Springer Proceedings in Mathematics & Statistics

Volume 139

Springer Proceedings in Mathematics & Statistics

This book series features volumes composed of selected contributions from workshops and conferences in all areas of current research in mathematics and statistics, including operation research and optimization. In addition to an overall evaluation of the interest, scientific quality, and timeliness of each proposal at the hands of the publisher, individual contributions are all refereed to the high quality standards of leading journals in the field. Thus, this series provides the research community with well-edited, authoritative reports on developments in the most exciting areas of mathematical and statistical research today.

More information about this series at http://www.springer.com/series/10533

Ram N. Mohapatra · Dipanwita Roy Chowdhury
Debasis Giri
Editors

Mathematics and Computing

ICMC, Haldia, India, January 2015

Springer

Editors
Ram N. Mohapatra
Department of Mathematics
University of Central Florida
Orlando, FL
USA

Debasis Giri
Department of Computer Science
and Engineering
Haldia Institute of Technology
Haldia, West Bengal
India

Dipanwita Roy Chowdhury
Department of Computer Science
and Engineering
Indian Institute of Technology Kharagpur
Kharagpur, West Bengal
India

ISSN 2194-1009 ISSN 2194-1017 (electronic)
Springer Proceedings in Mathematics & Statistics
ISBN 978-81-322-2451-8 ISBN 978-81-322-2452-5 (eBook)
DOI 10.1007/978-81-322-2452-5

Library of Congress Control Number: 2015939433

Springer New Delhi Heidelberg New York Dordrecht London

Printed on acid-free paper

Springer (India) Pvt. Ltd. is part of Springer Science+Business Media (www.springer.com)

Preface

The Second International Conference on Mathematics and Computing (ICMC 2015) was held at the Haldia Institute of Technology, Haldia, from January 5 to 10, 2015. Haldia is a city and a municipality in Purba Medinipur in the Indian state of West Bengal, and Haldia Institute of Technology is a premier institution that gives training to engineers and computer scientists. It has gained reputation through its institutional dedication to teaching and research.

In response to the call for papers for ICMC 2015, a total of 69 papers were submitted for presentation and inclusion in the proceedings of the conference. These papers were evaluated and ranked on the basis of their significance, novelty, and technical quality by at least two reviewers per paper. After a careful and blind refereeing process, 34 papers were selected for inclusion in the conference proceedings. The papers cover current research in cryptography, abstract algebra, functional analysis, Pal and fractal approximation, fluid dynamics, fuzzy modeling and optimization, and statistics. ICMC 2015 saw eminent personalities both from India and abroad (USA, France, Russia, Japan, Turkey, China, and Indonesia) who delivered invited addresses, workshop lectures, and tutorial talks. Speakers from India were recognized leaders from government, industry, and academic institutions like Defense Research and Development Organization (DRDO), New Delhi; Indian Institute of Sciences (IISc), Bangalore; Indian Statistical Institute (ISI), Kolkata; Indian Institute of Technology (IIT) Kharagpur; and IIT Madras. All of them are involved in research dealing with the current issues of interest related to the theme of the conference. The conference hosted four tutorial talks by Prof. Peeyush Chandra (IIT Kanpur), Dr. Prasanta Kumar Srivastava (IIT Patna), Dr. Dhanonjoy Dey (DRDO, New Delhi), and Prof. Ram N. Mohapatra (University of Central Florida, USA). There were four workshops, ranging over 2 days by Prof. Ram N. Mohapatra, Prof. Manoranjan Maiti (Vidyasagar University), Prof. C. Pandurangan (IIT Madras), and Prof. Abhijit Das (IIT Kharagpur). In addition to these, there was one keynote talk by Prof. Heinrich Begehr (Freie Universitat Berlin, Germany) and 11 invited talks by Prof. C.E Venimadhavan (IISc, Bangalore), Prof. Ekrem Savaş (Istanbul Commerce University, Turkey), Dr. P.K. Saxena (DRDO, New Delhi), Prof. Peeyush Chandra, Dr. Christina Boura (Versailles Saint-Quentin-en-Yvelines

University, France), Prof. Birendra Nath Mandal (ISI, Kolkata), Dr. Mridul Nandi (ISI, Kolkata), Prof. G.P. Kapoor (IIT Kanpur), Prof. P.D. Srivastava (IIT Kharagpur), Prof. Abhijit Das (IIT Kharagpur), and Prof. B.C. Tripathy (Institute of Advanced Study in Science and Technology, India).

A conference of this kind would not be possible to organize without the full support from different people across different committees. All logistics and general organizational aspects were looked after by the organizing committee members who spent their time and energy in making the conference a reality. We also thank all the technical program committee members and external reviewers for thoroughly reviewing the papers submitted to the conference and sending their constructive suggestions within the deadlines. Our hearty thanks to Springer for agreeing to publish the proceedings.

We are indebted to DRDO, Department of Electronics and Information Technology (Ministry of Communication and Information Technology, the Government of India), Indian Space Research Organisation (ISRO), University of Central Florida, The International Society for Analysis, its Applications and Computation (ISAAC), Cryptology Research Society of India (CRSI), Science and Engineering Research Board (DST), and Haldia Institute of Technology for sponsoring/supporting the event. Their support has significantly helped in raising the profile of the conference.

Last but not least, our sincere thanks go to all authors who submitted papers to ICMC 2015 and to all speakers and participants. We sincerely hope that the readers will find the proceedings stimulating and inspiring.

Ram N. Mohapatra
Dipanwita Roy Chowdhury
Debasis Giri

Committee

Patron
Lakshman Seth, Chairman, Haldia Institute of Technology, Haldia, India

General Co-Chairs
P.K. Saxena, SAG, DRDO, New Delhi, India
P.D. Srivastava, IIT Kharagpur, India

Workshop Chair
Peeyush Chandra, IIT Kanpur, India

Program Co-Chairs
Ram N. Mohapatra, University of Central Florida, USA
Dipanwita Roy Chowdhury, IIT Kharagpur, India
Debasis Giri, Haldia Institute of Technology, India

Program Committee Members

PC Member	Organization
P.N. Agarwal	Indian Institute of Technology Roorkee, India
Ravi P. Agarwal	Texas A&M University, Kingsville, USA
Rafikul Alam	Indian Institute of Technology Guwahati, India
Rana Barua	Indian Statistical Institute Kolkata, India
Heinrich Begehr	Freie Universitat Berlin, Germany
Amitava Bhattacharya	Tata Institute of Fundamental Research, Mumbai, India
Somnath Bhattacharyya	Indian Institute of Technology Kharagpur, India
Jaydeb Bhaumik	Haldia Institute of Technology, India

(continued)

PC Member	Organization
Leonid Bokut	Sobolev Institute of Mathematics, Novosibirsk, Russia
Ioana Boureanu	Akamai Technologies Limited, UK
Teodor Bulboaca	Babes-Bolyai University, Cluj-Napoca, Romania
Sucheta Chakrabarty	DRDO, Delhi, India
Kalyan Chakraborty	Harish-Chandra Research Institute, Allahabad, India
Peeyush Chandra	Indian Institute of Technology Kanpur, India
Janka Chlebikova	University of Portsmouth, UK
Dipanwita Roy Chowdhury	Indian Institute of Technology Kharagpur, India
Rifat Colak	Firat University, Elazig, Turkey
Abhijit Das	Indian Institute of Technology Kharagpur, India
Ashok Kumar Das	IIIT Hyderabad, India
Kinkar Das	Sungkyunkwan University, Korea
Manik Lal Das	DAIICT, India
B.K. Dass	University of Delhi, India
Biswa Datta	Northern Illinois University, USA
Anilkumar Devarapu	Albany State University, GA, USA
Dhananjoy Dey	DRDO, Delhi, India
Jana Dittmann	University of Magdeburg, Germany
Philippe Gaborit	University of Limoges, France
Sugata Gangopadhyay	Indian Institute of Technology Roorkee, India
Praveen Gauravaram	Tata Consultancy Services, India
Debasis Giri	Haldia Institute of Technology, Haldia 721657, India
Narendra Govil	Auburn University, Alabama, USA
Shay Gueron	University of Haifa, Israel
Indivar Gupta	DRDO, Delhi, India
U.C. Gupta	Indian Institute of Technology Kharagpur, India
Tian-Xiao He	Illinois Weslyan University, USA
Aoi Honda	Kyushu Institute of Technology, Japan
Don Hong	Middle Tennessee State University, USA
Honggang Hu	University of Science and Technology of China, China
N.J. Hunag	Sichuan University, Chengdu, Republic of China
Peter Johnson	Auburn University, Alabama, USA
Mohan S. Kankanhalli	National University of Singapore, Singapore
G.P. Kapoor	Indian Institute of Technology Kanpur, India
Sandip Karmakar	IIIT Guwahati, India
K.V. Krishna	Indian Institute of Technology. Guwahati, India
Somesh Kumar	Indian Institute of Technology Kharagpur, India
Shanta Laishram	Indian Statistical Institute, Delhi, India
Duan Li	The Chinese University of Hong Kong, Hong Kong
Shijun Liao	Shanghai Jiao Tong University, China

(continued)

PC Member	Organization
Jiqiang Lu	Institute for Infocomm Research (I2R), Singapore
Manoranjan Maiti	Vidyasagar University, India
Soumen Maity	IISER, Pune, India
Birendra Nath Manda	Indian Statistical Institute Kolkata, India
Keith Martin	Royal Holloway, University of London
Edgar Martinez-Moro	University of Valladolid, Spain
Miodrag Mihaljevic	Serbian Academy of Science and Arts, Belgrade, Serbia
P.R. Mishra	DRDO, Delhi, India
Tapan Misra	Space Applications Centre, ISRO, Ahmedabad, India
Ram N. Mohapatra	University of Central Florida, USA
Shyamal Mondal	Vidyasagar University, India
Debdeep Mukhopadhyay	IIT Kharagpur, India
M. Mursaleen	Aligarh Muslim University, India
Sukumar Nandi	Indian Institute of Technology Guwahati, India
Vilem Novak	University of Ostrava, Czech Republic
S. Padhi	Birla Institute of Technology Mesra, India
Madhumangal Pal	Vidyasagar University, India
Saibal K. Pal	DRDO, Delhi, India
Kolin Paul	Indian Institute of Technology Delhi, India
Svetla Petkova-Nikova	KU Leuven, ESAT-COSIC, Belgium
Rajesh Pillai	DRDO, Delhi, India
V. Sree Hari Rao	IDRBT, India
Bimal Roy	Indian Statistical Institute, Kolkata, India
Kouichi Sakurai	Kyushu University, Fukuoka, Japan
Suman Sanyal	Marshall University, West Virginia
Santanu Sarkar	Chennai Mathematical Institute, India
P.K. Saxena	DRDO, Delhi, India
G.P. Raja Sekhar	Indian Institute of Technology Kharagpur, India
S. Samarendra Singh	Manipur University, India
Seong Han Shin	National Institute of Advanced Industrial Science and Technology, Japan
P.D. Srivastava	Indian Institute of Technology Kharagpur, India
S. Sundar	Indian Institute of Technology Madras, India
Pamini Thangarajah	Mount Royal University, Calgary, Alberta, Canada
P. Vellaisamy	Indian Institute of Technology Bombay, India
C.E. Veni Madhavan	Indian Institute of Science, Bangalore, India
Ram U. Verma	Texas A&M University-Kingsville, USA
V. Vetrivel	Indian Institute of Technology Madras, India
Konstantin Volkov	Kingston University, UK
Bixiang Wang	New Mexico Institute of Mining and Technology, USA
Yao Zhao	Beijing Jiao Tong University, China

Additional Reviewers

Reviewer	Organization
Abhijit Datta Banik	Indian Institute of Technology Bhubaneswar, India
Rajesh Deo	Institute of Information Security, India
Manish Kant Dubey	SAG, DRDO, Delhi, India
Kishan Gupta	Indian Statistical Institute Kolkata, India
Dipak Kumar Jana	Haldia Institute of Technology, India
Louis Yang Liu	Michigan State University, USA
Mukund Madhav Mishra	University of Delhi, India
Seiichiro Mizoguchi	Kyushu University, Japan
Kirill Morozov	Kyushu University, Japan
Chandal Nahak	Indian Institute of Technology Kharagpur, India
Harika Narumanchi	Jawaharlal Nehru Technological University, India
Ameeya Nayak	Indian Institute of Technology Roorkee, India
Gayatri Ramesh	University of Central Florida, USA
David Rollins	University of Central Florida, USA
Sujit Samanta	National Institute of Technology Raipur, India
Joe Sawada	University of Guelph, Canada
Shalabh	Indian Institute of Technology Kanpur, India
Juan Jacobo Simon Pinero	Universidad de Murcia, Espaa
Yogesh Tripathi	Indian Institute of Technology Patna, India

Organizing Committee

Asish Lahiri
A.K. Dey
Anjan Mishra
Debasis Giri
Jaydeb Bhaumik
Sk. Sahnawaj
Sudipta Kumar Basu
Debasis Das
Soumen Paul
Tarun Kumar Ghosh
Apratim Mitra
Subhabrata Barman
Sourav Mandal

Subhankar Joardar
Susmit Maity
Palash Roy
Sk. Arif Ahmed
Bidesh Chakraborty
Mrinmoy Sen
Jayeeta Majumder
Mihir Baran Bera

Message from the General Chairs

As we all are aware mathematics has always been a discipline of interest not only to theoreticians but also to all practitioners, irrespective of their specific profession. Be it science, technology, economics, commerce, or even sociology, new mathematical principles and models have been emerging and helping in new research and in drawing inferences from practical data as well as through logic. The past few decades have seen enormous growth in applications of mathematics in different areas multidisciplinary in nature. Cryptography and signal processing are such areas, which have got more focus recently due to the need for securing communication while connecting with others. With emerging computing facilities and speeds, a phenomenal growth has happened in the problem solving area. Earlier, some observations were made and conjectures were drawn which remained conjectures till somebody could either prove it theoretically or found counter examples. But today, we can write algorithms and use computers for long calculations, verifications, or for generation of huge amounts of data. With available computing capabilities, we can find factors of very large integers of the size of hundreds of digits; we can find inverses of very large size matrices and solve a large set of linear equations, and so on. Thus, mathematics and computations have become more integrated areas of research these days, and it was thought to organize an event where thoughts may be shared by researchers and new challenging problems could be deliberated for solving these.

Apart from many other interdisciplinary areas of research, cryptography has emerged as one of the most important areas of research with discrete mathematics as a base. Several research groups are actively pursuing research on different aspects of cryptology not only in terms of new crypto-primitives and algorithms, but a whole lot of concepts related to authentication, integrity, and security proofs/protocols, many times with open and competitive evaluation mechanism to evolve standards.

As conferences, seminars, and workshops are the mechanisms to share knowledge and new research results that give us a chance to get new innovative ideas for futuristic needs as threats and computational capabilities of adversaries are ever-increasing, it was thought appropriate to organize the present conference focused on mathematics and computations covering theoretical as well as practical aspects of research, cryptography being one of them.

Eminent personalities working in mathematical sciences and related areas were invited from abroad as well as from within the country to deliver invited talks and tutorials for participants. The talks by these speakers covered a wide spectrum, namely number theoretic concepts, cryptography, algebraic concepts, and applications. The conference was spread over 6 days (January 5–10, 2015) with the first 2 days dedicated to workshops and the next one day dedicated to tutorials. The main conference was planned with special talks by experts and paper presentations in each session.

We hope that the conference met the aspirations of the participants and its objective of ideas and current research being shared and new targets/problems identified in the domain of cryptography, computational number theory, algebra, frame theory, optimizations, fuzzy logic, stochastic processes, compressive sensing, functional analysis, complex variables, etc., so that researchers and students would get new directions to pursue their future research.

P.K. Saxena
P.D. Srivastava

Message from the Program Chairs

It is a great pleasure for us to organize the Second International Conference on Mathematics and Computing 2015 held from January 5 to 10, 2015 at the Haldia Institute of Technology, Purba Medinipur, West Bengal, India. Our main goal was to provide an opportunity to the participants to learn about contemporary research in mathematics and computing and exchange ideas among themselves and with experts present in the conference as workshop presenters, tutorial presenters, and the plenary as well as invited speakers. With this aim in mind, we carefully selected the invited speakers and the speakers for the workshops and tutorials. It is our sincere hope that the conference would help the participants in their research and training and open new avenues for work for those who are either starting their research or are looking for extending their area of research to a different area of current research in mathematics and computing.

The below table shows the workshops held on January 5–6, 2015.

Title of the workshop	Name of the speaker	Duration of the workshop
Signcryption in standard model	Prof. C. Pandu Rangan	3 h and 45 min
Elliptic curve cryptography	Prof. Abhijit Das	3 h and 45 min
Riesz bases and frames	Prof. Ram N. Mohapatra	3 h and 45 min
Conventional and metaheuristic optimization techniques	Prof. Manoranjan Maiti	3 h and 45 min

The below table shows the tutorials held on January 7, 2015.

Title of the tutorial	Name of the speaker	Duration of the tutorial
Mathematical epidemiology	Prof. Peeyush Chandra	2 h
Mathematical epidemiology	Dr. Prasanta K. Srivastava	1 h and 45 min
A gentle introduction to block ciphers: design and analysis	Dr. Dhananjoy Dey	2 h
Some mathematical snapshots	Prof. Ram N. Mohapatra	1 h and 45 min

The conference began after a formal opening ceremony on January 8. There was one keynote 90-min talk by Prof. Heinrich Begehr and seven invited 1-h talks by Prof. Peeyush Chandra, Prof. C.E. Venni Madhavan, Dr. P.K. Saxena, Prof. Ekrem Savaş, Prof. Abhijit Das, Dr. Christina Boura, Dr. Mridul Nanadi, Prof. Birendra Nath Mandal, Prof. P.D. Srivastava, Prof. G.B. Kapoor, and Prof. Binod Chandra Tripathy. There were 32 contributed half-hour talks. Our speakers/contributors came from Germany, France, Japan, Turkey, Indonesia, India, China, and USA. After an initial call for papers, 69 papers were submitted for presentation at the conference. All submitted papers were sent to external referees and after refereeing, 34 papers were recommended for presentation. The proceedings of the conference contains 34 papers published by Springer. We are grateful to the speakers, participants, referees, organizers, sponsors, and funding agencies (from DRDO, University of Central Florida, DeitY-DIT, ISRO, CRSI, ISAAC, SERB-DST, Haldia Institute of Technology) for their support and help, without which it would have been impossible to organize the conference, the workshops, and the tutorials. We owe our gratitude to the volunteers who worked behind the scene tirelessly in taking care of the details in making this conference a success.

Ram N. Mohapatra
Dwipanita Roy Chowdhury
Debasis Giri

Contents

Editors and Contributors

About the Editors

Ram N. Mohapatra is Professor of Mathematics, University of Central Florida, Orlando, USA. He received his PhD degree from the University of Jabalpur, India, in 1968. Earlier, he taught at Sambalpur University in India, American University in Beirut, Lebanon, University of Alberta, and York University, Canada, prior to coming to Orlando. His area of research is Mathematical Analysis and he is the author of two books, two edited monographs, and over 120 research papers. He referees articles for professional journals and serves as a member of the editorial board of a number of journals.

Dipanwita Roy Chowdhury is Professor at the Department of Computer Science and Engineering, Indian Institute of Technology (IIT) Kharagpur, India. She is also the chairman of Kalpana Chawla Space Technology Cell, IIT Kharagpur. Professor Roy Chowdhury is PhD from IIT Kharagpur and M.Tech. in Computer Science from the University of Calcutta. Her topics of research interest are cryptography, error-correcting code, and cellular automata. She has published more than 150 research papers in several international journals and conference proceedings. Professor Roy Chowdhury is the recipient of INSA Young Scientist Award and associate of Indian Academy of Science. She is a fellow of Indian National Academy of Engineers (INAE).

Debasis Giri is Professor at the Department of Computer Science and Engineering, Haldia Institute of Technology, India. His topics of interest are discrete mathematics, cryptography, information security, coding theory, advanced algorithms, design and analysis of algorithms, and formal languages and automata theory. His research interests include cryptography, network security, security in wireless sensor networks, and security in VANETs. Dr. Giri had delivered several talks and guest lectures at various universities and conferences. He is supervisor of three PhD research scholars, and has guided many B.Tech. and M.Tech. students. He is associate editor of the *Journal of Security and Communication Networks* (Wiley), and the *Journal of Electrical and Computer Engineering Innovations*. Further, he is the editorial board member and reviewer of many reputed international journals. He is also a program committee member of many international conferences. He is a life member of Cryptology Research Society of India. He received his PhD on "Cryptanalysis and improvement of protocols for digital signature, smartcard authentication and access control" from Indian Institute of Technology (IIT) Kharagpur. He did both his M.Tech. and M.Sc. from IIT Kharagpur. He secured 10th position in all India rank with percentile score 98.42 in the Graduate Aptitude Test in Engineering (GATE) examination in 1999. Dr. Giri has published more than 30 technical papers in several international journals and proceedings.

Contributors

Avishek Adhikari Department of Pure Mathematics, University of Calcutta, Kolkata, India

Yasmeen Akhtar Indian Institute of Science Education and Research, Pune, India

S. Amutha Ramanujan Centre for Higher Mathematics, Alagappa University, Karaikudi, TamilNadu, India

Sudeshna Banerjea Department of Mathematics, Jadavpur University, Kolkata, India

Heinrich Begehr Mathematical Institute, Free University Berlin, Berlin, Germany

Pramod Kumar Behera Department of Mathematics, Veer Surendra Sai University of Technology, Burla, India

S. Bera Department of Mathematics, Indian Institute of Technology Kharagpur, Kharagpur, India

Dambaru Bhatta The University of Texas, Edinburg, TX, USA

S. Bhattacharyya Department of Mathematics, Indian Institute of Technology Kharagpur, Kharagpur, India

Sushil Kumar Bhuiya Department of Mathematics, Indian Institute of Technology Kharagpur, Kharagpur, India

Klara Ajeng Canyarasmi Institut Teknologi Bandung, Bandung, Indonesia

Debjani Chakraborty Department of Mathematics, Indian Institute of Technology Kharagpur, Kharagpur, India

A.K.B. Chand Department of Mathematics, Indian Institute of Technology Madras, Chennai, Tamil Nadu, India

Reshma C. Chandrasekharan Indian Institute of Science Education and Research, Pune, India

Chen-Mou Cheng Kyushu University, Fukuoka, Japan; National Taiwan University, Taipei, Taiwan

P. Chinnaraj Department of Mathematics, Park College of Engineering and Technology, Coimbatore, Tamilnadu, India

Dipanwita Roy Chowdhury Department of Computer Science and Engineering, Indian Institute of Technology Karagpur, Kharagpur, India

A.K. Das SQC & OR Unit, Indian Statistical Institute, Kolkata, India

Angsuman Das Department of Mathematics, St. Xavier's College, Kolkata, India

Ashok Kumar Das Center for Security, Theory and Algorithmic Research, International Institute of Information Technology, Hyderabad, India

Deepmala SQC & OR Unit, Indian Statistical Institute, Kolkata, India

Harpreet Dhillon Department of Mathematics, Jadavpur University, Kolkata, India

Navneet Gaba Scientific Analysis Group, Defence Research and Development Organization, Delhi, India

A. Garai Memari College, Memari, Burdwan, West Bengal, India

S.H. Ghate Department of Mathematics, R.T.M. Nagpur University, Nagpur, India

Shamit Ghosh Department of Computer Science and Engineering, Indian Institute of Technology Karagpur, Kharagpur, India

R.G. Ghumde Department of Mathematics, Ramdeobaba College of Engineering and Management, Nagpur, India

Debasis Giri Department of Computer Science and Engineering, Haldia Institute of Technology, Haldia, India

Indivar Gupta Scientific Analysis Group, Defence Research and Development Organization, Delhi, India

Prasun Hazra Paladion Networks, Bangalore, India

Don Hong Computational Science Program, Middle Tennessee State University, Murfreesboro, TN, USA; College of Sciences, North China University of Technology, Beijing, China

Biswapati Jana Department of Computer Science, Vidyasagar University, West Bengal, India

Mrinal Jana Department of Mathematics, Indian Institute of Technology Kharagpur, Kharagpur, India

S. Jana Department of Mathematics, Indian Institute of Technology Kharagpur, Kharagpur, India

Shashi Kant Department of Applied Mathematics, Delhi Technological University, Delhi, India

S.K. Katiyar Department of Mathematics, Indian Institute of Technology Madras, Chennai, India

Pankaj Kumar Department of Mathematics, Indian Institute of Technology Kharagpur, Kharagpur, India

Vivek Kumar Department of Applied Mathematics, Delhi Technological University, Delhi, India

Lakshmi Kuppusamy Society For Electronic Transactions and Security, MGR Knowledge City, Chennai, Tamilnadu, India

Jingsai Liang Computational Science Program, Middle Tennessee State University, Murfreesboro, TN, USA

Soumen Maity Indian Institute of Science Education and Research, Pune, India

Prasanna Raghaw Mishra Scientific Analysis Group, Defence Research and Development Organization, Delhi, India

R.N. Mohapatra Department of Mathematics, University of Central Florida, Orlando, FL, USA

Shyamal Kumar Mondal Department of Applied Mathematics with Oceanology and Computer Programming, Vidyasagar University, Midnapore, India

C. Nahak Department of Mathematics, Indian Institute of Technology Kharagpur, Kharagpur, India

Saroj Kumar Padhan Department of Mathematics, Veer Surendra Sai University of Technology, Burla, India

Geetanjali Panda Department of Mathematics, Indian Institute of Technology Kharagpur, Kharagpur, India

Udjianna Sekteria Pasaribu Institut Teknologi Bandung, Bandung, Indonesia

Kartik Patra Department of Mathematics, Sikkim Manipal Institute of Technology, Sikkim Manipal University, East Sikkim, India

R. Perumal Department of Mathematics, Kumaraguru College of Technology, Coimbatore, Tamilnadu, India

Gayatri Ramesh University of Central Florida, Orlando, FL, USA

Jothi Rangasamy Society For Electronic Transactions and Security, MGR Knowledge City, Chennai, Tamilnadu, India

S. Ray Visva Bharati, Santiniketan, West Bengal, India

Daniel N. Riahi The University of Texas, Edinburg, TX, USA

Partha Sarathi Roy Department of Pure Mathematics, University of Calcutta, Kolkata, India

Kouichi Sakurai Institute of Systems, Information Technologies and Nanotechnologies, Fukuoka, Japan; Kyushu University, Fukuoka, Japan

Amita Samanta Department of Computer Science, Vidyasagar University, West Bengal, India

Ekrem Savaş Istanbul Commerce University, Department of Mathematics, Istanbul, Turkey

V.P. Singh Department of Mathematics, Indian Institute of Technology Kharagpur, Kharagpur, India

Aaron Carl Smith Department of Mathematics, University of Central Florida, Orlando, FL, USA

N. Sridharan Department of Mathematics, Alagappa University, Karaikudi, TamilNadu, India

Satoshi Tanaka Institute of Systems, Information Technologies and Nanotechnologies, Fukuoka, Japan; Kyushu University, Fukuoka, Japan

K.R. Tyada Department of Mathematics, Indian Institute of Technology Madras, Chennai, India

Ram U. Verma Department of Mathematics, Texas State University, San Marcos, TX, USA

N. Vijender Department of Mathematics, VIT University, Chennai, India

Fengqing (Zoe) Zhang Department of Psychology, Drexel University, Philadelphia, PA, USA

Jiancheng Zou Computational· Science Program, Middle Tennessee State University, Murfreesboro, TN, USA; College of Sciences, North China University of Technology, Beijing, China

Integral Representations Related to Complex Partial Differential Operators

Heinrich Begehr

Abstract Integral representations are an essential tool for treating differential equations. They serve to solve initial and boundary value problems and to guarantee smoothness properties for solutions. Well known are the Green representation formulas for harmonic functions and the Cauchy formula for analytic functions. This survey concentrates on representation formulas in plane domains for the polyanalytic and the polyharmonic operators. They generalize the Cauchy-Riemann and the Laplace operator, respectively, to higher order partial differential operators. The kernels of these operators are the sets of polyanalytic and polyharmonic functions. Having constructed the fundamental solutions to these particular model operators, higher order Pompeiu area integral operators, providing particular solutions to the related inhomogeneous equations, serve to treat any higher order linear partial differential equation, the leading term of which is a product of the mentioned model operators.

Keywords Polyanalytic and polyharmonic equations · Integral representations · Hybrid polyharmonic Green functions · Iterated polyharmonic Green and Neumann functions · Polyharmonic Green-Almansi functions

Mathematics Subject Classifications: 31A25 · 31A30 · 30E25 · 35J05 · 35J08 · 35J30 · 35G15 · 35C15

1 Basic Integral Representations

The Gauss divergence theorem is the origin of a variety of integral representations related to partial differential operators. It is in fact the main theorem of calculus in higher dimensions. While the main theorem of calculus in the case of one real variable immediately leads to an integral representation formula for continuously dif-

H. Begehr (✉)
Mathematical Institute, Free University Berlin, Arnimallee 3, 14195 Berlin, Germany
e-mail: begehrh@zedat.fu-berlin.de

R.N. Mohapatra et al. (eds.), *Mathematics and Computing*,
Springer Proceedings in Mathematics & Statistics 139,
DOI 10.1007/978-81-322-2452-5_1

ferentiable functions and by an iteration process directly to the Taylor representation formula, in higher dimensions on one hand different kinds of differential operators are available and on the other a modification of the divergence theorem is needed in order to create representation formulas. The Gauss theorem is at first applied to the product of two admissible functions and then the resulting formula applied when one of the two is chosen as the fundamental solution to the respective differential operator. A representation formula then arises from the point singularity of the fundamental solution. As in the one real variable case, these representation formulas are proper for being iterated leading as well to a hierarchy of representation formulas for higher order differential operators, arbitrary powers of the original one, and on the other hand at the same time to fundamental solutions of these higher order differential operators.

Fundamental solutions to higher order differential operators can also be constructed from the ones for lower order operators. A fundamental solution to the product of two differential operators ∂_1, ∂_2 is found from a fundamental solution to ∂_1, say f_1, as a primitive with respect to ∂_2 of f_1, $\partial_2^{-1} f_1$. As f_1 satisfies $\partial_1 f_1 = \delta$ with the Dirac δ-operator, then $\partial_1 \partial_2 \partial_2^{-1} f_1 = \delta$. This process is used in [25] to construct fundamental solutions to complex partial differential operators of arbitrary order $\partial_z^m \partial_{\overline{z}}^n$ for $m, n \in \mathbb{N}$ in the complex plane $\mathbb{C}$. Here $\partial_{\overline{z}}$, ∂_z are the Cauchy-Riemann differential operator and its complex conjugate, given by $2\partial_{\overline{z}} = \partial_x + i\partial_y$, $2\partial_z = \partial_x - i\partial_y$, where $z = x + iy \in \mathbb{C}$, $x, y \in \mathbb{R}$. The initial fundamental solution to $\partial_{\overline{z}}$ is up to the factor $-\frac{1}{\pi}$ the Cauchy kernel $\frac{1}{z}$. Taking continued primitives against $\partial_{\overline{z}}$ leads to $\frac{\overline{z}^{n-1}}{[n-1]!z}$ for the polyanalytic differential operator $\partial_{\overline{z}}^n$, $n \in \mathbb{N}$. Similarly, a primitive to the Cauchy kernel with respect to ∂_z is $\log z$, a better one because symmetric in z and $\overline{z}$ is $\log z + \log \overline{z} = \log |z|^2$, the fundamental solution to $\partial_z \partial_{\overline{z}}$, the complex form of the Laplace differential operator $\Delta_z = 4\partial_z \partial_{\overline{z}}$. Continuing, $\frac{z^{m-1}\overline{z}^{n-1}}{(m-1)!(n-1)!} \log |z|^2$ turns out as a fundamental solution to $\partial_z^m \partial_{\overline{z}}^n$. A better one is

$$\frac{z^{m-1}\overline{z}^{n-1}}{(m-1)!(n-1)!}\left[\log |z|^2 + \sum_{\mu=1}^{m-1} \frac{1}{\mu} + \sum_{\nu=1}^{n-1} \frac{1}{\nu}\right].$$

Applying differential operators ∂_z, $\partial_{\overline{z}}$, respectively, keeps its form with m and n, respectively, lowered by 1. Based on these fundamental solutions, integral representation formulas are available. They are, however, not proper to solve boundary value problems. For this purpose, the fundamental solutions have to be adjusted to certain boundary behavior. For the polyanalytic operator, there is the polyanalytic Schwarz kernel, see [1, 2, 8, 25, 27, 38, 42] related to the Schwarz problem. Also a Neumann, a particular Robin [11, 22, 61] and many mixed problems [60] are treatable.

For the polyharmonic operator $(\partial_z \partial_{\overline{z}})^n$, there are a variety of different boundary conditions, the higher the order n the more possible boundary conditions exist. Exactly, n conditions may be posed and this may be done in many different ways [6, 7, 10, 12–14, 16, 17, 19–21, 24, 28–30, 39, 44, 45, 47, 49–51, 56, 59, 63, 65–70, 72, 73]. For the harmonic operator, $n = 1$, there are three main boundary conditions, the Dirichlet, the Neumann, and the Robin condition. The related fundamental solutions are well know in the literature. They are called the harmonic Green, the harmonic Neumann, and the harmonic Robin function. Explicit formulas and methods to find them are given, e.g., in [29, 30, 33, 34, 36, 48]. For a certain class of plane domains, the parqueting reflection principle [18, 23, 24, 30–33, 35, 37, 69, 72] is effective.

One possibility to determine such fundamental solutions for the polyharmonic operator is to iteratively convolute the three types of harmonic fundamental solutions. This process leads to hybrid polyharmonic Green functions [3–5, 14, 29, 30]. The procedure of iterating representation formulas [11, 12] can be used to deduce the Poisson representation formula and the one for the Bitsadze operator from the Cauchy-Pompeiu formulas just on the basis of the Gauss divergence theorem. In a natural way, even the idea of Green and Neumann functions arises here.

The theory of hybrid polyharmonic representations is far from being complete. Calculating particular samples demands a lot of area integral evaluations. Also not all convolutions seem to be obvious, for the Robin function, e.g., iterations are only performed for particular parameters [23] and the modified Robin function [36] is not yet iterated.

The iteration process for constructing certain fundamental solutions and related integral representation formulas for higher order differential operators does work also in higher dimensional cases as in quarternionic, in octonionic, and in Clifford analysis [9, 40, 41, 43, 64, 71]. Some other complex differential equations were treated in a similar way in [26, 46, 50].

Inhomogeneous polyanalytic and polyharmonic equations are just model equations for arbitrary higher order elliptic equations. But their potentials serve to treat more general equations, the leading term of which is the operator of one of the model equations. Such general linear equations can be investigated on the basis of the Fredholm alternative for singular integral equations [2–6]. The general model differential operator

$$\partial_z^m \partial_{\overline{z}}^n = (\partial_z \partial_{\overline{z}})^m \partial_{\overline{z}}^{(n-m)}, \ (m \leq n)$$

can be written as a product of a polyanalytic and a polyharmonic operator. An iteration of the representation formulas for these two factors will lead to one for this general operator.

1.1 Cauchy-Pompeiu Representations

A direct consequence of the Gauss divergence theorem for bounded domains D of the complex plane $\mathbb{C}$ with piecewise smooth boundary ∂D are the Cauchy-Pompeiu formulas [11]. Under proper growth restrictions, they also hold for unbounded domains [56].

Theorem 1 *Any $w \in C^1(D;\mathbb{C}) \cap C(\overline{D};\mathbb{C})$ can be represented as*

$$w(z) = \frac{1}{2\pi i}\int_{\partial D} w(\zeta)\frac{d\zeta}{\zeta - z} - \frac{1}{\pi}\int_D w_{\overline{\zeta}}(\zeta)\frac{d\xi d\eta}{\zeta - z},$$

$$w(z) = -\frac{1}{2\pi i}\int_{\partial D} w(\zeta)\frac{d\overline{\zeta}}{\overline{\zeta - z}} - \frac{1}{\pi}\int_D w_{\zeta}(\zeta)\frac{d\xi d\eta}{\overline{\zeta - z}}.$$

These two representation formulas are the basis for an iteration process leading to all subsequent higher order representations. Second-order formulas are attained by representing $w_{\overline{z}}$ by the first formula and inserting this expression into the original formula. Also, applying the second formula to $w_{\overline{z}}$ and inserting the result into the first one leads to another representation formula. On this way, four second-order formulas are attained. Two of them are as follows [11].

Theorem 2 *Any $w \in C^2(D;\mathbb{C}) \cap C^1(\overline{D};\mathbb{C})$ can be represented as*

$$w(z) = \frac{1}{2\pi i}\int_{\partial D} w(\zeta)\frac{d\zeta}{\zeta - z} - \frac{1}{2\pi i}\int_{\partial D} w_{\overline{\zeta}}(\zeta)\frac{\overline{\zeta - z}}{\zeta - z}d\zeta$$

$$+\frac{1}{\pi}\int_D w_{\overline{\zeta}\,\overline{\zeta}}(\zeta)\frac{\overline{\zeta - z}}{\zeta - z}d\xi d\eta$$

and as

$$w(z) = \frac{1}{2\pi i}\int_{\partial D} w(\zeta)\frac{d\zeta}{\zeta - z} + \frac{1}{2\pi i}\int_{\partial D} w_{\overline{\zeta}}(\zeta)\log|\zeta - z|^2 d\overline{\zeta}$$

$$+\frac{1}{\pi}\int_D w_{\zeta\overline{\zeta}}(\zeta)\log|\zeta - z|^2 d\xi d\eta.$$

The second-order differential operators $\partial_{\overline{z}}^2$ and $\partial_z\partial_{\overline{z}}$ are the bianalytic or Bitsadze and the harmonic or Laplace operator, respectively. The representation formula for the Laplace operator requires alteration [11]. Because harmonic functions are known to be determined just by their boundary values, the boundary integral with first-order derivatives of the function involved is redundant.

1.2 Green Representations

There are different ways to change the representation formula for the Laplace operator. Just applying the Gauss theorem appropriately in the case of the unit disk, the concepts of the Green and the Neumann functions arise [11]. Using them for proper domains, Green and Neumann representation formulas are attained adjusted to certain boundary behavior of the function represented. Using the concept of Robin functions, both these representation formulas can be combined into one formula [36].

Definition 1 For $\alpha, \beta \in \mathbb{R}$, $0 < \alpha^2+\beta^2$, a real-valued function $R_{1;\alpha,\beta}(z,\zeta)$, $z,\zeta \in D$, $z \neq \zeta$, is called Robin function if for any $\zeta \in D$ it has the properties

- $R_{1;\alpha,\beta}(\cdot\,,\zeta)$ is harmonic in $D \setminus \{\zeta\}$ and continuously differentiable in $\overline{D} \setminus \{\zeta\}$,
- $h(z,\zeta) = R_{1;\alpha,\beta}(z,\zeta) + \log|\zeta - z|^2$ is harmonic for $z \in D$,
- $\alpha R_{1;\alpha,\beta}(z,\zeta) + \beta\partial_{\nu_z} R_{1;\alpha,\beta}(z,\zeta) = \beta\sigma(s)$ for $z = z(s) \in \partial D$, where the density function σ is a real-valued, piecewise constant function of s, the arc length parameter on ∂D, with finite mass $\int\limits_{\partial D} \sigma(s)ds$,
- $\beta \int\limits_{\partial D} \sigma(s_z) R_{1;\alpha,\beta}(z,\zeta)ds_z = 0$ *(normalization condition).*

This Robin function can be shown to satisfy

$$R_{1;\alpha,\beta}(z,\zeta) = R_{1;\alpha,\beta}(\zeta,z) \ \ (symmetry).$$

For $\beta = 0$ it is the Green function for the domain D, $G_1(z,\zeta) = R_{1;\alpha,0}(z,\zeta)$, for $\alpha = 0$ it is the Neumann function, $N_1(z,\zeta) = R_{1;0,\beta}(z,\zeta)$. For the Green function, there appears no normalization condition and the density function σ is not needed. Both functions differ mainly in their boundary behavior.

Manipulating the representation formula for the Laplacian in introducing the Robin function some formulas appear interpolating the Green and the Neumann representation.

Theorem 3 *Any function $w \in C^2(D;\mathbb{C}) \cap C^1(\overline{D};\mathbb{C})$ can be represented as*

$$w(z) = \omega_k(z) - \frac{1}{\pi}\int\limits_D \partial_\zeta\partial_{\overline{\zeta}} w(\zeta) R_{1;\alpha,\beta}(\zeta,z)d\xi d\eta, \quad k = 1,2,3,$$

$$\omega_1(z) = -\frac{1}{4\pi}\int\limits_{\partial D} \{w(\zeta)\partial_{\nu_\zeta} R_{1;\alpha,\beta}(\zeta,z) - \partial_\nu w(\zeta) R_{1;\alpha,\beta}(\zeta,z)\}ds_\zeta,$$

$$4\pi\beta\omega_2(z) = \int\limits_{\partial D} \{\alpha w(\zeta) + \beta\partial_\nu w(\zeta)\} R_{1;\alpha,\beta}(\zeta,z)ds_\zeta - \beta\int\limits_{\partial D} \sigma w(\zeta)ds_\zeta \ \ (\beta \neq 0),$$

$$4\pi\alpha\omega_3(z) = -\int_{\partial D} \{\alpha w(\zeta) + \beta \partial_\nu w(\zeta)\} \partial_{\nu_\zeta} R_{1;\alpha,\beta}(\zeta, z) ds_\zeta + \beta \int_{\partial D} \sigma \partial_\nu w(\zeta) ds_\zeta \quad (\alpha \neq 0).$$

The two subcases for $\beta \neq 0$ and $\alpha \neq 0$ are proper for representing functions w satisfying the Robin boundary value problem.

Definition 2 A solution of the Poisson equation

$$\partial_z \partial_{\overline{z}} w = f \ \ in \ \ D$$

satisfying the Robin boundary condition

$$\alpha w + \beta \partial_\nu w = \gamma \ \ on \ \ \partial D$$

is called a solution to the Robin problem.

For $\beta = 0$ this is the Dirichlet, for $\alpha = 0$ the Neumann problem. In general, the Robin problem is conditionally solvable. The conditions for solvability depend on the domain. Here is the result for the unit disk $\mathbb{D}$ [36].

Theorem 4 *For $f \in L_p(\mathbb{D}; \mathbb{C}), 2 < p, \ \gamma \in C(\partial\mathbb{D}; \mathbb{C})$, the Robin problem*

$$\partial_z \partial_{\overline{z}} w = f \ in \ \mathbb{D}, \ \ \alpha w + \beta \partial_\nu w = \gamma \ on \ \partial\mathbb{D},$$

(i) if $\beta \neq 0$ is solvable if and only if

$$\frac{1}{2\pi i} \int_{\partial\mathbb{D}} \gamma(\zeta) \frac{d\zeta}{\zeta} - \frac{\alpha}{2\pi i} \int_{\partial\mathbb{D}} w(\zeta) \frac{d\zeta}{\zeta} = \frac{2\beta}{\pi} \int_{\mathbb{D}} f(\zeta) d\xi d\eta,$$

the solution being then

$$w(z) = \frac{1}{4\pi i\beta} \int_{\partial\mathbb{D}} \gamma(\zeta) R_{1;\alpha,\beta}(\zeta, z) \frac{d\zeta}{\zeta} + \frac{1}{2\pi i} \int_{\partial\mathbb{D}} w(\zeta) \frac{d\zeta}{\zeta} - \frac{1}{\pi} \int_{\mathbb{D}} f(\zeta) R_{1;\alpha,\beta}(\zeta, z) d\xi d\eta,$$

(ii) if $\alpha \neq 0$ is solvable if and only if

$$\frac{\beta}{2\pi i} \int_{\partial\mathbb{D}} \partial_\nu w(\zeta) \frac{d\zeta}{\zeta} = \frac{2\beta}{\pi} \int_{\mathbb{D}} f(\zeta) d\xi d\eta,$$

the solution then being

$$w(z) = -\frac{1}{4\pi i\alpha} \int_{\partial\mathbb{D}} \gamma(\zeta) \partial_{\nu_\zeta} R_{1;\alpha,\beta}(\zeta, z) \frac{d\zeta}{\zeta} - \frac{\beta}{2\pi i\alpha} \int_{\partial\mathbb{D}} \partial_\nu w(\zeta) \frac{d\zeta}{\zeta}$$

$$-\frac{1}{\pi}\int_{\mathbb{D}} f(\zeta)R_{1;\alpha,\beta}(\zeta,z)d\xi d\eta.$$

Remark 1 If $\alpha = 0$ then $\gamma = \beta\partial_\nu w$ and the solvability condition is the known one for the Neumann problem, see, e.g., [11]. In case of $\beta = 0$ there is no solvability condition!

2 Polyharmonic Representations

The iteration of integral representation formulas can also be applied to get representations for poly-Poisson equations $(\partial_z\partial_{\bar{z}})^n w = f$, $n \in \mathbb{N}$. As this equation can be decomposed into a system of n, Poisson equations $\partial_z\partial_{\bar{z}}w_\mu = w_{\mu+1}, 0 \le \mu \le n-1$, $w_0 = w$, $w_n = f$, and each of these Poisson equations allows one boundary condition, there exists a variety of boundary value problems for the poly-Poisson equation. The classical ones are [57, 66] the Dirichlet, the Neumann, and the Riquier problem.

Definition 3 A solution of the inhomogeneous polyharmonic equation

$$(\partial_z\partial_{\bar{z}})^n w = f \quad in \quad D$$

is a solution to the **polyharmonic Dirichlet problem** if it satisfies

$$\partial_\nu^\mu w = \gamma_\mu, \quad 0 \le \mu \le n-1 \quad \text{on} \quad \partial D, \tag{1}$$

solves the **polyharmonic Neumann problem** if

$$\partial_\nu^\mu w = \gamma_\mu, \quad 1 \le \mu \le n \quad \text{on} \quad \partial D \tag{2}$$

are satisfied, and solves the **polyharmonic Riquier problem** if

$$(\partial_z\partial_{\bar{z}})^\mu w = \gamma_\mu, \quad 0 \le \mu \le n-1 \quad \text{on} \quad \partial D \tag{3}$$

hold.

Some other boundary value problems for the n-**Poisson** equation are
Problem 1

$$\partial_\nu(\partial_z\partial_{\bar{z}})^\mu w = \gamma_\mu, \quad 0 \le \mu \le n-1, \quad \text{on} \quad \partial D,$$

Problem 2

$$(\partial_z\partial_{\bar{z}})^\mu w = \gamma_{0\mu}, \quad 0 \le 2\mu \le n-1,$$

$$\partial_\nu(\partial_z\partial_{\bar{z}})^\mu w = \gamma_{1\mu}, \quad 0 \le 2\mu \le n-2, \quad \text{on} \quad \partial D,$$

Problem 3

$$(\partial_z \partial_{\overline{z}})^{2\mu} w = \gamma_{0\mu}, \quad 0 \le 2\mu \le n-1,$$

$$\partial_\nu (\partial_z \partial_{\overline{z}})^{2\mu+1} w = \gamma_{1\mu}, \quad 0 \le 2\mu \le n-2, \quad \text{on } \partial D,$$

Problem 4

$$(\partial_z \partial_{\overline{z}})^{2\mu+1} w = \gamma_{0\mu}, \quad 0 \le 2\mu \le n-2,$$

$$\partial_\nu (\partial_z \partial_{\overline{z}})^{2\mu} w = \gamma_{1\mu}, \quad 0 \le 2\mu \le n-1, \quad \text{on } \partial D.$$

Prescribing arbitrarily boundary conditions for the functions w_μ of the decomposition of w produces a variety of boundary conditions, and iterating the representation formulas for these functions results in a family of hybrid polyharmonic Green functions [13–17, 28–30, 39, 47, 49, 51, 56, 59, 69, 72].

2.1 Hybrid Polyharmonic Green Functions

Choosing just Dirichlet conditions for all the w_μ leads to a certain iterated polyharmonic Green function

$$\widehat{G}_\mu(z,\zeta) = -\frac{1}{\pi} \int_D G_1(z,\widetilde{\zeta}) \widehat{G}_{\mu-1}(\widetilde{\zeta},\zeta) d\widetilde{\xi} d\widetilde{\eta}, \quad 2 \le \mu \le n,$$

where $\widehat{G}_1(z,\zeta) = G_1(z,\zeta)$ for $n = 1$.

Definition 4 A real-valued function $K_m(z,\zeta)$ on $D \times D$, $z \ne \zeta$, with the properties

- *$K_m(\cdot,\zeta)$ is polyharmonic of order m in $D \setminus \{\zeta\}$,*
- *$K_m(z,\zeta) + \frac{|\zeta - z|^{2(m-1)}}{(m-1)!^2} \log |\zeta - z|^2$ is polyharmonic of order m in D for any $\zeta \in D$,*
- *$K_m(z,\zeta)$ satisfies the (boundary and side) conditions B_m for $z \in \partial D$, $\zeta \in D$,*

is called polyharmonic Green function of order m with boundary behavior B_m.

Definition 5 For two polyharmonic Green functions of order m and n and with boundary behavior B_m and $\widehat{B}_n$, K_m and $\widehat{K}_n$, respectively, the convolution

$$K_m \widehat{K}_n(z,\zeta) = -\frac{1}{\pi} \int_D K_m(z,\widetilde{\zeta}) \widehat{K}_n(\widetilde{\zeta},\zeta) d\widetilde{\xi} d\widetilde{\eta} \tag{4}$$

is called hybrid polyharmonic Green function.

Theorem 5 *$K_m\widehat{K}_n(z,\zeta)$ satisfies as a function of z for any $\zeta \in D$ the boundary value problem*

$$(\partial_z\partial_{\overline{z}})^m K_m\widehat{K}_n(z,\zeta) = \widehat{K}_n(z,\zeta) \text{ in } D, \tag{5}$$

$$B_m(K_m\widehat{K}_n(z,\zeta)) = -\frac{1}{\pi}\int_D (B_m(K_m(z,\widetilde{\zeta})))\widehat{K}_n(\widetilde{\zeta},\zeta)d\widetilde{\xi}d\widetilde{\eta} \text{ on } \partial D, \tag{6}$$

and as a function of ζ for any fixed $z \in D$

$$(\partial_\zeta\partial_{\overline{\zeta}})^n(K_m\widehat{K}_n(z,\zeta)) = K_m(z,\zeta) \text{ in } D, \tag{7}$$

$$\widehat{B}_n(K_m\widehat{K}_n(z,\zeta)) = -\frac{1}{\pi}\int_D K_m(z,\widetilde{\zeta})\widehat{B}_n(\widehat{K}_n(\widetilde{\zeta},\zeta))d\widetilde{\xi}d\widetilde{\eta} \text{ on } \partial D. \tag{8}$$

In cases where $K_m(z,\zeta)$ and $\widehat{K}_n(z,\zeta)$ are both symmetric then

$$K_m\widehat{K}_n(z,\zeta) = \widehat{K}_n K_m(\zeta, z)$$

follows.

Of particular interest are the cases when only Dirichlet and only Neumann conditions are involved leading to the iterated polyharmonic Green and Neumann functions $\widehat{G}_n(z,\zeta)$ and $N_n(z,\zeta)$, respectively, [6, 29, 39].

2.2 Riquier Problem, Iterated Polyharmonic Green Function

The iterated polyharmonic Green function $\widehat{G}_\mu$ of order μ, $2 \le \mu$, is a solution to the Dirichlet problem for the Poisson equation,

$$\partial_z\partial_{\overline{z}}\widehat{G}_\mu(z,\zeta) = \widehat{G}_{\mu-1}(z,\zeta) \text{ in } D,$$

$$\widehat{G}_\mu(z,\zeta) = 0, \quad \text{on } \partial D.$$

It has for any $\zeta \in D$ the properties

- $\widehat{G}_\mu(\cdot,\zeta)$ is a polyharmonic function of order μ in $D\backslash\{\zeta\}$,
- $\widehat{G}_\mu(z,\zeta) + \frac{|\zeta-z|^{2(\mu-1)}}{(\mu-1)!^2}\log|\zeta-z|^2$ is polyharmonic of order μ in D,
- $(\partial_z\partial_{\overline{z}})^\nu\widehat{G}_\mu(z,\zeta) = 0, 0 \le \nu \le \mu-1$, on ∂D,
- $\widehat{G}_\mu(z,\zeta) = \widehat{G}_\mu(\zeta,z)$, $z,\zeta \in D, z \ne \zeta$ (symmetry).

The normal derivatives of these polyharmonic Green functions are needed for the solution to the Riquier problem (3).

Definition 6 For $1 \le \mu$

$$g_\mu(z,\zeta) = -\frac{1}{2}\partial_{\nu_\zeta}\widehat{G}_\mu(z,\zeta),\, z \in D,\, \zeta \in \partial D$$

is called polyharmonic Poisson kernel of order μ.

Theorem 6 *The unique solution to the Riquier problem (3) is*

$$w(z) = \frac{1}{2\pi}\sum_{\mu=0}^{n-1}\int_{\partial D}\gamma_\mu(\zeta)g_{\mu+1}(z,\zeta)ds_\zeta - \frac{1}{\pi}\int_D f(\zeta)\widehat{G}_n(z,\zeta)d\xi d\eta.$$

2.3 Polyharmonic Poisson Kernels

Obviously, explicit knowledge of the polyharmonic Poisson kernels is important. Unfortunately, the procedure of repeatedly iterating the Green function is involved. Even for the unit disk polyharmonic Green functions are explicitly known only up to order 4 [29, 30, 49, 72]. But the polyharmonic Poisson kernels [45] are found in [47, 51, 52] on the basis of fife characteristic properties without explicit knowledge of the iterated polyharmonic Green functions.

Theorem 7 *The sequence of polyharmonic Poisson kernels* $\{g_n(z,\zeta)\}$ *for the unit disk* $\mathbb{D}$ *is uniquely determined by*

- $\partial_z\partial_{\bar z}g_1(z,\zeta) = 0,\ \partial_z\partial_{\bar z}g_n(z,\zeta) = g_{n-1}(z,\zeta),\quad 2 \le n,$
- $\displaystyle\lim_{z\to t,|z|<1,|t|=1}\frac{1}{2\pi i}\int_{\partial\mathbb{D}}\gamma(\zeta)g_1(z,\zeta)\frac{d\zeta}{\zeta} = \gamma(t),\quad \text{for } \gamma \in C(\partial\mathbb{D};\mathbb{C}),$
- $\displaystyle\lim_{z\to t,|z|<1,|t|=1}\frac{1}{2\pi i}\int_{\partial\mathbb{D}}\gamma(\zeta)g_2(z,\zeta)\frac{d\zeta}{\zeta} = 0 \ \text{ for } \ \gamma \in C(\partial\mathbb{D};\mathbb{C}),$
- $\displaystyle\lim_{z\to t,|z|<1,|t|=1} g_n(z,\zeta) = 0 \ \text{ for } \ 2 < n \ \text{ and } \ |\zeta| = 1,$
- $g_n(\cdot,\zeta) \in C^{2n}(\mathbb{D};\mathbb{C})$ for any $\zeta \in \partial\mathbb{D}$,

  $g_n(z,\zeta),\ \partial_z g_n(z,\zeta),\ \partial_{\bar z}g_n(z,\zeta) \in C(\mathbb{D}\times\partial\mathbb{D};\mathbb{C}),\quad n \in \mathbb{N}.$

The polyharmonic Poisson kernels for the upper half plane are constructed in [53]; in [54, 55] they are calculated for the n-dimensional unit ball and a half space.

2.4 Iterated Polyharmonic Neumann Function

Iterating the harmonic Neumann function gives the iterated polyharmonic Neumann functions N_n [39]. According to (4)

$$N_n(z,\zeta) = -\frac{1}{\pi}\int\limits_D N_1(z,\widetilde{\zeta})N_{n-1}(\widetilde{\zeta},\zeta)d\widetilde{\xi}d\widetilde{\eta},\ \ 2 \le n.$$

Theorem 8 *The polyharmonic Neumann function N_n for the unit disk $\mathbb{D}$ satisfies*

- $N_n(\cdot,\zeta)$ *is polyharmonic of order n in $\mathbb{D}\backslash\{\zeta\}$,*
- $N_n(z,\zeta) + \frac{|\zeta-z|^{2(n-1)}}{(n-1)!^2}\log|\zeta-z|^2$ *is polyharmonic of order n in $\mathbb{D}$ for any $\zeta \in \mathbb{D}$,*
- $\partial_{\nu_z}N_n(z,\zeta) = -\frac{2}{(n-1)!^2}(|\zeta|^2-1)^{n-1}$

$$+\sum_{\mu=[\frac{n}{2}]}^{n-2}\frac{\mu!^2}{(n-1)!(n-1-\mu)!^2(2\mu-n+1)!}\partial_{\nu_z}N_{\mu+1}(z,\zeta)\ \text{ for }\ z\in\partial\mathbb{D},\zeta\in\mathbb{D},$$

- $\frac{1}{2\pi i}\int_{\partial\mathbb{D}}N_n(z,\zeta)\frac{dz}{z} = 0$ for $\zeta\in\mathbb{D}$ *(normalization),*
- $N_n(z,\zeta) = N_n(\zeta,z)$ for $z,\zeta\in\mathbb{D}, z\neq\zeta$ *(symmetry).*

Moreover, for $1 < n$

$$\partial_z\partial_{\overline{z}}N_n(z,\zeta) = N_{n-1}(z,\zeta)\ \ in\ \ \mathbb{D}.$$

Explicit formulas for the first three N_n, $n = 1, 2, 3$, are given in [16, 49, 72].

The polyharmonic Neumann problem (2) with certain additional normalization conditions is conditionally uniquely solvable [39].

2.5 Polyharmonic Green-Almansi Function

The classical polyharmonic Green-Almansi function [7, 10, 56, 70] is not an iterated Green function.

Definition 7 A real-valued function $G_n(z,\zeta)$, $z,\zeta\in D$, $z\neq\zeta$, satisfying

- $G_n(\cdot,\zeta)$ is polyharmonic of order n in $D\backslash\{\zeta\}$,
- $G_n(z,\zeta) + \frac{|\zeta-z|^{2(n-1)}}{(n-1)!^2}\log|\zeta-z|^2$ is polyharmonic of order n in D for any $\zeta\in D$,
- $\partial^{\mu}_{\nu_z}G_n(z,\zeta) = 0$ for $z\in\partial D,\zeta\in D,\ \ 0\le\mu\le n-1$,
- $G_n(z,\zeta) = G_n(\zeta,z)$ for $z,\zeta\in D, z\neq\zeta$ (symmetry),

is called polyharmonic Green-Almansi function.

Moreover, G_n satisfies the additional boundary conditions

- $(\partial_z\partial_{\overline{z}})^{\mu}G_n(z,\zeta) = 0,\ 0\le 2\mu\le n-1$, for $z\in\partial D,\ \zeta\in D$,
- $\partial_{\nu_z}(\partial_z\partial_{\overline{z}})^{\mu}G_n(z,\zeta) = 0,\ 0\le 2\mu\le n-2$, for $z\in\partial D,\ \zeta\in D$.

For the unit disk $\mathbb{D}$ [10, 70]

$$G_n(z,\zeta) = \frac{|\zeta - z|^{2(n-1)}}{(n-1)!^2} \log\left|\frac{1 - z\overline{\zeta}}{\zeta - z}\right|^2$$

$$-\sum_{\mu=1}^{n-1} \frac{1}{\mu} |\zeta - z|^{2(n-1-\mu)} (1 - |z|^2)^\mu (1 - |\zeta|^2)^\mu, \quad z, \zeta \in \mathbb{D}, \quad z \neq \zeta.$$

For the upper half plane $\mathbb{H}$ [56]

$$G_n(z,\zeta) = \frac{|\zeta - z|^{2(n-1)}}{(n-1)!^2} \log\left|\frac{\overline{\zeta} - z}{\zeta - z}\right|^2$$

$$+\sum_{n=1}^{n-1} \frac{1}{\mu} |\zeta - z|^{2(n-1-\mu)} (\zeta - \overline{\zeta})^\mu (z - \overline{z})^\mu, \quad z, \zeta \in \mathbb{H}^+, \quad z \neq \zeta.$$

The polyharmonic Green-Almansi function serves for a representation formula.

Theorem 9 *Any $w \in C^{2n}(D;\mathbb{C}) \cap C^{2n-1}(\overline{D};\mathbb{C})$, $n \in \mathbb{N}$, is representable by*

$$w(z) = -\sum_{\mu=0}^{[\frac{n}{2}]-1} \frac{1}{4\pi} \int_{\partial D} \partial_{\nu_\zeta} (\partial_\zeta \partial_{\overline{\zeta}})^{n-\mu-1} G_n(z,\zeta) (\partial_\zeta \partial_{\overline{\zeta}})^\mu w(\zeta) ds_\zeta$$

$$+\sum_{\mu=0}^{[\frac{n-1}{2}]} \frac{1}{4\pi} \int_{\partial D} (\partial_\zeta \partial_{\overline{\zeta}})^{n-\mu-1} G_n(z,\zeta) \partial_{\nu_\zeta} (\partial_\zeta \partial_{\overline{\zeta}})^\mu w(\zeta) ds_\zeta \tag{9}$$

$$-\frac{1}{\pi} \int_D G_n(z,\zeta) (\partial_\zeta \partial_{\overline{\zeta}})^n w(\zeta) d\xi d\eta.$$

On the basis of this representation, in [15, 20, 44] proper polyharmonic Dirichlet problems (see Problem 2 in Sect. 2) are treated for the n-Poisson equation in the unit disk and in the upper half plane. The unique solutions are given in explicit form.

A polyharmonic Green-Almansi function for the n-dimensional unit ball is constructed in [58, 62].

2.6 A Particular Hybrid Tetraharmonic Green Function

Taking $m = n = 2$ in (4) and choosing K_2 and $\widehat{K}_2$ as G_2 and $\widehat{G}_2$, respectively, the hybrid tetraharmonic Green function [29]

$$H_4(z,\zeta) = G_2\widehat{G}_2(z,\zeta) = -\frac{1}{\pi} \int_D G_2(z,\widetilde{\zeta}) \widehat{G}_2(\widetilde{\zeta},\zeta) d\widetilde{\xi} d\widetilde{\eta}$$

is attained. As function of z it is a solution to the Dirichlet problem for the biharmonic Poisson equation

$$(\partial_z \partial_{\overline{z}})^2 H_4(z, \zeta) = \widehat{G}_2(z, \zeta) \text{ in } D \text{ for any } \zeta \in D,$$

$$H_4(z, \zeta) = 0, \ \partial_{\nu_z} H_4(z, \zeta) = 0 \text{ for } z \in \partial D, \ \zeta \in D.$$

Moreover, as a function of ζ it solves the Requier problem for the biharmonic Poisson equation

$$(\partial_\zeta \partial_{\overline{\zeta}})^2 H_4(z, \zeta) = G_2(z, \zeta) \text{ in } D \text{ for any } \zeta \in D,$$

$$H_4(z, \zeta) = 0, \ \partial_\zeta \partial_{\overline{\zeta}} H_4(z, \zeta) = 0 \text{ for } \zeta \in \partial D, \ z \in D.$$

Its properties are

- $H_4(z, \zeta)$ is tetraharmonic for $z \in D \backslash \{\zeta\}$ and for $\zeta \in D \backslash \{z\}$,
- $H_4(z, \zeta) + \frac{|\zeta - z|^6}{3!^2} \log |\zeta - z|^2$ is tetraharmonic for $z, \zeta \in D$,
- $H_4(z, \zeta) = 0, \ \partial_{\nu_z} H_4(z, \zeta) = 0, \ (\partial_z \partial_{\overline{z}})^2 H_4(z, \zeta) = 0, \ (\partial_z \partial_{\overline{z}})^3 H_4(z, \zeta) = 0$ for $z \in \partial D, \ \zeta \in D$,
- $H_4(z, \zeta) = 0, \ \partial_\zeta \partial_{\overline{\zeta}} H_4(z, \zeta) = 0, \ (\partial_\zeta \partial_{\overline{\zeta}})^2 H_4(z, \zeta) = 0, \ \partial_{\nu_\zeta} (\partial_\zeta \partial_{\overline{\zeta}})^2 H_4(z, \zeta) = 0$ for $\zeta \in \partial D, \ z \in D$.

Obviously, $H_4(z, \zeta)$ is not symmetric in its variables. Therefore two different representation formulas are available, proper for different tetraharmonic boundary value problems.

Theorem 10 *Any $w \in C^8(D; \mathbb{C}) \cap C^7(\overline{D}; \mathbb{C})$ can be represented as*

$$w(z) = -\frac{1}{4\pi} \int_{\partial D} \Big\{ \partial_{\nu_\zeta} (\partial_\zeta \partial_{\overline{\zeta}})^3 H_4(z, \zeta) w(\zeta) - (\partial_\zeta \partial_{\overline{\zeta}})^3 H_4(z, \zeta) \partial_{\nu_\zeta} w(\zeta)$$

$$+ \partial_{\nu_\zeta} \partial_\zeta \partial_{\overline{\zeta}} H_4(z, \zeta) (\partial_\zeta \partial_{\overline{\zeta}})^2 w(\zeta) + \partial_{\nu_\zeta} H_4(z, \zeta) (\partial_\zeta \partial_{\overline{\zeta}})^3 w(\zeta) \Big\} ds_\zeta$$

$$- \frac{1}{\pi} \int_D H_4(z, \zeta) (\partial_\zeta \partial_{\overline{\zeta}})^4 w(\zeta) d\xi d\eta, \tag{10}$$

$$w(z) = -\frac{1}{4\pi} \int_{\partial D} \Big\{ \partial_{\nu_\zeta} (\partial_\zeta \partial_{\overline{\zeta}})^3 H_4(\zeta, z) w(\zeta) + \partial_{\nu_\zeta} (\partial_\zeta \partial_{\overline{\zeta}})^2 H_4(\zeta, z) \partial_\zeta \partial_{\overline{\zeta}} w(\zeta)$$

$$+ \partial_{\nu_\zeta} \partial_\zeta \partial_{\overline{\zeta}} H_4(\zeta, z) (\partial_\zeta \partial_{\overline{\zeta}})^2 w(\zeta) - \partial_\zeta \partial_{\overline{\zeta}} H_4(\zeta, z) \partial_{\nu_\zeta} (\partial_\zeta \partial_{\overline{\zeta}})^2 w(\zeta) \Big\} ds_\zeta$$

$$- \frac{1}{\pi} \int_D H_4(\zeta, z) (\partial_\zeta \partial_{\overline{\zeta}})^4 w(\zeta) d\xi d\eta. \tag{11}$$

H_4 is just one of a whole variety of hybrid polyhamonic Green functions. Calculating them for simple domains will demand many area integral evaluations via the Gauss and the Cauchy integral theorems. This would be some task for computer algebraists. Also the evaluation of iterated polyharmonic Green functions could be done with computer algebra help.

References

1. Akel, M.S., Hussein, H.S.: Two basic boundary value problems for the Cauchy-Riemann equation in an infinite sector. Adv. Pure Appl. Math. **3**, 315–328 (2012)
2. Aksoy, Ü.: Schwarz problem for complex partial differential equations. Ph.D. thesis, METU, Ankara (2007)
3. Aksoy, Ü., Celebi, A.O.: Neumann problem for generalized n-Poisson equation. J. Math. Anal. Appl. **357**, 438–446 (2009)
4. Aksoy, Ü., Celebi, A.O.: Dirichlet problems for generalized n-Poisson equation. Oper. Th. Adv. Appl. **205**, 129–142 (2010)
5. Aksoy, Ü., Celebi, A.O.: Mixed boundary value problems for higher-order complex partial differential equations. Analysis **30**, 157–169 (2010)
6. Aksoy, Ü., Celebi, A.O.: A survey on boundary value problems for complex partial differential equations. Adv. Dyn. Syst. Appl. **5**, 133–158 (2010)
7. Almansi, E.: Sull'integrazione dell'equazione differenziale $\triangle^{2n} u = 0$. Ann. Math. **3**(2), 1–59 (1899)
8. Begehr, H.: Complex analytic methods for partial differential equations. An introductory text. World Scientific, Singapore (1994)
9. Begehr, H.: Iterated integral operators in Clifford analysis. ZAA **18**, 361–377 (1999)
10. Begehr, H.: Orthogonal decomposition of the function space $L_2(\overline{\mathbb{D}}; \mathbb{C})$. J. Reine Angew. Math. **549**, 191–219 (2002)
11. Begehr, H.: Boundary value problems in complex Analysis; I. II. Bol. Asoc. Mat. Venezolana **XII** 65–85; 217–250 (2005)
12. Begehr, H.: The main theorem of calculus in complex analysis. Ann. EAS **2005**, 184–210 (2006)
13. Begehr, H.: Biharmonic Green functions. Le Matematiche **LXI**, 395–405 (2006)
14. Begehr, H.: Hybrid Green functions and related boundary value problems. In: Rezapour, S. (ed.) Extended Abstracts of AIMC 37, 37th Annual Iranian Mathamatical Conference, pp. 275–278 (2006)
15. Begehr, H.: A particular polyharmonic Dirichlet problem. In: Aliyev Azeroglu, T., Tamrazov, P.M. (eds.) Complex Analysis and Potential Theory, Proceedings Conference Satellite ICM 2006. World Scientific, New Jersey, pp. 84–115 (2007)
16. Begehr, H.: Six biharmonic Dirichlet problems in complex analysis. In: Le, H.S. et al. (eds.) Function Spaces in Complex and Clifford Analysis. In: Proceedings of 14th International Conference Finite Infinite Dimensional Complex Analysis and Applications, Hue University, National University Publishers, Vietnam, Hanoi, pp. 243–252 (2008)
17. Begehr, H.: Iterated polyharmonic Green functions for plane domains. Acta Math. Vietnamica **36**, 169–181 (2011)
18. Begehr, H.: Green function for a hyperbolic strip and a class of related plane domains. Appl. Anal., **93**, 2370–2385 (2004). http://www.tandfonline.com/doi/abs/10.1080/00036811.2014.926336
19. Begehr, H.: The parqueting-reflection principle. In: Mityushev, V.V., Ruzhansky, M. (eds.) Current Trends in Analysis and its Applications, Trends in Mathematics, Springer, Switzerland, Basel, pp. 77–84 (2015)

20. Begehr, H., Gaertner, E.: A Dirichlet problem for the inhomogeneous polyharmonic equation in the upper half plane. Georg. Math. J. **14**, 33–52 (2007)
21. Begehr, H., Gilbert, R.P.: Transformations, transmutations, and kernel functions; I, II. Longman, Harlow (1992; 1993)
22. Begehr, H., Harutyunyan, G.: Robin boundary value problem for the Cauchy-Riemann operator. Complex Var. Theor. Appl. **50**, 1125–1136 (2005)
23. Begehr, H., Harutyunyan, G.: Robin boundary value problem for the Poisson equation. J. Anal. Appl. **4**, 201–213 (2006)
24. Begehr, H., Harutyunyan, G.: Neumann problem for the Beltrami operator and for second order operators with Poisson/Bitsadze operator as main part. Complex Var. Ell. Eqs. **54**, 1129–1150 (2009)
25. Begehr, H., Hile, G.N.: A hierarchy of integral operators. Rocky Mt. J. Math. **27**, 669–706 (1997)
26. Begehr, H., Kumar, A.: Boundary value problems for bi-polyanalytic functions. Appl. Anal. **85**, 1045–1077 (2006)
27. Begehr, H., Schmersau, D.: The Schwarz problem for polyanalytic functions. ZAA **24**, 341–351 (2005)
28. Begehr, H., Vaitekhovich, T.: Green functions in complex plane domains. Uzbek Math. J. **4**, 29–34 (2008)
29. Begehr, H., Vaitekhovich, T.: Iterated Dirichlet problem for higher order Poisson equation. Le Matematiche **LXIII**, 139–154 (2008)
30. Begehr, H., Vaitekhovich, T.: Polyharmonic Green functions for particular plane domains. In: Beznea, L. et al. (eds) Proceedings of 6th Congress of Romanian Mathematicians, Vol. 1. Bucharest 2007, Publ. House Romanian Acad. Sci., Bucharest, pp. 119–126 (2009)
31. Begehr, H., Vaitekhovich, T.: Harmonic boundary value problems in half disc and half ring. Funct. Approx. Comment Math. **40**, 251–282 (2009)
32. Begehr, H., Vaitekhovich, T.: Some harmonic Robin functions in the complex plane. Adv. Pure Appl. Math. **1**, 19–34 (2010)
33. Begehr, H., Vaitekhovich, T.: Green function, reflections, and plane parqueting. Eurasian Math. J. **1**, 17–31 (2010); **2**, 139–142 (2011)
34. Begehr, H., Vaitekhovich, T.: How to find harmonic Green functions in the plane. Complex Var. Ell. Eqs. **56**, 1169–1181 (2011)
35. Begehr, H., Vaitekhovich, T.: Harmonic Dirichlet problems for some equilateral triangle. Complex Var. Ell. Eqs. **57**, 185–196 (2012)
36. Begehr, H., Vaitekhovich, T.: Modified harmonic Robin functions. Complex Var. Ell. Eqs. **58**, 483–496 (2013)
37. Begehr, H., Vaitekhovich, T.: The parqueting-reflection principle for constructing Green function. In: Rogosin, S.V., Dubatovskaya, M.V. (eds.) Analytic Methods of Analysis and Differential Equations: AMADE-2012, Cambridge Scientific Publishers, Cottenham, 11–20 (2013)
38. Begehr, H., Vaitekhovich, T.: Schwarz problem in lens and lune. Complex Var. Ell. Eqs. **59**, 76–84 (2014)
39. Begehr, H., Vanegas, C.J.: Iterated Neumann problem for the higher order Poisson equation. Math. Nachr. **279**, 38–57 (2006)
40. Begehr, H., Gackstatter, F., Krausz, A.: Integral representations in octonionic analysis. In: Kajiwara, J., Kim, K.W., Shon, K.H. (eds.) Proceedings of 10th International Conference Finite, Infinite Dimensional Complex Analysis and Applications, Complex Analysis, Busan, Korea, pp. 1–7 (2002)
41. Begehr, H., Du, J., Zhang, Z.: On higher order Cauchy-Pompeiu formula in Clifford analysis and its application. Gen. Math. **11**, 5–26 (2003)
42. Begehr, H., Kumar, A., Schmersau, D., Vanegas, J.C.: Mixed complex boundary value problems in complex analysis. In: Kazame, H., Morimoto, M., Yang, C.C. (eds.) Proceedings of 12th International Conference Finite, Infinite Complex Analysis and Applications, Kyushu University Press, Fukuoka, pp. 25–40 (2005)

43. Begehr, H., Otto, H., Zhang, Z.-H.: Differential operators, their fundamental solutions, and related integral representations in Clifford analysis. Complex Var Theor. Appl. **51**, 407–427 (2006)
44. Begehr, H., Vu, T.N.H., Zhang, Z.-X.: Polyharmonic Dirichlet problems. Proc. Steklov Inst. Math. **255**, 13–34 (2006)
45. Begehr, H., Du, J., Wang, Y.: A Dirichlet problem for polyharmonic functions. Ann. Math. Pura Appl. **187**, 435–457 (2008)
46. Begehr, H., Chaudhary, A., Kumar, A.: Bi-polyanalytic functions on the upper half plane. Complex Var. Ell. Eqs. **55**, 305–316 (2010)
47. Begehr, H., Du, Z., Wang, N.: Dirichlet problems for inhomogeneous complex mixed differential equations in the unit disc: new view. Oper. Th. Adv. Appl. **205**, 101–128 (2010)
48. Begehr, H., Costache, M.-R., Tappert, S., Vaitekhovich, T: Harmonic Green and Neumann representations in a triangle, quarter disc, and octo plane. In: Ruzhansky, M., Wirth, J. (eds.) Progress in Analysis. Proceedings of 7th International ISAAC Congress, London, 2009, World Scientific, Singapore, pp. 74–80 (2010)
49. Burgumbayeva, S.: Boundary value problems for triharmonic functions in the unit disc. Ph.D. thesis, FU Berlin, 2009. http://www.diss.fu-berlin.de/diss/receive/FUDISS_thesis_000000012636
50. Chaudhary, A.: Complex boundary value problems in unbounded domains. Ph.D. thesis, University of Delhi (2009)
51. Du, Z.: Boundary value problems for higher order complex partial differential equations. Ph.D. thesis, FU Berlin (2008). http://www.diss.fu-berlin.de/diss/receive/FUDISS_thesis_000000003677
52. Du, Z., Kou, K.L., Wang, J.: L^p polyharmonic Dirichlet problems in regular domains. I: The unit disc. Complex Var. Ell. Equ. **58**, 1387–1405 (2013)
53. Du, Z., Qian, T., Wang, J.: L^p polyharmonic Dirichlet problems in regular domains. II: the upper half plane. J. Differ. Equ. **252**, 1789–1812 (2013)
54. Du, Z., Qian, T., Wang, J.: L^p polyharmonic Dirichlet problems in regular domains. IV: the upper-half space. J. Differ. Equ. **255**, 779–795 (2013)
55. Du, Z., Qian, T., Wang, J.: L^p polyharmonic Dirichlet problems in regular domains. III: The unit ball. Complex Var. Ell. Equ. **59**, 947–965 (2014)
56. Gaertner, E.: Basic complex boundary value problems in the upper half plane. Ph.D. thesis, FU Berlin, (2006). http://www.diss.fu-berlin.de/diss/receive/FUDISS_thesis_000000002129
57. Gazzola, F., Grunau, H.-Ch., Sweers, G.: Polyharmonic boundary value problems. Positivity preserving and nonlinear higher order elliptic equations in bounded domains. Lecture Notes Mathamatics 1991, Springer, Berlin (2010)
58. Kalmenov., TSh, Koshanov, B.D., Nemchenko, M.Y.: Green function representation for the Dirichlet problem of the polyharmonic equation in a sphere. Complex Var. Ell. Equ. **53**, 177–183 (2008)
59. Krausz, A.: Integraldarstellungen mit Greenschen Funktionen höherer Ordnung in Gebieten und Polygebieten, Ph.D. thesis, FU Berlin, 2005. http://www.diss.fu-berlin.de/diss/receive/FUDISS_thesis_000000001659
60. Kumar, A., Prakash, R.: Iterated boundary value problems for the inhomogeneous polyanalytic equation. Complex Var. Ell. Equ. **52**, 921–932 (2007)
61. Linares, Y.R., Vanegas, C.J.: A Robin boundary value problem in the upper half plane for the Bitsadze equation. J. Math. Anal. Appl. **419**, 200–217 (2014)
62. Nemchenko, M.: Explicit construction of a Green function for a poly-harmonic equation in the case of an even space dimension. Ph.D. thesis, Kazakh National al-Farabi University, Almaty, 2009 (Russian)
63. Nicolesco, M.: Les fonctions polyharmoniques. Actual Sci. Ind. **331**, Hermann, Paris (1936)
64. Otto, H.A.M.: Cauchy-Pompeiusche Integraldarstellungen in der Clifford Analysis. Ph.D. thesis, FU Berlin, 2006. http://www.diss.fu-berlin.de/diss/receive/FUDISS_thesis_000000002246
65. Prakash, R.: Boundary value in complex analysis. Ph.D. thesis, University of Delhi (2006)

66. Riquier, C.H.: Sur quelques problems relatifs à l'équation aux derivées patielle $(\frac{\partial^2}{\partial x^2} + \frac{\partial^2}{\partial y^2})^n$ $u = 0$. J. de Math. **9**(5), 297–393 (1926)
67. Shupeyeva, B.: Some basic boundary value problems for complex partial differential equations in quarter ring and half hexagon. Ph.D. thesis, FU Berlin, 2013. http://www.diss.fu-berlin.de/diss/receive/FUDISS_thesis_000000094596
68. Shupeyeva, B.: Boundary value problems and method of reflection for quarter ring and half hexagon. In: Mityushev, V.V., Ruzhansky, M. (eds.) Current Trends in Analysis and its Applications, Trends in Mathematics, Springer, Switzerland, Basel, pp. 59–66 (2015)
69. Vaitsiakhovich, T.: Boundary value problems for complex partial differential equations in a ring domain. Ph.D. thesis, FU Berlin, (2008). www.diss.fu-berlin.de/diss/receive/FUDISS_thesis_000000003859
70. Vekua, I.N.: New methods for solving elliptic equations. North Holland Publishing Company, Wiley, Amsterdam, New York (1967)
71. Vu, T.N.H.: Intregral representations in quaternionic analysis related to the Helmholtz operator. Ph.D. thesis, FU Berlin, 2005. http://www.diss.fu-berlin.de/diss/receive/FUDISS_thesis_000000001591
72. Wang, Y. Boundary value problems for complex partial differential equations in fan-shaped domains. Ph.D. thesis, FU Berlin, (2011). http://www.diss.fu-berlin.de/diss/receive/FUDISS_thesis_000000021359
73. Wang, Y., Du, J.: Harmonic Dirichlet problem in a ring sector. In: Mityushev, V.V., Ruzhansky, M. (eds.) Current Trends in Analysis and its Applications, Trends in Mathematics, Springer Switzerland, Basel, pp. 67–75 (2015)

Higher Order Hybrid Invexity Frameworks and Discrete Multiobjective Fractional Programming Problems

Ram U. Verma

Abstract Based on the higher order hybrid $(\Phi, \rho, \eta, \zeta, \theta)-$invexities, first some parametrically generalized sufficient efficiency conditions for multiobjective fractional programming are developed and then efficient solutions to the multiobjective fractional programming problems are established. Furthermore, the obtained results on sufficient efficiency conditions are generalized to the case of the $\varepsilon-$efficient solutions. The results thus obtained generalize and unify a wide spectrum of investigations on the theory and applications to the multiobjective fractional programming based on the hybrid $(\Phi, \rho, \eta, \zeta, \theta)-$invexity frameworks.

Keywords Higher order hybrid invexity · Multiobjective fractional programming · Efficient solutions

AMS Subject Classification 90C32 · 90C45

1 Introduction

Mangasarian [8] investigated second order duality for a conventional nonlinear programming problem, where the approach is based on constructing a second order dual problem by taking linear and quadratic approximations of the objective and constraint functions for an arbitrary but fixed point leading to the Wolfe dual model for the approximated problem, while letting the fixed point to vary. Recently, Verma [22] investigated a general framework for a class of $(\rho, \eta, \theta)-$invex functions to examine some parametric sufficient efficiency constraints for multiobjective fractional programming problems leading to weakly $\varepsilon-$efficient solutions. Motivated by these research developments, we first introduce the higher order hybrid $(\Phi, \rho, \eta, \zeta, \theta)-$invexities, second, introduce some parametrically sufficient efficiency conditions for multiobjective fractional programming, and finally, explore the

R.U. Verma (✉)
Department of Mathematics, Texas State University, San Marcos, TX 78666, USA
e-mail: verma99@msn.com

R.N. Mohapatra et al. (eds.), *Mathematics and Computing*,
Springer Proceedings in Mathematics & Statistics 139,
DOI 10.1007/978-81-322-2452-5_2

efficient solutions to multiobjective fractional programming problems. The results established in this paper, not only generalize and unify the results on general sufficient efficiency conditions for multiobjective fractional programming problems based on the hybrid invexity of functions, but also generalize second order invexity results in more general settings.

We consider, based on the higher order hybrid $(\Phi, \rho, \eta, \zeta, \theta)$−invexities of functions, the following multiobjective fractional programming problem:
(P)

$$Minimize\left(\frac{f_1(x)}{g_1(x)}, \frac{f_2(x)}{g_2(x)}, \cdots, \frac{f_p(x)}{g_p(x)}\right)$$

subject to $x \in Q = \{x \in X : H_j(x) \leq 0, j \in \{1, 2, \cdots, m\}\}$, where X is an open convex subset of $\Re^n$ (n-dimensional Euclidean space), f_i and g_i for $i \in \{1, \cdots, p\}$ and H_j for $j \in \{1, \cdots, m\}$ are real-valued functions defined on X such that $f_i(x) \geq 0$, $g_i(x) > 0$ for $i \in \{1, \cdots, p\}$ and for all $x \in Q$. Here Q denotes the feasible set of (P).

Next, we observe that problem (P) is equivalent to the nonfractional programming problem:
(Pλ)

$$Minimize\Big(f_1(x) - \lambda_1 g_1(x), \cdots, f_p(x) - \lambda_p g_p(x)\Big)$$

subject to $x \in Q$ with

$$\lambda = \Big(\lambda_1, \lambda_2, \cdots, \lambda_p\Big) = \left(\frac{f_1(x^*)}{g_1(x^*)}, \frac{f_2(x^*)}{g_2(x^*)}, \cdots, \frac{f_p(x^*)}{g_p(x^*)}\right),$$

where x^* is an efficient solution to (P).
General mathematical programming problems offer a wide range of applications to other fields, such as applications to game theory, statistical analysis, engineering design (including design of control systems, design of earthquake-resistant structures, digital filters, and electronic circuits), random graphs, boundary value problems, wavelet analysis, environmental protection planning, decision and management sciences, optimal control problems, continuum mechanics, and others. Recently, Pitea and Postolache [18] introduced and studied a new class of multitime multiobjective variational problems for minimizing a vector of functionals of curvilinear integral type relating to Mond-Weir-Zalmai type duality based on the notion of (ρ, b)-quasiivexity. They also established some weak duality theorems showing the value of the objective function of the primal cannot exceed the value of the dual. On the other hand, there are accelerated advances on duality models for a class of multiobjective control problems with generalized invexity, especially the work of Zhian and Qingkai [41], where they have discussed the duality models for multiobjective control problems using the generalized invexity. For more details on generalized efficiency and efficiency results and applications, we recommend the reader [1–41].

This submission is organized as follows: the introductory section deals with a brief historical development for the multiobjective fractional mathematical programming, while emphasizing the roles of the generalized invex functions. In Sect. 2, the hybrid $(\Phi, \rho, \eta, \zeta, \theta)-$invex functions of higher orders are introduced, and Sect. 3 deals with sufficient efficiency conditions leading to the solvability of the problem (P) using the hybrid $(\Phi, \rho, \eta, \zeta, \theta)-$invexities.

2 Hybrid Invexities

In this section, we introduce the notion of higher order $(\Phi, \rho, \eta, \zeta, \theta)-$invexities, which encompass most of the existing generalized invexities in the current literature. Let X be an open convex subset of $\Re^n$ (n-dimensional Euclidean space). Let $\langle \cdot, \cdot \rangle$ denote the inner product, and let $z \in \Re^n$. Suppose that $f : X \rightarrow \Re$ is a real-valued twice continuously differentiable function defined on X, and that $\triangledown f(y)$ and $\nabla^2 f(y)$ denote, respectively, the gradient and Hessian of f at y.

Definition 2.1 A twice differentiable function $f : X \rightarrow \Re$ is said to be hybrid $(\Phi, \rho, \eta, \zeta, \theta)-$invex at x^* of second order if there exists a function $\Phi : \Re \rightarrow \Re$ such that for each $x \in X$, $\rho : X \times X \rightarrow \Re$, $\eta, \theta, \zeta : X \times X \rightarrow \Re^n$, and $z \in \Re^n$,

$$\Phi\Big(f(x) - f(x^*)\Big) \geq \langle \triangledown f(x^*) + \nabla^2 f(x^*)z, \eta(x, x^*)\rangle - \frac{1}{2}\big\langle \nabla^2 f(x^*)z, \zeta(x, x^*)\big\rangle + \rho(x, x^*)\|\theta(x, x^*)\|^2.$$

Definition 2.2 A twice differentiable function $f : X \rightarrow \Re$ is said to be hybrid $(\Phi, \rho, \eta, \zeta, \theta)-$pseudo-invex at x^* of second order if there exists a function $\Phi : \Re \rightarrow \Re$ such that for each $x \in X$, $\rho : X \times X \rightarrow \Re$, $\eta, \zeta, \theta : X \times X \rightarrow \Re^n$, and $z \in \Re^n$,

$$\langle \triangledown f(x^*) + \nabla^2 f(x^*)z, \eta(x, x^*)\rangle - \frac{1}{2}\big\langle \nabla^2 f(x^*)z, \zeta(x, x^*)\big\rangle + \rho(x, x^*)\|\theta(x, x^*)\|^2 \geq 0$$
$$\Rightarrow \Phi\Big(f(x) - f(x^*)\Big) \geq 0.$$

Definition 2.3 A twice differentiable function $f : X \rightarrow \Re$ is said to be strictly hybrid $(\Phi, \rho, \eta, \zeta, \theta)-$ pseudo-invex at x^* of second order if there exists a function $\Phi : \Re \rightarrow \Re$ such that for each $x \in X$, $\rho : X \times X \rightarrow \Re$, $\eta, \theta : X \times X \rightarrow \Re^n$, and $z \in \Re^n$,

$$\langle \triangledown f(x^*) + \nabla^2 f(x^*)z, \eta(x, x^*)\rangle - \frac{1}{2}\big\langle \nabla^2 f(x^*)z, \zeta(x, x^*)\big\rangle + \rho(x, x^*)\|\theta(x, x^*)\|^2 \geq 0$$
$$\Rightarrow \Phi\Big(f(x) - f(x^*)\Big) > 0.$$

Definition 2.4 A twice differentiable function $f : X \rightarrow \Re$ is said to be prestrictly hybrid $(\Phi, \rho, \eta, \zeta, \theta)-$pseudo-invex at x^* of second order if there exists a function

$\Phi : \Re \to \Re$ such that for each $x \in X$, $\rho : X \times X \to \Re$, $\eta, \zeta, \theta : X \times X \to \Re^n$, and $z \in \Re^n$,

$$\langle \nabla f(x^*) + \nabla^2 f(x^*)z, \eta(x, x^*)\rangle - \frac{1}{2}\langle \nabla^2 f(x^*)z, \zeta(x, x^*)\rangle + \rho(x, x^*)\|\theta(x, x^*)\|^2 > 0$$
$$\Rightarrow \Phi\Big(f(x) - f(x^*)\Big) \geq 0.$$

Definition 2.5 A twice differentiable function $f : X \to \Re$ is said to be hybrid $(\Phi, \rho, \eta, \zeta, \theta)-$quasi-invex at x^* of second order if there exists a function $\Phi : \Re \to \Re$) such that for each $x \in X$, $\rho : X \times X \to \Re$, $\eta, \zeta, \theta : X \times X \to \Re^n$, and $z \in \Re^n$,

$$\Phi\big(f(x) - f(x^*)\big) \leq 0$$
$$\Rightarrow \langle \nabla f(x^*) + \nabla^2 f(x^*)z, \eta(x, x^*)\rangle - \frac{1}{2}\langle \nabla^2 f(x^*)z, \zeta(x, x^*)\rangle + \rho(x, x^*)\|\theta(x, x^*)\|^2 \leq 0.$$

Definition 2.6 A twice differentiable function $f : X \to \Re$ is said to be strictly hybrid $(\Phi,\ \rho,\ \eta,\ \zeta,\ \theta)-$quasi-invex at x^* of second order if there exists a function $\Phi : \Re \to \Re$ such that for each $x \in X$, $\rho : X \times X \to \Re$, $\eta, \zeta, \theta : X \times X \to \Re^n$, and $z \in \Re^n$,

$$\Phi\big(f(x) - f(x^*)\big) \leq 0$$
$$\Rightarrow \langle \nabla f(x^*) + \nabla^2 f(x^*)z, \eta(x, x^*)\rangle - \frac{1}{2}\langle \nabla^2 f(x^*)z, \zeta(x, x^*)\rangle + \rho(x, x^*)\|\theta(x, x^*)\|^2 < 0.$$

Definition 2.7 A twice differentiable function $f : X \to \Re$ is said to be prestrictly hybrid $(\Phi, \rho, \eta, \zeta, \theta)-$quasi-invex at x^* of second order if there exists a function $\Phi : \Re \to \Re$ such that for each $x \in X$, $\rho : X \times X \to \Re$, $\eta, \zeta, \theta : X \times X \to \Re^n$, and $z \in \Re^n$,

$$\Phi\big(f(x) - f(x^*)\big) < 0$$
$$\Rightarrow \langle \nabla f(x^*) + \nabla^2 f(x^*)z, \eta(x, x^*)\rangle - \frac{1}{2}\langle \nabla^2 f(x^*)z, \zeta(x, x^*)\rangle + \rho(x, x^*)\|\theta(x, x^*)\|^2 \leq 0,$$

equivalently,

$$\langle \nabla f(x^*) + \nabla^2 f(x^*)z, \eta(x, x^*)\rangle - \frac{1}{2}\langle \nabla^2 f(x^*)z, \zeta(x, x^*)\rangle + \rho(x, x^*)\|\theta(x, x^*)\|^2 > 0$$
$$\Rightarrow \Phi\big(f(x) - f(x^*)\big) \geq 0.$$

Definition 2.8 A point $x^* \in Q$ is an efficient solution to (P) if there exists no $x \in Q$ such that

$$\frac{f_i(x)}{g_i(x)} \leq \frac{f_i(x^*)}{g_i(x^*)} \forall i = 1, \cdots, p,$$

$$\frac{f_j(x)}{g_j(x)} < \frac{f_j(x^*)}{g_j(x^*)} \text{ for some } j \in \{1, \cdots, p\}.$$

Next to this context, we have the following auxiliary problem:
(P$\bar{\lambda}$)

$$minimize_{x \in Q}(f_1(x) - \bar{\lambda}_1 g_1(x), \cdots, f_p(x) - \bar{\lambda}_p g_p(x)),$$

subject to $x \in Q$,
where $\bar{\lambda}_i$ for $i \in \{1, \cdots, p\}$ are parameters, and $\bar{\lambda}_i = \frac{f_i(x^*)}{g_i(x^*)}$.

Example 2.1 Consider a twice differentiable function $f : X \to \Re$ such that there exist functions $\Phi : \Re \to \Re$, $\rho : X \times X \to \Re$, $\eta, \theta, \zeta : X \times X \to \Re^n$. Then f is hybrid $(\Phi, \rho, \eta, \zeta, \theta)-$invex at x^* of second order since for each $x \in X$, and $z \in \Re^n$,

$$\Phi\Big(f(x) - f(x^*)\Big) \geq \langle \nabla f(x^*) + \frac{1}{2}\nabla^2 f(x^*)z, \eta(x, x^*)\rangle + \rho(x, x^*)\|\theta(x, x^*\|^2.$$

Example 2.2 Consider a differentiable function $f : X \to \Re$ such that there exist functions $\Phi : \Re \to \Re$, $\rho : X \times X \to \Re$, $\eta, \theta, \zeta : X \times X \to \Re^n$. Then f is hybrid $(\Phi, \rho, \eta, \theta)-$invex at x^* of first order since for each $x \in X$, and $z \in \Re^n$,

$$\Phi\Big(f(x) - f(x^*)\Big) \geq \langle \nabla f(x^*), \eta(x, x^*)\rangle + \rho(x, x^*)\|\theta(x, x^*\|^2.$$

Next, we introduce the efficiency solvability conditions for (P$\bar{\lambda}$) problem.

Definition 2.9 A point $x^* \in Q$ is an efficient solution to (P$\bar{\lambda}$) if there does not exist an $x \in Q$ such that

$$f_i(x) - \bar{\lambda}_i g_i(x) \leq f_i(x^*) - \bar{\lambda}_i g_i(x^*) \,\forall\, i = 1, \cdots, p,$$

$$f_j(x) - \bar{\lambda}_j g_j(x) < f_j(x^*) - \bar{\lambda}_j g_j(x^*) \text{ for some} j \in \{1, \cdots, p\},$$

where $\bar{\lambda}_i = \frac{f_i(x^*)}{g_i(x^*)}$ for $i = 1, \cdots, p$.
Next, we recall the following result (Verma [24]) that provides a set of necessary efficiency conditions for problem (P) for developing some sufficient efficiency conditions for the next section based on second order $(\Phi, \rho, \eta, \zeta, \theta)-$invexities.

Theorem 2.1 *[24] Let $x^* \in \mathbb{F}$ and $\lambda^* = \max_{1 \leq i \leq p} f_i(x^*)/g_i(x^*)$ for each $i \in \underline{p}$, and let f_i and g_i be twice continuously differentiable at x^* for each $i \in \underline{p}$. For each $j \in \underline{q}$, let the function $z \to G_j(z, t)$ be twice continuously differentiable at x^* for all $t \in T_j$, and for each $k \in \underline{r}$, let the function $z \to H_k(z, s)$ be twice continuously differentiable at x^* for all $s \in S_k$. If x^* is an efficient solution of (P), if the second order generalized Abadie constraint qualification holds at x^*, and if for any critical direction y, the set cone*

$$\{\Big(\nabla G_j(x^*,t),\langle y,\nabla^2 G_j(x^*,t)y\rangle\Big):t\in\hat{T}_j(x^*),\ j\in\underline{q}\}$$
$$+\ span\{\Big(\nabla H_k(x^*,s),\langle y,\nabla^2 H_k(x^*,s)y\rangle\Big):s\in S_k,k\in\underline{r}\},$$

where $\hat{T}_j(x^*)\equiv\{t\in T_j:G_j(x^*,t)=0\}$, *is closed, then there exist* $u^*\in U\equiv\{u\in\mathbb{R}^p:u\geq 0,\sum_{i=1}^p u_i=1\}$ *and integers* ν_0^* *and* ν^* *with* $0\leq\nu_0^*\leq\nu^*\leq n+1$ *such that there exist* ν_0^* *indices* j_m *with* $1\leq j_m\leq q$ *together with* ν_0^* *points* $t^m\in\hat{T}_{j_m}(x^*)$, $m\in\underline{\nu_0^*}$, $\nu^*-\nu_0^*$ *indices* k_m *with* $1\leq k_m\leq r$ *together with* $\nu^*-\nu_0^*$ *points* $s^m\in S_{k_m}$ *for* $m\in\underline{\nu^*}\backslash\underline{\nu_0^*}$, *and* ν^* *real numbers* v_m^* *with* $v_m^*>0$ *for* $m\in\underline{\nu_0^*}$ *with the property that*

$$\sum_{i=1}^p u_i^*[\nabla f_i(x^*)-\lambda^*(\nabla g_i(x^*)]+\sum_{m=1}^{\nu_0^*}v_m^*[\nabla G_{j_m}(x^*,t^m)$$
$$+\sum_{m=\nu_0^*+1}^{\nu^*}v_m^*\nabla H_k(x^*,s^m)=0, \tag{2.1}$$

$$\langle y,\Big[\sum_{i=1}^p u_i^*[\nabla^2 f_i(x^*)-\lambda^*\nabla^2 g_i(x^*)]+\sum_{m=1}^{\nu_0^*}v_m^*\nabla^2 G_{j_m}(x^*,t^m)$$
$$+\sum_{m=\nu_0^*+1}^{\nu^*}v_m^*\nabla^2 H_k(x^*,s^m)\Big]y\rangle\geq 0, \tag{2.2}$$

$$u_i^*[f_i(x^*)-\lambda^* g_i(x^*)]=0,\quad i\in\underline{p}, \tag{2.3}$$

where $\underline{\nu}\setminus\underline{\nu_0}$ *is the complement of the set* $\underline{\nu_0}$ *relative to the set* $\underline{\nu}$.

3 Sufficient Efficiency Conditions for Problem (P)

This section deals with some parametrically sufficient efficiency conditions for problem (P) under the hybrid frameworks for $(\Phi,\rho,\eta,\zeta,\theta)-$invexities. We begin with real-valued functions $E_i(.,x^*,u^*)$ and $B_j(.,v)$ defined by

$$E_i(x,x^*,u^*)=u_i[f_i(x)-\Big(\frac{f_i(x^*)}{g_i(x^*)}\Big)g_i(x)],\ i\in\{1,\cdots,p\}$$

and

$$B_j(.,v)=v_jH_j(x),\ \ j=1,\cdots,m.$$

Theorem 3.1 *Let $x^* \in Q$, f_i, g_i for $i \in \{1, \cdots, p\}$ with $\frac{f_i(x^*)}{g_i(x^*)} \geq 0$, $g_i(x^*) > 0$ and H_j for $j \in \{1, \cdots, m\}$ be twice continuously differentiable at $x^* \in Q$, and let there exist $u^* \in U = \{u \in \Re^p : u > 0, \Sigma_{i=1}^p u_i = 1\}$ and $v^* \in \Re_+^m$ such that*

$$\Sigma_{i=1}^p u_i^*[\nabla f_i(x^*) - (\frac{f_i(x^*)}{g_i(x^*)}) \nabla g_i(x^*)] + \Sigma_{j=1}^m v_j^* \nabla H_j(x^*) = 0, \tag{3.1}$$

$$\Big\langle\Big[\sum_{i=1}^p u_i^*[\nabla^2 f_i(x^*) - (\frac{f_i(x^*)}{g_i(x^*)})\nabla^2 g_i(x^*)] + \sum_{j=1}^m v_j^*\nabla^2 H_j(x^*)\Big]z, \eta(x, x^*)\Big\rangle \geq 0, \tag{3.2}$$

$$-\frac{1}{2}\Big\langle\Big[\sum_{i=1}^p u_i^*[\nabla^2 f_i(x^*) - (\frac{f_i(x^*)}{g_i(x^*)})\nabla^2 g_i(x^*)] + \sum_{j=1}^m v_j^*\nabla^2 H_j(x^*)\Big]z, \zeta(x, x^*)\Big\rangle \geq 0, \tag{3.3}$$

and

$$v_j^* H_j(x^*) = 0, \ j \in \{1, \cdots, m\}. \tag{3.4}$$

Suppose, in addition, that any one of the following assumptions holds (for $\rho(x, x^*) \geq 0$):

(i) $E_i(.\,; x^*, u^*)$ $\forall i \in \{1, \cdots, p\}$ are hybrid $(\Phi, \rho, \eta, \zeta, \theta)-$pseudo-invex at x^* with $\bar{\Phi}(a) \geq 0 \Rightarrow a \geq 0$, and $B_j(.\,, v^*)$ $\forall j \in \{1, \cdots, m\}$ are hybrid $(\tilde{\Phi}, \bar{\rho}, \eta, \zeta, \theta)-$quasi-invex at x^* for $\tilde{\Phi}$ increasing and $\tilde{\Phi}(0) = 0$.

(ii) $E_i(.\,; x^*, u^*)$ $\forall i \in \{1, \cdots, p\}$ are prestrictly hybrid $(\Phi, \rho, \eta, \zeta, \theta)-$pseudo-invex at x^* for $\Phi(a) \geq 0 \Rightarrow a \geq 0$, and $B_j(.\,, v^*)$ $\forall j \in \{1, \cdots, m\}$ are strictly hybrid $(\tilde{\Phi}, \rho, \eta, \zeta, \theta)-$quasi-invex at x^* for $\tilde{\Phi}$ increasing and $\tilde{\Phi}(0) = 0$.

(iii) $E_i(.\,; x^*, u^*)$ $\forall i \in \{1, \cdots, p\}$ are prestrictly hybrid $(\Phi, \rho, \eta, \zeta, \theta)-$quasi-invex at x^* for $\Phi(a) \geq 0 \Rightarrow a \geq 0$, and $B_j(.\,, v^*)$ $\forall j \in \{1, \cdots, m\}$ are strictly hybrid $(\tilde{\Phi}, \bar{\rho}, \eta, \zeta, \theta)-$quasi-invex at x^* for $\tilde{\Phi}$ increasing and $\tilde{\Phi}(0) = 0$.

(iv) For each $i \in \{1, \cdots, p\}$, f_i is hybrid $(\Phi, \rho_1, \eta, \theta)-$invex and $-g_i$ is hybrid $(\Phi, \Psi, \rho_2, \eta, \theta)-$invex at x^* for $\Phi(a) \geq 0 \Rightarrow a \geq 0$, $H_j(.\,, v^*)$ $\forall j \in \{1, \cdots, m\}$ is hybrid $(\bar{\Phi}, \rho_3, \eta, \theta)-$quasi-invex at x^* for $\bar{\Phi}$ increasing and $\bar{\Phi}(0) = 0$, and $\Sigma_{j=1}^m v_j^* \rho_3(x, x^*) + \rho^*(x, x^*) \geq 0$ for $\rho^* = \Sigma_{i=1}^p u_i^*(\rho_1(x, x^*) + \phi(x^*)\rho_2(x, x^*))$ and for $\phi(x^*) = \frac{f_i(x^*)}{g_i(x^*)}$.

Then x^* is an efficient solution to (P).

Proof If (i) holds, and if $x \in Q$, then it follows from (3.1)–(3.3) that

$$\Big\langle \Sigma_{i=1}^p u_i^*[\nabla f_i(x^*) - (\frac{f_i(x^*)}{g_i(x^*)}) \nabla g_i(x^*)] + \sum_{i=1}^p u_i^*[\nabla^2 f_i(x^*)z - (\frac{f_i(x^*)}{g_i(x^*)})\nabla^2 g_i(x^*)z], \eta(x, x^*)\Big\rangle$$

$$
\begin{aligned}
&-\frac{1}{2}\Big\langle\sum_{i=1}^{p}u_i^*[\nabla^2 f_i(x^*)z-(\frac{f_i(x^*)}{g_i(x^*)})\nabla^2 g_i(x^*)z],\zeta(x,x^*)\Big\rangle\\
&+\Big\langle\Sigma_{j=1}^{m}v_j^*\nabla H_j(x^*)+\Sigma_{j=1}^{m}v_j^*\nabla^2 H_j(x^*)z,\eta(x,x^*)\Big\rangle\\
&-\frac{1}{2}\Big\langle\Sigma_{j=1}^{m}v_j^*\nabla^2 H_j(x^*)z,\zeta(x,x^*)\Big\rangle\geq 0. \qquad (3.5)
\end{aligned}
$$

Since $v^*\geq 0$, $x\in Q$ and (3.4) holds, we have

$$\Sigma_{j=1}^{m}v_j^*H_j(x)\leq 0=\Sigma_{j=1}^{m}v_j^*H_j(x^*),$$

and in light of assumptions on $\tilde{\Phi}$, we find

$$\tilde{\Phi}\Big(\Sigma_{j=1}^{m}v_j^*H_j(x)-\Sigma_{j=1}^{m}v_j^*H_j(x^*)\Big)\leq 0,$$

which applying the hybrid $(\tilde{\Phi},\bar{\rho},\eta,\zeta,\theta)$−quasi-invexity of $B_j(.,v^*)$ at x^*, results in

$$
\begin{aligned}
&\Big\langle\Sigma_{j=1}^{m}v_j^*\nabla H_j(x^*)+\Sigma_{j=1}^{m}v_j^*\nabla^2 H_j(x^*)z,\eta(x,x^*)\Big\rangle\\
&-\frac{1}{2}\Big\langle\Sigma_{j=1}^{m}v_j^*\nabla^2 H_j(x^*)z,\zeta(x,x^*)\Big\rangle+\bar{\rho}(x,x^*)\|\theta(x,x^*)\|^2\leq 0. \qquad (3.6)
\end{aligned}
$$

It follows from (3.5) and (3.6) that

$$
\begin{aligned}
&\Big\langle\Sigma_{i=1}^{p}u_i^*[\nabla f_i(x^*)-(\frac{f_i(x^*)}{g_i(x^*)})\nabla g_i(x^*)]\\
&+\sum_{i=1}^{p}u_i^*[\nabla^2 f_i(x^*)z-(\frac{f_i(x^*)}{g_i(x^*)})\nabla^2 g_i(x^*)z],\eta(x,x^*)\Big\rangle\\
&-\frac{1}{2}\Big\langle\sum_{i=1}^{p}u_i^*[\nabla^2 f_i(x^*)z-(\frac{f_i(x^*)}{g_i(x^*)})\nabla^2 g_i(x^*)z],\zeta(x,x^*)\Big\rangle\\
&\geq\bar{\rho}(x,x^*)\|\theta(x,x^*)\|^2\geq-\rho(x,x^*)\|\theta(x,x^*)\|^2. \qquad (3.7)
\end{aligned}
$$

Since $\rho(x,x^*)\geq 0$, applying the hybrid $(\Phi,\rho,\eta,\zeta,\theta)$−pseudo-invexity at x^* to (3.7) and assumptions on Φ, we have

$$\Phi\Big(\Sigma_{i=1}^{p}u_i^*[f_i(x)-(\frac{f_i(x^*)}{g_i(g^*)})g_i(x)]-\Sigma_{i=1}^{p}u_i^*[f_i(x^*)-(\frac{f_i(x^*)}{g_i(x^*)})g_i(x^*)]\Big)\geq 0,$$

which implies

$$\begin{aligned}&\Sigma_{i=1}^{p}u_i^*[f_i(x)-(\frac{f_i(x^*)}{g_i(x^*)})g_i(x)]\\ &\geq \Sigma_{i=1}^{p}u_i^*[f_i(x^*)-(\frac{f_i(x^*)}{g_i(x^*)})g_i(x^*)])\\ &=0.\end{aligned}$$

Thus, we have

$$\Sigma_{i=1}^{p}u_i^*[f_i(x)-(\frac{f_i(x^*)}{g_i(x^*)})g_i(x)]\geq 0. \tag{3.8}$$

Since $u_i^* > 0$ for each $i \in \{1, \cdots, p\}$, we conclude that there does not exist an $x \in Q$ such that

$$\frac{f_i(x)}{g_i(x)}-(\frac{f_i(x^*)}{g_i(x^*)})\leq 0\ \forall i=1,\cdots,p,$$

$$\frac{f_j(x)}{g_j(x)}-(\frac{f_j(x^*)}{g_j(x^*)})<0 \quad \text{for some } j\in\{1,\cdots,p\}.$$

Hence, x^* is an efficient solution to (P).

Next, If (ii) holds, and if $x \in Q$, then it follows from (3.1)–(3.3) that

$$\begin{aligned}&\Big\langle \Sigma_{i=1}^{p}u_i^*[\nabla f_i(x^*)-(\frac{f_i(x^*)}{g_i(x^*)})\nabla g_i(x^*)]\\ &+\sum_{i=1}^{p}u_i^*[\nabla^2 f_i(x^*)z-(\frac{f_i(x^*)}{g_i(x^*)})\nabla^2 g_i(x^*)z],\eta(x,x^*)\Big\rangle\\ &-\frac{1}{2}\Big\langle\sum_{i=1}^{p}u_i^*[\nabla^2 f_i(x^*)z-(\frac{f_i(x^*)}{g_i(x^*)})\nabla^2 g_i(x^*)z],\zeta(x,x^*)\Big\rangle\\ &+\Big\langle\Sigma_{j=1}^{m}v_j^*\nabla H_j(x^*)+\Sigma_{j=1}^{m}v_j^*\nabla^2 H_j(x^*)z,\eta(x,x^*)\Big\rangle\\ &-\frac{1}{2}\Big\langle\Sigma_{j=1}^{m}v_j^*\nabla^2 H_j(x^*)z,\zeta(x,x^*)\Big\rangle\geq 0.\end{aligned} \tag{3.9}$$

Since $v^* \geq 0$, $x \in Q$ and (3.3) holds, we have

$$\Sigma_{j=1}^{m}v_j^*H_j(x)\leq 0=\Sigma_{j=1}^{m}v_j^*H_j(x^*),$$

which results (using assumptions on $\tilde{\Phi}$) in

$$\tilde{\Phi}\Big(\Sigma_{j=1}^{m}v_j^*H_j(x)-\Sigma_{j=1}^{m}v_j^*H_j(x^*)\Big)\leq 0.$$

Now, in light of the strictly hybrid $(\tilde{\Phi}, \bar{\rho}, \eta, \zeta, \theta)-$quasi-invexity of $B_j(., v^*)$ at x^*, we find

$$\Big\langle \Sigma_{j=1}^m v_j^* \bigtriangledown H_j(x^*) + \Big\langle \Sigma_{j=1}^m v_j^* \nabla^2 H_j(x^*)z, \eta(x, x^*)\Big\rangle$$
$$-\frac{1}{2}\Big\langle \Sigma_{j=1}^m v_j^* \nabla^2 H_j(x^*)z, \zeta(x, x^*)\Big\rangle + \bar{\rho}(x, x^*)\|\theta(x, x^*)\|^2 < 0. \qquad (3.10)$$

It follows from (3.9) and (3.10) that

$$\Big\langle \Sigma_{i=1}^p u_i^*[\bigtriangledown f_i(x^*) - (\frac{f_i(x^*)}{g_i(x^*)}) \bigtriangledown g_i(x^*)]$$
$$+\sum_{i=1}^p u_i^*[\nabla^2 f_i(x^*)z - (\frac{f_i(x^*)}{g_i(x^*)})\nabla^2 g_i(x^*)z], \eta(x, x^*)\Big\rangle$$
$$-\frac{1}{2}\Big\langle \sum_{i=1}^p u_i^*[\nabla^2 f_i(x^*)z - (\frac{f_i(x^*)}{g_i(x^*)})\nabla^2 g_i(x^*)z], \zeta(x, x^*)\Big\rangle$$
$$> \bar{\rho}(x, x^*)\|\theta(x, x^*)\|^2 > -\rho(x, x^*)\|\theta(x, x^*)\|^2. \qquad (3.11)$$

As a result, since $\rho(x, x^*) \geq 0$, applying the prestrictly hybrid $(\Phi, \rho, \eta, \zeta, \theta)-$pseudo-invexity at x^* to (3.11) and assumptions on Φ, we have

$$\Phi\Big(\Sigma_{i=1}^p u_i^*[f_i(x) - (\frac{f_i(x^*)}{g_i(g^*)})g_i(x)] - \Sigma_{i=1}^p u_i^*[f_i(x^*) - (\frac{f_i(x^*)}{g_i(x^*)})g_i(x^*)]\Big) \geq 0,$$

which implies

$$\Sigma_{i=1}^p u_i^*[f_i(x) - (\frac{f_i(x^*)}{g_i(x^*)})g_i(x)]$$
$$\geq \Sigma_{i=1}^p u_i^*[f_i(x^*) - (\frac{f_i(x^*)}{g_i(x^*)})g_i(x^*)])$$
$$= 0.$$

Thus, we have

$$\Sigma_{i=1}^p u_i^*[f_i(x) - (\frac{f_i(x^*)}{g_i(x^*)})g_i(x)] \geq 0. \qquad (3.12)$$

Since $u_i^* > 0$ for each $i \in \{1, \cdots, p\}$, we conclude that there does not exist an $x \in Q$ such that

$$\frac{f_i(x)}{g_i(x)} - (\frac{f_i(x^*)}{g_i(x^*)}) \leq 0 \; \forall i = 1, \cdots, p,$$

$$\frac{f_j(x)}{g_j(x)} - (\frac{f_j(x^*)}{g_j(x^*)}) < 0 \text{ for some } j \in \{1, \cdots, p\}.$$

Hence, x^* is an efficient solution to (P).

The important aspect of the proof applying (iii) is that we use the equivalent form for Definition 2.7 instead. Since $B_j(\cdot, v_{j^*})$ is strictly hybrid $(\tilde{\Phi}, \bar{\rho}, \eta, \zeta, \theta)-$quasi-invex at x^* for $\tilde{\Phi}$ increasing and $\tilde{\Phi}(0) = 0$, we have

$$\Big\langle \Sigma_{j=1}^{m} v_j^* \nabla H_j(x^*) + \Sigma_{j=1}^{m} v_j^* \nabla^2 H_j(x^*) z, \eta(x, x^*)\Big\rangle$$
$$- \frac{1}{2}\Big\langle \Sigma_{j=1}^{m} v_j^* \nabla^2 H_j(x^*) z, \zeta(x, x^*)\Big\rangle + \bar{\rho}(x, x^*)\|\theta(x, x^*)\|^2 < 0. \quad (3.13)$$

Next, applying (3.13)–(3.15), we arrive at

$$\Big\langle \Sigma_{i=1}^{p} u_i^* [\nabla f_i(x^*) - (\frac{f_i(x^*)}{g_i(x^*)}) \nabla g_i(x^*)]$$
$$+ \sum_{i=1}^{p} u_i^* [\nabla^2 f_i(x^*) z - (\frac{f_i(x^*)}{g_i(x^*)}) \nabla^2 g_i(x^*) z], \eta(x, x^*)\Big\rangle$$
$$- \frac{1}{2}\Big\langle \sum_{i=1}^{p} u_i^* [\nabla^2 f_i(x^*) z - (\frac{f_i(x^*)}{g_i(x^*)}) \nabla^2 g_i(x^*) z], \zeta(x, x^*)\Big\rangle$$
$$> \bar{\rho}(x, x^*)\|\theta(x, x^*)\|^2 > -\rho(x, x^*)\|\theta(x, x^*)\|^2. \quad (3.14)$$

At this stage, since $E_i(.\,; x^*, u^*)$ $\forall i \in \{1, \cdots, p\}$ are prestrictly hybrid $(\Phi, \rho, \eta, \zeta, \theta)-$quasi-invex at x^* for $\Phi(a) \geq 0 \Rightarrow a \geq 0$, we have

$$\Phi\Big(\Sigma_{i=1}^{p} u_i^* [f_i(x) - (\frac{f_i(x^*)}{g_i(g^*)}) g_i(x)] - \Sigma_{i=1}^{p} u_i^* [f_i(x^*) - (\frac{f_i(x^*)}{g_i(x^*)}) g_i(x^*)]\Big) \geq 0,$$

which implies

$$\Sigma_{i=1}^{p} u_i^* [f_i(x) - (\frac{f_i(x^*)}{g_i(x^*)}) g_i(x)]$$
$$\geq \Sigma_{i=1}^{p} u_i^* [f_i(x^*) - (\frac{f_i(x^*)}{g_i(x^*)}) g_i(x^*)]) = 0.$$

Thus, we have

$$\Sigma_{i=1}^{p} u_i^* [f_i(x) - (\frac{f_i(x^*)}{g_i(x^*)}) g_i(x)] \geq 0. \quad (3.15)$$

Since $u_i^* > 0$ for each $i \in \{1, \cdots, p\}$, we conclude that there does not exist an $x \in Q$ such that

$$\frac{f_i(x)}{g_i(x)} - (\frac{f_i(x^*)}{g_i(x^*)}) \leq 0 \;\forall i = 1, \cdots, p,$$

$$\frac{f_j(x)}{g_j(x)} - (\frac{f_j(x^*)}{g_j(x^*)}) < 0 \text{ for some } j \in \{1, \cdots, p\}.$$

Hence, x^* is an efficient solution to (P).

Finally, we prove using (iv) as follows: since $x \in Q$, it follows that $H_j(x) \leq H_j(x^*)$, which implies $\bar{\Phi}\Big(H_j(x) - H_j(x^*)\Big) \leq 0$. Then applying the hybrid $(\bar{\Phi}, \rho_3, \eta, \zeta, \theta)-$quasi-invexity of H_j at x^* and $v^* \in R^m_+$, we have

$$\Big\langle \Sigma_{j=1}^m v_j^* \nabla H_j(x^*) + \Sigma_{j=1}^m v_j^* \nabla^2 H_j(x^*) z, \eta(x, x^*)\Big\rangle$$
$$- \frac{1}{2}\Big\langle \Sigma_{j=1}^m v_j^* \nabla^2 H_j(x^*) z, \zeta(x, x^*)\Big\rangle + \bar{\rho}(x, x^*)\|\theta(x, x^*)\|^2 \leq 0. \quad (3.16)$$

Since $u^* \geq 0$ and $\frac{f_i(x^*)}{g_i(x^*)} \geq 0$, it follows from the hybrid $(\Phi, \rho_3, \eta, \zeta, \theta)-$invexity assumptions that

$$\begin{aligned}
&\Phi\Big(\Sigma_{i=1}^p u_i^*[f_i(x) - (\frac{f_i(x^*)}{g_i(x^*)})g_i(x)]\Big)\\
&= \Phi\Big(\Sigma_{i=1}^p u_i^*\{[f_i(x) - f_i(x^*)] - (\frac{f_i(x^*)}{g_i(x^*)})[g_i(x) - g_i(x^*)]\}\Big)\\
&\geq \Sigma_{i=1}^p u_i^*\{\langle \nabla f_i(x^*) - (\frac{f_i(x^*)}{g_i(x^*)}) \nabla g_i(x^*)\\
&+ \Sigma_{i=1}^p u_i^*[\nabla^2 f_i(x^*)z - (\frac{f_i(x^*)}{g_i(x^*)})\nabla^2 g_i(x^*)z, \eta(x, x^*)\rangle]\\
&- \frac{1}{2}\langle \Sigma_{i=1}^p u_i^*[\nabla^2 f_i(x^*)z - (\frac{f_i(x^*)}{g_i(x^*)})\nabla^2 g_i(x^*)z, \zeta(x, x^*)\rangle]\\
&+ \Sigma_{i=1}^p u_i^*[\rho_1(x, x^*) + \phi(x^*)\rho_2(x, x^*)]\|\theta(x, x^*)\|^2\\
&\geq -\Big[\langle \Sigma_{j=1}^m v_j^* \nabla H_j(x^*) + \Sigma_{j=1}^m v_j^* \nabla^2 H_j(x^*)z, \eta(x, x^*)\Big\rangle\\
&- \frac{1}{2}\langle \Sigma_{j=1}^m v_j^* \nabla^2 H_j(x^*)z, \zeta(x, x^*)\rangle\Big]\\
&+ \Sigma_{i=1}^p u_i^*[\rho_1(x, x^*) + \phi(x^*)\rho_2(x, x^*)]\|\theta(x, x^*)\|^2\\
&\geq (\Sigma_{j=1}^m v_j^* \rho_3(x, x^*) + \Sigma_{i=1}^p u_i^*[\rho_1(x, x^*) + \phi(x^*)\rho_2(x, x^*)])\|\theta(x, x^*)\|^2\\
&= (\Sigma_{j=1}^m v_j^* \rho_3 + \rho^*(x, x^*))\|\theta(x, x^*)\|^2\\
&\geq 0,
\end{aligned}$$

where $\phi(x^*) = \frac{f_i(x^*)}{g_i(x^*)}$ and $\rho^* = \Sigma_{i=1}^p u_i^*(\rho_1(x, x^*) + \phi(x^*)\rho_2(x, x^*))$.
This implies that

$$\Phi\Big(\Sigma_{i=1}^p u_i^*[f_i(x) - (\frac{f_i(x^*)}{g_i(x^*)})g_i(x)]\Big) \geq 0.$$

Next we consider the case when the functions are first-order differentiable, Theorem 3.1 reduces to the result which is similar to the results of Zalmai ([35], Theorems 3.1, 3.2), and Zalmai and Zhang [38].

Theorem 3.2 *For $x^* \in Q$, let f_i, g_i for $i \in \{1, \cdots, p\}$ with $\frac{f_i(x^*)}{g_i(x^*)} \geq 0$, $g_i(x^*) > 0$ and H_j for $j \in \{1, \cdots, m\}$ be differentiable at $x^* \in Q$, and let there exist $u^* \in U = \{u \in \Re^p : u > 0, \Sigma_{i=1}^p u_i = 1\}$ and $v^* \in \Re_+^m$ such that*

$$\left\langle \Sigma_{i=1}^p u_i^*[\nabla f_i(x^*) - (\frac{f_i(x^*)}{g_i(x^*)})\nabla g_i(x^*)] + \Sigma_{j=1}^m v_j^* \nabla H_j(x^*), \eta(x, x^*)) \right\rangle \geq 0 \quad (3.17)$$

and

$$v_j^* H_j(x^*) = 0, \; j \in \{1, \cdots, m\}. \quad (3.18)$$

Suppose, in addition, that any one of the following assumptions holds (for $\rho(x, x^*) \geq 0$):

(i) *$E_i(.; x^*, u^*)$ $\forall i \in \{1, \cdots, p\}$ are first-order hybrid $(\Phi, \rho, \eta, \theta)-$pseudo-invex at x^* for $\Phi(a) \geq 0 \Rightarrow a \geq 0$, and $B_j(., v^*)$ $\forall j \in \{1, \cdots, m\}$ are first-order hybrid $(\bar{\Phi}, \bar{\rho}, \eta, \theta)-$quasi-invex at x^* for $\bar{\Phi}$ increasing and $\bar{\Phi}(0) = 0$.*

(ii) *$E_i(.; x^*, u^*)$ $\forall i \in \{1, \cdots, p\}$ are first-order hybridprestrictly $(\Phi, \rho, \eta, \theta)-$pseudo-invex at x^* for $\Phi(a) \geq 0 \Rightarrow a \geq 0$, and $B_j(., v^*)$ $\forall j \in \{1, \cdots, m\}$ are first-order strictly hybrid $(\bar{\Phi}, \bar{\rho}, \eta, \theta)-$quasi-invex at x^* for $\bar{\Phi}$ increasing and $\bar{\Phi}(0) = 0$.*

(iii) *$E_i(.; x^*, u^*)$ $\forall i \in \{1, \cdots, p\}$ are first-order prestrictly hybrid $(\Phi, \rho, \eta, \theta)-$quasi-invex at x^* $\Phi(a) \geq 0 \Rightarrow a \geq 0$, and $B_j(., v^*)$ $\forall j \in \{1, \cdots, m\}$ are first-order strictly hybrid $(\bar{\Phi}, \bar{\rho}, \eta, \theta)-$quasi-invex at x^* for $\bar{\Phi}$ increasing and $\bar{\Phi}(0) = 0$.*

(iv) *For each $i \in \{1, \cdots, p\}$, f_i is first-order hybrid $(\Phi, \rho_1, \eta, \theta)-$invex and $-g_i$ is first-order hybrid $(\Phi, \rho_2, \eta, \theta)-$invex at x^* for $\Phi(a) \geq 0 \Rightarrow a \geq 0$. $H_j(., v^*)$ $\forall j \in \{1, \cdots, m\}$ is hybrid $(\bar{\Phi}, \bar{\rho_3}, \eta, \theta)-$quasi-invex at x^*, and $\Sigma_{j=1}^m v_j^* \rho_3(x, x^*) + \rho^*(x, x^*) \geq 0$ for $\bar{\Phi}$ increasing and $\bar{\Phi}(0) = 0$, $\rho^*(x, x^*) = \Sigma_{i=1}^p u_i^*(\rho_1(x, x^*) + \phi(x^*)\rho_2(x, x^*))$ and for $\phi(x^*) = \frac{f_i(x^*)}{g_i(x^*)}$.*

Then x^* is an efficient solution to (P).

We observe that Theorem 3.1 can be further generalized to the case of the $\varepsilon-$efficient conditions based on the hybrid $(\Phi, \rho, \eta, \zeta, \theta)-$invexity frameworks. As a matter of fact, we generalize the $\varepsilon-$efficient solvability conditions for problem (P) based on the work of Verma [22], and Kim, Kim and Lee [6], where they have investigated the $\varepsilon-$efficiency as well as the weak $\varepsilon-$efficiency conditions for multiobjective fractional programming problems under constraint qualifications. We recall some auxiliary concepts (for the hybrid $(\Phi, \rho, \eta, \zeta, \theta)-$invexity) crucial to the problem on hand.

Definition 3.1 A point $x^* \in Q$ is an $\varepsilon-$efficient solution to (P) if there does not exist an $x \in Q$ such that

$$\frac{f_i(x)}{g_i(x)} \leq \frac{f_i(x^*)}{g_i(x^*)} - \varepsilon_i \; \forall\, i = 1, \cdots, p,$$

$$\frac{f_j(x)}{g_j(x)} < \frac{f(_j x^*)}{g_j(x^*)} - \varepsilon_j \text{ for some} j \in \{1, \cdots, p\},$$

where ε_i=$(\varepsilon_1, \cdots, \varepsilon_p)$ is with $\varepsilon_i \geq 0$ for $i = 1, \cdots, p$.
For $\varepsilon = 0$, Definition 3.1 reduces to the case that $x^* \in Q$ is an efficient solution to (P).
Next, we start with real-valued functions $E_i(., x^*, u^*)$ and $B_j(., v)$ defined by

$$E_i(x, x^*, u^*) = u_i[f_i(x) - \Big(\frac{f_i(x^*)}{g_i(x^*)} - \varepsilon_i\Big)g_i(x)],\; i \in \{1, \cdots, p\}$$

and

$$B_j(., v^*) = v_j^* H_j(x),\;\; j = 1, \cdots, m.$$

Theorem 3.3 *Let* $x^* \in Q$, f_i, g_i *for* $i \in \{1, \cdots, p\}$ *with* $f_i(x^*) \geq \varepsilon_i g_i(x^*)$, $g_i(x^*) > 0$ *and* H_j *for* $j \in \{1, \cdots, m\}$ *be twice continuously differentiable at* $x^* \in Q$*, and let there exist* $u^* \in U = \{u \in \Re^p : u > 0, \Sigma_{i=1}^p u_i = 1\}$, $v^* \in \Re_+^m$ *and* $z \in \Re^n$ *such that*

$$\Sigma_{i=1}^p u_i^*[\nabla f_i(x^*) - \Big(\frac{f_i(x^*)}{g_i(x^*)} - \varepsilon_i\Big) \bigtriangledown g_i(x^*)] + \Sigma_{j=1}^m v_j^* \bigtriangledown H_j(x^*) = 0, \quad (3.19)$$

$$\Big\langle\Big[\sum_{i=1}^p u_i^*[\nabla^2 f_i(x^*) - (\frac{f_i(x^*)}{g_i(x^*)} - \varepsilon_i)\nabla^2 g_i(x^*)] + \sum_{j=1}^m v_j^* \nabla^2 H_j(x^*)\Big]z, \eta(x, x^*)\Big\rangle \geq 0, \quad (3.20)$$

$$-\frac{1}{2}\Big\langle\Big[\sum_{i=1}^p u_i^*[\nabla^2 f_i(x^*) - (\frac{f_i(x^*)}{g_i(x^*)} - \varepsilon_i)\nabla^2 g_i(x^*)] + \sum_{j=1}^m v_j^* \nabla^2 H_j(x^*)\Big]z, \zeta(x, x^*)\Big\rangle \geq 0, \quad (3.21)$$

and

$$v_j^* H_j(x^*) = 0,\; j \in \{1, \cdots, m\}. \quad (3.22)$$

Suppose, in addition, that any one of the following assumptions holds (for $\rho(x, x^*) \geq 0$*):*

(i) $E_i(.; x^*, u^*)$ $\forall i \in \{1, \cdots, p\}$ *are hybrid* $(\Phi, \rho, \eta, \theta)$*−pseudo-invex at* x^* *for* $\Phi(a) \geq 0 \Rightarrow a \geq 0$*, and* $B_j(., v^*)$ $\forall j \in \{1, \cdots, m\}$ *are hybrid* $(\tilde{\Phi}, \bar{\rho}, \eta, \theta)$*−quasi-invex at* x^* *for* $\tilde{\Phi}$ *increasing and* $\tilde{\Phi}(0) = 0$.
(ii) $E_i(.; x^*, u^*)$ $\forall i \in \{1, \cdots, p\}$ *are prestrictly hybrid* $(\Phi, \rho, \eta, \theta)$*−pseudo-invex at* x^* *for* $\Phi(a) \geq 0 \Rightarrow a \geq 0$*, and* $B_j(., v^*)$ $\forall j \in \{1, \cdots, m\}$ *are strictly hybrid* $(\Phi, \rho, \eta, \theta)$*−quasi-invex at* x^* *for* $\bar{\Phi}$ *increasing and* $\bar{\Phi}(0) = 0$.

(iii) $E_i(.\,; x^*, u^*)$ $\forall\, i \in \{1, \cdots, p\}$ *are prestrictly hybrid* $(\Phi, \rho, \eta, \theta)-$*quasi-invex at* x^* *for* $\Phi(a) \geq 0 \Rightarrow a \geq 0$, *and* $B_j(.\,, v^*)$ $\forall\, j \in \{1, \cdots, m\}$ *are strictly hybrid* $(\bar{\Phi}, \bar{\rho}, \eta, \theta)-$*quasi-invex at* x^* *for* $\bar{\Phi}$ *increasing and* $\bar{\Phi}(0) = 0$.

(iv) *For each* $i \in \{1, \cdots, p\}$, f_i *is hybrid* $(\Phi, \rho_1, \eta, \theta)-$*invex and* $-g_i$ *is* $(\Phi, \rho_2, \eta, \theta)-$*invex at* x^* *for* $\Phi(a) \geq 0 \Rightarrow a \geq 0$, *and* $H_j(.\,, v^*)$ $\forall\, j \in \{1, \cdots, m\}$ *is hybrid* $(\bar{\Phi}, \rho_3, \eta, \theta)-$*quasi-invex at* x^* *for* $\bar{\Phi}$ *increasing and* $\bar{\Phi}(0) = 0$ *and* $\Sigma_{j=1}^{m} v_j^* \rho_3(x, x^*) + \rho^*(x, x^*) \geq 0$ *for* $\rho^* = \Sigma_{i=1}^{p} u_i^*(\rho_1(x, x^*) + \phi(x^*)\rho_2(x, x^*))$, *where* $\phi(x^*) = \frac{f_i(x^*)}{g_i(x^*)} - \varepsilon_i$.

Then x^* *is an* $\varepsilon-$*efficient solution to (P).*

Proof The proofs are similar to that of Theorem 3.1.

4 Concluding Remarks

We observe that the higher order hybrid $(\Phi, \rho, \eta, \zeta, \theta)-$ invexities can effectively be applied generalizing/unifying the first-order sufficient efficiency condition results [35], first-order parametric duality model results [36] as well as second order duality model results (Zalmai [37]) on Hanson-Antczak-type generalized V-invex functions in semi-infinite multiobjective fractional programming. Based on new duality models and suitable constraint structures, the weak, strong, and strict converse duality theorems can be established using appropriate hybrid $(\Phi, \rho, \eta, \zeta, \theta)-$ invexities.

References

1. Ben-Israel, A., Mond, B.: What is the invexity? J. Aust. Math. Soc. Ser. B **28**, 1–9 (1986)
2. Caiping L., Xinmin Y.: Generalized $(\rho, \theta, \eta)-$invariant monotonicity and generalized $(\rho, \theta, \eta)-$invexity of non-differentiable functions. J. Inequal. Appl., Article ID # 393940, p 16 (2009)
3. Hanson, M.A.: On sufficiency of the Kuhn-Tucker conditions. J. Math. Anal. Appl. **80**, 545–550 (1981)
4. Jeyakumar, V.: Strong and weak invexity in mathematical programming. Methods Oper. Res. **55**, 109–125 (1985)
5. Kawasaki, H.: Second-order necessary conditions of the Kuhn-Tucker type under new constraint qualifications. J. Optim. Theory Appl. **57**(2), 253–264 (1988)
6. Kim, M.H., Kim, G.S., Lee, G.M.: On $\varepsilon-$optimality conditions for multiobjective fractional optimization problems, Fixed Point Theory Appl. **6** (2011). doi:10.1186/1687-1812-2011-6
7. Liu, J.C.: Second order duality for minimax programming. Utilitas Math. **56**, 53–63 (1999)
8. Mangasarian, O.L.: Second- and higher-order duality theorems in nonlinear programming. J. Math. Anal. Appl. **51**, 607–620 (1975)
9. Mishra, S.K.: Second order generalized invexity and duality in mathematical programming. Optimization **42**, 51–69 (1997)
10. Mishra, S.K.: Second order symmetric duality in mathematical programming with F-convexity. European J. Oper. Res. **127**, 507–518 (2000)

11. Mishra, S.K., Rueda, N.G.: Higher-order generalized invexity and duality in mathematical programming. J. Math. Anal. Appl. **247**, 173–182 (2000)
12. Mishra, S.K., Rueda, N.G.: Second-order duality for nondifferentiable minimax programming involving generalized type I functions. J. Optim. Theory Appl. **130**, 477–486 (2006)
13. Mishra, S.K., Laha, V., Verma, R.U.: Generalized vector variational-like inequalities and nonsmooth vector optimization of radially continuous functions. Adv. Nonlinear Variational Inequal. **14**(2), 1–18 (2011)
14. Mond, B., Weir, T.: Generalized convexity and higher-order duality, J. Math. Sci. 16–18 (1981–1983), 74–94
15. Mond, B., Zhang, J.: Duality for multiobjective programming involving second-order V-invex functions. In: Glover, M., Jeyakumar, V. (eds.) Proceedings of the Optimization Miniconference II (B, pp. 89–100. University of New South Wales, Sydney, Australia (1995)
16. Mond B., Zhang, J.: Higher order invexity and duality in mathematical programming. In: Crouzeix, J. P., et al. (eds.) Generalized convexity, generalized monotonicity : recent results, Kluwer Academic Publishers, Netherlands, 1998, pp. 357–372
17. Patel, R.B.: Second order duality in multiobjective fractional programming. Indian J. Math. **38**, 39–46 (1997)
18. Pitea, A., Postolache, M.: Duality theorems for a new class of multitime multiobjective variational problems. J. Global Optim. **54**, 47–58 (2012)
19. Srivastava, K.K., Govil, M.G.: Second order duality for multiobjective programming involving (F, ρ, σ)-type I functions. Opsearch **37**, 316–326 (2000)
20. Verma, R.U.: Parametric duality models for multiobjective fractional programming based on new generation hybrid invexities. J. Appl. Funct. Anal. **10**(3–4), 234–253 (2015)
21. Verma, R.U.: Multiobjective fractional programming problems and second order generalized hybrid invexity frameworks statistics. Optim. Inform. Comput. **2**(4), 280–304 (2014)
22. Verma, R.U.: Weak $\varepsilon-$ efficiency conditions for multiobjective fractional programming. Appl. Math. Comput. **219**, 6819–6827 (2013)
23. Verma, R.U.: New $\varepsilon-$optimality conditions for multiobjective fractional subset programming problems. Trans. Math. Program. Appl. **1**(1), 69–89 (2013)
24. Verma, R.U.: Second-order $(\Phi, \eta, \rho, \theta)-$invexities and parameter-free $\varepsilon-$efficiency conditions for multiobjective discrete minmax fractional programming problems. Adv. Nonlinear Variational Inequal. **17**(1), 27–46 (2014)
25. Yang, X.M.: Second order symmetric duality for nonlinear programs. Opsearch **32**, 205–209 (1995)
26. Yang, X.M.: On second order symmetric duality in nondifferentiable multiobjective programming. J. Ind. Manag. Optim. **5**, 697–703 (2009)
27. Yang, X.M., Hou, S.H.: Second-order symmetric duality in multiobjective programming. Appl. Math. Lett. **14**, 587–592 (2001)
28. Yang, X.M., Teo, K.L., Yang, X.Q.: Higher-order generalized convexity and duality in nondifferentiable multiobjective mathematical programming. J. Math. Anal. Appl. **29**, 48–55 (2004)
29. Yang, X.M., Yang, X.Q., Teo, K.L.: Nondifferentiable second order symmetric duality in mathematical programming with F-convexity. European J. Oper. Res. **144**, 554–559 (2003)
30. Yang, X.M., Yang, X.Q., Teo, K.L.: Huard type second-order converse duality for nonlinear programming. Appl. Math. Lett. **18**, 205–208 (2005)
31. Yang, X.M., Yang, X.Q., Teo, K.L.: Higher-order symmetric duality in multiobjective programming with invexity. J. Ind. Manag. Optim. **4**, 385–391 (2008)
32. Yang, X.M., Yang, X.Q., Teo, K.L., Hou, S.H.: Second order duality for nonlinear programming. Indian J. Pure Appl. Math. **35**, 699–708 (2004)
33. Yokoyama, K.: Epsilon approximate solutions for multiobjective programming problems. J. Math. Anal. Appl. **203**(1), 142–149 (1996)
34. Zalmai, G.J.: Global parametric sufficient optimality conditions for discrete minmax fractional programming problems containing generalized (θ, η, ρ)-V-invex functions and arbitrary norms. J. Appl. Math. Comput. **23**(1–2), 1–23 (2007)

35. Zalmai, G.J.: Hanson-Antczak-type generalized $(\alpha, \beta, \gamma, \xi, \eta, \rho, \theta)$-V-invex functions in semi-infinite multiobjective fractional programming I : sufficient efficiency conditions. Adv. Nonlinear Variational Inequal. **16**(1), 91–114 (2013)
36. Zalmai, G.J.: Hanson-Antczak-type generalized $(\alpha, \beta, \gamma, \xi, \eta, \rho, \theta)$-V-invex functions in semi-infinite multiobjective fractional programming II : first-order parametric duality models. Adv. Nonlinear Variational Inequal. **16**(2), 61–90 (2013)
37. Zalmai, G.J.: Hanson-Antczak-type generalized $(\alpha, \beta, \gamma, \xi, \eta, \rho, \theta)$-V-invex functions in semi-infinite multiobjective fractional programming, III: second-order parametric duality models. Adv. Nonlinear Variational Inequal. **16**(2), 91–126 (2013)
38. Zalmai, G.J., Zhang, Q.: Global nonparametric sufficient optimality conditions for semi-infinite discrete minmax fractional programming problems involving generalized $(\rho, \theta)-$invex functions. Numer. Funct. Anal. Optim. **28**(1–2), 173–209 (2007)
39. Zhang, J., Mond, B.: Second order b-invexity and duality in mathematical programming. Utilitas Math. **50**, 19–31 (1996)
40. Zhang, J., Mond, B.: Second order duality for multiobjective nonlinear programming involving generalized convexity. In: Glover, B.M., Craven, B.D., Ralph, D. (eds.) Proceedings of the Optimization Miniconference III , University of Ballarat, pp. 79–95, (1997)
41. Zhian, L., Qingkai, Y.: Duality for a class of multiobjective control problems with generalized invexity. J. Math. Anal. Appl. **256**, 446–461 (2001)

A Study of Generalized Invex Functions on Riemannian Manifold

S. Jana and C. Nahak

Abstract In this article, we introduce (p, r)-invex, $\rho-(p, r)$-invex, and semistrictly geodesic η-prequasi invex functions in the setting of Riemannian manifolds. We construct counter examples to show that these functions generalize the notion of invexity. We also study the optimality conditions of a minimization problem under these generalized invexities on Riemannian manifolds.

Keywords Riemannian manifold · Generalized invexity · KKT conditions · Nonconvex optimization

1 Introduction

Convexity plays a vital role in the theory of optimization but it is often not enjoyed by real problems. Therefore, several generalizations have been developed for the classical properties of convexity. An important and significant generalization of convexity is invexity which was introduced by Hanson [6], in the year 1981. Later on Zalmai [19] generalized the class of invex functions into $\rho-(\eta, \theta)$-invex functions. In 2001, Antczak [3] introduced (p, r)-invex sets and functions. Mandal and Nahak [10] introduced $(p, r)-\rho-(\eta, \theta)$-invexity which is a generalization of the results of both Zalmai [19] and Antczak [3]. Yang and Li [17] introduced semistrictly preinvex functions on Euclidean spaces.

Rapcsak [14] introduced a generalization of convexity called geodesic convexity and extended many results of convex analysis and optimization theory from linear spaces to Riemannian manifolds. Udriste [15] established duality results for a convex programming problem on Riemannian manifolds. Pini [13] introduced

S. Jana (✉) · C. Nahak
Department of Mathematics, Indian Institute of Technology Kharagpur,
Kharagpur 721302, India
e-mail: shreyasi.iitkgp@gmail.com

C. Nahak
e-mail: cnahak@maths.iitkgp.ernet.in

R.N. Mohapatra et al. (eds.), *Mathematics and Computing*,
Springer Proceedings in Mathematics & Statistics 139,
DOI 10.1007/978-81-322-2452-5_3

the notion of invex functions on a manifold. Motivated by Pini [13], Mititelu [12] generalized invexity by defining (ρ, η)-invex, (ρ, η)-pseudoinvex, and (ρ, η)-quasiinvex functions. Mititelu [12] also established the necessary and sufficient conditions of Karush-Kuhn-Tucker type for a vector programming problem defined on a differentiable manifold. Mond–Weir-type duality for vector programming problems on differentiable manifolds was developed by Ferrara and Mititelu [5]. The concepts of geodesic invex sets, geodesic invex, and geodesic preinvex functions were introduced by Barani and Pouryayevali [4] on Riemannian manifolds. Ahmad et al. [2] extended these results by introducing geodesic η-pre-pseudo invex functions and geodesic η-prequasi invex functions. Recently, Iqbal et al. [7] defined geodesic E-convex sets and geodesic E-convex functions. Further, Agarwal et al. [1] introduced geodesic α-invex sets, geodesic α-invex, and α-preinvex functions. Agarwal et al. [1] extended the results of Yang and Li by introducing semistrictly geodesic η-preinvex functions over a Riemannian manifold.

Motivated by the above concepts, we extend (p, r)-invex, $\rho - (p, r)$-invex functions from Euclidean spaces to Riemannian manifolds. We introduce the notion of semistrictly geodesic η-prequasi invex functions on Riemannian manifolds which extend semistrictly quasi invex functions introduced by Yang et al. [18]. We have studied optimality conditions of the nonlinear programming problems on Riemannian manifolds under these generalized invexity assumptions.

2 Preliminaries

In this section, we recall some definitions concerning Riemannian geometry which will be used throughout the article. These standard materials can be found in [15, 16]. A Riemannian manifold M is a C^∞ smooth manifold endowed with a Riemannian metric $< .,. >$ on the tangent space $T_x M$ and the corresponding norm is denoted by $\|.\|_x$, where the subscript x will be omitted. The tangent bundle of M is denoted by $TM = \cup_{x \in M} T_x M$, which is naturally a manifold. The length of a piecewise smooth curve $\gamma : [a, b] \to M$ joining x to y such that $\gamma(a) = x$ and $\gamma(b) = y$, is defined by $L(\gamma) = \int_a^b \| \dot{\gamma}(t) \|_{\gamma(t)} \, dt$. We define the distance d between any two points $x, \ y \in M$ by

$$d(x, y) = \inf\{L(\gamma) : \gamma \ is \ a \ piecewise \ C^1 \ curve \ joining \ x \ to \ y\}.$$

Then d is a distance which induces the original topology on M. On every Riemannian manifold, there exists exactly one covariant derivation called Levi-Civita connection denoted by $\nabla_X Y$ for any vector fields X, Y on M. We recall that a geodesic is a C^∞ smooth path γ whose tangent is parallel along the path γ, that is γ satisfies the equation $\nabla_{\frac{d\gamma(t)}{dt}} \frac{d\gamma(t)}{dt} = 0$. A geodesic joining x to y in M is said to be a minimal geodesic if its length equals $d(x, y)$.

A Riemannian manifold is complete if for any $x \in M$ all geodesics emanating from x are defined for all $t \in \mathbb{R}$. By the Hopf-Rinow theorem; we know that if M is complete then any pair of points in M can be joined by a minimal geodesic. Moreover, (M, d) is a complete metric space and bounded closed subsets are compact.

Definition 1 ([5]) Let $F : M \to \mathbb{R}$ be a differentiable function. The differential of F at x, namely $\mathrm{d}F_x : T_x M \to T_{F(x)}\mathbb{R} \equiv \mathbb{R}$, is introduced by $\mathrm{d}F_x(v) = \mathrm{d}F(x)v$, $v \in T_x M$.

Definition 2 ([14]) A subset K of M is said to be geodesic convex if and only if for any two points $x, y \in K$, the geodesic joining x to y is contained in K. That is if $\gamma : [0, 1] \to M$ is a geodesic with $x = \gamma(0)$ and $y = \gamma(1)$, then $\gamma(t) \in K, \ for \ 0 \leq t \leq 1$.

Definition 3 ([14]) A real-valued function $f : M \to \mathbb{R}$ defined on a geodesic convex set K is said to be geodesic convex function if and only if for $0 \leq t \leq 1$,

$$f(\gamma(t)) \leq (1-t)f(\gamma(0)) + tf(\gamma(1)).$$

We consider now a map $\eta : M \times M \to TM$ such that $\eta(x, y) \in T_y M$ for every $x, y \in M$.

Definition 4 ([4]) Let M be an n-dimensional Riemannian manifold and $\eta : M \times M \to TM$ be a function such that for every $x, y \in M$, $\eta(x, y) \in T_y M$. A nonempty subset S of M is said to be geodesic invex set with respect to η if for every $x, y \in S$, there exists a unique geodesic $\gamma_{x,y} : [0, 1] \to M$ such that

$$\gamma_{x,y}(0) = y, \ \ \gamma'_{x,y}(0) = \eta(x, y), \ \ \gamma_{x,y}(t) \in S, \ \forall t \in [0, 1].$$

Definition 5 ([4]) Let M be an n-dimensional Riemannian manifold and S be an open subset of M which is geodesic invex with respect to $\eta : M \times M \to TM$. A function $f : S \to \mathbb{R}$ is said to be geodesic η-preinvex if $\forall x, y \in S$, we have

$$f(\gamma_{x,y}(t)) \leq tf(x) + (1-t)f(y) \quad \forall t \in [0, 1].$$

If f be differentiable on S. We say that f is geodesic η-invex on S if the following holds

$$f(x) - f(y) \geq df_y(\eta(x, y)), \ \ \forall x, y \in S.$$

Later on Mititelu [12] generalized the above definition as follows.

Definition 6 The differentiable function f is said to be (ρ, η)-invex at y if there exist an $\eta : M \times M \to TM$ and $\rho \in \mathbb{R}$ such that

$$\forall x \in M : f(x) - f(y) \geq \mathrm{d}f_y(\eta(x, y)) + \rho \mathrm{d}^2(x, y).$$

Definition 7 ([4]) A closed η-path joining the points y and $u = \alpha_{x,y}(1)$ is a set of the form $P_{yu} = \{v : v = \alpha(t) : t \in [0, 1]\}$.

Definition 8 ([2]) Let M be an n-dimensional Riemannian manifold and S be an open subset of M which is geodesic invex with respect to $\eta : M \times M \to TM$. A function $f : S \to \mathbb{R}$ is said to be geodesic η-prequasi invex on S if

$$f(\gamma_{x,y}(t)) \leq \max\{f(x), f(y)\} \, \forall x, y \in S, \; \forall t \in [0, 1].$$

3 Main Results

3.1 (p,r)-Invexity

In the year 2001, Antczak [3] introduced (p, r)-invex function over $\mathbb{R}^n$ which generalizes the notion of invexity.

Definition 9 (*Antczak (2001)*) Let $f : \mathbb{R}^n \to \mathbb{R}$ be a differentiable function and p, r be arbitrary real numbers. The function f is said to be (p, r)-invex with respect to $\eta : \mathbb{R}^n \times \mathbb{R}^n \to \mathbb{R}^n$ at u, if any one of the following conditions holds

$$\frac{1}{r}e^{rf(x)} \geq \frac{1}{r}e^{rf(u)}[1 + \frac{r}{p}\nabla f(u)(e^{p\eta(x,u)} - \mathbf{1})], \; for \; p \neq 0, \; r \neq 0, \tag{1}$$

$$\frac{1}{r}e^{rf(x)} \geq \frac{1}{r}e^{rf(u)}[1 + r\nabla f(u)\eta(x, u)], \; for \; p = 0, \; r \neq 0, \tag{2}$$

$$f(x) - f(u) \geq \frac{1}{p}\nabla f(u)(e^{p\eta(x,u)} - \mathbf{1}), \; for \; p \neq 0, \; r = 0, \tag{3}$$

$$f(x) - f(u) \geq \nabla f(u)\eta(x, u), \; for \; p = 0, \; r = 0. \tag{4}$$

We introduce the (p, r)-invex function on a Riemannian manifold M. Using (p, r)-invexity assumptions, we derive optimality conditions for optimization problems on these spaces.

Definition 10 ([8]) Let M be an n-dimensional Riemannian manifold and $f : M \to \mathbb{R}$ be a differentiable function. Let η be a map $\eta : M \times M \to TM$ such that $\eta(x, u) \in T_uM$ for all $x, u \in M$. The exponential map on M is a map $exp_u : T_uM \to M$ and the differential of the exponential map $(\mathrm{d}exp_u)_a : T_a(T_uM) \cong T_uM \to T_cM$, where $a = t_0\eta(x, u)$, $t_0 \in [0, 1]$, and $c \in P_{xu}$ where P_{xu} is a closed path joining the point x and u. Let p, r be arbitrary real numbers. If for all $x \in M$, the relations

$$\frac{1}{r}(e^{r(f(x)-f(u))} - 1) \geq \frac{1}{p}\mathrm{d}f_c([(\mathrm{d}exp_u)_a(p\eta(x, u))] - \mathbf{I}), \; for \; p \neq 0, \; r \neq 0, \tag{5}$$

$$\frac{1}{r}(e^{r(f(x)-f(u))}-1) \geq \mathrm{d}f_u(\eta(x,u)), \, for\ p=0,\ r\neq 0, \tag{6}$$

$$f(x)-f(u) \geq \frac{1}{p}\mathrm{d}f_c([(\mathrm{d}exp_u)_a(p\eta(x,u))]-\mathbf{I}), \, for\ p\neq 0,\ r=0, \tag{7}$$

$$f(x)-f(u) \geq \mathrm{d}f_u(\eta(x,u)), \, for\ p=0,\ r=0, \tag{8}$$

hold, then f is said to be (p, r)-invex function at u on M. Here $\mathbf{I} \in T_c M$ such that for a co-ordinate chart ϕ, $\phi(\mathbf{I}) = \mathbf{1}$, where $\mathbf{1} = (1, 1, ..., 1) \in \mathbb{R}^n$.

Remark 1 We denote the exponential map on the manifold by $\exp(x)$ for $x \in M$ and e^x for $x \in \mathbb{R}$.

Example 1 The circle S can be thought of as the set $\{(x, y) \in \mathbb{R}^2 : x^2 + y^2 = 1\}$ of the Euclidean space $\mathbb{R}^2$. In the case of the circle S the possible co-ordinate charts are

$U_1 = \{(x, y) : x > 0\}$ $\quad \phi_1(x, y) = y$
$U_2 = \{(x, y) : x < 0\}$ $\quad \phi_2(x, y) = y$
$U_3 = \{(x, y) : y > 0\}$ $\quad \phi_3(x, y) = x$
$U_4 = \{(x, y) : y < 0\}$ $\quad \phi_4(x, y) = x.$

We define a differentiable function on S. Let $x = (x_1, x_2) \in S$, and $f : S \to \mathbb{R}$ be defined by $f(x) = x_1 + \sin x_2$. Let $u = (u_1, u_2) \in S$.
The tangent space of S at u is the set $T_u S = \{v \in \mathbb{R}^2 : u \cdot v = 0\}$.
We choose $\eta : S \times S \to T_u S$ as $\eta(x, u) = (-u_2, u_1) \in T_u S$.
Let $a = \eta(x, u) = (-u_2, u_1)$.
We now find $\mathrm{d}f_u(a)$. We take a chart $\phi_3(-u_2, u_1) = \phi(-u_2, u_1) = u_2$ at a and the identity mapping as a chart ψ at $f(a)$. Here both S and $\mathbb{R}$ are of dimension 1. We now find the Jacobian matrix $\psi o f o \phi^{-1}$ at $\phi(a)$.

$$\mathrm{d}f_a(\frac{\partial}{\partial\phi})(\psi) = \frac{\partial}{\partial\phi}(\psi o f) = \frac{\partial}{\partial\phi}(f(-u_2, u_1)) = \frac{\partial}{\partial u_2}(-u_2 + \sin u_1) = -1.$$

i.e., $\mathrm{d}f_u(\eta(x, u)) = -1$.
Now $e^{f(x)-f(u)} - 1 - \mathrm{d}f_u(\eta(x, u)) = e^{f(x)-f(u)} - 1 + 1 = e^{f(x)-f(u)} \geq 0, \forall x, u \in S$.
Hence f is (0, 1)-invex on S.
But if we take $x = (1/\sqrt{2}, -1\sqrt{2}) \in S$, $u = (1/\sqrt{2}, 1\sqrt{2}) \in S$.
Then $f(x) - f(u) = 1/\sqrt{2} - \sin 1/\sqrt{2} - 1/\sqrt{2} - \sin 1/\sqrt{2} = -1.299$ and $df_u(\eta(x, u)) = -1$. Hence $f(x) - f(u) \leq \mathrm{d}f_u(\eta(x, u))$, i.e., f is not invex on S.

Sufficient Optimality Conditions

In recent years, many traditional optimization methods have been successfully generalized to minimize objective functions on manifolds. Mititelu [12] established necessary and sufficient conditions of Karush-Kuhn-Tucker (KKT) [11] type for a vector programming problem on differentiable manifolds. Consider the following primal optimization problem on a Riemannian manifold M

(P) *Minimize* $f(x)$

subject to $g_i(x) \leq 0, \quad i = 1, ..., m,$

where $f : M \to \mathbb{R},\ g_i : M \to \mathbb{R},\ i = 1, ..., m,$ are differentiable functions. Let D denote the set of all feasible solutions of (**P**).

Let $\bar{x} \in$ D be an optimal solution of (**P**) and we define the set $J^\circ = \{j \in 1, ..., m : g_j(\bar{x}) = 0\}$. Suppose that the domain D satisfies the following constraint qualification at $\bar{x}$:

$$R(\bar{x}) : \exists v \in TM : \mathrm{d}(g_{J^\circ})_{\bar{x}}(v) \leq 0.$$

Here $\mathrm{d}(g_{J^\circ})_{\bar{x}}(v)$ is the vector components of $\mathrm{d}(g_j)_{\bar{x}}(v), \forall j \in J^\circ$, taken in increasing order of j.

Lemma 1 (Necessary Karush-Kuhn-Tucker (KKT) condition) *[12] If a feasible point $\bar{x} \in M$ is an optimal solution of the problem (**P**) and satisfies the constraint qualification $R(\bar{x})$, then there exists multiplier $\xi = (\xi_1, ..., \xi_m)^T \in \mathbb{R}^m$, such that the following conditions hold*

$$\mathrm{d}f_{\bar{x}} + \xi^T \mathrm{d}g_{\bar{x}} = 0, \tag{9}$$

$$\xi^T g(\bar{x}) = 0, \tag{10}$$

$$\xi \geq 0, \quad i = 1, 2, ..., m, \tag{11}$$

here $g = (g_1, g_2, ..., g_m)^T$.

Theorem 1 (Sufficient Optimality Condition) *Assume that a point $\bar{x} \in M$ is feasible for problem (**P**), and let the KKT conditions* (9)–(11) *be satisfied at $(\bar{x}, \xi)$. If the objective function f and the function $\xi^T g$ are (p, r)-invex with respect to the same function η at $\bar{x}$ on D, then $\bar{x}$ is a global minimum point of the problem (**P**).*

Proof Let x be a feasible point for the problem (**P**). Since f and $\xi^T g$ are (p, r)-invex with respect to the same function η at $\bar{x}$ on D, $\forall x, \bar{x} \in$ D, we have,

$$\frac{1}{r}(e^{r(f(x)-f(\bar{x}))} - 1) \geq \frac{1}{p}\mathrm{d}f_{\bar{x}}(\mathrm{d}(exp_{\bar{x}}(p\eta(x, \bar{x}))) - \mathbf{I}), \tag{12}$$

$$\frac{1}{r}(e^{r(\xi^T g(x)-\xi^T g(\bar{x}))} - 1) \geq \frac{\xi^T}{p}\mathrm{d}g_{\bar{x}}(\mathrm{d}(exp_{\bar{x}}(p\eta(x, \bar{x}))) - \mathbf{I}). \tag{13}$$

Adding (12) and (13), we have

$$\frac{1}{r}[(e^{r(f(x)-f(\bar{x}))} - 1 + e^{r(\xi^T g(x)-\xi^T g(\bar{x}))} - 1] \geq \frac{1}{p}(\mathrm{d}f_{\bar{x}} + \xi^T \mathrm{d}g_{\bar{x}})(\mathrm{d}(exp_{\bar{x}}(p\eta(x, \bar{x}))) - \mathbf{I}),$$

and by KKT condition (9), we have

$$\frac{1}{r}(e^{r(f(x)-f(\bar{x}))}-1) \geq \frac{1}{r}(1-e^{r(\xi^T g(x)-\xi^T g(\bar{x}))}),$$

or, by KKT condition (10)

$$\frac{1}{r}(e^{r(f(x)-f(\bar{x}))}-1) \geq \frac{1}{r}(1-e^{r(\xi^T g(x))}).$$

Without loss of generality, let $r > 0$ (in the case when $r < 0$ the proof is analogous; one should change only the direction of some inequalities below to the opposite one). Since x is a feasible point of (**P**), then $g(x) \leq 0$ and $\xi \geq 0$ imply that $1-e^{r\xi^T g(x)} \geq 0$ and $e^{r(f(x)-f(\bar{x}))} \geq 1$.
Hence $f(x) \geq f(\bar{x})$. Therefore, $\bar{x}$ is an optimal solution of the problem (**P**).

3.2 ρ-(p,r)-Invexity

Definition 11 (*Mandal and Nahak (2011)*) Let $f : \mathbb{R}^n \to \mathbb{R}$ be a differentiable function and p, r be arbitrary real numbers, $\rho \in \mathbb{R}$. The function f is said to be $(p, r) - \rho - (\eta, \theta)$-invex with respect to $\eta, \theta : \mathbb{R}^n \times \mathbb{R}^n \to \mathbb{R}^n$ at u, if any one of the following conditions holds

$$\frac{1}{r}(e^{r(f(x)-f(u))}-1) \geq \frac{1}{p}\nabla f(u)(e^{p\eta(x,u)}-\mathbf{1})]+\rho\|\theta(x,u)\|^2, \ for\ p \neq 0,\ r \neq 0, \tag{14}$$

$$\frac{1}{r}(e^{r(f(x)-f(u))}-1) \geq \nabla f(u)\eta(x,u)+\rho\|\theta(x,u)\|^2,\ for\ p=0,\ r\neq 0, \tag{15}$$

$$f(x)-f(u) \geq \frac{1}{p}\nabla f(u)(e^{p\eta(x,u)}-\mathbf{1})+\rho\|\theta(x,u)\|^2,\ for\ p\neq 0,\ r=0, \tag{16}$$

$$f(x)-f(u) \geq \nabla f(u)\eta(x,u)+\rho\|\theta(x,u)\|^2,\ for\ p=0,\ r=0. \tag{17}$$

Here the exponentials appearing on the right-hand sides of inequalities above are understood to be taken componentwise and $\mathbf{1} = (1, 1, ..., 1)$.
Motivated by the $(p, r)-\rho-(\eta, \theta)$-invex function, we introduce the $\rho-(p, r)$-invex function and study the sufficient optimality conditions for optimization problems defined on a Riemannian manifold.

Definition 12 ([9]) Let M be an n-dimensional Riemannian manifold and $f : M \to \mathbb{R}$ be a differentiable function. Let η be a map $\eta : M \times M \to TM$ such that $\eta(x, u) \in T_u M$ for all $x, u \in M$. The exponential map on M is a map $exp_u : T_u M \to M$ and the differential of the exponential map $(\mathrm{d}exp_u)_a : T_a(T_u M) \cong T_u M \to T_c M$, where

$a = t_0\eta(x, u)$, $t_0 \in [0, 1]$, and $c \in P_{xu}$ where P_{xu} is a closed path joining the point x and u. Let p, r, and ρ be arbitrary real numbers. If for all $x \in M$, the relations

$$\frac{1}{r}(e^{r(f(x)-f(u))} - 1) \geq \frac{1}{p}\mathrm{d}f_c([(\mathrm{d}exp_u)_a(p\eta(x, u))] - \mathbf{I}) + \rho\mathrm{d}^2(x, u), \; for\; p \neq 0, r \neq 0, \tag{18}$$

$$\frac{1}{r}(e^{r(f(x)-f(u))} - 1) \geq \mathrm{d}f_u(\eta(x, u)) + \rho\mathrm{d}^2(x, u), \; for\; p = 0,\; r \neq 0, \tag{19}$$

$$f(x) - f(u) \geq \frac{1}{p}\mathrm{d}f_c([(\mathrm{d}exp_u)_a(p\eta(x, u))] - \mathbf{I}) + \rho\mathrm{d}^2(x, u), \; forp \neq 0, r = 0, \tag{20}$$

$$f(x) - f(u) \geq \mathrm{d}f_u(\eta(x, u)) + \rho\mathrm{d}^2(x, u), \; for\; p = 0,\; r = 0, \tag{21}$$

hold, then f is said to be $\rho - (p, r)$-invex function at u on M. Here $\mathbf{I} \in T_cM$ such that for a co-ordinate chart ϕ, $\phi(\mathbf{I}) = \mathbf{1}$, where $\mathbf{1} = (1, 1, ..., 1)$.

Note

1. If $\rho > 0$, then we call the functions as "strongly $\rho - (p, r)$-invex" functions.
2. If $\rho = 0$, then the functions reduce to "(p, r)-invex" functions.
3. If $\rho < 0$, then we call the functions as "weakly $\rho - (p, r)$-invex" functions.

It is clear that every strongly $\rho - (p, r)$-invex function is (p, r)-invex but weakly $\rho - (p, r)$-invex function is not (p, r)-invex in general. We construct the following counter example.

Example 2 We consider the circle $S = \{(x, y) \in \mathbb{R}^2 : x^2 + y^2 = 18^2\}$ of the Euclidean space $\mathbb{R}^2$. In the case of the circle S the possible co-ordinate charts are

$U_1 = \{(x, y) : x > 0\}$ $\phi_1(x, y) = y$
$U_2 = \{(x, y) : x < 0\}$ $\phi_2(x, y) = y$
$U_3 = \{(x, y) : y > 0\}$ $\phi_3(x, y) = x$
$U_4 = \{(x, y) : y < 0\}$ $\phi_4(x, y) = x$.

Let $x = (x_1, x_2) \in S$ and we define a differentiable function f on S by $f : S \to \mathbb{R}$, $f(x) = -x_1 + \cos x_2$. Let $u = (u_1, u_2) \in S$ and the angle between x and u is $\theta°$, $(\theta \geq 1)$.

Hence $\mathrm{d}(x, u) = \frac{2\pi \times 18\theta}{360} = \frac{11\theta}{35} = .3143\theta$.

The tangent space of S at u is the set $T_uS = \{v \in \mathbb{R}^2 : u \cdot v = 0\}$.

We choose $\eta : S \times S \to T_uS$ as $\eta(x, u) = (-u_2, u_1) \in T_uS$.

Let $a = \eta(x, u) = (-u_2, u_1)$.

We now find $\mathrm{d}f_u(a)$. We take a chart $\phi_3(-u_2, u_1) = \phi(-u_2, u_1) = u_2$ at a and the identity mapping as a chart ψ at $f(a)$. Here both S and $\mathbb{R}$ are of dimension 1. We now find the Jacobian matrix $\psi o f o \phi^{-1}$ at $\phi(a)$.

$$\mathrm{d}f_a(\frac{\partial}{\partial\phi})(\psi) = \frac{\partial}{\partial\phi}(\psi o f) = \frac{\partial}{\partial\phi}(f(-u_2, u_1)) = \frac{\partial}{\partial u_2}(u_2 + \cos u_1) = 1.$$

i.e., $\mathrm{d}f_u(\eta(x,u)) = 1$.
Now $e^{f(x)-f(u)} - 1 - \mathrm{d}f_u(\eta(x,u)) - \rho \mathrm{d}^2(x,u) = e^{f(x)-f(u)} - 1 - 1 - (0.3143)^2 \rho\theta^2 > -2 - .0987\rho\theta^2$ (since $e^{f(x)-f(u)} > 0$).
If we take $\rho = -50$, then $e^{f(x)-f(u)} - 1 - \mathrm{d}f_u(\eta(x,u)) - \rho \mathrm{d}^2(x,u) > -2 + 4.935\theta^2 > 0$, $\forall x, u \in S$ (we take $\theta \geq 1$). Hence f is ((-50)-(0,1))-invex on S, i.e., f is weakly $50 - (0, 1)$-invex.
But if we take $x = (15, 3\sqrt{11}) \in S$, $u = (15, -3\sqrt{11}) \in S$,
then $e^{f(x)-f(u)} - 1 - \mathrm{d}f_u(\eta(x,u)) = 1 - 1 - 1 = -1 < 0$, i.e., f is not (0,1)-invex on S.

Sufficient Optimality Conditions

We now consider the optimization problem (P) and prove the sufficient optimality conditions.

Theorem 2 (Sufficient Optimality Condition) *Assume that a point $\bar{x} \in M$ is feasible for problem (**P**), and let the KKT conditions* (9)–(11) *be satisfied at $(\bar{x}, \xi)$. If the objective function f and the function $\xi^T g$ are $\rho_1 - (p, r)$-invex and $\rho_2 - (p, r)$-invex, respectively, at $\bar{x}$ on D with respect to the same function η with $(\rho_1 + \rho_2) \geq 0$, then $\bar{x}$ is an optimal solution of the problem (**P**).*

Proof Let x be a feasible point for the problem (**P**). Since f and $\xi^T g$ are $\rho_1 - (p, r)$-invex and $\rho_2 - (p, r)$-invex, respectively, at $\bar{x}$ on D with respect to the same function η, $\forall x \in$ D, we have

$$\frac{1}{r}(e^{r(f(x)-f(\bar{x}))} - 1) \geq \frac{1}{p}\mathrm{d}f_{\bar{x}}(\mathrm{d}(exp_{\bar{x}}(p\eta(x,\bar{x}))) - \mathbf{I}) + \rho_1 \mathrm{d}^2(x,\bar{x}), \tag{22}$$

$$\frac{1}{r}(e^{r(\xi^T g(x)-\xi^T g(\bar{x}))} - 1) \geq \frac{\xi^T}{p}\mathrm{d}g_{\bar{x}}(\mathrm{d}(exp_{\bar{x}}(p\eta(x,\bar{x}))) - \mathbf{I}) + \rho_2 \mathrm{d}^2(x,\bar{x}). \tag{23}$$

Adding (22) and (25) we have

$$\frac{1}{r}[(e^{r(f(x)-f(\bar{x}))} - 1 + e^{r(\xi^T g(x)-\xi^T g(\bar{x}))} - 1)] \geq \frac{1}{p}(\mathrm{d}f_{\bar{x}} + \xi^T \mathrm{d}g_{\bar{x}})(\mathrm{d}(exp_u(p\eta(x,\bar{x}))) - \mathbf{I}) + (\rho_1 + \rho_2)\mathrm{d}^2(x,\bar{x}).$$

By KKT conditions and as $(\rho_1 + \rho_2) \geq 0$, we have

$$\frac{1}{r}[(e^{r(f(x)-f(\bar{x}))} - 1] \geq \frac{1}{r}(1 - e^{r(\xi^T g(x))}). \tag{24}$$

Without loss of generality, let $r > 0$ (in the case when $r < 0$ the proof is analogous; one should change only the direction of some inequalities below to the opposite one).

Since x is a feasible solution of the problem (**P**), then $g(x) \leq 0$ and $\xi \geq 0$ imply that $(1 - e^{r(\xi^T g(x))}) \geq 0$. From which we get $e^{r(f(x)-f(\bar{x}))} \geq 1$.
Hence $f(x) \geq f(\bar{x})$ holds for all feasible $x \in$ D of the problem (**P**). Therefore, $\bar{x}$ is an optimal solution of the problem (**P**).

3.3 Semistrictly Geodesic η-prequasi Invex Functions

In this section, the notion of semistrictly geodesic η-prequasi invex functions is introduced and their properties are studied. Throughout the section, M denotes a finite dimensional Riemannian manifold.

Definition 13 Let S be an open subset of M which is geodesic invex with respect to $\eta : M \times M \to TM$. A function $f : S \to \mathbb{R}$ is said to be semistrictly geodesic η-prequasi invex if $\forall x, y \in S,\ f(x) \neq f(y)$, we have

$$f(\gamma_{x,y}(t)) < \max\{f(x), f(y)\} \quad \forall t \in (0, 1).$$

We show by an example that semistrictly geodesic η-prequasi invex function need not be geodesic η-prequasi invex function [2].

Example: Let $M = \{e^{i\theta} = | -\pi \leq \theta < \pi\}$ and $S = \{e^{i\theta} = | -\frac{\pi}{2} < \theta < \frac{\pi}{2}\}$. Then S is an open set in M. Let $x, y \in S$, where $x = e^{i\theta_1}$, $y = e^{i\theta_2}$, and $\eta(x, y) = (\theta_2 - \theta_1)(\sin\theta_2, -\cos\theta_2)$.
We define a geodesic on M as $\gamma_{x,y} : [0, 1] \to M$ such that $\gamma_{x,y}(t) = (\cos((1 - t)\theta_2 + t\theta_1), \sin((1 - t)\theta_2 + t\theta_1))$. Clearly

$$\gamma_{x,y}(0) = y, \quad \gamma'_{x,y}(0) = \eta(x, y), \quad \gamma_{x,y}(t) \in S, \ \forall t \in [0, 1].$$

Hence S is a geodesic invex set in M.
Now we define $f : S \to \mathbb{R}$ by

$$f(x) = \begin{cases} 1, & \text{if } \theta = 0, \\ 0, & \text{otherwise.} \end{cases}$$

Then $\forall x, y \in S,\ f(x) \neq f(y)$, we have

$$f(\gamma_{x,y}(t)) < \max\{f(x), f(y)\} \quad \forall t \in (0, 1).$$

i.e., f is semistrictly geodesic η-prequasi invex function.
Let $\theta_1 = \frac{\pi}{4},\ \theta_2 = -\frac{\pi}{4},\ t = \frac{1}{2}$, then
$f(\gamma_{x,y}(t)) = f(\cos(\frac{1}{2}\theta_2 + \frac{1}{2}\theta_1), \sin(\frac{1}{2}\theta_2 + \frac{1}{2}\theta_1)) = f(\cos 0, \sin 0) = 1$
$\nleq \max\{f(e^{\frac{i\pi}{4}}), f(e^{-\frac{i\pi}{4}})\} = 0$.
Hence f is not geodesic η-prequasi invex function.

Theorem 3 *Let S be a nonempty geodesic invex subset of M with respect to $\eta : M \times M \rightarrow TM$ and $f : M \rightarrow \mathbb{R}$ be a semistrictly geodesic η-prequasi invex function. If $\bar{x} \in S$ is a local optimal solution to the problem*

$$(MP) \quad \min_{x \in S} f(x)$$

then $\bar{x}$ is a global minimum of (MP).

Proof let $\bar{x} \in S$ be a local minimum of (MP). Then there is a neighborhood $N_\varepsilon(\bar{x})$ of $\bar{x}$ such that

$$f(\bar{x}) \leq f(x) \ \forall x \in S \cap N_\varepsilon(\bar{x}). \tag{25}$$

If possible let $\bar{x}$ is not a global minimum of f then there exists a point $x^* \in S$ such that $f(x^*) \leq f(\bar{x})$.
Since S is a geodesic invex set with respect to η, there exists exactly one geodesic $\gamma_{x^*,\bar{x}}$ joining x^*, $\bar{x}$ such that

$$\gamma_{x^*,\bar{x}}(0) = \bar{x}, \ \gamma'_{x^*,x}(0) - \eta(x^*, \bar{x}), \ \gamma_{x^*,\bar{x}}(t) \in S \ \forall t \in [0, 1].$$

Let us choose $\varepsilon > 0$ small enough such that $d(\gamma_{x^*,\bar{x}}(t), \bar{x}) < \varepsilon$, then $\gamma_{x^*,\bar{x}}(t) \in N_\varepsilon(\bar{x})$.
Since f is semistrictly geodesic η-prequasi invex function, we have

$$f(\gamma_{x^*,\bar{x}}(t)) < \max\{f(x^*), f(\bar{x})\} \quad \forall t \in (0, 1).$$

i.e., for all $\gamma_{x^*,\bar{x}}(t) \in S \cap N_\varepsilon(\bar{x})$, we have $f(\gamma_{x^*,\bar{x}}(t)) < f(\bar{x})$, which is a contradiction to (25).
Hence $\bar{x}$ is a global minimum of (MP).

4 Conclusions

The notions of (p, r)-invex and $\rho - (p, r)$-invex functions on Riemannian manifolds are introduced in this paper which generalizes invex functions. We establish optimality conditions under these generalized invexity assumptions for a general nonlinear programming problem that is built upon on Riemannian manifolds. We extend the notion of semistrictly prequasi invex functions from Euclidean spaces to Riemannian manifolds by introducing semistrictly geodesic η-prequasi invex functions. We hope that our results give a new direction to the researchers in this interesting area of research. Variational and control problems on Riemannian manifolds under geodesic η-invexity will orient the future study of the authors.

Acknowledgments The authors wish to thank the referee for his valuable comments.

References

1. Agarwal, R.P., Ahmad, I., Iqbal, A., Ali, S.: Generalized invex sets and preinvex functions on Riemannian manifolds. Taiwanese J. Math. **16**, 1719–1732 (2012)
2. Ahmad, I., Iqbal, A., Ali, S.: On properties of geodesic η-preinvex functions. Adv. Oper. Res. **2009**, Article ID 381831, pp 10. doi:10.1155/2009/381831
3. Antczak, T.: (p, r)-invex sets and functions. J. Math. Anal. Appl. **263**, 355–379 (2001)
4. Barani, A., Pouryayevali, M.R.: Invex sets and preinvex functions on Riemannian manifolds. J. Math. Anal. Appl. **328**, 767–779 (2007)
5. Ferrara, M., Mititelu, S.: Mond-Weir duality in vector programming with generalized invex functions on differentiable manifolds. Balkan J. Geom. Appl. **11**, 80–87 (2006)
6. Hanson, M.A.: On sufficiency of the Kuhn-Tucker conditions. J. Math. Anal. Appl. **80**, 545–550 (1981)
7. Iqbal, A., Ali, S., Ahmad, I.: On geodesic E-convex sets, geodesic E-convex functions and E-epigraphs. J. Optim. Theor. Appl. **155**, 239–251 (2012)
8. Jana, S., Nahak, C.: Optimality conditions and duality results of the nonlinear programming problems under (p, r)-invexity on differentiable manifolds. BSG Proc. **21**, 84–95 (2014)
9. Jana, S., Nahak, C.: Optimality conditions and duality results of the nonlinear programming problems under $\rho - (p, r)$-invexity on differentiable manifolds. J. Appl. Math. Inf. **32**(3–4), 491–502 (2014)
10. Mandal, P., Nahak, C.: $(p, r) - \rho - (\eta, \theta)$-invexity in multiobjective programming problems. Int. J. Optim. Theor. Methods Appl. **2**, 273–282 (2010)
11. Mangasarian, O.L.: Nonlinear Programming. McGraw-Hill Book Company, New York (1969)
12. Mititelu, S.: Generalized invexity and vector optimization on differentiable manifolds. Differ. Geom. Dyn. Syst. **3**, 21–31 (2001)
13. Pini, R.: Convexity along curves and invexity. Optimization **29**, 301–309 (1994)
14. Rapcsak, T.: Geodesic convexity in nonlinear optimization. J. Optim. Theor. Appl. **69**, 169–183 (1991)
15. Udriste, C.: Convex Functions and Optimization Methods on Riemannian Manifolds, Mathematics and Applications, vol. 297. Kluwer Academic Publishers, Providence (1994)
16. Willmore, T.J.: An Introduction to Differential Geometry. Oxford University Press, Oxford (1959)
17. Yang, X.M., Li, D.: Semistrictly preinvex functions. J. Math. Anal. Appl. **258**, 287–308 (2001)
18. Yang, X.M., Yang, X.Q., Teo, K.L.: Characterizations and applications of prequasi-invex functions. J. Optim. Theor. Appl. **110**(3), 645–668 (2001)
19. Zalmai, G.J.: Generalized sufficiency criteria in continuous-time programming with application to a class of variational-type inequalities. J. Math. Anal. Appl. **153**, 331–355 (1990)

Second-order Symmetric Duality and Variational Problems

Saroj Kumar Padhan, Pramod Kumar Behera and R.N. Mohapatra

Abstract The concept of second-order symmetric duality of the variational problem is studied in the present investigation. Appropriate duality results for a pair of second-order symmetric variational problems are established under generalized invexity assumptions. It is observed that some of the known results in the literature are the particular cases of our work.

Keywords Second-order symmetric duality · Variational problem · Generalized invexity · Weak duality · Strong duality · Converse duality

Mathematics Subject Classifications: 65K05 · 65K10 · 65K99

1 Introduction

In general nonlinear programming problems, the dual of a dual need not be the original primal. So, the concept of symmetric duality in general nonlinear programming was introduced by Dorn [3], where the dual of the dual is always the original primal. Gulati et al. [4] established the Mond-Weir type symmetric duality for multiobjective variational problems and proved the desired duality results under generalized convexity assumptions. Kim and Lee [8] formulated a pair of multiobjective generalized nonlinear symmetric dual variational problems involving vector valued functions which unify the Wolfe and Mond-Weir type duals. They also observed that various

S.K. Padhan (✉) · P.K. Behera
Department of Mathematics, Veer Surendra Sai University
of Technology, Burla 768018, India
e-mail: sarojpadhan@gmail.com

P.K. Behera
e-mail: pramoda.1977@rediffmail.com

R.N. Mohapatra
Department of Mathematics, University of Central Florida, Orlando, FL 32816, USA
e-mail: ramm1627@gmail.com

R.N. Mohapatra et al. (eds.), *Mathematics and Computing*,
Springer Proceedings in Mathematics & Statistics 139,
DOI 10.1007/978-81-322-2452-5_4

known results are the particular cases of their work. Symmetric dual for the multiobjective fractional variational problems with partial invexity was studied by Xiuhong [17]. Kassem [7] introduced a new class of generalized cone-pseudo convex functions and strongly cone-pseudo convex functions and established different duality theorems. A variety of duality theorems for variational problems and symmetric mathematical programming problems have appeared in the literature (Husain and Zabeen [6], Husain et al. [5], Lotfi et al. [10], Padhan and Nahak [11, 12], Suneja and Louhan [15], Verma [16]). There are many problems in real-life situations for which the first-order dual has no solution, while the second-order dual has a solution. Again, convexity assumptions make the solution of an optimization problem relatively easy and assure global optimal results. But there are many optimization problems which contain nonconvex objective functions. To minimize a vector of functionals of curvilinear integral type, Pitea and Postolache [13] considered a new class of multitime multiobjective variational problems. They established duals of Mond-Weir type and generalized Mond-Weir-Zalmai type, based on the normal efficiency conditions for multitime multiobjective variational problems. Desired duality results were also studied under (ρ, b)-quasiinvexity assumptions. Recently, Ahmed et al. [1] proved results on mixed type symmetric duality for multiobjective variational problems with cone constraints. Moreover, the variational problems have been given special attention in the optimization theory which is concerned with problems involving infinite dimensional spaces. To the best of our knowledge, no one has studied the second-order symmetric duality for variational problems.

In this paper, we develop the concept of second-order symmetric duality of the variational problem and obtained duality results for a pair of symmetric variational problem under generalized invexity assumptions. The results proved by Smart and Mond [14] and Kim and Lee [9] are the particular cases of this paper.

2 Notation and Preliminaries

Let $I = [a, b]$ be an interval (through out) and $f : I \times \mathbb{R}^n \times \mathbb{R}^n \times \mathbb{R}^m \times \mathbb{R}^m \to \mathbb{R}$. Consider the real valued function $f(t, x, \dot{x}, y, \dot{y})$, where $t \in I$, $x : I \to \mathbb{R}^n$, $y : I \to \mathbb{R}^m$ and $\dot{x}$, $\dot{y}$ denote the derivatives of x and y, respectively, with respect to t. Assume that f has continuous fourth-order partial derivatives with respect to x, y, $\dot{x}$ and $\dot{y}$. Notational distinction is not considered between row and column vectors. Denote the first partial derivatives of f with respect to x, and $\dot{x}$ by f_x and $f_{\dot{x}}$, respectively, that is,

$$f_x = \begin{pmatrix} \frac{\partial f}{\partial x_1} \\ \frac{\partial f}{\partial x_2} \\ \cdot \\ \cdot \\ \cdot \\ \frac{\partial f}{\partial x_n} \end{pmatrix}, \quad f_{\dot{x}} = \begin{pmatrix} \frac{\partial f}{\partial \dot{x}_1} \\ \frac{\partial f}{\partial \dot{x}_2} \\ \cdot \\ \cdot \\ \cdot \\ \frac{\partial f}{\partial \dot{x}_n} \end{pmatrix}.$$

f_{xx}, the $n \times n$ Hessian matrix of f with respect to x, is defined as

$$f_{xx} = \begin{pmatrix} \frac{\partial^2 f}{\partial x_1 \partial x_1} & \frac{\partial^2 f}{\partial x_1 \partial x_2} & \cdots & \frac{\partial^2 f}{\partial x_1 \partial x_n} \\ \frac{\partial^2 f}{\partial x_2 \partial x_1} & \frac{\partial^2 f}{\partial x_2 \partial x_2} & \cdots & \frac{\partial^2 f}{\partial x_2 \partial x_n} \\ \cdot & \cdot & \cdots & \cdot \\ \cdot & \cdot & \cdots & \cdot \\ \cdot & \cdot & \cdots & \cdot \\ \frac{\partial^2 f}{\partial x_n \partial x_1} & \frac{\partial^2 f}{\partial x_n \partial x_2} & \cdots & \frac{\partial^2 f}{\partial x_n \partial x_n} \end{pmatrix}.$$

Similarly $f_{x\dot{x}}$, $f_{\dot{x}\dot{x}}$, f_y, f_{yy}, $f_{\dot{y}}$, $f_{y\dot{y}}$, and $f_{\dot{y}\dot{y}}$ are also defined.
Consider the variational problem

$$\text{(VP)} \ \min \int_a^b f\,(t, x(t), \dot{x}(t))\,dt, \tag{1}$$

subject to

$$g\,(t, x(t), \dot{x}(t)) \leq 0, \tag{2}$$

$$x(a) = \gamma_1, x(b) = \gamma_2;\ \dot{x}(a) = \delta_1, \dot{x}(b) = \delta_2, \tag{3}$$

where f and g are twice continuously differentiable functions from $I \times \mathbb{R}^n \times \mathbb{R}^n$ into $\mathbb{R}$ and $\mathbb{R}^m$, respectively.

Lemma 2.1 *[2] If (VP) attains a local (or global) minimum at $\bar{x} \in S$, then there exist Lagrange multiplier $\tau \in \mathbb{R}$ and piecewise smooth $\lambda : I \to \mathbb{R}^m$ such that,*

$$\begin{aligned} &\tau f_x(t, \bar{x}(t), \dot{\bar{x}}(t)) + g_x(t, \bar{x}(t), \dot{\bar{x}}(t))^T \lambda(t) \\ &= \frac{d}{dt}[\tau f_{\dot{x}}(t, \bar{x}(t), \dot{\bar{x}}(t)) + g_{\dot{x}}(t, \bar{x}(t), \dot{\bar{x}}(t))^T \lambda(t)], t \in I \end{aligned} \tag{4}$$

$$\lambda(t)^T g(t, \bar{x}(t), \dot{\bar{x}}(t)) = 0, t \in I \tag{5}$$

$$(\tau, \lambda(t)^T) \geq 0, t \in I. \tag{6}$$

Remark 2.1 Equations (4), (5) and (6) give the Fritz-John necessary conditions for (VP), and they become Kuhn-Tucker conditions if $\tau = 1$.

Definitions of invexity:

Definition 2.1 The functional $\int_a^b f(t, ., ., ., .)dt$ is said to be second-order $\rho_1 - (\eta, \theta)-$invex in x and $\dot{x}$ if for each $v : I \to \mathbb{R}^m$, with $\dot{v}$ is piecewise smooth, there exist functions $\eta : I \times \mathbb{R}^n \times \mathbb{R}^n \times \mathbb{R}^n \times \mathbb{R}^n \to \mathbb{R}^n, \theta : I \times \mathbb{R}^n \times \mathbb{R}^n \times \mathbb{R}^m \times \mathbb{R}^m \to \mathbb{R}$ and $\rho_1 \in \mathbb{R}$, such that

$$\begin{aligned}\int_a^b f(t, x, \dot{x}, v, \dot{v})dt &- \int_a^b f(t, u, \dot{u}, v, \dot{v})dt \\ &\geq \int_a^b \{\eta(t, x, \dot{x}, u, \dot{u})\big[f_x(t, u, \dot{u}, v, \dot{v}) - \frac{d}{dt} f_{\dot{x}}(t, u, \dot{u}, v, \dot{v})\big] \\ &+ \frac{1}{2}\eta(t, x, \dot{x}, u, \dot{u})\big[f_{xx}(t, u, \dot{u}, v, \dot{v}) - 2\frac{d}{dt} f_{x\dot{x}}(t, u, \dot{u}, v, \dot{v}) \\ &+ \frac{d^2}{dt^2} f_{\dot{x}\dot{x}}(t, u, \dot{u}, v, \dot{v})\big]u(t) + \rho_1|\theta(t, x, u, y, v)|^2\}dt,\end{aligned}$$

for all $x : I \to \mathbb{R}^n$, $u : I \to \mathbb{R}^n$ with $(\dot{x}, \dot{u})$ piecewise smooth on I.

Definition 2.2 The functional $-\int_a^b f(t, ., ., ., .)dt$ is said to be second-order $\rho_2 - (\xi, \theta)-$invex in y and $\dot{y}$ if for each $x : I \to \mathbb{R}^n$, with $\dot{x}$ is piecewise smooth, there exist functions $\xi : I \times \mathbb{R}^m \times \mathbb{R}^m \times \mathbb{R}^m \times \mathbb{R}^m \to \mathbb{R}^m, \theta : I \times \mathbb{R}^n \times \mathbb{R}^n \times \mathbb{R}^m \times \mathbb{R}^m \to \mathbb{R}$ and $\rho_2 \in \mathbb{R}$, such that

$$\begin{aligned}-\int_a^b &\big[f(t, x, \dot{x}, v, \dot{v}) - f(t, x, \dot{x}, y, \dot{y})\big]dt \\ &\geq -\int_a^b \{\xi(t, v, \dot{v}, y, \dot{y})\big[f_y(t, x, \dot{x}, y, \dot{y}) - \frac{d}{dt} f_{\dot{y}}(t, x, \dot{x}, y, \dot{y})\big] \\ &+ \frac{1}{2}\xi(t, v, \dot{v}, y, \dot{y})\big[f_{yy}(t, x, \dot{x}, y, \dot{y}) - 2\frac{d}{dt} f_{y\dot{y}}(t, x, \dot{x}, y, \dot{y}) \\ &+ \frac{d^2}{dt^2} f_{\dot{y}\dot{y}}(t, x, \dot{x}, y, \dot{y})\big]y(t) + \rho_2|\theta(t, x, u, y, v)|^2\}dt,\end{aligned}$$

for all $v : I \to \mathbb{R}^m$, $y : I \to \mathbb{R}^m$ with $(\dot{v}, \dot{y})$ piecewise smooth on I.

Note: Throughout the paper, $\rho_1 - (\eta, \theta)-$invex means invexity with respect to x and $\dot{x}$. Again $\rho_2 - (\xi, \theta)-$invex means invexity with respect to y and $\dot{y}$.

3 Symmetric Duality

We formulate the following pair of second-order symmetric nonlinear variational primal problems, where $(\dot{x}(t), \dot{y}(t))$ is piecewise smooth.

$$(SSVP)\ \min \int_a^b \Big\{ f(t,x,\dot{x},y,\dot{y}) - y(t) f_y(t,x,\dot{x},y,\dot{y}) + y(t)\frac{d}{dt} f_{\dot{y}}(t,x,\dot{x},y,\dot{y}) - \frac{1}{2} y(t)[f_{yy}(t,x,\dot{x},y,\dot{y}) - 2\frac{d}{dt} f_{y\dot{y}}(t,x,\dot{x},y,\dot{y}) + \frac{d^2}{dt^2} f_{\dot{y}\dot{y}}(t,x,\dot{x},y,\dot{y})]y(t) \Big\} dt$$

$$\text{subject to} \quad x(a) = x_0,\ x(b) = x_1,\ y(a) = y_0,\ y(b) = y_1, \tag{7}$$

$$f_y(t,x,\dot{x},y,\dot{y}) - \frac{d}{dt} f_{\dot{y}}(t,x,\dot{x},y,\dot{y}) + [f_{yy}(t,x,\dot{x},y,\dot{y}) - 2\frac{d}{dt} f_{y\dot{y}}(t,x,\dot{x},y,\dot{y}) + \frac{d^2}{dt^2} f_{\dot{y}\dot{y}}(t,x,\dot{x},y,\dot{y})]y(t) \le 0, \tag{8}$$

$$x(t) \ge 0. \tag{9}$$

$$(SSVD)\ \max \int_a^b \Big\{ f(t,u,\dot{u},v,\dot{v}) - u(t) f_x(t,u,\dot{u},v,\dot{v}) + u(t)\frac{d}{dt} f_{\dot{x}}(t,u,\dot{u},v,\dot{v}) - \frac{1}{2} u(t)[f_{xx}(t,u,\dot{u},v,\dot{v}) - 2\frac{d}{dt} f_{x\dot{x}}(t,u,\dot{u},v,\dot{v}) + \frac{d^2}{dt^2} f_{\dot{x}\dot{x}}(t,u,\dot{u},v,\dot{v})]u(t) \Big\} dt$$

$$\text{subject to} \quad u(a) = u_0,\ u(b) = u_1,\ v(a) = v_0,\ v(b) = v_1, \tag{10}$$

$$f_x(t,u,\dot{u},v,\dot{v}) - \frac{d}{dt} f_{\dot{x}}(t,u,\dot{u},v,\dot{v}) + [f_{xx}(t,u,\dot{u},v,\dot{v}) - 2\frac{d}{dt} f_{x\dot{x}}(t,u,\dot{u},v,\dot{v}) + \frac{d^2}{dt^2} f_{\dot{x}\dot{x}}(t,u,\dot{u},v,\dot{v})]u(t) \ge 0, \tag{11}$$

$$v(t) \ge 0, \tag{12}$$

where inequalities (3) and (11) may not satisfy at the corners of $(\dot{x}(t), \dot{y}(t))$ and $(\dot{u}(t), \dot{v}(t))$, respectively, but must be satisfied for unique right- and left- hand limits.

Theorem 3.1 (Weak Duality) *Let (x, y) and (u, v) be the feasible solutions of (SSVP) and (SSVD), respectively. Suppose $\int_a^b f(t, ., ., ., .)dt$ and $-\int_a^b f(t, ., ., ., .)dt$ are second-order $\rho_1 - (\eta, \theta)-$invex and second-order $\rho_2 - (\xi, \theta)-$invex functions, respectively, with respect to the same function θ, and $\rho_2 - \rho_1 \ge 0$. Also assume that $\eta(t, x, \dot{x}, u, \dot{u}) + u(t) \ge 0$ and $\xi(t, v, \dot{v}, y, \dot{y}) + y(t) \ge 0$ (except perhaps at the corner of $(\dot{x}(t), \dot{y}(t))$ and $(\dot{u}(t), \dot{v}(t)$, respectively, but must be satisfied for unique right- and left- hand limits). Then the following inequality holds between the primal (SSVP) and the dual (SSVD),*

$$\int_a^b \{f(t,x,\dot{x},y,\dot{y}) - y(t)f_y(t,x,\dot{x},y,\dot{y}) + y(t)\frac{d}{dt}f_{\dot{y}}(t,x,\dot{x},y,\dot{y})$$
$$-\frac{1}{2}y(t)[f_{yy}(t,x,\dot{x},y,\dot{y})$$
$$-2\frac{d}{dt}f_{y\dot{y}}(t,x,\dot{x},y,\dot{y}) + \frac{d^2}{dt^2}f_{\dot{y}\dot{y}}(t,x,\dot{x},y,\dot{y})]y(t)\}dt$$
$$\geq \int_a^b \{f(t,u,\dot{u},v,\dot{v}) - u(t)f_x(t,u,\dot{u},v,\dot{v})$$
$$+u(t)\frac{d}{dt}f_{\dot{x}}(t,u,\dot{u},v,\dot{v}) - \frac{1}{2}u(t)[f_{xx}(t,u,\dot{u},v,\dot{v})$$
$$-2\frac{d}{dt}f_{x\dot{x}}(t,u,\dot{u},v,\dot{v}) + \frac{d^2}{dt^2}f_{\dot{x}\dot{x}}(t,u,\dot{u},v,\dot{v})]u(t)\}dt.$$

Proof By the assumptions of second-order $\rho_1-(\eta,\theta)-$invexity of $\int_a^b f(t,.,.,.,.)dt$ and second-order $\rho_2-(\xi,\theta)-$invexity of $-\int_a^b f(t,.,.,.,.)dt$, we have

$$\int_a^b f(t,x,\dot{x},y,\dot{y})dt - \int_a^b f(t,u,\dot{u},v,\dot{v})dt$$
$$\geq \int_a^b \{\eta(t,x,\dot{x},u,\dot{u})\big[f_x(t,u,\dot{u},v,\dot{v}) - \frac{d}{dt}f_{\dot{x}}(t,u,\dot{u},v,\dot{v})\big]$$
$$+\frac{1}{2}\eta(t,x,\dot{x},u,\dot{u})[f_{xx}(t,u,\dot{u},v,\dot{v})$$
$$-2\frac{d}{dt}f_{x\dot{x}}(t,u,\dot{u},v,\dot{v}) + \frac{d^2}{dt^2}f_{\dot{x}\dot{x}}(t,u,\dot{u},v,\dot{v})]u(t)\}dt$$
$$-\xi(t,v,\dot{v},y,\dot{y})\big[f_y(t,x,\dot{x},y,\dot{y})$$
$$-\frac{d}{dt}f_{\dot{y}}(t,x,\dot{x},y,\dot{y})\big] - \frac{1}{2}\xi(t,v,\dot{v},y,\dot{y})[f_{yy}(t,x,\dot{x},y,\dot{y})$$
$$-2\frac{d}{dt}f_{y\dot{y}}(t,x,\dot{x},y,\dot{y})$$
$$+\frac{d^2}{dt^2}f_{\dot{y}\dot{y}}(t,x,\dot{x},y,\dot{y})]y(t) + (\rho_1-\rho_2)|\theta(t,x,u,y,v)|^2\}dt \tag{13}$$

Now using the assumptions $\eta(t,x,\dot{x},u,\dot{u}) + u(t) \geq 0$, $\xi(t,v,\dot{v},y,\dot{y}) + y(t) \geq 0$ and $\rho_2 - \rho_1 \geq 0$, inequality (13) becomes

$$\int_a^b f(t,x,\dot{x},y,\dot{y})dt - \int_a^b f(t,u,\dot{u},v,\dot{v})dt$$

$$\geq \int_a^b \{-u(t)\big[f_x(t,u,\dot{u},v,\dot{v}) - \frac{d}{dt}f_{\dot{x}}(t,u,\dot{u},v,\dot{v})\big] - \frac{1}{2}u(t)[f_{xx}(t,u,\dot{u},v,\dot{v})$$

$$-2\frac{d}{dt}f_{x\dot{x}}(t,u,\dot{u},v,\dot{v}) + \frac{d^2}{dt^2}f_{\dot{x}\dot{x}}(t,u,\dot{u},v,\dot{v})]u(t)\}dt + y(t)\big[f_y(t,x,\dot{x},y,\dot{y})$$

$$-\frac{d}{dt}f_{\dot{y}}(t,x,\dot{x},y,\dot{y})\big] + \frac{1}{2}y(t)[f_{yy}(t,x,\dot{x},y,\dot{y}) - 2\frac{d}{dt}f_{y\dot{y}}(t,x,\dot{x},y,\dot{y})$$

$$+\frac{d^2}{dt^2}f_{\dot{y}\dot{y}}(t,x,\dot{x},y,\dot{y})]y(t)\}dt$$

$$\Rightarrow \int_a^b \{f(t,x,\dot{x},y,\dot{y}) - y(t)f_y(t,x,\dot{x},y,\dot{y}) + y(t)\frac{d}{dt}f_{\dot{y}}(t,x,\dot{x},y,\dot{y}) - \frac{1}{2}y(t)[f_{yy}(t,x,\dot{x},y,\dot{y})$$

$$-2\frac{d}{dt}f_{y\dot{y}}(t,x,\dot{x},y,\dot{y}) + \frac{d^2}{dt^2}f_{\dot{y}\dot{y}}(t,x,\dot{x},y,\dot{y})]y(t)\}dt$$

$$\geq \int_a^b \{f(t,u,\dot{u},v,\dot{v}) - u(t)f_x(t,u,\dot{u},v,\dot{v}) + u(t)\frac{d}{dt}f_{\dot{x}}(t,u,\dot{u},v,\dot{v}) - \frac{1}{2}u(t)[f_{xx}(t,u,\dot{u},v,\dot{v})$$

$$-2\frac{d}{dt}f_{x\dot{x}}(t,u,\dot{u},v,\dot{v}) + \frac{d^2}{dt^2}f_{\dot{x}\dot{x}}(t,u,\dot{u},v,\dot{v})]u(t)\}dt.$$

□

Theorem 3.2 (Strong Duality) *Let (x, y) be an optimal solution of (SSVP). Suppose $\int_a^b f(t,.,.,.,.)dt$ and $-\int_a^b f(t,.,.,.,.)dt$ are second-order $\rho_1-(\eta,\theta)-$ invex and second-order $\rho_2-(\xi,\theta)-$ invex functions, respectively, with respect to the same functions θ, and $\rho_2-\rho_1 \geq 0$. Also the weak duality Theorem 3.1 holds between (SSVP) and (SSVD). Then $(x(t), y(t) = 0)$ is an optimal solution of (SSVD), and the optimal values of (SSVP) and (SSVD) are equal.*

Proof (x, y) is an optimal solution of (*SSVP*). Using Lemma 2.1, it can be easily shown that $(x(t), y(t) = 0)$ satisfies all the constraints of (SSVD). Again weak duality Theorem 3.1 shows that $(x(t), y(t) = 0)$ is an optimal solution of (*SSVD*). □

Theorem 3.3 (Converse Duality) *Let (u, v) be an optimal solution of (SSVD). Suppose $\int_a^b f(t,.,.,.,.)dt$ and $-\int_a^b f(t,.,.,.,.)dt$ are second-order $\rho_1-(\eta,\theta)-$invex and second-order $\rho_2-(\xi,\theta)-$invex functions, respectively, with respect to the same functions θ, and $\rho_2-\rho_1 \geq 0$. Again the weak duality Theorem 3.1 holds between (SSVP) and (SSVD). Further, assume that*

(i) $u(t) \leq 0$ and $f_{uu}(t,u,\dot{u},v,\dot{v}) - 2\frac{d}{dt}f_{u\dot{u}}(t,u,\dot{u},v,\dot{v}) + \frac{d^2}{dt^2}f_{\dot{u}\dot{u}}(t,u,\dot{u},v,\dot{v}) \leq 0$,

(ii) $v(t) \leq 0$ and $f_{vv}(t,u,\dot{u},v,\dot{v}) - 2\frac{d}{dt}f_{v\dot{v}}(t,u,\dot{u},v,\dot{v}) + \frac{d^2}{dt^2}f_{\dot{v}\dot{v}}(t,u,\dot{u},v,\dot{v}) \leq 0$.

Then (u, v) is an optimal solution of (SSVP).

Proof Suppose (u, v) is not an optimal solution of (*SSVP*). Then there exists a feasible solution (x, y) such that

$$\int_a^b \{f(t,x,\dot{x},y,\dot{y}) - y(t)f_y(t,x,\dot{x},y,\dot{y}) + y(t)\frac{d}{dt}f_{\dot{y}}(t,x,\dot{x},y,\dot{y}) - \frac{1}{2}y(t)[f_{yy}(t,x,\dot{x},y,\dot{y}) - 2\frac{d}{dt}f_{y\dot{y}}(t,x,\dot{x},y,\dot{y}) + \frac{d^2}{dt^2}f_{\dot{y}\dot{y}}(t,x,\dot{x},y,\dot{y})]y(t)\}dt$$
$$< \int_a^b \{f(t,u,\dot{u},v,\dot{v}) - v(t)f_v(t,u,\dot{u},v,\dot{v}) + v(t)\frac{d}{dt}f_{\dot{x}}(vt,u,\dot{u},v,\dot{v}) - \frac{1}{2}v(t)[f_{vv}(t,u,\dot{u},v,\dot{v}) - 2\frac{d}{dt}f_{v\dot{v}}(t,u,\dot{u},v,\dot{v}) + \frac{d^2}{dt^2}f_{\dot{v}\dot{v}}(t,u,\dot{u},v,\dot{v})]v(t)\}dt. \quad (14)$$

Now by the assumptions of second-order $\rho_1-(\eta,\theta)-$invexity of $\int_a^b f(t,.,.,.,.)dt$, second-order $\rho_2-(\xi,\theta)-$invexity of $-\int_a^b f(t,.,.,.,.)dt$ and weak duality Theorem 3.1, we have

$$\int_a^b \{f(t,x,\dot{x},y,\dot{y}) - y(t)f_y(t,x,\dot{x},y,\dot{y}) + y(t)\frac{d}{dt}f_{\dot{y}}(t,x,\dot{x},y,\dot{y}) - \frac{1}{2}y(t)[f_{yy}(t,x,\dot{x},y,\dot{y}) - 2\frac{d}{dt}f_{y\dot{y}}(t,x,\dot{x},y,\dot{y}) + \frac{d^2}{dt^2}f_{\dot{y}\dot{y}}(t,x,\dot{x},y,\dot{y})]y(t)\}dt$$
$$- \int_a^b \{f(t,u,\dot{u},v,\dot{v}) - v(t)f_v(t,u,\dot{u},v,\dot{v}) + v(t)\frac{d}{dt}f_{\dot{x}}(vt,u,\dot{u},v,\dot{v}) - \frac{1}{2}v(t)[f_{vv}(t,u,\dot{u},v,\dot{v}) - 2\frac{d}{dt}f_{v\dot{v}}(t,u,\dot{u},v,\dot{v}) + \frac{d^2}{dt^2}f_{\dot{v}\dot{v}}(t,u,\dot{u},v,\dot{v})]v(t)\}dt.$$
$$\geq \int_a^b \{f(t,u,\dot{u},v,\dot{v}) - u(t)f_u(t,u,\dot{u},v,\dot{v}) + u(t)\frac{d}{dt}f_{\dot{u}}(t,u,\dot{u},v,\dot{v}) - \frac{1}{2}u(t)[f_{uu}(t,u,\dot{u},v,\dot{v}) - 2\frac{d}{dt}f_{u\dot{u}}(t,u,\dot{u},v,\dot{v}) + \frac{d^2}{dt^2}f_{\dot{u}\dot{u}}(t,u,\dot{u},v,\dot{v})]u(t)\}dt$$
$$- \int_a^b \{f(t,u,\dot{u},v,\dot{v}) - v(t)f_v(t,u,\dot{u},v,\dot{v}) + v(t)\frac{d}{dt}f_{\dot{x}}(vt,u,\dot{u},v,\dot{v}) - \frac{1}{2}v(t)[f_{vv}(t,u,\dot{u},v,\dot{v}) - 2\frac{d}{dt}f_{v\dot{v}}(t,u,\dot{u},v,\dot{v}) + \frac{d^2}{dt^2}f_{\dot{v}\dot{v}}(t,u,\dot{u},v,\dot{v})]v(t)\}dt.$$
$\geq 0.$(By (i), (ii) and (u, v), $(x, y)$are optimal solutions of $(SSVD)$ and $(SSVP)$)

Which is a contradiction and hence the result. □

4 Concluding Remarks

1. For the first time we have established the second-order symmetric duality for variational problems.
2. When $\rho_1 = 0 = \rho_2$ and eliminate all second-order partial derivatives, the invexity and duality defined by Smart and Mond [14], and Kim and Lee [9] are the particular cases of our work.

Acknowledgments The authors wish to thank the referees for their valuable suggestions that improved the presentation of the paper.

References

1. Ahmad, I., Husain, Z., Al-Homidan, S.: Second-order duality in nondifferentiable fractional programming. Int. Trans. Oper. Res. **21**, 291–310 (2014)
2. Chen, X.: Second order duality for the variational problems. J. Math. Anal. Appl. **286**, 261–270 (2003)
3. Dorn, W.S.: A symmetric dual theorem for quadratic programs. J. Opes. Res. Soc. Japan **2**, 93–97 (1960)
4. Gulati, T.R., Husain, I., Ahmed, A.: Symmetric duality for multiobjective variational problems. J. Math. Anal. Appl. **210**, 22–38 (1997)
5. Husain, I., Ahmed, A., Masoodi, M.: Second-order duality for variational problems. Eur. J. Pure Appl. Math. **2**, 278–295 (2009)
6. Husain, I., Jabeen, Z.: On variational problems involving higher order derivatives. J. Appl. Math. Comput. **17**, 433–455 (2005)
7. Kassem, A.: Multiobjective nonlinear second order symmetric duality with (K, F)-pseudoconvexity. Appl. Math. Comput. **219**, 2142–2148 (2012)
8. Kim, D.S., Lee, W.J.: Generalized symmetric duality for multiobjective variational problems with invexity. J. Math. Anal. Appl. **234**, 147–164 (1999)
9. Kim, D.S., Lee, W.J.: Symmetric duality for multiobjective variational problems with invexity. J. Math. Anal. Appl. **218**, 34–48 (1998)
10. Lotfi, F.H., Noora, A.A., Jahanshahloo, G.R., Khodabakhshi, M., Payan, A.: Alinear programming approach to test efficiency in multi-objective linear fractional programming problems. Appl. Math. Model. **34**, 4179–4183 (2010)
11. Padhan, S.K., Nahak, C.: Second order duality for the variational problems under $\rho - (\eta, \theta)$-invexity. Comput. Math. Appl. **60**, 3072–3081 (2010)
12. Padhan, S.K., Nahak, C.: Higher-order symmetric duality in multiobjective programming problems under higher-order invexity. Appl. Math. Comput. **218**, 1705–1712 (2011)
13. Pitea, A., Postolache, M.: Duality theorems for a new class of multitime multiobjective variational problems. J. Glob. Optim. **54**, 47–58 (2011)
14. Smart, I., Mond, B.: Symmetric duality with invexity in variational problems. J. Math. Anal. Appl. **152**, 536–545 (1996)
15. Suneja, S.K., Louhan, P.: Higher-order symmetric duality under cone-invexity and other related concepts. J. Comput. Appl. Math. **255**, 825–836 (2014)
16. Verma, R.U.: Weak $\epsilon-$efficiency conditions for multiobjecitve fractional programming. Appl. Math. Comput. **219**, 6819–6827 (2013)
17. Xiuhong, C.: Symmetric duality for multiobjective fractional variational problems with partial invexity. J. Math. Anal. Appl. **245**, 105–123 (2000)

Efficient Portfolio for Interval Sharpe Ratio Model

Mrinal Jana, Pankaj Kumar and Geetanjali Panda

Abstract In this paper a problem related to portfolio optimization model is proposed to maximize the Sharpe ratio of the portfolio with varying parameters. The Sharpe ratio model is an interval fractional programming problem in which the function in objective and in constraints are interval-valued function. A methodology is developed to solve the Sharpe ratio model. This model is transformed into a general optimization problem and relation between the original problem and the transformed problem is established.

Keywords Portfolio optimization · Efficient portfolio · Fractional programming · Interval-valued function · Interval inequalities

1 Introduction

Balancing reward against risk is the base of a general mean-variance portfolio optimization problem. Reward is measured by the portfolio expected return and risk is measured by the portfolio variance. In the most basic form, portfolio optimization model determines the proportion of the total investment x_j of jth asset of a portfolio $x = (x_1, x_2, ..., x_n)$, where $\sum_{j=1}^{n} x_j = 1$. A common criterion for this assessment is the expected return-to-risk trade-off which is known as the Sharpe ratio of the portfolio [7, 8]. An important objective of selecting a portfolio is to maximize the Sharpe ratio [2, 5, 9]. In general the portfolio optimization model which maximizes the Sharpe ratio, is a fractional programming problem, when the coefficients in the

M. Jana (✉) · P. Kumar · G. Panda
Department of Mathematics, Indian Institute of Technology Kharagpur,
Kharagpur 721302, India
e-mail: mrinal.jana88@gmail.com

P. Kumar
e-mail: maths.pk@gmail.com

G. Panda
e-mail: geetanjali@maths.iitkgp.ernet.in

R.N. Mohapatra et al. (eds.), *Mathematics and Computing*,
Springer Proceedings in Mathematics & Statistics 139,
DOI 10.1007/978-81-322-2452-5_5

objective function and constraints (return, risk, and other parameters) are usually crisp values. Due to uncertainty in financial market, return of the asset is not fixed. Usually expected return is estimated from historical data, which in term affects the risk and performances of the portfolio. Recently many researchers have tried to select efficient portfolios by solving portfolio optimization models and addressing the uncertain parameters like return, risk, etc., using probability theory and fuzzy set theory. These methods have certain limitations while selecting suitable probability distribution functions and membership functions, respectively. Selection of suitable distribution function and membership function can be avoided if we consider the lower and upper level of the return from historical data. In that case the return of an asset will lie in closed interval which can include all types of market uncertainties. But, if the return is considered as a closed interval then the risk and performance of the portfolio have to be expressed in terms of intervals, which may be treated as interval-valued functions in mathematical sense. In these situations the portfolio optimization model cannot be handled by general optimization techniques. As a result of which, the parameters of the Sharpe ratio model become intervals and the portfolio selection model becomes an interval fractional programming model. Formulation of such a model is discussed in detail in Sect. 2. To study this model, we focus on its solution methodology. In the proposed methodology, the interval Sharpe ratio model is transformed to a general optimization problem which is free from interval uncertainty and relation between the original problem and the transformed problem is established. The theoretical developments are illustrated through a numerical example.

Hladík [4] consider a generalized linear fractional programming problem with interval data and present a method for computing the range of optimal values. Interval nonlinear fractional programming problem has been discussed by [1]. In [1], the denominator of the objective function is interval-valued affine function and in our paper the denominator of the objective function is an interval-valued quadratic function. In this paper we concentrate on nonlinear interval fractional programming problem and studied the existence of its solution. Then the Sharpe ratio model in Sect. 2 is solved by the proposed methodology. Throughout this paper (**SOI**) denotes an interval Sharpe ratio model.

The paper is divided in six major sections. Section 2 describes the proposed Sharpe ratio model as an interval fractional programming problem. Section 3 describes solution methodology for solving (**SOI**) with a numerical example. Section 4 includes an application of methodology to portfolio selection, and Sect. 5 provides some concluding remarks.

2 Sharpe Ratio Model

As we discussed earlier, an important objective of selecting a portfolio, based on risk-adjusted performance, is to maximize the Sharpe ratio. The higher portfolio's Sharpe ratio, the better its risk-adjusted performance has been. A negative Sharpe ratio indicates that a riskless asset would perform better than the security/risky asset.The parameters like return, variance, etc., are estimated from historical data. However, in the financial market, several types of uncertainties are affecting while estimating these parameters. For this reason, it is reasonable to consider the upper and lower bound of these parameters. We propose a Sharpe ratio model with parameters varies into closed intervals. Assuming that an investor invests his/her money into n-number of risky securities in such a way that the portfolio performance measure Sharpe ratio become maximum. Following are the notations and assumptions:

Λ_k	$\{1, 2, ..., k\}$.
T	Total number of time period.
p_{jt}	Rate of return of jth risky asset in kth time period, $j \in \Lambda_n,\ t \in \Lambda_T$.
p_j	Expected return of jth asset and is equal to $\frac{1}{T}\sum_{k=1}^{T} r_{jt}$.
$p_j^L(p_j^R)$	Lower (upper) bound of expected return of jth asset such that $r_j^L \leq r_j \leq r_j^R$.
p_f	Return of riskless asset.
s_{ij}	The covariance between ith and jth assets returns.
$s_{ij}^L(s_{ij}^R)$	Lower(upper) bound of covariance of return between ith and jth assets, i.e., $s_{ij}^L \leq s_{ij} \leq s_{ij}^R$, this implies $s_{ij} \in [s_{ij}^L, s_{ij}^R]$.
s_j	standard deviation of jth asset.
$s_j^L(s_j^R)$	Lower(upper) bound of standard deviation of jth asset.
x_j	The proportion of the total funds invested on jth assets.

Total expected return of portfolio $(x_1, x_2, \ldots, x_n)$ is $\sum_{j=1}^{n} p_j x_j$. Since $p_j \in [p_j^L, p_j^R]$, therefore portfolio return becomes $\sum_j^n [p_j^L, p_j^R] x_j$. Since p_f is the return of riskless assets, so total expected return is

$$\sum_{j}^{n}[p_j^L, p_j^R]x_j \ominus [p_f, p_f] = \sum_{j}^{n}[p_j^L - p_f, p_j^R - p_f]x_j.$$

Since the return vary in intervals so they will affect the variance of the portfolio. Hence the variance of the portfolio is

$$\sum_{i}^{n}\sum_{j=1}^{n}[s_{ij}^L, s_{ij}^R]x_i x_j.$$

Consequently, the Sharpe ratio becomes an interval-valued function

$$\sum_{j}^{n}[p_j^L - p_f, p_j^R - p_f]x_j \oslash \sqrt{\sum_{i}^{n}\sum_{j=1}^{n}[s_{ij}^L, s_{ij}^R]x_i x_j},$$

where $\sqrt{\sum_{i=1}^{n}\sum_{j=1}^{n} s_{ij}^L x_i x_j} > 0$.

In addition to this, the sum of the proportion of total investment for each stock of the portfolio is equal to one, i.e., $\sum_{j=1}^{n} x_j = 1$.

Sometimes the solution of the model, considering the above constraint only, typically leads to extreme portfolio weights, particularly when the number of assets is large. One can take the naive $\frac{1}{n}$ portfolio as the benchmark and impose the constraint: $\left|x_j - \frac{1}{n}\right| \leq \alpha$, where α is a positive constant. To get better portfolio output we impose constraints on each asset to improve the portfolio performance. The main idea of these constraints is that the estimation error of return, variance for all the assets from the historical data is not same for all the assets. The estimation errors are larger for the assets with larger sample variances. The constraint imposed on the weight of a given asset is inversely proportional to its standard deviation. We replace the constraint $\left|x_j - \frac{1}{n}\right| \leq \alpha$ by

$$\left|x_j - \frac{1}{n}\right| [s_j^L, s_j^R] \preceq \frac{\alpha}{n} \sum_{j=1}^{n} [s_j^L, s_j^R],\ j = 1, 2, \ldots, n,$$

where $[s_j^l, s_j^R]$ is the standard deviation of jth asset. This constraint implies that the higher $[s_j^L, s_j^R]$ (relative to the average standard deviation) the tighter the constraint imposed on the weight of stock j. Note that if all stocks have the same standard deviation, these constraints reduce to the homogeneous constraint $\left|x_j - \frac{1}{n}\right| \leq \alpha$. Hence the maximization of Sharpe ratio model with interval parameter is formulated as follows:

3 Solving Sharpe Ratio Model

The feasible region of the model (**SOI**) is

$$S = \left\{ (x_1, x_2, \ldots, x_n) : \sum_{j=1}^{n} x_j = 1,\ \left|x_j - \frac{1}{n}\right| [s_j^L, s_j^R] \preceq \frac{\alpha}{n} \sum_{j=1}^{n} [s_j^L, s_j^R],\ x_j \geq 0,\ j \in \Lambda_n \right\}.$$

The objective function is an interval-valued function. So the conditions for existence of feasible and optimal solution of (**SOI**) is not similar as classical optimization problem. A partial ordering is required to prove this result. $I(\mathbb{R})$ is not a totally ordered set. Several partial orderings in $I(\mathbb{R})$ exist in literature (see [3, 6]). Order relations between two intervals $\hat{a}$ and $\hat{b}$ can be explained in two ways; first one is an extension of $<$ on real line, that is, $\hat{a} < \hat{b}$ iff $a^R < b^L$, and the other is an extension of the concept of set inclusion, that is, $\hat{a} \subseteq \hat{b}$ iff $a^L \geq b^L$ and $a^R \leq b^R$. These order relations cannot explain ranking between two overlapping intervals. We introduce the following order relations $\preceq_\chi$ and $\succeq_\chi$ in $I(\mathbb{R})$ which describes partial ordering for overlapped intervals and helps to justify the existence of solution of (**SOI**) at later stage. We call this partial order as χ-partial order.

3.1 χ-Partial Order Relation

Two intervals may overlap, one interval may lie behind another interval or one interval may include another interval. To describe this concept mathematically, associate a function $\chi : I(\mathbb{R}) \times I(\mathbb{R}) \to [0, 1]$ as follows. For two intervals $\hat{a}$ and $\hat{b}$,

$$\chi(\hat{a}, \hat{b}) = \begin{cases} 1, & a^R \leq b^L \\ 0, & a^L \geq b^R \\ \frac{b^R - a^L}{(b^R - b^L) + (a^R - a^L)} \in (0, 1), & a^L < b^R \text{ and } a^R > b^L. \end{cases}$$

$\chi(\hat{a}, \hat{b})$ represents degree of closeness of $\hat{a}$ with $\hat{b}$. One may observe here that χ is continuous and belongs to $[0, 1]$. Moreover $\chi(\hat{a}, \hat{b}) + \chi(\hat{b}, \hat{a}) = 1$.

Based upon this concept of closeness of two intervals, we define order relation $\succeq_\chi$ between two intervals as follows:

Definition 1 For two intervals $\hat{a},\ \hat{b} \in I(\mathbb{R})$,

$$\begin{aligned} \hat{a} \succeq_\chi \hat{b} \ \text{ iff } \mu(\hat{a}) \leq \mu(\hat{b}) \text{ and } \chi(\hat{b}, \hat{a}) \in [\tfrac{1}{2}, 1], \\ \hat{a} \succ_\chi \hat{b} \ \text{ iff } \mu(\hat{a}) \leq \mu(\hat{b}) \text{ and } \ \chi(\hat{b}, \hat{a}) = 1, \\ \hat{a} = \hat{b} \ \text{ iff } \mu(\hat{a}) = \mu(\hat{b}) \text{ and } \chi(\hat{a}, \hat{b}) = 1/2. \end{aligned}$$

$\chi(\hat{b}, \hat{a}) \in [\frac{1}{2}, 1]$ means $\chi(\hat{a}, \hat{b}) \leq \chi(\hat{b}, \hat{a})$.

For example, $\mu([1, 4]) < \mu([0, 4])$ and $\chi([0, 4], [1, 4]) = \frac{4}{7}$. So $[1, 4] \succeq_\chi [0, 4]$ with degree of closeness $\frac{4}{7}$;

$\chi([3, 6], [2, 8]) = \frac{5}{9}$ but $[2, 8] \succeq_\chi [3, 6]$ is not true, since $\mu([2, 8]) \nleq \mu([3, 6])$.

$\mu([4, 5]) < \mu([2, 3])$ and $\chi([2, 3], [4, 5]) = 1$. So $[4, 5] \succ_\chi [2, 3]$ with degree of closeness 1; $\mu([-3, 0]) = \mu([1, 4])$ but $\chi([1, 4], [-3, 0]) = 0$, so $[-3, 0] \succeq_\chi [1, 4]$ is not true.

It is easy to prove that $\succeq_\chi$ is a partial order.

$I(\mathbb{R})^n$ is not a totally ordered set. To compare the interval vectors in $I(\mathbb{R})^n$, we define the following partial ordering $\succeq_\chi^n$.

Definition 2 For $\hat{a}_v = (\hat{a}_1\ \hat{a}_2\ ...\ \hat{a}_n)^T$ and $\hat{b}_v = (\hat{b}_1\ \hat{b}_2\ ...\ \hat{b}_n)^T$ in $I(\mathbb{R})^n$,

$$\hat{a}_v \succeq_\chi^n \hat{b}_v \ \text{ iff } \ \hat{a}_i \succeq_\chi \hat{b}_i,\ \forall i \in \Lambda_n.$$

Using the concept of closeness between two intervals, degree of closeness between two interval vectors $\hat{a}_v$ and $\hat{b}_v$ of dimension n can be defined as

$$\chi(\hat{a}_v, \hat{b}_v) = \min_{i \in \Lambda_n} \{\chi(\hat{a}_i, \hat{b}_i)\}. \tag{1}$$

Consider the interval vectors $\hat{a}_v = \begin{pmatrix}[0,4]\\ [2,3]\end{pmatrix}$ and $\hat{b}_v = \begin{pmatrix}[-2,3]\\ [1,3]\end{pmatrix}$. $\chi([-2,3],[0,4]) = \frac{2}{3}$ and $\chi([1,3],[2,3]) = \frac{2}{3}$.

$\chi(\hat{b}_v, \hat{a}_v) = \min\{\chi([-2,3],[0,4]), 1-\chi([1,3],[2,3])\} = \frac{2}{3}$. Hence $\hat{a}_v \succeq_\chi^{\frac{2}{3}} \hat{b}_v$ with degree of closeness $\frac{2}{3}$.

In similar way, we define order relation $\preceq_\chi$ between two intervals as follows:

Definition 3 For two intervals $\hat{a},\ \hat{b} \in I(\mathbb{R})$,

$$\begin{aligned}
&\hat{a} \preceq_\chi \hat{b} \ \text{ iff } \mu(\hat{a}) \le \mu(\hat{b}) \text{ and } \chi(\hat{a},\hat{b}) \in [1/2, 1],\\
&\hat{a} \prec_\chi \hat{b} \ \text{ iff } \mu(\hat{a}) \le \mu(\hat{b}) \text{ and } \ \chi(\hat{a},\hat{b}) = 1,\\
&\hat{a} = \hat{b} \ \text{ iff } \mu(\hat{a}) = \mu(\hat{b}) \text{ and } \chi(\hat{a},\hat{b}) = 1/2.
\end{aligned}$$

$\chi(\hat{a},\hat{b}) \in [\frac{1}{2}, 1]$ means $\chi(\hat{a},\hat{b}) \ge \chi(\hat{b},\hat{a})$.

For example, $\mu([1,4]) < \mu([0,5])$ and $\chi([1,4],[0,5]) = \frac{1}{2}$. So $[1,4] \preceq_\chi [0,5]$ with degree of closeness $\frac{1}{2}$;
$\chi([1,5],[3,6]) = \frac{5}{7}$ but $[1,5] \preceq_\chi [3,6]$ is not true, so $\mu([1,5]) \not\le \mu([3,6])$.
$\mu([1,4]) < \mu([5,9])$ and $\chi([1,4],[5,9]) = 1$. So $[1,4] \prec_\chi [5,9]$ with degree of closeness 1;

$\mu([1,4]) = \mu([-3,0])$ but $\chi([1,4],[-3,0]) = 0$, so $[1,4] \preceq_\chi [-3,0]$ is not true.

It is easy to prove that $\preceq_\chi$ is a partial order.

$I(\mathbb{R})^n$ is not a totally ordered set. To compare the interval vectors in $I(\mathbb{R})^n$, we define the following partial ordering $\preceq_\chi^n$.

Definition 4 For $\hat{a}_v = (\hat{a}_1\ \hat{a}_2 \ldots \hat{a}_n)^T$ and $\hat{b}_v = (\hat{b}_1\ \hat{b}_2 \ldots \hat{b}_n)^T$ in $I(\mathbb{R})^n$,

$$\hat{a}_v \preceq_\chi^n \hat{b}_v \ \text{ iff } \ \hat{a}_i \preceq_\chi \hat{b}_i,\ \forall i \in \Lambda_n.$$

Using the concept of closeness between two intervals, degree of closeness between two interval vectors $\hat{a}_v$ and $\hat{b}_v$ of dimension n can be defined as

$$\chi(\hat{a}_v, \hat{b}_v) = \min_{i \in \Lambda_n}\{\chi(\hat{a}_i, \hat{b}_i)\}. \tag{2}$$

Consider the interval vectors $\hat{a}_v = \left([0,2]\ [2,3]\right)^{\mathbf{T}}$ and $\hat{b}_v = \left([1,4]\ [2,4]\right)^{\mathbf{T}}$. Here $\chi([0,2],[1,4]) = \frac{4}{5}$ and $\chi([2,3],[2,4]) = \frac{2}{3}$.

$\chi(\hat{a}_v, \hat{b}_v) = \min\{\chi([0,2],[1,4]), \chi([2,3],[2,4])\} = \frac{2}{3}$. Hence we say $\hat{a}_v \preceq_\chi^{\frac{2}{3}} \hat{b}_v$ with degree of closeness 66 %.

In this present work we follow $\succeq_\chi$ partial ordering to interpret the meaning of minimization in the problem and $\preceq_\chi$ partial ordering to interpret the feasible region. Here χ-partial ordering is considered. Similar methodology in the light of the developments of this section can be established with respect to any other type partial ordering in the set of intervals. In that case the construction of (**SOI**) and proof of the corresponding theorem may be derived accordingly following the steps of this section, which is beyond the scope of the present work.

It is clear from the above discussion that two types of uncertainties are present in (**SOI**), one in objective function and another in constraints. These two uncertainties can be addressed separately in the following subsections to find the solution of (**SOI**).

3.2 Addressing the Uncertainty Present in Constraints of (SOI)

It can be observed that uncertainty is associated with n interval inequalities $\left|x_j - \frac{1}{n}\right| [s_j^L, s_j^R] \preceq \frac{\alpha}{n} \sum_{j=1}^n [s_j^L, s_j^R]$, $j \in \Lambda_n$. Every x in S satisfying the interval inequalities $\left|x_j - \frac{1}{n}\right| [s_j^L, s_j^R] \preceq \frac{\alpha}{n} \sum_{j=1}^n [s_j^L, s_j^R]$, $j \in \Lambda_n$ can be less or more acceptable for a decision maker. That is, every point x in S is associated with certain degree of feasibility/closeness between the interval vectors $\begin{pmatrix} \left|x_1 - \frac{1}{n}\right| [s_1^L, s_1^R] \\ \left|x_2 - \frac{1}{n}\right| [s_2^L, s_2^R] \\ \vdots \\ \left|x_n - \frac{1}{n}\right| [s_n^L, s_n^R] \end{pmatrix}$ and $\begin{pmatrix} \frac{\alpha}{n}[\sum_{j=1}^n s_j^L, \sum_{j=1}^n s_j^R] \\ \frac{\alpha}{n}[\sum_{j=1}^n s_j^L, \sum_{j=1}^n s_j^R] \\ \vdots \\ \frac{\alpha}{n}[\sum_{j=1}^n s_j^L, \sum_{j=1}^n s_j^R] \end{pmatrix}$. Using the concept of closeness of two interval vectors discussed in (1), if $\chi_j^F : I(\mathbb{R}) \times I(\mathbb{R}) \rightarrow [0, 1]$ given by

$$\chi_j^F \left(\left|x_j - \frac{1}{n}\right| [s_j^L, s_j^R], \frac{\alpha}{n} \sum_{j=1}^n [s_j^L, s_j^R] \right) = \begin{cases} 1, & \left|x_j - \frac{1}{n}\right| s_j^R \leq \frac{\alpha}{n} \sum_{j=1}^n s_j^L \\ 0, & \left|x_j - \frac{1}{n}\right| s_j^L \geq \frac{\alpha}{n} \sum_{j=1}^n s_j^R \\ \dfrac{\frac{\alpha}{n} \sum_{j=1}^n s_j^R - \left|x_j - \frac{1}{n}\right| s_j^L}{\left|x_j - \frac{1}{n}\right| \left(s_j^R - s_j^L\right) + \frac{\alpha}{n} \left(\sum_{j=1}^n s_j^R - \sum_{j=1}^n s_j^L\right)} & \text{elsewhere} \end{cases} \tag{3}$$

describes the closeness of $\left|x_j - \frac{1}{n}\right| [s_j^L, s_j^R]$ with $\frac{\alpha}{n} \sum_{j=1}^n [s_j^L, s_j^R]$ for every j, then degree of feasibility of x is $\min_{1 \leq j \leq n} \left\{ \chi_j^F \left(\left|x_j - \frac{1}{n}\right| [s_j^L, s_j^R], \frac{\alpha}{n} \sum_{j=1}^n [s_j^L, s_j^R] \right) \right\}$. Define a set

$$S_a = \left\{ (x : \tau) : \tau = \min_{1 \leq j \leq n} \left\{ \chi_j^F \left(\left|x_j - \frac{1}{n}\right| [s_j^L, s_j^R], \frac{\alpha}{n} \sum_{j=1}^n [s_j^L, s_j^R] \right) \right\}, \sum_{j=1}^n x_j = 1, \right.$$
$$\left. x_j \geq 0 \right\}.$$

For $(x : \tau) \in S_a$, we say x is a feasible point with acceptable degree τ and S_a is the acceptable feasible region.

Feasible solution of (**SOI**) associated with degree of closeness can be defined as follows:

Definition 5 $x \in \mathbb{R}^n$ is said to an acceptable feasible portfolio of (**SOI**) with acceptable degree τ if $x \in S_a$ with degree of closeness/acceptability τ, where $\tau \in [\frac{1}{2}, 1]$. We denote an acceptable feasible portfolio as $(x : \tau)$ henceforth.

3.3 Addressing Uncertainty Present in the Objective Function

Since the objective function is in interval form, the minimization in the problem (**SOI**) can be defined with respect to some partial ordering. In this present work we follow $\preceq_\chi$ partial ordering (defined in Sect. 3.1) to interpret the meaning of minimization in the problem. The objective function of (**SOI**) is a ratio of two interval valued functions, which can be expressed as a interval valued function as follows:

$$\sum_{j=1}^{n}[p_j^L, p_j^R]x_j \ominus [p_f, p_f] \oslash \sqrt{\sum_{i=1}^{n}\sum_{j=1}^{n}[s_{ij}^L, s_{ij}^R]x_i x_j}$$

$$= \left[\sum_{j=1}^{n} p_j^L x_j - p_f, \sum_{j=1}^{n} p_j^R x_j - p_f\right] \oslash \left[\sqrt{\sum_{i=1}^{n}\sum_{j=1}^{n} s_{ij}^L x_i x_j}, \sqrt{\sum_{i=1}^{n}\sum_{j=1}^{n} s_{ij}^R x_i x_j}\right]$$

$$= \left[\sum_{j=1}^{n} p_j^L x_j - p_f, \sum_{j=1}^{n} p_j^R x_j - p_f\right] \otimes \left[\frac{1}{\sqrt{\sum_{i=1}^{n}\sum_{j=1}^{n} s_{ij}^R x_i x_j}}, \frac{1}{\sqrt{\sum_{i=1}^{n}\sum_{j=1}^{n} s_{ij}^L x_i x_j}}\right]$$

$$= \left[\sum_{j=1}^{n} p_j^L x_j - p_f, \sum_{j=1}^{n} p_j^R x_j - p_f\right] \times q$$

$$\text{where } \frac{1}{\sqrt{\sum_{i=1}^{n}\sum_{j=1}^{n} s_{ij}^R x_i x_j}} \leq q \leq \frac{1}{\sqrt{\sum_{i=1}^{n}\sum_{j=1}^{n} s_{ij}^L x_i x_j}}$$

$$= \left[\sum_{j=1}^{n} p_j^L (x_j q) - p_f q, \sum_{j=1}^{n} p_j^R (x_j q) - p_f q\right]$$

$$\text{with } \sum_{i=1}^{n}\sum_{j=1}^{n} s_{ij}^L (x_i q)(x_j q) \leq 1 \text{ and } \sum_{i=1}^{n}\sum_{j=1}^{n} s_{ij}^R (x_i q)(x_j q) \geq 1$$

$$= \left[\sum_{j=1}^{n} p_j^L y_j - p_f q, \sum_{j=1}^{n} p_j^R y_j - p_f q\right]$$

$$\text{with } \sum_{i=1}^{n}\sum_{j=1}^{n} s_{ij}^L y_i y_j \leq 1, \ \sum_{i=1}^{n}\sum_{j=1}^{n} s_{ij}^R y_i y_j \geq 1 \text{ and } x_j q = y_j \ \forall j \in \Lambda_n.$$

Hence the transformed objective function is $\left[\sum_{j=1}^{n} p_j^L y_j - p_f q, \sum_{j=1}^{n} p_j^R y_j - p_f q\right]$ with $s_{ij}^L y_i y_j \leq 1$, $s_{ij}^R y_i y_j \geq 1$ and $x_j q = y_j$ $\forall j \in \Lambda_n$. The transformed objective function is an interval-valued function. So the conditions for existence of feasible and optimal solution (**SOI**) is not similar as classical optimization problem. To compare two intervals we have considered $\succeq_\chi$ partial order relation. Following the definition of χ-partial ordering, we define solution of (**SOI**) as follows.

Definition 6 An acceptable feasible portfolio $(x^* : \tau^*)$ with degree of feasibility τ^* of (**SOI**) is said to be a χ-efficient portfolio of (**SOI**) if there does not exist any acceptable feasible portfolio $(x : \tau)$ with $\tau > \tau^*$ of (**SOI**) such that

$$\sum_{j=1}^{n}[p_j^L, p_j^R]x_j \ominus [p_f, p_f] \oslash \sqrt{\sum_{i=1}^{n}\sum_{j=1}^{n}[s_{ij}^L, s_{ij}^R]x_i x_j}$$
$$\succ_\chi \sum_{j=1}^{n}[p_j^L, p_j^R]x_i^* \ominus [p_f, p_f] \oslash \sqrt{\sum_{i=1}^{n}\sum_{j=1}^{n}[s_{ij}^L, s_{ij}^R]x_i^* x_j^*}.$$

Considering the transformed objective function we construct the (**SOI**)(**y**) model as

$$\begin{aligned}
(\mathbf{SOI})(\mathbf{y}): \quad & \max \left[\sum_{j=1}^{n} p_j^L y_j - p_f q, \sum_{j=1}^{n} p_j^R y_j - p_f q\right] \\
\text{subject to} \quad & \sum_{i,j}^{n} s_{ij}^L y_i y_j \le 1, \\
& \sum_{i,j}^{n} s_{ij}^R y_i y_j \ge 1, \\
& \left|\frac{y_j}{q} - \frac{1}{n}\right| [s_j^L, s_j^R] \preceq \frac{\alpha}{n}\sum_{j=1}^{n}[s_j^L, s_j^R], \ \forall j \in \Lambda_n, \\
& \sum_{j=1}^{n} y_j = q, \\
& y_j \ge 0, \ \forall j \in \Lambda_n.
\end{aligned}$$

Using the discussion in Sect. 3.2, we get the acceptable feasible region of the interval optimization problem (**SOI**)(**y**) as

$$S_a(y) = \left\{ (y:\tau) : \tau = \min_{1\le j\le n} \left\{ \chi_j^F \left(\left| \frac{y_j}{q} - \frac{1}{n} \right| [s_j^L, s_j^R], \frac{\alpha}{n} \sum_{j=1}^{n} [s_j^L, s_j^R] \right) \right\}, \right.$$
$$\left. \sum_{i,j}^{n} s_{ij}^L y_i y_j \le 1,\ \sum_{i,j}^{n} s_{ij}^R y_i y_j \ge 1,\ \sum_{j=1}^{n} y_j = q,\ y_j \ge 0 \right\}.$$

For $(y:\tau) \in S_a(y)$, we say y is a feasible point with acceptable degree τ and $S_a(y)$ is the acceptable feasible region of (**SOI**). In the light of the definition of χ-efficient portfolio of (**SOI**), χ-efficient portfolio of (**SOI**)(y) can be defined as follows.

Definition 7 An acceptable feasible portfolio $(y^*, q^* : \tau^*)$ with degree of feasibility τ^* of (**SOI**)(**y**) is said to be a χ-efficient portfolio of (**SOI**) if there does not exist any acceptable feasible portfolio $(y, q : \tau)$ with $\tau > \tau^*$ of (**SOI**)(**y**) such that

$$\left[\sum_{j=1}^{n} p_j^L y_j - p_f q, \sum_{j=1}^{n} p_j^R y_j - p_f q \right] \succ_\chi \left[\sum_{j=1}^{n} p_j^L y_j^* - p_f q^*, \sum_{j=1}^{n} p_j^R y_j^* - p_f q^* \right]$$

One may observe that solution of (**SOI**) is related to the solution of (**SOI**)(**y**) by the relation $y_j = x_j q,\ \forall j$.

To address the uncertainty present in objective function in (**SOI**)(**y**), we will assign goal to the objective function. Goal can be provided by decision-makers. Let $[l, u]$ is preassigned goal of the objective function. That is,
$\left[\sum_{j=1}^{n} p_j^L y_j - p_f q, \sum_{j=1}^{n} p_j^R y_j - p_f q\right] \succeq_\chi [l, u]$. For every $(y, q : \tau) \in S_a(y)$, deviation of the objective function from the goal $[l, u]$ may be more or less acceptable for the decision-maker. This implies that every interval-valued objective function is associated with certain degree of flexibility from its goal.

This logic is similar to the discussion in Sect. 3.1 for the closeness between two intervals. Using the closeness between two intervals, we get the degree of closeness between two intervals
$\left[\sum_{j=1}^{n} p_j^L y_j - p_f q, \sum_{j=1}^{n} p_j^R y_j - p_f q\right]$ and $[l, u]$ as

$$\chi^O \left([l, u], \left[\sum_{j=1}^{n} p_j^L y_j - p_f q, \sum_{j=1}^{n} p_j^R y_j - p_f q \right] \right)$$
$$= \begin{cases} 1, & \text{if } u \le \left(\sum_{j=1}^{n} p_j^L y_j - p_f q\right) \\ 0, & \text{if } l \ge \left(\sum_{j=1}^{n} p_j^R y_j - p_f q\right) \\ \frac{(\sum_{j=1}^{n} p_j^R y_j - p_f q) - l}{(u-l) + \left((\sum_{j=1}^{n} p_j^R y_j - p_f q) - (\sum_{j=1}^{n} p_j^L y_j - p_f q)\right)}, & \text{elsewhere.} \end{cases} \quad (4)$$

3.4 Conversion of the Sharpe Ratio Model into Deterministic Form

The objective functions are characterized by their degree of flexibility $\chi^O\left([l,u],\left[\sum_{j=1}^n p_j^L y_j - p_f q, \sum_{j=1}^n p_j^R y_j - p_f q\right]\right)$, and the constraints are characterized by their degree of feasibility $\chi_j^F\left(\left|x_j - \frac{1}{n}\right| [s_j^L, s_j^R], \frac{\alpha}{n}\sum_{j=1}^n [s_j^L, s_j^R]\right)$. So in this uncertain environment a decision y for $(\mathbf{SOI})(\mathbf{y})$ is the selection of activities that simultaneously satisfies the objective function and constraints. Hence degree of acceptability of this y is

$$\min\left\{\chi^O\left(\left[[l,u],\sum_{j=1}^n p_j^L y_j - p_f q, \sum_{j=1}^n p_j^R y_j - p_f q\right]\right);\ (y:\tau)\in S_a(y)\right\}$$

$$= \min_{(y:\tau)\in S_a(y)} \frac{(\sum_{j=1}^n p_j^L y_j - p_f q) - l}{(u-l)+\left((\sum_{j=1}^n p_j^R y_j - p_f q) - (\sum_{j=1}^n p_j^L y_j - p_f q)\right)}. \tag{5}$$

This problem is equivalent to the max–min problem

$(\overline{\mathbf{SOI}})(y):\quad \max\ \theta$

$$\text{subject to } \theta \le \frac{(\sum_{j=1}^n p_j^R y_j - p_f q) - l}{(u-l)+\left((\sum_{j=1}^n p_j^R y_j - p_f q) - (\sum_{j=1}^n p_j^L y_j - p_f q)\right)},$$

$$\theta \le \frac{\frac{\alpha}{n}\sum_{j=1}^n s_j^R - \left|\frac{y_j}{q} - \frac{1}{n}\right| s_j^L}{\left|\frac{y_j}{q} - \frac{1}{n}\right|\left(s_j^R - s_j^L\right) + \frac{\alpha}{n}\left(\sum_{j=1}^n s_j^R - \sum_{j=1}^n s_j^L\right)},$$

$$\sum_{i,j}^n s_{ij}^L y_i y_j \le 1,$$

$$\sum_{i,j}^n s_{ij}^R y_i y_j \ge 1,$$

$$\sum_{j=1}^n y_j = q,$$

$$y_j \ge 0,\ q > 0,$$

$$\frac{1}{2} \le \theta \le 1.$$

It can be observed that $(\overline{\mathbf{SOI}})(y)$ a general non linear programming problem. Solution of this problem can be found using any nonlinear programming technique. Let (y^*, q^*, θ^*) be a solution of the problem $(\overline{\mathbf{SOI}})(y)$. The following theorem states the relation between the solution of $(\mathbf{SOI})(y)$ and $(\overline{\mathbf{SOI}})(y)$.

Theorem 1 *If* (y^*, q^*, θ^*) *be an optimal solution of the problem* $(\overline{\mathbf{SOI}})(y)$*, then* (y^*, q^*) *is an* χ*-efficient portfolio of the problem* $(\mathbf{SOI})(\mathbf{y})$.

Above methodology is explained in the following numerical example first. Then the methodology is applied for real data from BSE, India.

Example 1 Consider the following problem

$$\begin{aligned}
\max \quad & \hat{P}(x) \oslash \hat{Q}(x) \\
\text{subject to} \quad & x_1 + x_2 = 1, \\
& \left| x_1 - \frac{1}{2} \right| [0.144, 0.1764] \preceq \frac{2}{2} \odot [0.2809, 0.3445], \\
& \left| x_2 - \frac{1}{2} \right| [0.1369, 0.1681] \preceq \frac{2}{2} \odot [0.2809, 0.3445], \\
& x_1, x_2 \geq 0,
\end{aligned}$$

where

$$\hat{P}(x) = [0.12, 0.14]x_1 \oplus [0.18, 0.22]x_2 \ominus [0.04, 0.04]$$

and $$\hat{Q}(x) = \sqrt{[0.144, 0.1764]x_1^2 \oplus 2[0.0703, 0.0681]x_1x_2 \oplus [0.1369, 0.1681]x_2^2}.$$

Solution The transformed interval optimization problem $(\mathbf{SOI})(\mathbf{y})$ is given by

$$\begin{aligned}
(\mathbf{SOI})(\mathbf{y}): \quad \max \quad & [0.12y_1 + 0.18y_2 - 0.04q, 0.14y_1 + 0.22y_2 - 0.04q] \\
\text{subject to} \quad & 0.144y_1^2 + 0.1406y_1y_2 + 0.1369y_2^2 \leq 1, \\
& 0.176y_1^2 + 0.1722y_1y_2 + 0.1681y_2^2 \geq 1, \\
& \left| \frac{y_1}{q} - \frac{1}{2} \right| [0.144, 0.1764] \preceq \frac{2}{2} \odot [0.2809, 0.3445], \\
& \left| \frac{y_2}{q} - \frac{1}{2} \right| [0.1369, 0.1681] \preceq \frac{2}{2} \odot [0.2809, 0.3445], \\
& y_1 + y_2 = q, \\
& y_1, y_2 \geq 0, \\
& q > 0.
\end{aligned}$$

The acceptable feasible region of the interval optimization problem $(\mathbf{SOI})(\mathbf{y})$ is given by

$$S_a(y) = \Bigg\{(y:\tau) : \tau = \min\Bigg\{\chi_1^F\left(\left|\frac{y_1}{q} - \frac{1}{2}\right|[0.144, 0.1764], \frac{2}{2}\odot[0.2809, 0.3445]\right),$$
$$\chi_2^F\left(\left|\frac{y_2}{q} - \frac{1}{2}\right|[0.1369, 0.1681], \frac{2}{2}\odot[0.2809, 0.3445]\right)\Bigg\},$$
$$0.144y_1^2 + 0.1406y_1y_2 + 0.1369y_2^2 \le 1,\ 0.176y_1^2 + 0.1722y_1y_2 + 0.1681y_2^2 \ge 1,\ y_1 + y_2 = q,\ y_1,\ y_2 \ge 0\Bigg\}.$$

where

$$\chi_1^F\left(\left|\frac{y_1}{q} - \frac{1}{2}\right|[0.144, 0.1764], \frac{2}{2}\odot[0.2809, 0.3445]\right) = \begin{cases} 1, & \left|\frac{y_1}{q} - \frac{1}{2}\right| 0.1764 \le 0.2809 \\ 0, & \left|\frac{y_1}{q} - \frac{1}{2}\right| 0.144 \ge 0.3445 \\ \frac{0.3445 - 0.144\left|\frac{y_1}{q} - \frac{1}{2}\right|}{0.0636 + 0.0324\left|\frac{y_1}{q} - \frac{1}{2}\right|}, & \text{elsewhere} \end{cases} \tag{6}$$

$$\chi_2^F\left(\left|\frac{y_2}{q} - \frac{1}{2}\right|[0.1369, 0.1681], \frac{2}{2}\odot[0.2809, 0.3445]\right) = \begin{cases} 1, & \left|\frac{y_2}{q} - \frac{1}{2}\right| 0.1681 \le 0.2809 \\ 0, & \left|\frac{y_2}{q} - \frac{1}{2}\right| 0.3974 \ge 0.3445 \\ \frac{0.3445 - 0.1369\left|\frac{y_2}{q} - \frac{1}{2}\right|}{0.0636 + 0.0312\left|\frac{y_2}{q} - \frac{1}{2}\right|}, & \text{elsewhere.} \end{cases} \tag{7}$$

Let goal of the objective function of $(\mathbf{SOI})(\mathbf{y})$ is given by $[0.3, 0.5]$, then degree of flexibility of the objective function is given by

$$\chi^O([0.3, 0.5], [0.12y_1 + 0.18y_2 - 0.04q, 0.14y_1 + 0.22y_2 - 0.04q])$$
$$= \begin{cases} 1, & \text{if } 0.5 \le (0.12y_1 + 0.18y_2 - 0.04q) \\ 0, & \text{if } 0.3 \ge (0.14y_1 + 0.22y_2 - 0.04q) \\ \frac{(0.14y_1 + 0.22y_2 - 0.04q) - 0.3}{(0.2) + ((0.14y_1 + 0.22y_2 - 0.04q) - (0.12y_1 + 0.18y_2 - 0.04q))}, & \text{elsewhere.} \end{cases} \tag{8}$$

Hence the deterministic model is

$$(\overline{\mathbf{SOI}})(y): \quad \max\ \theta$$
$$\text{subject to } \theta \le \frac{(0.14y_1 + 0.22y_2 - 0.04q) - 0.3}{(0.2) + (0.02y_1 + 0.04y_2)},$$
$$\theta \le \frac{0.3445 - 0.144\left|\frac{y_1}{q} - \frac{1}{2}\right|}{0.0636 + 0.0324\left|\frac{y_1}{q} - \frac{1}{2}\right|},$$
$$\theta \le \frac{0.3445 - 0.1369\left|\frac{y_2}{q} - \frac{1}{2}\right|}{0.0636 + 0.0312\left|\frac{y_2}{q} - \frac{1}{2}\right|},$$
$$0.144y_1^2 + 0.1406y_1y_2 + 0.1369y_2^2 \le 1,$$
$$0.176y_1^2 + 0.1722y_1y_2 + 0.1681y_2^2 \ge 1,$$
$$y_1 + y_2 = q,$$
$$y_1,\ y_2 \ge 0, q > 0,$$
$$\frac{1}{2} \le \theta \le 1.$$

χ-efficient portfolio of the problem $(\mathbf{SOI})(\mathbf{y})$ is $((y_1^*, y_2^*), q^*) = ((0.1736, 2.6091), 2.7828)$ with degree of acceptability 0.6075. Objective value is [0.4592, 0.5670]. Accordingly the χ-efficient portfolio of the original problem $(\mathbf{SOI})$ is given by $(x_1^*, x_2^*) = (0.0624, 0.9376)$.

In order to show the applicability of our proposed model, we apply this model to real data set taken from Indian stocks market National Stock Exchange, India.

4 Empirical Result

In this section we present an illustration of proposed Sharpe ratio model considering fifteen stocks from the Bombay Stock Exchange, India. The code of all fifteen stocks are given in Table 1. We also consider a riskless asset with rate of return 0.2 % per month. We collect the weekly opening-, maximum-, minimum-, and closing price from April 1, 2010 to December 30, 2013. Rate of return of each week and average rate of return of each stocks are calculated. To estimate the bounds of expected return of each stocks, we find the maximum and minimum average return from the average rate of returns corresponding to all the prices, which represent upper and lower bound, respectively. The lower and upper bound of all the fifteen stocks are given in Table 2. Further, we estimated the bounds of elements of covariance matrix based on the obtained bound of expected rate of return of stocks. The lower bound and upper bound of elements of covariance are given in Tables 3 and 4, respectively.

Based on the above information, next we obtain an efficient portfolio. In order to obtain the efficient portfolio we apply our proposed methodology established in section for targeted value of the objective function [0.02, 0.03]. We choose $\alpha = 0.97$. Using LINGO 11, we solve $(\overline{\mathbf{SOI}})(y)$, and obtain the value of $y_1 = 0.041$, $y_2 = 8.038$, $y_3 = 8.930$, $y_4 = 6.118$, $y_5 = 8.111$, $y_6 = 0.060$, $y_7 = 0.000$, $y_8 = 7.174$, $y_9 = 2.119$, $y_{10} = 5.038$, $y_{11} = 6.815$, $y_{12} = 0.174$, $y_{13} = 8.037$, $y_{14} = 0.098$, $y_{15} = 0.081$, value of $\theta = 1$ and $Q = 60.831$ for $\alpha = 0.97$. Hence the χ-efficient portfolio is given by $x_1 = 0.001$, $x_2 = 0.132$, $x_3 = 0.147$, $x_4 = 0.101$, $x_5 = 0.133$, $x_6 = 0.001$, $x_7 = 0.000$, $x_8 = 0.118$, $x_9 = 0.035$, $x_{10} = 0.083$, $x_{11} = 0.112$, $x_{12} = 0.003$, $x_{13} = 0.132$, $x_{14} = 0.002$ $x_{15} = 0.001$.

5 Concluding Remarks

This paper presents a Sharpe ratio in which all the coefficients of the objective function and constraints are intervals. Concept of existence of χ-efficient portfolio is introduced. A methodology is developed to determine an χ-efficient portfolio based on a partial order relation in the set of interval. The methodology is illustrated through a numerical example. Finally, the portfolio selection model, based on the real data from Bombay Stock Exchange, India, is solved by the developed methodology to find a χ-efficient portfolio.

Table 1 Code of fifteen stocks

Stock	Stock 1	Stock 2	Stock 3	Stock 4	Stock 5	Stock 6
Code	ACC	AMBUJACEM	ASIANPAINT	AXISBANK	BAJAJ-AUTO	BANKBARODA
Stock	Stock 7	Stock 8	Stock 9	Stock 10	Stock 11	Stock 12
Code	BHEL	BPCL	BHARTIARTL	CAIRN	CIPLA	DLF
Stock	Stock 13	Stock 14	Stock 15			
Code	DRREDDY	GAIL	GRASIM			

Table 2 Bounds of expected return of stocks

Stock	Stock 1	Stock 2	Stock 3	Stock 4	Stock 5	Stock 6	Stock 7	
Lower bound	0.0013	0.0029	0.0006	0.0014	0.0012	0.0009	−0.0080	
Upper bound	0.0018	0.0034	0.0072	0.0019	0.0046	0.0015	−0.0026	
Stock	Stock 8	Stock 9	Stock 10	Stock 11	Stock 12	Stock 13	Stock 14	Stock 15
Lower bound	0.0000	0.0011	0.0007	0.0012	−0.0017	0.0038	−0.0005	0.0005
Upper bound	0.0034	0.0014	0.0011	0.0101	−0.0010	0.0041	0.0000	0.0007

Table 3 Lower bound of covariance of stocks

Stock	Stock 1	Stock 2	Stock 3	Stock 4	Stock 5	Stock 6	Stock 7	Stock 8	Stock 9	Stock 10	Stock 11	Stock 12	Stock 13	Stock 14	Stock 15
Stock 1	0.00107	0.00056	0.00024	0.00046	0.00041	0.00047	0.00035	0.00021	0.00022	0.00011	0.00007	0.00052	0.00006	0.00029	0.00039
Stock 2	0.00056	0.00143	0.00017	0.00032	0.00025	0.00035	0.00038	0.00030	0.00029	0.00017	0.00018	0.00055	0.00016	0.00034	0.00050
Stock 3	0.00024	0.00017	0.00115	0.00007	0.00026	0.00039	0.00037	0.00025	0.00012	0.00012	0.00004	0.00032	0.00013	0.00013	0.00030
Stock 4	0.00046	0.00032	0.00007	0.00183	0.00003	0.00027	−0.00033	0.00020	0.00007	−0.00014	0.00011	0.00006	−0.00005	0.00025	0.00028
Stock 5	0.00041	0.00025	0.00026	0.00003	0.00141	0.00044	0.00052	0.00042	0.00005	0.00024	0.00009	0.00060	−0.00004	0.00021	0.00017
Stock 6	0.00047	0.00035	0.00039	0.00027	0.00044	0.00175	0.00087	0.00051	0.00032	0.00021	0.00013	0.00084	0.00014	0.00037	0.00034
Stock 7	0.00035	0.00038	0.00037	−0.00033	0.00052	0.00087	0.00272	0.00050	0.00032	0.00049	0.00001	0.00111	0.00007	0.00037	0.00044
Stock 8	0.00021	0.00030	0.00025	0.00020	0.00042	0.00051	0.00050	0.00220	0.00024	−0.00002	0.00004	0.00084	0.00014	0.00039	0.00028
Stock 9	0.00022	0.00029	0.00012	0.00007	0.00005	0.00032	0.00032	0.00024	0.00174	0.00017	0.00002	0.00061	0.00002	0.00023	0.00022
Stock 10	0.00011	0.00017	0.00012	−0.00014	0.00024	0.00021	0.00049	−0.00002	0.00017	0.00084	−0.00008	0.00044	0.00011	0.00017	0.00016
Stock 11	0.00007	0.00018	0.00004	0.00011	0.00009	0.00013	0.00001	0.00004	0.00002	−0.00008	0.00068	0.00031	−0.00010	0.00015	0.00010
Stock 12	0.00052	0.00055	0.00032	0.00006	0.00060	0.00084	0.00111	0.00084	0.00061	0.00044	0.00031	0.00273	0.00012	0.00058	0.00065
Stock 13	0.00006	0.00016	0.00013	−0.00005	−0.00004	0.00014	0.00007	0.00014	0.00002	0.00011	−0.00010	0.00012	0.00071	0.00017	0.00016
Stock 14	0.00029	0.00034	0.00013	0.00025	0.00021	0.00037	0.00037	0.00039	0.00023	0.00017	0.00015	0.00058	0.00017	0.00080	0.00034
Stock 15	0.00039	0.00050	0.00030	0.00028	0.00017	0.00034	0.00044	0.00028	0.00022	0.00016	0.00010	0.00065	0.00016	0.00034	0.00122

Table 4 Upper Bound of covariance of stocks

Stock	Stock 1	Stock 2	Stock 3	Stock 4	Stock 5	Stock 6	Stock 7	Stock 8	Stock 9	Stock 10	Stock 11	Stock 12	Stock 13	Stock 14	Stock 15
Stock 1	0.00116	0.00098	0.00024	0.00046	0.00051	0.00047	0.00035	0.00035	0.00032	0.00018	0.00043	0.00078	0.00021	0.00040	0.00072
Stock 2	0.00098	0.00165	0.00017	0.00032	0.00058	0.00045	0.00047	0.00031	0.00041	0.00024	0.00038	0.00076	0.00020	0.00046	0.00068
Stock 3	0.00024	0.00017	0.00115	0.00007	0.00026	0.00039	0.00037	0.00025	0.00019	0.00013	0.00004	0.00032	0.00013	0.00013	0.00032
Stock 4	0.00046	0.00032	0.00007	0.00263	0.00003	0.00027	−0.00033	0.00020	0.00007	−0.00014	0.00084	0.00006	−0.00005	0.00025	0.00028
Stock 5	0.00051	0.00058	0.00026	0.00003	0.00141	0.00049	0.00052	0.00042	0.00047	0.00028	0.00009	0.00086	0.00022	0.00024	0.00048
Stock 6	0.00047	0.00045	0.00039	0.00027	0.00049	0.00197	0.00087	0.00073	0.00042	0.00040	0.00017	0.00143	0.00026	0.00055	0.00056
Stock 7	0.00035	0.00047	0.00037	−0.00033	0.00052	0.00087	0.00272	0.00082	0.00041	0.00057	0.00038	0.00129	0.00028	0.00049	0.00063
Stock 8	0.00035	0.00031	0.00025	0.00020	0.00042	0.00073	0.00082	0.00220	0.00042	0.00019	0.00042	0.00112	0.00014	0.00058	0.00037
Stock 9	0.00032	0.00041	0.00019	0.00007	0.00047	0.00042	0.00041	0.00042	0.00176	0.00020	0.00002	0.00069	0.00016	0.00034	0.00039
Stock 10	0.00018	0.00024	0.00013	−0.00014	0.00028	0.00040	0.00057	0.00019	0.00020	0.00120	−0.00008	0.00081	0.00026	0.00022	0.00025
Stock 11	0.00043	0.00038	0.00004	0.00084	0.00009	0.00017	0.00038	0.00042	0.00002	−0.00008	0.03470	0.00070	−0.00010	0.00016	0.00055
Stock 12	0.00078	0.00076	0.00032	0.00006	0.00086	0.00143	0.00129	0.00112	0.00069	0.00081	0.00070	0.00387	0.00043	0.00077	0.00090
Stock 13	0.00021	0.00020	0.00013	−0.00005	0.00022	0.00026	0.00028	0.00014	0.00016	0.00026	−0.00010	0.00043	0.00082	0.00017	0.00023
Stock 14	0.00040	0.00046	0.00013	0.00025	0.00024	0.00055	0.00049	0.00058	0.00034	0.00022	0.00016	0.00077	0.00017	0.00097	0.00049
Stock 15	0.00072	0.00068	0.00032	0.00028	0.00048	0.00056	0.00063	0.00037	0.00039	0.00025	0.00055	0.00090	0.00023	0.00049	0.00143

References

1. Bhurjee, A.K., Panda, G.: Nonlinear fractional programming problem with inexact parameter. J. Appl. Math. Inf. **31**(5), 853–867 (2013)
2. Brennan, M.J., Xia, Y.: Assessing asset pricing anomalies. Rev. Financ. Stud. **14**(4), 905–942 (2001)
3. Hansen, E., Walster, G.W.: Global Optimization Using Interval Analysis. Marcel Dekker Inc, New York (2004)
4. Hladík, M.: Generalized linear fractional programming under interval uncertainty. Eur. J. Oper. Res. **205**(1), 42–46 (2010)
5. Jorion, P.: Bayes-stein estimation for portfolio analysis. J. Financ. Quant. Anal. **21**(03), 279–292 (1986)
6. Moore, R.E., Kearfott, R.B., Cloud, M.J.: Introduction to interval analysis. Society for Industrial Mathematics, Philadelphia (2009)
7. Sharpe, W.F.: The sharpe ratio. J. Portfolio Mgmt. **21**(1), 49–58 (1994)
8. Sharpe, W.F., Sharpe, W.: Portfolio Theory and Capital Markets, vol. 217. McGraw-Hill, New York (1970)
9. Xia, Y.: Learning about predictability: the effects of parameter uncertainty on dynamic asset allocation. J. Finance **56**(1), 205–246 (2001)

On Solvability for Certain Functional Equations Arising in Dynamic Programming

Deepmala and A.K. Das

Abstract In this paper, we study the existence, uniqueness, and iterative approximations of solutions for the functional equations arising in dynamic programming under Banach spaces and complete metric spaces. Our results unify the results of Bellman [1], Bhakta and Mitra [3], Bhakta and Choudhury [4], Liu [8], Liu and Ume [10], Liu et al. [11], Liu et al. [13], Liu and Kang [9], and Jiang et al. [7]. Examples are provided to support our results.

Keywords Dynamic programming · Multistage decision process · Functional equations · Fixed point · Banach space · Metric space

1 Introduction

The existence of solutions for various functional equations arising in dynamic programming is important in both theory and practice. In optimization, dynamic programming is an interesting field of research because of its applicability in multistage decision processes. For details, see [1, 6, 14, 15] and references therein. Bellman [1, 2] worked on the existence of solutions for some classes of functional equations arising in dynamic programming. Bellman and Lee [2] considered the functional equation in dynamic programming of multistage decision process as

$$f(x) = \text{opt}_{y \in D} H(x, y, f(T(x, y))), \quad \forall x \in S.$$

In the past two decades, Bhakta and Mitra [3], Bhakta and Chuodhary [4], Liu and Ume [10], Liu et al. [11, 13], Liu and Kang [9] and Jiang et al. [7] established the

Deepmala (✉) · A.K. Das
SQC & OR Unit, Indian Statistical Institute, 203 B. T. Road, Kolkata 700108, India
e-mail: dmrai23@gmail.com

A.K. Das
e-mail: akdas@isical.ac.in

R.N. Mohapatra et al. (eds.), *Mathematics and Computing*,
Springer Proceedings in Mathematics & Statistics 139,
DOI 10.1007/978-81-322-2452-5_6

existence of solutions for the following functional equation (1)–(9) in Banach spaces and complete metric spaces.

$$f(x) = \inf_{y\in D} \max\{r(x,y), s(x,y), f(b(x,y))\}, \quad x \in S. \tag{1}$$

$$f(x) = \inf_{y\in D} \max\{r(x,y), f(b(x,y))\}, \quad x \in S. \tag{2}$$

$$f(x) = \sup_{y\in D}\{p(x,y) + A(x,y,f(a(x,y)))\}, \quad x \in S. \tag{3}$$

$$f(x) = \sup_{y\in D}\{p(x,y) + f(a(x,y))\}, \quad x \in S. \tag{4}$$

$$f(x) = \operatorname*{opt}_{y\in D}\{a\,[u(x,y) + f(T(x,y))] + b\ \operatorname{opt}[v(x,y), f(T(x,y))]\}, \quad x \in S,\ a+b=1. \tag{5}$$

$$f(x) = \operatorname*{opt}_{y\in D}\{u(x,y) + \operatorname{opt}\{p_i(x,y) + q_i(x,y) f_i(a_i(x,y)) : i = 1,2\}\}, \quad x \in S. \tag{6}$$

$$\begin{aligned} f(x) = {} & \operatorname*{opt}_{y\in D}\{p(x,y) + q(x,y) f(a(x,y)) \\ & + \operatorname{opt}\{r(x,y), s(x,y) f(b(x,y)), t(x,y) f(c(x,y))\}\}. \end{aligned} \tag{7}$$

$$f(x) = \inf_{y\in D} \max\{p(x,y), f(a(x,y)), q(x,y) + f(b(x,y))\}, \qquad \forall\, x \in S. \tag{8}$$

$$\begin{aligned} f(x) = {} & \operatorname*{opt}_{y\in D} \operatorname{opt}\{p(x,y), q(x,y) f(a(x,y)), r(x,y) f(b(x,y)), s(x,y) f(c(x,y))\}, \\ & forall\ x \in S. \end{aligned} \tag{9}$$

We introduce the following generalized functional equation related to dynamic programming of multistage decision processes:

$$\begin{aligned} f(x) = {} & \operatorname*{opt}_{y\in D}\{u(x,y) + r(x,y) f(s(x,y)) \\ & + \operatorname{opt}\{v(x,y),\ p_i(x,y) + q_i(x,y) f(a_i(x,y)) : i = 1,2,3\}\}, \forall x \in S \end{aligned} \tag{10}$$

where opt denotes the sup or inf, x and y stands for the state and decision vectors, respectively, a_i represents the transformation of the processes, $f(x)$ denotes the optimal return function with initial state x. We establish the existence and uniqueness of solutions for the proposed functional equation.

In Sect. 2, we recall some of the basic concepts and results to be used in this paper. In Sect. 3, we utilize the fixed point theorem due to Boyd and Wong [5] to establish the existence, uniqueness, and iterative approximation of solution for this generalized functional equation in Banach spaces. In Sect. 4, we obtain the existence, uniqueness, and iterative approximations of solutions for that functional equation in the complete metric spaces. We construct some nontrivial examples to explain our results. The results presented here unify the results of Bellman [1], Bhakta and Mitra [3], Bhakta and Choudhury [4], Liu and Ume [10], Liu et al. [11], Liu et al. [13], Liu and Kang [9], Jiang et al. [7], and Liu [8].

2 Preliminaries

We begin by introducing some notations, definitions, and results that will be used in this paper. Let $\mathbb{R} = (-\infty, +\infty)$, $R^+ = [0, +\infty)$ and $R^- = (-\infty, 0]$. For any $t \in \mathbb{R}$, $[t]$ denotes the largest integer not exceeding t and $(X, \|.\|)$ and $(Y, \|.\|')$ be real Banach spaces. $S \subseteq X$ and $D \subseteq Y$ denote state space and decision space, respectively. Define

$\Phi_1 = \{\varphi | \varphi : R^+ \longrightarrow R^+ \text{ is nondecreasing}\}$,
$\Phi_2 = \{(\varphi, \psi) : \varphi, \psi \in \Phi_1, \psi(t) > 0 \text{ and } \sum_{n=0}^{\infty} \psi(\varphi^n(t)) < \infty \text{ for } t > 0\}$,
$\Phi_3 = \{(\varphi, \psi) : \varphi, \psi \in \Phi_1, \psi(t) > 0 \text{ and } \lim_{n \longrightarrow \infty} \psi(\varphi^n(t)) = 0 \text{ for } t > 0\}$,
$\Phi_4 = \{\varphi : \varphi \in \Phi_1 \text{ and } \sum_{n=0}^{\infty} \varphi^n(t) < \infty \text{ for } t > 0\}$,
$B(S) = \{f | f : S \longrightarrow \mathbb{R} \text{ is bounded}\}$,
$BC(S) = \{f : f \in B(S) \text{ is continuous}\}$,
$BB(S) = \{f | f : S \longrightarrow \mathbb{R} \text{ is bounded on bounded subsets of S}\}$.

Define norm $\|f\|_1 = \sup_{x \in S} |f(x)|$, then $(B(S), \|.\|_1)$ and $(BC(S), \|.\|_1)$ are Banach spaces. Put,

$$d_k(f, g) = \sup \{|f(x) - g(x)| : x \in \overline{B}(0, k)\},$$

$$d(f, g) = \sum_{k=1}^{\infty} \frac{1}{2^k} \cdot \frac{d_k(f, g)}{1 + d_k(f, g)},$$

for any positive integer k, $\overline{B}(0, k) = \{x : x \in S \text{ and } \|x\| \leq k\}$ and $f, g \in BB(S)$. Thus, $(BB(S), d)$ is a complete metric space. A metric space (M, ρ) is said to metrically convex if for each $x, y \in M$, there is a $z \neq x, y$ for which $\rho(x, y) = \rho(x, z) + \rho(z, y)$ and any Banach space is metrically convex.

Lemma 1 [5] *Suppose (M, ρ) is a completely metrically convex metric space and $f : M \longrightarrow M$ satisfies*

$$\rho(f(x), f(y)) \leq \varphi(\rho(x, y)) \qquad for\ x, y \in M, \tag{11}$$

where $\varphi : \overline{P} \longrightarrow R^+$ satisfies $\varphi(t) < t$ for $t \in \overline{P} - \{0\}$, $P = \{\rho(x, y) : x, y \in M\}$ and $\overline{P}$ denotes the closure of P. Then f has a fixed point $u \in M$ and $lim_{n \longrightarrow \infty} f^n(x) = u$, for each $x \in M$.

Lemma 2 [7] $\{a_i, b_i : 1 \leq i \leq n\} \subseteq \mathbb{R}$. *Then*

$$|\operatorname{opt}\{a_i : 1 \leq i \leq n\} - \operatorname{opt}\{b_i : 1 \leq i \leq n\}| \leq \max \{|a_i - b_i| : 1 \leq i \leq n\}. \tag{12}$$

Lemma 3 [7]

(i) Assume that $A : S \times D \rightarrow \mathbb{R}$ is a mapping such that $\operatorname{opt}_{y \in D} A(x_0, y)$ *is bounded for some $x_0 \in S$. Then*

$$|\operatorname{opt}_{y\in D} A(x_0, y)| \le \sup_{y\in D} |A(x_0, y)|; \tag{13}$$

(ii) Assume that $A, B : S \times D \to \mathbb{R}$ *is a mapping such that* $\mathrm{opt}_{y\in D}A(x_1, y)$ *and* $\mathrm{opt}_{y\in D}B(x_2, y)$ *is bounded for some* $x_1, x_2 \in S$. *Then*

$$|\operatorname{opt}_{y\in D} A(x_1, y) - \operatorname{opt}_{y\in D} B(x_2, y)| \le \sup_{y\in D} |A(x_1, y) - B(x_2, y)|. \tag{14}$$

3 Main Results

First of all, we show the existence and uniqueness of solutions in $BC(S)$ and $B(S)$.

Theorem 1 *Let* S *be compact. Let* $u, v, r, p_i, q_i : S\times D \longrightarrow \mathbb{R}$ *and* $s, a_i : S\times D \longrightarrow S$ *for* $i = 1, 2, 3$, *and satisfy the following conditions:*

(C_1) u, v *and* p_i *are bounded for* $i = 1, 2, 3$.
(C_2) *for each* $x_0 \in S$, $u(x, y) \longrightarrow u(x_0, y)$, $r(x, y) \longrightarrow r(x_0, y)$, $s(x, y) \longrightarrow s(x_0, y)$, $v(x, y) \longrightarrow v(x_0, y)$, $p_i(x, y) \longrightarrow p_i(x_0, y)$, $q_i(x, y) \longrightarrow q_i(x_0, y)$, $a_i(x, y) \longrightarrow a_i(x_0, y)$ *as* $x \longrightarrow x_0$ *uniformly for* $y \in D$ *and* $i = 1, 2, 3$.
(C_3) $|r(x, y)|+ \max\{|q_i(x, y)| : i = 1, 2, 3\} \le \alpha$, *for some* $\alpha \in (0, 1)$*and* $(x, y) \in S \times D$.

Then the functional equation (10) possesses a unique solution $w \in BC(S)$ *and* $\{H^n h\}_{n\ge 1}$ *converges to* w *for each* $h \in BC(S)$, *where* H *is defined as*

$$\begin{aligned} Hh(x) = {} & \operatorname{opt}_{y\in D}\{u(x, y) + r(x, y)h(s(x, y)) \\ & + \operatorname{opt}\{v(x, y),\ p_i(x, y) + q_i(x, y)h(a_i(x, y)) : i = 1, 2, 3\}\},\ \forall x \in S. \end{aligned} \tag{15}$$

Proof Let $h \in BC(S)$ and $x_0 \in S$ and $\varepsilon > 0$, by (C_1), (C_2) and compactness of S, there exist a constant $M > 0$, $\delta > 0$ and $\delta_1 > 0$ such that

$$\max\{|u(x, y)|, |v(x, y)|, |p_i(x, y)| : i = 1, 2, 3\} \le M,\ \forall\, (x, y) \in S \times D. \tag{16}$$

$$\max\{|h(x)|, |h(s(x, y))|, |h(a_i(x, y))| : i = 1, 2, 3\} \le M,\ \forall\, (x, y) \in S \times D. \tag{17}$$

$$\sup_{y\in D}\{|u(x, y) - u(x_0, y)|\} \le \varepsilon/4,\ \ with\ \|x - x_0\| < \delta. \tag{18}$$

$$\max\{|p_i(x, y) - p_i(x_0, y)| : i = 1, 2, 3\} \le \varepsilon/4,\ \ with\ \|x - x_0\| < \delta. \tag{19}$$

$$\max\{|r(x, y) - r(x_0, y)|\} \le \varepsilon/4M,\ \ with\ \|x - x_0\| < \delta. \tag{20}$$

$$\max\{|q_i(x, y) - q_i(x_0, y)| : i = 1, 2, 3\} \le \varepsilon/4M,\ \ with\ \|x - x_0\| < \delta. \tag{21}$$

$$|h(x_1) - h(x_2)| < \varepsilon/4, \quad \forall\, x_1, x_2 \in S, \quad with\ \|x_1 - x_2\| < \delta_1. \tag{22}$$

$$\max\{\|a_i(x, y) - a_i(x_0, y)\| : i = 1, 2, 3\} < \delta_1, \quad \forall\, (x, y) \in S \times D\ with\ \|x - x_0\| < \delta. \tag{23}$$

$$\begin{aligned}|Hh(x)| &\le \sup_{y\in D}|\{u(x, y) + r(x, y)h(s(x, y)) \\ &\quad + \text{opt}\{v(x, y),\ p_i(x, y) + q_i(x, y)\, h(a_i(x, y)) : i = 1, 2, 3\}\}| \\ &\le \sup_{y\in D}\{|u(x, y)| + |r(x, y)|\, |h(s(x, y))| + \\ &\quad \max\{|v(x, y)|,\ |p_i(x, y)| + |q_i(x, y)|\, |h(a_i(x, y))| : i = 1, 2, 3\}\} \\ &\le 2M + \alpha M,\end{aligned}$$

This implies H is bounded.
From (C_3), (15), (17)–(23) and Lemmas 2 and 3, we obtain that for all $(x, y) \in S\times D$ with $\|x - x_0\| < \delta$,

$$\begin{aligned}|Hh(x) - Hh(x_0)| &= |\operatorname*{opt}_{y\in D}\{u(x, y) + r(x, y)h(s(x, y)) \\ &\quad + \text{opt}\{v(x, y), p_i(x, y) + q_i(x, y)h(a_i(x, y)) : i = 1, 2, 3\}\} \\ &\quad - \operatorname*{opt}_{y\in D}\{u(x_0, y) + r(x_0, y)h(s(x_0, y)) \\ &\quad + \text{opt}\{v(x_0, y), p_i(x_0, y) + q_i(x_0, y)h(a_i(x_0, y)) : i = 1, 2, 3\}\}| \\ &\le sup_{y\in D}\{|u(x, y) - u(x_0, y)| + |r(x, y)h(s(x, y)) - r(x_0, y)h(s(x_0, y))| \\ &\quad + \max\{|v(x, y) - v(x_0, y)|,\ |p_i(x, y) - p_i(x_0, y)| \\ &\quad + |q_i(x, y)h(a_i(x, y)) - q_i(x_0, y)h(a_i(x_0, y))| : i = 1, 2, 3\}\} \\ &\le sup_{y\in D}\{|u(x, y) - u(x_0, y)| + |r(x, y) - r(x_0, y)|\, |h(s(x, y))| \\ &\quad + |r(x_0, y)|\, |h(s(x, y)) - h(s(x_0, y))| \\ &\quad + \max\{|v(x, y) - v(x_0, y)|,\ |p_i(x, y) - p_i(x_0, y)| \\ &\quad + |q_i(x, y)h(a_i(x, y)) - q_i(x_0, y)h(a_i(x_0, y))| : i = 1, 2, 3\}\} \\ &\le \frac{\varepsilon}{4} + \frac{\varepsilon}{4M}M + \alpha\frac{\varepsilon}{4} + \max\{\frac{\varepsilon}{4}, \frac{\varepsilon}{4} + \frac{\varepsilon}{4M}M + \alpha\, \frac{\varepsilon}{4}\} \\ &< \varepsilon,\end{aligned} \tag{24}$$

which ensure that Hh is continuous at x_0. So, H is a self-mapping on $BC(S)$.
Given $\varepsilon > 0$, $x \in S$ and $h, g \in BC(S)$. Suppose $\text{opt}_{y\in D} = \sup_{y\in D}$. Then $\exists y, z \in D$ such that

$$\begin{aligned}Hh(x) &< u(x, y) + r(x, y)h(s(x, y)) + \text{opt}\{v(x, y),\ p_i(x, y) + q_i(x, y)h(a_i(x, y)) : i = 1, 2, 3\} + \varepsilon, \\ Hg(x) &< u(x, z) + r(x, z)g(s(x, z)) + \text{opt}\{v(x, z),\ p_i(x, z) + q_i(x, z)g(a_i(x, z)) : i = 1, 2, 3\} + \varepsilon, \\ Hh(x) &\ge u(x, z) + r(x, z)h(s(x, z)) + \text{opt}\{v(x, z),\ p_i(x, z) + q_i(x, z)h(a_i(x, z)) : i = 1, 2, 3\}, \\ Hg(x) &\ge u(x, y) + r(x, y)g(s(x, y)) + \text{opt}\{v(x, y),\ p_i(x, y) + q_i(x, y)g(a_i(x, y)) : i = 1, 2, 3\}.\end{aligned} \tag{25}$$

By (25), we get

$$\begin{aligned}Hh(x)-Hg(x) &< r(x,y)(h(s(x,y))-g(s(x,y)))\\&\quad+\text{opt}\{v(x,y),\ p_i(x,y)+q_i(x,y)h(a_i(x,y)):i=1,2,3\}.\\&\quad-\text{opt}\{v(x,y),\ p_i(x,y)+q_i(x,y)g(a_i(x,y)):i=1,2,3\}+\varepsilon,\\&\le |r(x,y)|\ |(h(s(x,y))-g(s(x,y)))|\\&\quad+\max\{|q_i(x,y)|\ |h(a_i(x,y))-g(a_i(x,y))|:i=1,2,3\}+\varepsilon,\\&\le [\ |r(x,y)|+\max\{|q_i(x,y)|:i=1,2,3\}]\\&\quad\max\{|(h(s(x,y))-g(s(x,y)))|,|h(a_i(x,y))-g(a_i(x,y))|:i=1,2,3\}+\varepsilon,\\&\le \alpha\ \max\{|(h(s(x,y))-g(s(x,y)))|,|h(a_i(x,y))-g(a_i(x,y))|:i=1,2,3\}+\varepsilon.\end{aligned}$$

and

$$\begin{aligned}Hh(x)-Hg(x) &> r(x,z)(h(s(x,z))-g(s(x,z)))\\&\quad+\text{opt}\{v(x,z),\ p_i(x,z)+q_i(x,z)h(a_i(x,z)):i=1,2,3\}.\\&\quad-\text{opt}\{v(x,z),\ p_i(x,z)+q_i(x,z)g(a_i(x,z)):i=1,2,3\}-\varepsilon,\\&\ge -|r(x,z)|\ |(h(s(x,z))-g(s(x,z)))|\\&\quad\max\{|q_i(x,z)|\ |h(a_i(x,z))-g(a_i(x,z))|:i=1,2,3\}-\varepsilon,\\&\ge [\ -|r(x,z)|+\max\{|q_i(x,z)|:i=1,2,3\}]\\&\quad\max\{|(h(s(x,z))-g(s(x,z)))|,|h(a_i(x,z))-g(a_i(x,z))|:i=1,2,3\}-\varepsilon,\\&\ge -\alpha\ \max\{|(h(s(x,z))-g(s(x,z)))|,|h(a_i(x,z))-g(a_i(x,z))|:i=1,2,3\}-\varepsilon.\end{aligned}$$

which implies that

$$|Hh(x)-Hg(x)|\le\alpha\|h-g\|_1+\varepsilon.$$

by which we get

$$\|Hh-Hg\|_1\le\varphi(\|h-g\|_1)+\varepsilon,\ \forall\ h,g\in BC(S)$$

where $\varphi(\lambda)=\alpha\lambda,\ \lambda\in R^+$. Letting $\varepsilon\to 0^+$, we get

$$\|Hh-Hg\|_1\le\varphi(\|h-g\|_1),\ \forall\ h,g\in BC(S) \tag{26}$$

In a similar way, we conclude that (26) holds for $\text{opt}_{y\in D}=\inf_{y\in D}$. Lemma 1 ensures that H has a unique fixed point $w\in BC(S)$ and $\{H^n h\}_{n\ge1}$ converges to w for each $h\in BC(S)$. It is obvious that w is also a unique solution of the functional equation (10) in $BC(S)$. This completes the proof.

If we remove the condition of compactness of S and (C_2) in Theorem 1, we obtain the below result.

Theorem 2 *Let $u,v,r,p_i,q_i:S\times D\longrightarrow\mathbb{R}$ and $s,a_i:S\times D\longrightarrow S$ for $i=1,2,3$, and satisfies conditions (C_1) and (C_3). Then the functional equation (10) possesses a unique solution $w\in B(S)$ and $\{H^n h\}_{n\ge1}$ converges to w for each $h\in B(S)$, where H is defined by (15).*

Remark 1 If $u(x, y) = r(x, y) = p_i(x, y) = 0$, for $i = 1, 2, 3$ and $\forall(x, y) \in S \times D$, then Theorems 1 and 2 reduce to the results of Jiang et al. [7]. The following example shows that Theorems 1 and 2 unify substantially the results in [7].

Example 1 Let $X = Y = \mathbb{R},\ S = [1, 2],\ D = R^+$, then Theorem 2 ensures that the functional equation given below possesses a unique solution in $B(S)$.

$$\begin{aligned} f(x) = \operatorname*{opt}_{y \in D} \Bigg\{ & \frac{x}{x+y^2} + \left(\frac{x+y}{1+3(x+y)}\right) f\left(\frac{2x^2+y^2}{x^2+y^2}\right) + \operatorname{opt}\Bigg\{ \frac{x^2}{1+xy}, \sin(x+2y+1) \\ & + \frac{1}{5}\sin(2x^2y+3) f\left(\frac{3x+y^3}{x+y^3}\right), \cos(x+y^2+1) + \frac{1}{7}\sin(2x+3y) f\left(\frac{7x^2+y}{x^2+y}\right), \\ & \frac{x^2}{x+y^2} + \frac{\sin(3x+5y+1)}{7+x^2+y} f\left(\frac{x^2 y \sin(1+xy)}{1+xy^2}\right)\Bigg\}\Bigg\}, \forall x \in S. \end{aligned} \tag{27}$$

Since,

$$\left|\frac{x+y}{1+3(x+y)}\right| + \max\left\{\frac{1}{5}|\sin(2x^2y+3)|, \frac{1}{7}|\sin(2x+3y)|, \frac{|\sin(3x+5y+1)|}{7+x^2+y}\right\} < 1.$$

However, the corresponding results in [7] are not applicable for the functional equation (27). Because,

$$\frac{x}{x+y^2} > 0,\ \forall\ (x, y) \in S \times D.$$

We point out that the functional equation (27) possesses also a unique solution in BC(S).

We discuss properties of solutions in $BB(S)$ in our next results.

Theorem 3 *Let $u, v, s, p_i, q_i : S \times D \longrightarrow \mathbb{R}$ and $s, a_i : S \times D \longrightarrow S$ for $i = 1, 2, 3$, and satisfy the following conditions:*

(B_1) *u, v and p_i are bounded on $\overline{B}(0, k) \times D$ for $k \geq 1$ and $i = 1, 2, 3$,*
(B_2) $\max\{\|s(x, y)\|, \|a_i(x, y)\| : i = 1, 2, 3\} \leq \|x\|$, *for $(x, y) \in S \times D$,*
(B_3) *there exists a constant α such that*
$\sup_{(x,y)\in S\times D}\{|r(x, y)| + \max|q_i(x, y)| : i = 1, 2, 3\} \leq \alpha < 1$,

then the functional equation (10) possesses a unique solution $w \in BB(S)$ and $\{H^n h\}_{n\geq1}$ converges to w for each $h \in BB(S)$, where H is defined by (15).

Proof For each $k \geq 1$ and $h \in BB(S)$, (B_1) and (B_2) imply that there exist $\beta(k) > 0$ and $\eta(k, h) > 0$ such that

$$\sup_{(x,y)\in\overline{B}(0,k)\times D} \{|u(x, y)|, |v(x, y)|, |p_i(x, y)| : i = 1, 2, 3\} \leq \beta(k) \tag{28}$$

$$\sup_{(x,y)\in\overline{B}(0,k)\times D} \{|h(s(x, y))|, |h(a_i(x, y))| : i = 1, 2, 3\} \leq \eta(k, h). \tag{29}$$

In view of (B_3), (28) and (29), we get

$$\begin{aligned}|Hh(x)| &\le \sup_{y\in D}\{|u(x,y)| + |r(x,y)|\,|h(s(x,y))| \\ &\qquad + \max\{|v(x,y)|, |p_i(x,y)| + |q_i(x,y)|\,|h(a_i(x,y))| : i = 1,2,3\}\}, \\ &\le 2\beta(k) + 2\eta(k,h).\end{aligned}$$

Thus Hh is bounded, i.e., H is a self-mapping on $BB(S)$.
Let $k \in N$, $x \in \overline{B}(0,k)$, $g, h \in BB(S)$ and $\varepsilon > 0$. Suppose $\mathrm{opt}_{y\in D} = \sup_{y\in D}$. Then there exist $y, z \in D$ satisfying

$$\begin{aligned}Hh(x) &< u(x,y) + r(x,y)h(s(x,y)) \\ &\qquad + \mathrm{opt}\{v(x,y),\ p_i(x,y) + q_i(x,y)h(a_i(x,y)) :\ i = 1,2,3\} + \varepsilon, \\ Hg(x) &< u(x,z) + r(x,z)g(s(x,z)) \\ &\qquad + \mathrm{opt}\{v(x,z),\ p_i(x,z) + q_i(x,z)g(a_i(x,z)) :\ i = 1,2,3\} + \varepsilon, \\ Hh(x) &\ge u(x,z) + r(x,z)h(s(x,z)) \\ &\qquad + \mathrm{opt}\{v(x,z),\ p_i(x,z) + q_i(x,z)h(a_i(x,z)) :\ i = 1,2,3\}, \\ Hg(x) &\ge u(x,y) + r(x,y)g(s(x,y)) \\ &\qquad + \mathrm{opt}\{v(x,y),\ p_i(x,y) + q_i(x,y)g(a_i(x,y)) :\ i = 1,2,3\}.\end{aligned}$$

Using (B_3), Lemma 2 and applying in the similar way as in Theorem 1 the above terms, we get

$$|Hh(x) - Hg(x)| \le \alpha d_k(h,g) + \varepsilon, \quad \forall h, g \in BB(S). \tag{30}$$

Similarly, we can show that (30) holds for $\mathrm{opt}_{y\in D} = \inf_{y\in D}$. It follows that

$$d_k(Hg, Hh) \le \varphi(d_k(h,g)) + \varepsilon, \tag{31}$$

where $\varphi(\lambda) = \alpha\lambda$ for $\lambda \in R^+$. As $\varepsilon \longrightarrow 0^+$ in (31), we get

$$d_k(Hg, Hh) \le \varphi(d_k(h,g)).$$

It follows from Theorem 2.2 in [4] that H has a unique fixed point $w \in BB(S)$ and $\{H^n h\}_{n\ge 1}$ converges to w for each $h \in BB(S)$. Obviously, w is also a unique solution of the functional equation (10). This completes the proof.

Remark 2

1. If $u = v = r = q_1 = q_3 = p_1 = p_2 = 0$, $\mathrm{opt}_{y\in D} = \inf_{y\in D}$ and opt $=$ max, then Theorem 3 reduces to Theorem 3.4 of Bhakta and Choudhary [4] and a result of Bellman [1].
2. If $v = r = p_3 = q_3 = 0$, then Theorem 3 reduces to a result of Liu et al. [11].
3. If $p_1 = p_2 = p_3 = q_3 = 0$, then Theorem 3 reduces to Theorem 4.1 of Liu et al. [13], which, in turn, generalizes the results in [1, 4].
4. Theorem 3.3 of Jiang et al. [7] is a particular case of Theorem 3.

The example below shows that Theorem 3 unifies the results in [1, 4, 7, 11, 13].

Example 2 Let $X = Y = \mathbb{R}$ and $S = D = R^+$. Consider the following functional equation:

$$\begin{aligned}
f(x) = \operatorname*{opt}_{y\in D}\Bigg\{&(1+|\sin(x+2y-1)|) + \frac{1+\sin(1+xy^2)}{5+5xy^2} f\left(\frac{x^2}{1+2x+3y}\right) \\
&+ \operatorname{opt}\Bigg\{\left(1+\frac{1}{2+x^2y}\right), \frac{3x^2}{1+x^2y^2} + \left(\frac{\sin(x^2+y^2)}{4+(x^2+y^2)}\right) f\left(\frac{x^3}{2+x^2+y}\right), \\
&x\cos(x+2y-1) + \left(\frac{1}{3+xy}\right) f\left(\frac{x^2}{1+x+y}\right), \; x\sin(2x+3y) \\
&+ \left(1+\frac{\sin(xy+2)}{7+x^3y}\right) f\left(\frac{x^3}{1+x^2+y}\right)\Bigg\}\Bigg\}, \quad \forall x \in S. \qquad (32)
\end{aligned}$$

Clearly,

$$\begin{aligned}
\sup_{(x,y)\in\overline{B}(0,k)\times D}\Bigg\{&(1+|\sin(x+2y-1)|), \; (1+\frac{1}{2+x^2y}), \; \frac{3x^2}{1+x^2y^2}, \\
&x|\cos(x+2y-1)|, \; x|\sin(2x+3y)|\Bigg\} \le 3k, \; k \ge 1,
\end{aligned}$$

$$\max\left\{\frac{x^2}{1+2x+3y}, \frac{x^3}{2+x^2+y}, \frac{x^2}{1+x+y}, \frac{x^3}{1+x^2+y}\right\} \le |x|, \; (x,y) \in S \times D$$

and

$$\sup_{(x,y)\in S\times D}\left\{\frac{1+|\sin(1+xy^2)|}{5+5xy^2} + \max\left\{\frac{|\sin(x^2+y^2)|}{4+(x^2+y^2)}, \frac{1}{3+xy}, \frac{1+|\sin(xy+2)|}{7+x^3y}\right\}\right\} < 1.$$

It follows from Theorem 3 that the functional equation (32) possesses a unique solution in $BB(S)$. However, the results in [1, 4, 7, 11, 13] are not applicable to the functional equation (32). Since

$$(1+|\sin(x+2y-1)|) > 0, \; \left(1+\frac{1}{2+x^2y}\right) > 0 \text{ and}$$
$$\left(1+\frac{|\sin(xy+2)|}{7+x^3y}\right) > 0, \; \forall\,(x,y) \in S \times D.$$

In the following theorem, we present the proof in line with the proof of Theorem 4.3 of Pathak and Deepmala [15].

Theorem 4 *Let $u, v, r, p_i, q_i : S\times D \longrightarrow \mathbb{R}$ and $s, a_i : S\times D \longrightarrow S$ for $i = 1, 2, 3$, and let (φ, ψ) be in (Φ_2) satisfying the following conditions:*

(B_4) $\max\{|u(x, y)| + |v(x, y)|, |p_i(x, y)| : i = 1, 2, 3\} \leq \psi(\|x\|), \forall (x, y) \in S \times D.$

(B_5) $\max\{\|s(x, y)\|, \|a_i(x, y)\| : i = 1, 2, 3\} \leq \varphi(\|x\|), \forall (x, y) \in S \times D.$

(B_6) $\sup_{y \in D}\{|r(x, y)| + \max\{|q_i(x, y)| : i = 1, 2, 3\}\} \leq 1, \forall x \in S.$

Then the functional equation (10) possesses a solution $w \in BB(S)$ that satisfies the following conditions:

(B_7) *The sequence $\{w_n\}_{n \geq 0}$ defined by*

$$w_0(x) = \text{opt}_{y \in D}\{u(x, y) + v(x, y) + \text{opt}\{p_i(x, y) : i = 1, 2, 3\}\}, \forall x \in S$$

$$w_n(x) = \text{opt}_{y \in D}\Big\{u(x, y) + r(x, y)w_{n-1}(s(x, y)) + \text{opt}\{v(x, y), p_i(x, y) + q_i(x, y)w_{n-1}(a_i(x, y)) : i = 1, 2, 3\}\Big\},$$

$\forall x \in S,\ n \geq 1$ *converges to w.*

(B_8) $\lim_{n \to \infty} w(x_n) = 0$ *for any $x_0 \in S$, $\{y_n\}_{n \geq 1} \subset D$ and $x_n \in \{a_i(x_{n-1}, y_{n-1}) : i = 1, 2, 3\}, \forall n \in N.$*

(B_9) *w is unique with respect to condition (B_8)*

Proof Since $(\varphi, \psi) \in \Phi_2$, thus

$$\varphi(t) < t \quad \text{for} \quad t < 0 \tag{33}$$

First, we show that the mapping H defined by (15) is nonexpansive on $BB(S)$, by (33) and (B_5), we get $\max\{\|s(x, y)\|, \|a_i(x, y)\| : i = 1, 2, 3\} \leq \varphi(\|x\|) < k$, for $(x, y) \in \overline{B}(0, k) \times D$, which implies that $\exists$ a constant $\theta(k, h) > 0$ with

$$\max\{|h(s(x, y))|, |h(a_i(x, y))| : i = 1, 2, 3\} \leq \theta(k, h),\ for (x, y) \in \overline{B}(0, k) \times D. \tag{34}$$

By virtue of (B_4), (B_6), (15), (34), Lemmas 2 and 3, we deduce that

$$\begin{aligned}|Hh(x)| &\leq \sup_{y \in D}\{|u(x, y)| + |r(x, y)||h(s(x, y))| \\ &\quad + \max\{|v(x, y)|, |p_i(x, y)| + |q_i(x, y)|\ |h(a_i(x, y))| : i = 1, 2, 3\}\} \\ &\leq \sup_{y \in D}\{|u(x, y)| + |v(x, y)| + \max\{|p_i(x, y)| : i = 1, 2, 3\} + [\ |r(x, y)| \\ &\quad + \max\{|q_i(x, y)| : i = 1, 2, 3\}\ ]\ \max\{|h(s(x, y))|, |h(a_i(x, y))| : i = 1, 2, 3\}\} \\ &\leq 2\psi(k) + \theta(k, h)\ for\ x \in \overline{B}(0, k).\end{aligned}$$

Thus H is a self-mapping on $BB(S)$.
By (B_6) and following the similar approach as in proof of Theorem 3, we conclude that for $h, g \in BB(S)$ and $k \geq 1$,

$$d_k(Hh, Hg) \leq d_k(h, g),$$

which implies that

$$d(Hh, Hg) = \sum_{k=1}^{\infty} \frac{1}{2^k} \frac{d_k(Hh, Hg)}{1 + d_k(Hh, Hg)} \leq \sum_{k=1}^{\infty} \frac{1}{2^k} \frac{d_k(h, g)}{1 + d_k(h, g)} = d(h, g)$$

for $h, g \in BB(S)$. That is, H is nonexpansive.
Now we assert that for each $n \geq 0$,

$$|w_n(x)| \leq 2 \sum_{j=0}^{n} \psi(\varphi^j(\|x\|)), \ x \in S \tag{35}$$

Now by (B_4) we see that

$$\begin{aligned}|w_0(x)| &\leq \sup_{y \in D}\{|u(x, y)| + |v(x, y)| + \max\{|p_i(x, y)| : i = 1, 2, 3\}\} \\ &\leq 2\psi(\|x\|), \ x \in S\end{aligned}$$

That is, (35) is true for $n = 0$. Suppose (35) holds for some $n \geq 0$. From (B_4)–(B_6) we know that

$$\begin{aligned}|w_{n+1}(x)| &= |\operatorname{opt}_{y \in D}\{u(x, y) + r(x, y)w_n(s(x, y)) \\ &\qquad + \operatorname{opt}\{v(x, y), p_i(x, y) + q_i(x, y)w_n(a_i(x, y)) : i = 1, 2, 3\}\}| \\ &\leq \sup_{y \in D}\{|u(x, y)| + |r(x, y)|\,|w_n(s(x, y))| \\ &\qquad + \max\{|v(x, y)|, |p_i(x, y)| + |q_i(x, y)|\,|w_n(a_i(x, y))| : i = 1, 2, 3\}\} \\ &\leq \sup_{y \in D}\{|u(x, y)| + |v(x, y)| + \max\{|p_i(x, y)| : i = 1, 2, 3\} \\ &\qquad + [\,|r(x, y)| + \max\{|q_i(x, y)| : i = 1, 2, 3\}] \\ &\qquad\quad \max\{|w_n(s(x, y))|, \ |w_n(a_i(x, y))| : i = 1, 2, 3\}\} \\ &\leq 2\psi(\|x\|) + 2\sum_{j=0}^{n} \psi(\varphi^j(\|x\|)) \\ &\leq 2\sum_{j=0}^{n+1} \psi(\varphi^j(\|x\|)),\end{aligned}$$

Hence (35) holds for $n \geq 0$.
Next we claim that $\{w_n\}_{n \geq 0}$ is a cauchy sequence in $BB(S)$. Given $k \geq 1$ and $x_0 \in \overline{B}(0, k)$. Let $\varepsilon > 0$, $n, m \in N$. Suppose $\operatorname{opt}_{y \in D} = \sup_{y \in D}$. Then we select $y, z \in D$ such that

$$\begin{aligned}w_n(x_0) &< u(x_0, y) + r(x_0, y)w_{n-1}(s(x_0, y)) \\ &\qquad + \operatorname{opt}\{v(x_0, y), \ p_i(x_0, y) + q_i(x_0, y)\, w_{n-1}(a_i(x_0, y)) : i = 1, 2, 3\} + \frac{\varepsilon}{2} \\ w_{n+m}(x_0) &< u(x_0, z) + r(x_0, z)w_{n+m-1}(s(x_0, z)) \\ &\qquad + \operatorname{opt}\{v(x_0, z), \ p_i(x_0, z) + q_i(x_0, z)\, w_{n+m-1}(a_i(x_0, z)) : i = 1, 2, 3\} + \frac{\varepsilon}{2} \\ w_n(x_0) &\geq u(x_0, z) + r(x_0, z)w_{n-1}(s(x_0, z)) \\ &\qquad + \operatorname{opt}\{v(x_0, z), \ p_i(x_0, z) + q_i(x_0, z)\, w_{n-1}(a_i(x_0, z)) : i = 1, 2, 3\},\end{aligned}$$

$$w_{n+m}(x_0) \geq u(x_0, y) + r(x_0, y)w_{n+m-1}(s(x_0, y))$$
$$+ \operatorname{opt}\{v(x_0, y),\ p_i(x_0, y) + q_i(x_0, y)\, w_{n+m-1}(a_i(x_0, y)) : i = 1, 2, 3\}. \tag{36}$$

Using (36), (B_6) and Lemma 2, we show that

$$\begin{aligned} |w_{n+m}(x_0) - w_n(x_0)| &< \max\{\max\{|w_{n+m-1}(a_i(x_0, z)) - w_{n-1}(a_i(x_0, z))| : i = 1, 2, 3\},\\ &\quad \max\{|w_{n+m-1}(a_i(x_0, y)) - w_{n-1}(a_i(x_0, y))| : i = 1, 2, 3\}\} + \varepsilon/2,\\ &= |w_{n+m-1}(x_1) - w_{n-1}(x_1)| + \varepsilon/2, \end{aligned} \tag{37}$$

for some $y_1 \in \{y, z\}$ and $x_1 \in \{a_i(x_0, y_1) : i = 1, 2, 3\}$.
Similarly, we conclude that the above inequality (37) holds for $\operatorname{opt}_{y\in D} = \inf_{y\in D}$. Proceeding in this way, we select $y_j \in D$ and $x_j \in \{a_i(x_{j-1}, y_j) : i = 1, 2, 3\}$ for $j = 2, 3, ..., n$ such that

$$\begin{aligned} |w_{n+m-1}(x_1) - w_{n-1}(x_1)| &< |w_{n+m-2}(x_2) - w_{n-2}(x_2)| + 2^{-2}\varepsilon\\ |w_{n+m-2}(x_2) - w_{n-2}(x_2)| &< |w_{n+m-3}(x_3) - w_{n-3}(x_3)| + 2^{-3}\varepsilon\\ &\cdots\cdots\cdots\\ |w_{m+1}(x_{n-1}) - w_1(x_{n-1})| &< |w_m(x_n) - w_0(x_n)| + 2^{-n}\varepsilon. \end{aligned} \tag{38}$$

It follows from (B_5), (33), (35), (37) and (38) that

$$\begin{aligned} |w_{n+m}(x_0) - w_n(x_0)| &< |w_m(x_n) - w_0(x_n)| + \textstyle\sum_{i=1}^{n} 2^{-i}\varepsilon\\ &< |w_m(x_n)| + |w_0(x_n)| + \varepsilon\\ &\leq 2\textstyle\sum_{i=0}^{m}\psi(\varphi^i(\|x_n\|)) + 2\psi(\|x_n\|) + \varepsilon\\ &\leq 2\textstyle\sum_{i=0}^{m}\psi(\varphi^{i+n}(\|x_0\|)) + 2\psi(\varphi^n(\|x_0\|)) + \varepsilon\\ &\leq 2\textstyle\sum_{j=n-1}^{\infty}\psi(\varphi^j(k)) + \varepsilon, \end{aligned}$$

which implies that

$$d_k(w_{n+m}, w_n) \leq 2\sum_{j=n-1}^{\infty} \psi(\varphi^j(k)) + \varepsilon \tag{39}$$

As $\varepsilon \to 0^+$ in the above inequality, we get $d_k(w_{n+m}, w_n) \leq 2\sum_{j=n-1}^{\infty} \psi(\varphi^j(k))$, which implies that $\{w_n\}_{n\geq 0}$ is a cauchy sequence in $(BB(S), d)$ since, $\sum_{i=0}^{\infty} \psi(\varphi^n(t)) < \infty$, for each $t > 0$. Suppose $\{w_n\}_{n\geq 0}$ converges to some $w \in BB(S)$. Since H is nonexpansive, it follows that

$$\begin{aligned} d(w, Hw) &\leq d(w, Hw_n) + d(Hw_n, Hw)\\ &\leq d(w, w_{n+1}) + d(w_n, w)\\ &\to 0 \text{ as } n \to \infty. \end{aligned}$$

That is, $Hw = w$. So, the functional equation (10) possesses a solution w.

Now we show that (B_8) holds. Let $\varepsilon > 0, x_0 \in S$, $\{y_n\}_{n\geq 1} \subset D$ and $x_n \in \{a_i(x, y), i = 1, 2, 3\}$ for $n \geq 1$. Put $k = [\|x_0\|] + 1$. Then there exists a positive integer m satisfying

$$d_k(w, w_n) + 2\sum_{j=n}^{\infty} \psi(\varphi^j(k)) < \varepsilon, \ for\ n > m. \tag{40}$$

By (35), (B_5) and (40), we show that for $n > m$,

$$\begin{aligned}|w(x_n)| &\leq |w(x_n) - w_n(x_n)| + |w_n(x_n)|\\ &\leq d_k(w, w_n) + 2\textstyle\sum_{j=0}^{\infty}\psi(\varphi^j(\|x_n\|))\\ &\leq d_k(w, w_n) + 2\textstyle\sum_{j=n}^{\infty}\psi(\varphi^j(k))\\ &\leq \varepsilon,\end{aligned}$$

which means that $\lim_{n\to\infty} w(x_n) = 0$.
Finally, we show that (B_9) holds. Suppose the functional equation (10) possesses another solution $h \in BB(S)$, which satisfies condition (B_8). Let $\varepsilon > 0$ and $x_0 \in S$. If $\text{opt}_{y\in D} = \sup_{y\in D}$, then there exist $y, z \in S$ such that

$$\begin{aligned}w(x_0) &< u(x_0, y) + r(x_0, y)w(s(x_0, y))\\ &\qquad + \text{opt}\{v(x_0, y),\ p_i(x_0, y) + q_i(x_0, y)w(a_i(x_0, y)) : i = 1, 2, 3\} + \frac{\varepsilon}{2},\\ h(x_0) &< u(x_0, z) + r(x_0, z)h(s(x_0, z))\\ &\qquad + \text{opt}\{v(x_0, z),\ p_i(x_0, z) + q_i(x_0, z)h(a_i(x_0, z)) : i = 1, 2, 3\} + \frac{\varepsilon}{2},\\ w(x_0) &\geq u(x_0, z) + r(x_0, z)w(s(x_0, z))\\ &\qquad + \text{opt}\{v(x_0, z),\ p_i(x_0, z) + q_i(x_0, z)w(a_i(x_0, z)) : i = 1, 2, 3\},\\ h(x_0) &\geq u(x_0, y) + r(x_0, y)w(h(x_0, y))\\ &\qquad + \text{opt}\{v(x_0, y),\ p_i(x_0, y) + q_i(x_0, y)h(a_i(x_0, y)) : i = 1, 2, 3\}.\end{aligned} \tag{41}$$

By Lemma 2, (B_6), and (41) we obtain

$$\begin{aligned}|w(x_0) - h(x_0)| &< \max\{\max\{|w(a_i(x_0, y)) - h(a_i(x_0, y))| : i = 1, 2, 3\},\\ &\qquad \max\{|w(a_i(x_0, z)) - h(a_i(x_0, z))| : i = 1, 2, 3\}\} + \frac{\varepsilon}{2},\\ &= |w(x_1) - h(x_1)| + \frac{\varepsilon}{2}.\end{aligned} \tag{42}$$

for some $y_1 \in \{y, z\}$ and $x_1 \in \{a_i(x_0, y_1) : i = 1, 2, 3\}$. Similarly, we conclude that (42) holds for $\text{opt}_{y\in D} = \inf_{y\in D}$. Proceeding in this way, we select $y_j \in D$ and $x_j \in \{a_i(x_{j-1}, y_j)\}$ for $j = 2, 3, ..., n$ satisfying

$$
\begin{aligned}
|w(x_1) - h(x_1)| &< |w(x_2) - h(x_2)| + 2^{-2}\varepsilon \\
|w(x_2) - h(x_2)| &< |w(x_3) - h(x_3)| + 2^{-3}\varepsilon \\
&\cdots\cdots\cdots \\
|w(x_{n-1}) - h(x_{n-1})| &< |w(x_n) - h(x_n)| + 2^{-n}\varepsilon
\end{aligned} \tag{43}
$$

Combining (42) and (43), we obtain

$$
|w(x_0) - h(x_0)| \;<\; |w(x_n) - h(x_n)| + \sum_{j=1}^{n} 2^{-j}\varepsilon \;<\; |w(x_n) - h(x_n)| + \varepsilon.
$$

Letting $n \to \infty$ in the above inequalities, by (B_8) we get

$$
|w(x_0) - h(x_0)| \le \varepsilon.
$$

As $\varepsilon \to 0^+$ in the above inequality, we know that $w(x_0) = h(x_0)$. This completes the proof.

Remark 3

1. Theorem 4 unifies the results of Liu et al. [11, 13].
2. In case, $u = r = p_1 = p_2 = p_3 = 0$ and $(\varphi, \psi) \in \Phi_3$, then Theorem 4 reduces to Theorem 3.4 of Jiang et al. [7], which, in turn, unifies the results of Bellman [1], Bhakta and Choudhary [4], Liu [8] and Liu and Ume [10].
3. If $r = p_1 = p_2 = p_3 = 0$, $q_1 = q_2 = q_3 = (1 - \lambda)$, $a_1 = a_2 = a_3 = a(x, y)$, $v(x, y) = (1 - \lambda)\, v_1(x, y)$, $u(x, y) = \lambda\, [u_1(x, y) + f(a(x, y))]$, $\psi(t) = Mt$ and $\varphi \in \Phi_4$, for $(x, y) \in S \times D$, $t \in R^+$, where λ is a constant in [0, 1] and M is a positive constant, then Theorem 4 reduces to Theorem 3.1 of Liu and Ume [10], which, in turn, unifies Theorem 3.5 in [4], Theorem 2.4 in [3] and a result in [1].
4. If we put $\text{opt}_{y\in D} = \sup_{y\in D}$, $u = r = p_1 = p_3 = q_3 = 0$, $q_1 = q_2 = 1$, then Theorem 4 reduces to Theorem 2.2 of Liu et al. [12].
5. Dropping condition (B_6) and replacing $\text{opt}_{y\in D} = \inf_{y\in D}$, opt $=$ max, $q_1 = q_2 = 1$, $u = r = p_1 = p_3 = q_3 = 0$, then Theorem 4 reduces to Theorem 3.4 of Liu and Kang [9], which, in turn, unifies the results in [9], [4] and [1].

The example given below demonstrates that Theorem 4 unifies the results in [1, 3, 4, 7–13]

Example 3 Let $X = Y = R$, $S = D = [1, \infty)$, $\psi(t) = 2t^2$, $\varphi(t) = \frac{t}{3}$, $\forall\, t \in R^+$. Consider the following functional equation:

$$
\begin{aligned}
f(x) = \operatorname*{opt}_{y\in D}\Bigg\{ & x^2\left(1 + \frac{1}{x + 2y}\right) + \frac{1}{3}\sin(x^2 - y^2) f\left(\frac{x\cos(2x + y)}{3 + xy}\right) + \operatorname{opt}\Bigg\{ \frac{1}{2 + \sin(2x + y)}, \\
& \frac{x^2}{1 + xy} + \frac{1}{2}\cos^2(x + 2y) f\left(\frac{x \sin x}{3 + x^2 y}\right), \frac{x^2}{1 + \sin(x + 3y)} + \frac{1}{3 + xy} f\left(\frac{x}{3 + \sin xy}\right), \\
& \frac{x}{1 + x^2 + y^2} + \frac{1}{5 + \sin(2x + y - 1)} f\left(\frac{x}{3 + xy^2}\right)\Bigg\}\Bigg\}, \ \forall\, x \in S.
\end{aligned} \tag{44}
$$

Note that

$$\begin{aligned}
|u(x,y)| &= x^2\left(1+\tfrac{1}{x+2y}\right) &\le 2x^2 = \psi(|x|),\\
|v(x,y)| &= \tfrac{1}{2+|\sin(2x+y)|} &\le 2x^2 = \psi(|x|),\\
|p_1(x,y)| &= \tfrac{x^2}{1+xy} &\le 2x^2 = \psi(|x|),\\
|p_2(x,y)| &= \tfrac{x^2}{1+|\sin(x+3y)|} &\le 2x^2 = \psi(|x|),\\
|p_3(x,y)| &= \tfrac{x}{1+x^2+y^2} &\le 2x^2 = \psi(|x|),
\end{aligned}$$

Also,

$$\max\left\{\frac{1}{3}|\sin(x^2-y^2)| + \max\left\{\frac{1}{2}|\cos^2(x+2y)|, \frac{1}{3+xy}, \frac{1}{5+|\sin(2x+y-1)|}\right\}\right\} < 1,$$

$$\begin{aligned}
|s(x,y)| &= \tfrac{x|\cos(2x+y)|}{3+xy} &\le \tfrac{x}{3} = \varphi(|x|),\\
|a_1(x,y)| &= \tfrac{x|\sin x|}{3+x^2y} &\le \tfrac{x}{3} = \varphi(|x|),\\
|a_2(x,y)| &= \tfrac{x}{3+|\sin xy|} &\le \tfrac{x}{3} = \varphi(|x|),\\
|a_3(x,y)| &= \tfrac{x}{3+xy^2} &\le \tfrac{x}{3} = \varphi(|x|).
\end{aligned}$$

It follows from Theorem 4 that the functional equation (44) possesses a solution $w \in BB(S)$. However, the corresponding results in [1, 3, 4, 7–13] are not applicable for the functional equation (44). Because,

$$\left|x^2\left(1+\frac{1}{x+2y}\right)\right| \le M|x|,$$

does not holds for $(x, y) = (1+M, 1) \in S \times D$, where M is a positive constant, also

$$x^2\left(1+\frac{1}{x+2y}\right) > 0 \text{ and } \frac{1}{2+|\sin(2x+y)|} > 0.$$

4 Conclusion

Dynamic programming is an important and challenging research field because of its applicability in multistage decision processes. For solving a class of functional equations arising in formulation of some real world problem, it is always demanding to show that the solution of these types of functional equations exists or not. If exists, then unique or multiple. Thus, we conclude that the results developed are useful for the researchers to show the existence and uniqueness of the solutions of the functional equations.

Acknowledgments This work was carried out under the project on Optimization and Reliability Modelling of Indian Statistical Institute. The authors wish to thank the unknown referees who patiently went through the article and whose suggestions considerably improved its presentation and readability.

References

1. Bellman, R.: Dynamic Programming. Princeton University Press, Princeton (1957)
2. Bellman, R., Lee, E.S.: Funct. Equat. Aris. Dyn. Program. Aequationes Math. **17**, 1–18 (1978)
3. Bhakta, P.C., Mitra, S.: Some existence theorems for functional equations arising in dynamic programming. J. Math. Anal. Appl. **98**, 348–362 (1984)
4. Bhakta, P.C., Choudhury, S.R.: Some existence theorems for functional equations arising in dynamic programming II. J. Math. Anal. Appl. **131**, 217–231 (1988)
5. Boyd, D.W., Wong, J.S.W.: On nonlinear contractions. Proc. Amer. Math. Soc. **20**, 458–464 (1969)
6. Deepmala: A study of fixed point theorems for nonlinear contractions and its applications. Ph.D. thesis, Pt. Ravishankar Shukla University Raipur (Chhatisgarh)-492010 India (2014)
7. Jiang, G., Kang, S. M., Kwun, Y.C.: Solvability and algorithms for functional equations originating from dynamic programming. Fixed Point Theor. Appl. **2011**(701519), 30 doi:10.1155/2011/701519
8. Liu, Z.: Existence theorems of solutions for certain classes of functional equations arising in dynamic programming. J. Math. Anal. Appl. **262**, 529–553 (2001)
9. Liu, Z., Kang, S.M.: Existence and uniqueness of solutions for two classes of functional equations arising in dynamic programming. Acta Math. Applicatae Sinica Engl. Ser. **23**(2), 195–208 (2007)
10. Liu, Z., Ume, J.S.: On properties of solutions for a class of functional equations arising in dynamic programming. J. Optim. Theor. Appl. **117**(3), 533–551 (2003)
11. Liu, Z., Agarwal, R.P., Kang, S.M.: On solvability of functional equations and system of functional equations arising in dynamic programming. J. Math. Anal. Appl. **297**, 111–130 (2004)
12. Liu, Z., Ume, J.S., Kang, S.M.: Some existence theorems for functional equations and system of functional equations arising in dynamic programming. Taiwanese J. Math. **14**(4), 1517–1536 (2010)
13. Liu, Z., Xu, Y., Ume, J.S., Kang, S.M.: Solutions to two functional equations arising in dynamic programming. J. Comput. Appl. Math. **192**, 251–269 (2006)
14. Pathak, H.K., Deepmala: Some common fixed point theorems for occasionally weakly compatible maps with applications in dynamic programming. Rev. Bull. Cal. Math. Soc. **19**(2), 209–218 (2011)
15. Pathak, H.K., Deepmala: Existence and uniqueness of solutions of functional equations arising in dynamic programming. Appl. Math. Comp. **218**(13), 7221–7230 (2012)

CASca:A CA Based Scalable Stream Cipher

Shamit Ghosh and Dipanwita Roy Chowdhury

Abstract This paper presents a scalable stream cipher based on Cellular Automata. The cipher uses linear and nonlinear cellular automata as crypto primitives. The properties of maximum length nonlinear cellular automata have been exploited to design the cipher. Rotational symmetric bent function is used in the final combiner of the cipher which is proven to be secured against certain kind of fault attacks. The scalability provides different security level for different applications. Finally the cipher is shown to be very hardware efficient.

Keywords Cellular automata · Stream cipher · Pseudo random sequence generator · Scalable stream cipher

1 Introduction

Stream Cipher is an important branch in symmetric key cryptography. The goal of a stream cipher design is that it must provide high-speed encryption and less design overhead in comparison with block ciphers. The conventional stream ciphers used linear feedback shift registers (LFSR) for randomness and sufficiently large period. However, attempts are made to replace LFSR with linear CA to get excellent random sequences with a high speed of execution. Nonlinearity is another essential property for security, which is typically introduced by nonlinear feedback shift registers (NFSR) in stream cipher designs. The challenge of designing a crypto-system is, in addition to providing the required security, the crypto-system should be easy to implement in both hardware and software together with high performance and minimal resource usage. Stream ciphers have gained a lot of attention in the past few years.

S. Ghosh (✉) · D.R. Chowdhury
Department of Computer Science and Engineering, Indian Institute of Technology Karagpur, Kharagpur 721302, India
e-mail: shamit.ghosh@cse.iitkgp.ernet.in

D.R. Chowdhury
e-mail: drc@cse.iitkgp.ernet.in

R.N. Mohapatra et al. (eds.), *Mathematics and Computing*,
Springer Proceedings in Mathematics & Statistics 139,
DOI 10.1007/978-81-322-2452-5_7

The eSTREAM [4] project was launched in 2004 in search of a good stream cipher. The eSTREAM project has been instrumental for this attention. The eSTREAM portfolio ciphers fall into two profiles. Profile 1 contains stream ciphers more suitable for software applications with high throughput requirements. The winners in this profile are HC-128 [11], Rabbit [6], Salsa20/12 [5], SOSEMANUK [3]. Profile 2 stream ciphers are particularly suitable for hardware applications with restricted resources such as limited storage, gate count, or power consumption. Grain v1 [10], MICKEY 2.0 [1] and Trivium [7] are the winner in this category.

The basic philosophy of a stream cipher is to generate pseudo-random sequences from a secret key or a seed. There is an optional provision for an initial value (IV) which provides security for multiple encryptions using the same secret key. This fact is the motivation of designing fast pseudo-random sequence generators. The cellular automata (CA) provide very good pseudo-random sequences which exhibit excellent statistical properties. A necessary requirement for such a sequence generator is large period. In our design, both linear and nonlinear part are maximum length sequence generator. Moreover, scalability is an important aspect any cryptographic design. Due to advancement of computing speed, the current security standard may be obsolete after a few years. Only for a scalable design, the new security standard can be achieved by increasing the key size without discarding the whole algorithm.

Our Contribution: In this paper we have designed a new CA-based scalable stream cipher CASca. The design specification and design rationale of the cipher is portrayed. Crypto properties of CASca is shown in detail which proves its security against all existing attacks. The design is shown to be suitable for constrained hardware environment.

The remainder of the paper is organized as follows. Section 2 draws the idea about some notions, definitions, and basic studies on CA. The design of CASca is depicted in Sect. 3. Section 4 shows the scalability and initialization of the cipher. Section 5 gives a security analysis of CASca against some popular cryptanalysis techniques. Finally Sect. 6 concludes the work.

2 Preliminaries

In this section we discuss some basic notions required for security analysis. Some definitions and properties related to CA are also highlighted. Based on these theoretical studies, we further proceed to our proposed scheme.

2.1 Notions

Throughout the paper, we use '+' to represent Boolean XOR operation in **GF**(2). In this subsection, some basic security properties for evaluating a cryptographic primitive are given. The entire theoretical studies and analysis of our scheme is done based on these properties.

Definition 1 Hamming Weight: Number of Boolean 1's in a Boolean function's truth table is called the Hamming weight of the function.

Definition 2 Affine Function in $GF(2)$: A Boolean function which can be expressed as XOR of some or all of its input variables and a Boolean constant is an affine function.

In this paper the term *Affine Function* simply refers to Affine Function in $GF(2)$.

Definition 3 Nonlinearity: Let, f be a Boolean function of variables, $x_1, x_2, \cdots, x_n$ and A be the set of all affine functions in $x_1, x_2, \cdots, x_n$. The minimum of all the Hamming distances between f and the Boolean functions in A is the nonlinearity of f.

Definition 4 Algebraic Normal Form: Any Boolean function can be expressed as XOR of conjunctions and a Boolean constant, True or False. This form of the Boolean function is called its Algebraic Normal Form (ANF).

Definition 5 Correlation Immunity : A function $f : \mathbb{F}_2^n \rightarrow \mathbb{F}_2$ is kth order correlation immune if for any independent n binary random variables $X_0 \ldots X_{n-1}$, the random variable $Z = f(X_0, \ldots, X_{n-1})$ is independent of any random vector $(X_{i_1} \ldots X_{i_k})$ with $0 \le i_1 < \ldots < i_k < n$.

Definition 6 Resiliency : A function $f : \mathbb{F}_2^n \rightarrow \mathbb{F}_2$ is kth order resistant if it is balanced and correlation immune of order k.

2.2 CA Basics

Cellular automata are studied as mathematical model for self-organizing statistical systems. CA can be one-dimensional or multi-dimensional. In this paper, we discuss only one-dimensional two state CA. They can be considered as an array of cells where each cell is a one-bit memory element.

The neighbor set $\mathbf{N}(i)$ is defined as the set of cells on which the ith cell is dependent on each iteration. The simplest class of CA are *elementary CA* or *three-neighborhood CA* where each cell evolves in every time step based on some combinatorial logic on the cell itself and its two nearest neighbors. More formally, for a three-neighborhood CA, $\mathbf{N}(i) = \{i-1, i, i+1\}$. So, if the value of ith cell at tth time step is $q_i(t)$, then

$$q_i(t+1) = f(q_{i-1}(t), q_i(t), q_{i+1}(t))$$

where f denotes some combinatorial logic. We call the set of all feedback functions as ruleset and express as $\mathcal{F}$. The state transition of one iteration of a CA is expressed as $\mathcal{S}_{t+1} = \mathcal{F}(\mathcal{S}_t)$ where $\mathcal{S}_t$ is the set of all cells in the CA at tth time step.

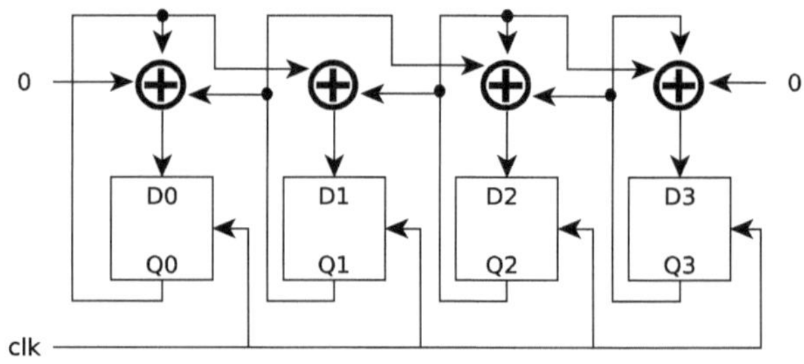

Fig. 1 4-cell 3-neighborhood null boundary LHCA with ruleset 1011

Since, a three-neighborhood CA having two states (0 or 1) can have $2^3 = 8$ possible binary states, there are a total of $2^{2^3} = 256$ possible rules. Each rule can be represented as an decimal integer from 0 to 255. If the combinatorial logic for the rules have only Boolean XOR operation, then it is called *linear* or *additive* rule. Some of the three-neighborhood additive CA rules are 0, 60, 90, 102, 150 etc. Moreover, if the combinatorial logic contains AND/OR operations, then it is called *nonlinear* rule.

An n cell CA with cells $\{x_0, x_1, \cdots, x_{n-1}\}$is called *null boundary* CA if $x_n = 0$ and $x_{-1} = 0$. Similarly, for a *periodic boundary* CA $x_n = x_0$.

A CA is called *uniform*, if all its cells follow the same rule. Otherwise, it is called *nonuniform* or *hybrid* CA. For a hybrid CA, the sequence of the rules followed by the cells in a particular order (MSB to LSB or vise versa). If all the ruleset of a hybrid CA is linear, then we call the CA linear hybrid cellular automata (LHCA), otherwise it is called nonlinear hybrid cellular automata (NHCA). In Fig. 1, a four-cell null boundary LHCA is shown.

The *shifting operation* [9] on an NHCA is defined as follows.

Definition 7 The one-cell shifting operation, denoted by $f_i \xrightarrow{P} f_{i\pm1}$ moves a set of ANF monomials P from ith cell of an NHCA to all the cells from $(i-1)$ to $(i+1)$-th cell, according to the dependency of the affected cells upon the ith cell. Each variable in P is changed by its previous state. Similarly, a k cell shifting is obtained by applying the one-cell shifting operation for k times upon the initial NHCA and symbolized as $f_i \xrightarrow{P} f_{i\pm k}$.

For example, we have a 5-bit 3 neighborhood CA with the following initial ruleset:

$$
\begin{aligned}
f_0 &= (x_1 \oplus x_0)\\
f_1 &= (x_2 \oplus x_1 \oplus x_0)\\
f_2 &= (x_3 \oplus x_2 \oplus x_1) \oplus x_4\\
f_3 &= (x_4 \oplus x_3 \oplus x_2)\\
f_4 &= x_3
\end{aligned}
$$

It is clear from the equations that x_3 is the previous state of x_4. Now applying the shifting $f_2 \xrightarrow{x_4} f_{2\pm 1}$, the new ruleset becomes:

$$
\begin{aligned}
f_0 &= x_1 \oplus x_0 \\
f_1 &= (x_2 \oplus x_1 \oplus x_0) \oplus x_3 \\
f_2 &= (x_3 \oplus x_2 \oplus x_1) \oplus x_3 \\
f_3 &= (x_4 \oplus x_3 \oplus x_2) \oplus x_3 \\
f_4 &= x_3
\end{aligned}
$$

3 Design of Scalable a Stream Cipher

The design of the proposed stream cipher consists of three parts, a maximum length sequence generator, a nonlinear sequence generator, and a final combiner function. Their design rationale and construction are discussed below. The overall scheme is depicted in Fig. 2.

3.1 Maximum Length Sequence Generator

The necessity of maximum length sequence is to prevent low period attack. We designed the maximum length sequence using linear hybrid cellular automata(LHCA) as it is widely known that linear functions provide good diffusion properties. To synthesize a maximum length LHCA rule a primitive polynomial is needed. From that primitive polynomial, a ruleset is generated using the algorithm described in [8]. In our design, the polynomial $x^{128} + x^{29} + x^{27} + x^2 + 1$ is used. We call this LHCA as $\mathcal{L}$. The individual bits of $\mathcal{L}$ is denoted by s_i where $i \in \{0, 127\}$. It is trivial that $\mathcal{L}$ is a null boundary CA.

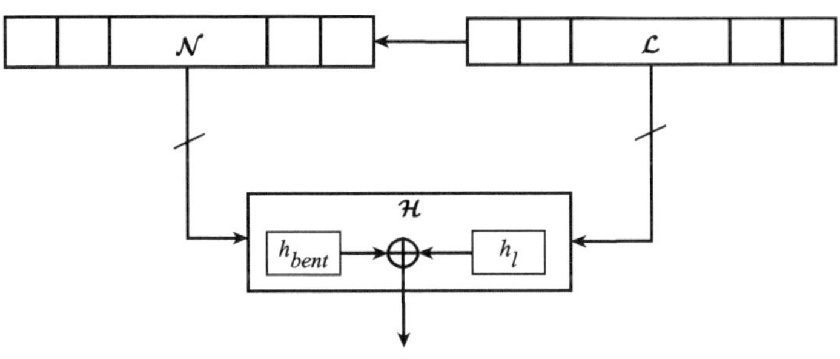

Fig. 2 Operations of cipher

3.2 Nonlinear Sequence Generator

Linear functions alone cannot provide cryptographic security as they can be easily cryptanalyzed. To introduce nonlinearity, a nonlinear sequence generator is needed. The design the nonlinear sequence generator should be in such a way that it has a long period. In this case, we use Algorithm 1 depicted in [9]. This algorithm takes a maximum length LHCA as input and injects required nonlinearity into some given positions of the CA while retaining the period of $2^n - 1$. In our design we synthesized the nonlinear sequence generator from an LHCA, the same as $\mathcal{L}$ and then injected nonlinearity at positions $\{20, 42, 79, 117\}$. For each position i, the nonlinear function $\mathbf{f_N}$, injected at ith position is, $(x_{i+2}.x_{i-2})$. We will call this nonlinear sequence generator as $\mathcal{N}$. The individual bits of $\mathcal{N}$ is denoted by b_i where $i \in \{0, 127\}$. The bit b_0 is bounded by null value, whereas b_{127} is bounded by s_0.

Algorithm [1]: NHCA Synthesize Algorithm

Input: A maximum length LHCA with ruleset $\mathcal{F}_L$, A position j to inject nonlinearity and the set of cells of the LHCA $\mathcal{S}$
Output: A maximum length NHCA ruleset $\mathcal{F}_N$

1: $\mathcal{F}_N \leftarrow \mathcal{F}_L$
2: Let $\mathcal{F}_N = \{f_{n-1}, \cdots, f_0\}$
3: $\mathcal{X} \subset \mathcal{S} : \forall x \in \mathcal{X}, x \notin \mathbf{N}(j)$ ⊳ select a subset from $\mathcal{S}$
4: $P \leftarrow \mathbf{f_N}(\mathcal{X})$ ⊳ $\mathbf{f_N}$ is non-linear function
5: $f_j \leftarrow f_j \oplus P$
6: $(f_j \xrightarrow{P} f_{j+1})$ ⊳ Apply shifting operation
7: $f_j \leftarrow f_j \oplus P$
8: **return** $\mathcal{F}_N$

3.3 Final Combiner Function

Some suitable tap bit positions are chosen from both $\mathcal{L}$ and $\mathcal{N}$. We call the set of these tap bit positions as $\mathcal{T}$. The final combiner function $\mathcal{H}$ is defined as $\mathcal{H} : 0, 1^{\{|\mathcal{T}|\}} \rightarrow \{0, 1\}$ where the input to $\mathcal{H}$ is $\mathcal{T}$. The construction of $\mathcal{H}$ has two primary parts, a bent function h_{bent} and a linear part h_l. The bent function provides high nonlinearity,[1] whereas the linear part increases the correlation immunity and resiliency of $\mathcal{H}$. The Boolean xor of l and b generates the required value of $\mathcal{H}$. The function b is defined as

[1]The nonlinearity of a bent function is the highest possible value among all Boolean functions of the same number of variables.

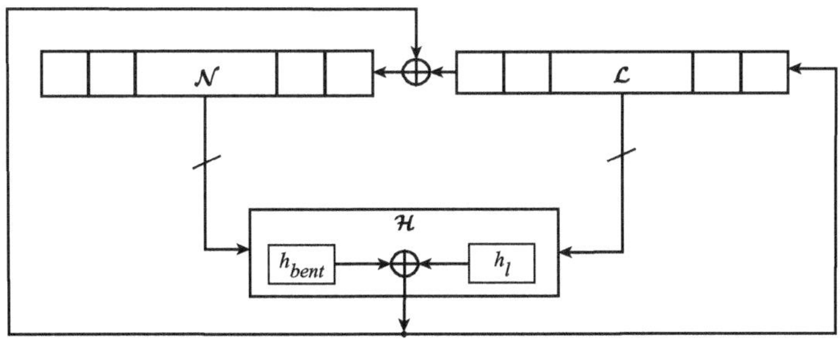

Fig. 3 Initialization of cipher

$$h_{bent}(x) = x_0x_2 + x_0x_6 + x_1x_3 + x_1x_7 + x_2x_4 + x_3x_5 + x_4x_6 + x_5x_7 + x_0x_2x_5 + x_0x_3x_5 + x_0x_3x_6 + x_1x_3x_6 + x_1x_4x_6 + x_1x_4x_7 + x_2x_4x_7 + x_2x_5x_7$$

where $x = \{b_{17}, s_{12}, s_{35}, s_{58}, s_{78}, s_{97}, s_{117}, b_{97}\}$. Similarly, l is defined as

$$h_l = \sum_{k \in A} b_k$$

where $A = \{21, 43, 80, 118\}$.

4 Scalability and Key Initialization

Before generating any keystream, the cipher is initialized with a key k and initial vector IV . The number of bits in IV is 96 and we denote the bits of IV as IV_i, $0 \le i \le 95$. The size of k is variable and can vary from 80 and 128. The size is chosen by the user according to the security parameter. Let n be the size of the key for a particular scheme where i^{th} bit of the key is denoted as k_i, $0 \le i \le n$. The 96 LSB bits of $\mathcal{L}$ is initialized with IV, $s_i = IV_i, 0 \le i \le 95$. The rest of the bits are set at 1. This ensures that $\mathcal{L}$ cannot be initialized with all in case of a chosen IV attack. Similarly, the first n bits of $\mathcal{N}$ is filled with k, $b_i = k_i, 0 \le i \le n$ and the remaining bits (if any) are set to 1. Next the cipher is clocked for 128 cycles without producing any keystream and the keystream is XORed with both the MSB of $\mathcal{L}$ and $\mathcal{N}$ as shown in Fig. 3.

5 Security

The principal design criteria of a stream cipher is to be secured against all existing cryptanalysis techniques. The best possible algorithm to recover the secret key is to be no less than exhaustive search of the key space. In this section we discuss some possible attacks and the corresponding design criteria in our cipher against them.

(a)

Itn.	b_{21}	b_{43}	b_{80}	b_{118}
1	4	2	4	2
2	16	0	16	0
3	0	4	0	16
4	0	0	32	64
5	256	0	128	128
6	0	256	64	128
7	256	0	256	0
8	4096	512	512	512
9	1024	1024	0	0
10	2048	2048	4096	2048

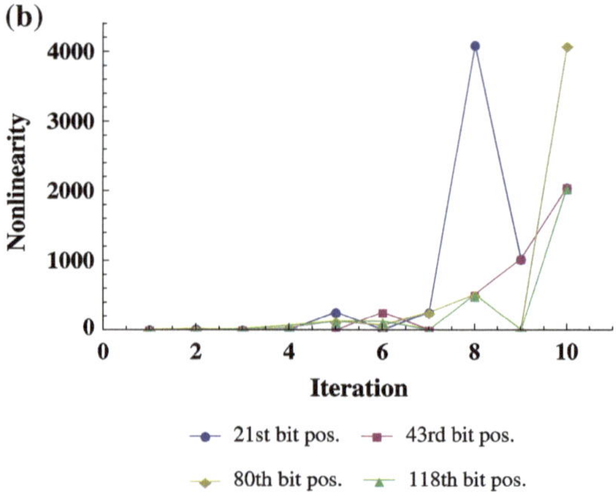

Fig. 4 Nonlinearity of the tap bits. **a** Nonlineartiy with iteration. **b** Nonlineartiy graph

5.1 Linear Cryptanalysis

Linear cryptanalysis tries to formulate a linear approximation of the cipher. High nonlinear value of the cipher protects the cipher against this attack. Table 4a shows the nonlinear values of the tap points with each iteration. After a few initial rounds the nonlinearity reaches a high value. So the cipher is expected to be secured against linear cryptanalysis techniques (Fig. 4).

5.2 Algebraic Attacks

Algebraic cryptanalysis techniques are very efficient in terms of finding loopholes in the design. Weak or careless design principal can cause such kinds of attacks. In our design, the $\mathcal{H}$ function provides a boolean function of degree three. With each iteration this degree increases if the output bit is expressed as a function of only the initial state bits. Hence, solving algebraic equations to cryptanalyze the system is computationally infeasible.

5.3 Correlation Attacks

Correlation immunity is an important aspect of designing stream ciphers which prevents the chosen *IV* attacks. The idea of the attack is to find any correlation between the *IV* and the output stream. Unbalanced output helps an adversary to find a correlation. Using only bent functions in $\mathcal{H}$ may cause vulnerability for finding correlation. Thus a linear function h_l is needed. The $\mathcal{H}$ function has correlation immunity 3

Table 1 Results of NIST statistical test suite

Test name	Status
Frequency (Monobit) test	Pass
Frequency test within a block	Pass
Runs test	Pass
Discrete fourier transform (Spectral) test	Pass
Non-overlapping template matching test	Pass
Overlapping template matching test	Pass
Serial test	Pass
Approximate entropy test	Pass
Cumulative sums (Cusum) test	Pass

and nonlinearity 1664. It is a balanced function, so it is also a 3 resilient function. If the input bits of $\mathcal{H}$ are represented as functions of the initial state bits, then with each iteration the nonlinearity as well as resiliency increases very fast. So correlation attack against CASca will not be faster than a brute force attack.

5.4 Statistical Analysis

Statistical analysis of the cipher is carried out using NIST Randomness Test Suite and Table 1 summarizes the result. The tests are performed by taking 10,000 bit keystream from a fixed 128 bit key and IV pair.

5.5 Fault Attacks

Fault attacks are the most powerful and popular cryptanalysis techniques. The easier way to inject fault into the cryptographic devices makes it highly feasible. Designing a cryptographic scheme that is fault attack resistant is a challenging task for researchers. Initially, an attacker injects single or multibit fault into the state of the cipher. The output difference of the fault-free and faulty ciphertext leaks some information about the state of the cipher. This leakage is indicated with some equation. This method is repeated multiple times until a probabilistic polynomial time algorithm can recover the secret key (or the state of the cipher). The fault locations are chosen by the attacker in such a fashion that the set of equation can be solved easily. In our design, the state of the cipher is implemented as CA. The high diffusion property of CA infects the state with the fault within a very few iteration. Hence, the algebraic degree and the number of unknown variables in the set of equations becomes so high that it is a hard problem to solve. Moreover, a fault attack based on the decomposition of the final combiner function in Grain v1 is discussed in [2]. This attack was possible as the combiner function can be decomposed into a form of $s.u + v$ where u is a function of only the nonlinear state bits and v is a function of the linear state bits. In our design, the h_{bent} function is a rotational symmetric bent function. This function hardly reveals any information about the fault positions for their symmetric property and cannot be decomposed like the above-mentioned attack. Thus the design is expected to be robust against fault attacks.

6 Conclusion

A new stream cipher CASca is proposed in this paper. The choice of design rationale of the cipher considers the existing cryptanalysis techniques. The size of IV is 96 bits but the key length has been kept as a variable one. The reason behind this is to provide scalability for the applications where the security parameter can vary. The design of CASca is very hardware efficient. It suitable for applications where low power consumption and low area overhead are required, such as mobile devices.

References

1. Babbage, S., Dodd, M.: The mickey stream ciphers. In: Robshaw, M., Billet O. (eds.) New Stream Cipher Designs, Lecture Notes in Computer Science, vol. 4986, pp. 191–209. Springer, Heidelberg (2008). doi:10.1007/978-3-540-68351-3_15. http://dx.doi.org/10.1007/978-3-540-68351-3_15
2. Banik, S., Maitra, S., Sarkar, S.: A differential fault attack on the grain family of stream ciphers. In: Prouff, E., Schaumont P. (eds.) Cryptographic Hardware and Embedded Systems - CHES 2012, Lecture Notes in Computer Science, vol. 7428, pp. 122–139. Springer, Heidelberg (2012). doi:10.1007/978-3-642-33027-8_8. http://dx.doi.org/10.1007/978-3-642-33027-8_8
3. Berbain, C., Billet, O., Canteaut, A., Courtois, N., Gilbert, H., Goubin, L., Gouget, A., Granboulan, L., Lauradoux, C., Minier, M., Pornin, T., Sibert, H.: Sosemanuk, a fast software-oriented stream cipher. In: Robshaw, M.J.B., Billet, O. (eds.) The eSTREAM Finalists, Lecture Notes in Computer Science, vol. 4986, pp. 98–118. Springer, Berlin (2008). http://dblp.uni-trier.de/db/series/lncs/lncs4986.html#BerbainBCCGGGLMPS08
4. Bernstein, D.J.: Notes on the ecrypt stream cipher project (estream). http://cr.yp.to/streamciphers.html
5. Bernstein, D.J.: New stream cipher designs. In: The Salsa20 Family of Stream Ciphers, pp. 84–97. Springer, Heidelberg (2008). doi:10.1007/978-3-540-68351-3_8. http://dx.doi.org/10.1007/978-3-540-68351-3_8
6. Boesgaard, M., Vesterager, M., Pedersen, T., Christiansen, J., Scavenius, O.: Rabbit: a new high-performance stream cipher. In: Fast Software Encryption, 10th International Workshop, FSE 2003, Lund, Sweden, 24–26 Feb 2003, Revised Papers, Lecture Notes in Computer Science, vol. 2887, pp. 307–329. Springer, Heidelberg (2003). doi:10.1007/978-3-540-39887-5_23. http://www.iacr.org/cryptodb/archive/2003/FSE/3049/3049.pdf
7. Cannire, C.: Trivium: a stream cipher construction inspired by block cipher design principles. In: Katsikas, S., Lpez, J., Backes, M., Gritzalis, S., Preneel, B. (eds.) Information Security, Lecture Notes in Computer Science, vol. 4176, pp. 171–186. Springer, Heidelberg (2006). doi:10.1007/11836810_13. http://dx.doi.org/10.1007/11836810_13

8. Cattell, K., Muzio, J.C.: Synthesis of one-dimensional linear hybrid cellular automata. IEEE Trans. Comput. Aided Des. Integr. Circuits Syst. **15**(3), 325–335 (1996). doi:10.1109/43.489103
9. Ghosh, S., Sengupta, A., Saha, D., Chowdhury, D.R.: A scalable method for constructing non-linear cellular automata with period $2^n - 1$. In: Cellular Automata—11th International Conference on Cellular Automata for Research and Industry, ACRI 2014, Krakow, Poland, 22–25 Sept 2014. Proceedings, pp. 65–74 (2014). doi:10.1007/978-3-319-11520-7_8. http://dx.doi.org/10.1007/978-3-319-11520-7_8
10. Hell, M., Johansson, T., Meier, W.: Grain: a stream cipher for constrained environments. Int. J. Wire. Mob. Comput. **2**(1), 86–93 (2007). doi:10.1504/IJWMC.2007.013798. http://dx.doi.org/10.1504/IJWMC.2007.013798
11. Wu, H.: The stream cipher hc-128. In: Robshaw, M.J.B., Billet, O. (eds.) The eSTREAM Finalists, Lecture Notes in Computer Science, vol. 4986, pp. 39–47. Springer, Heidelberg (2008). http://dblp.uni-trier.de/db/series/lncs/lncs4986.html#Wu08

Improved Cryptographic Puzzle Based on Modular Exponentiation

Lakshmi Kuppusamy and Jothi Rangasamy

Abstract Cryptographic puzzles are moderately hard–neither easy nor hard to solve—computational problems. They have been identified to be useful in mitigating a type of resource exhaustion attacks on Internet protocols. Puzzles based on modular exponentiation are interesting as they possess some desirable properties such as deterministic solving time, sequential (non-parallelizable) solving process and linear granularity. We propose a cryptographic puzzle based on modular exponentiation. Our puzzle is as efficient as the state-of-art puzzle of its kind and also overcomes the major limitation of the previous schemes.

Keywords Cryptographic puzzle · Proof-of-work · Denial-f-service protection · Unforgeability · Difficulty

1 Introduction

A cryptographic puzzle is a *moderately difficult* computational problem in which a prover(client) must demonstrate to a puzzle generator (verifier) that it has performed the required computational task. Cryptographic puzzles were first introduced as *proof-of-work* systems by Dwork and Naor [8] in 1992 for combating junk emails [8]. Rivest et al. (1996) used puzzles for realizing *time-release cryptography* [19]. Juels and Brainard (1999) considered puzzles to mitigate Denial-of-Service (DoS) attacks in network protocols. In server-client scenario, they are known as *client puzzle* protocols.

DoS attack is one of the most common real-world network security attacks and presents a severe threat to the Internet and e-commerce. In DoS attack, the attacker

L. Kuppusamy · J. Rangasamy (✉)
Society For Electronic Transactions and Security, MGR Knowledge City,
CIT Campus, Taramani, Chennai 600113, Tamilnadu, India
e-mail: jothi.rangasamy@gmail.com

L. Kuppusamy
e-mail: lakshdev21@gmail.com

R.N. Mohapatra et al. (eds.), *Mathematics and Computing*,
Springer Proceedings in Mathematics & Statistics 139,
DOI 10.1007/978-81-322-2452-5_8

targets to drain out the service provider's resources such as bandwidth, memory, and computational time so that the resources will become unavailable to process legitimate clients' requests. In recent years, major e-commerce sites including eBay, Yahoo!, Amazon, and Microsoft's name server [17] have faced huge financial loss due to DoS attacks. Very recently, a DoS attack on several thousands of *time keeping servers* distributed across the world to keep the time in sync by running the network time protocol (NTP) has been mounted. Two vulnerabilities in the NTP were exploited by the attackers to mount the DoS attack using IP spoofing technique. This attack has been described as the world's largest DoS attack to date by security researchers due to its amplification factor of 206x.

Cryptographic puzzles have been shown to be a promising and effective mechanism to deter the effect of malicious requests. When the server is under DoS attack, it generates a (client) puzzle instance and sends it as a reponse to the client's connection request. The server processes the client's request only if the client proves its legitimate intentions of getting the request by sending the correct puzzle solution. Generating and verifying a client puzzle must be computationally easy for the server. That is, it must add a little computational and memory overhead to the server. Otherwise, the client puzzle may introduce a resource exhaustion attack where an attacker triggers puzzle generation and verification process by sending a large number of pretended requests or a large number of fake puzzle solutions respectively.

On the other hand, finding a correct solution to the client puzzle must be moderately hard for the client. This property is called *puzzle difficulty* which is a property that every good puzzle must satisfy. That is, for a legitimate client, the computational burden for solving a client puzzle is not high, whereas for an attacker who makes multiple connection requests, finding solution for many client puzzles received through multiple requests must be a huge resource-consuming process.

1.1 Modular Exponentiation-Based Puzzle

Client puzzles are mostly either hash based [3, 9, 11] or modular exponentiation [12, 19] based puzzles. Though it is essential that all the client puzzles must satisfy the puzzle difficulty property, exponentiation-based client puzzles are known to achieve additional properties such as non-parallelizability, deterministic solving time, and finer granularity. In a non-parallelizable client puzzle, the solution finding time remains constant even if the attacker/client uses multiple machines to solve a single client puzzle. Unlike in the hash-based puzzles where the running time to find a puzzle solution is probabilistic, the modular exponentiation-based puzzles have the property that the minimum amount of work required to solve a puzzle can be determined. Moreover, these puzzles support linear granularity; the puzzle generator (server) has the ability to increase the puzzle difficulty level linearly. This property is useful since the puzzle issuing server will have more options for the difficulty level and can choose one accordingly.

Rivest et al. [19] gave the first modular exponentiation-based puzzle which achieves non-parallelizability, deterministic solving time, and finer granularity. A problem with Rivest et al. puzzle construction is that the server has to perform modular exponentiation in order to verify the puzzle solution. Karame and Čapkun [12] proposed two puzzle constructions. First, one works for the fixed-difficulty level and reduces the running time of the puzzle verification by a factor of $\frac{|n|}{2k}$ for a given RSA modulus n, where k is the security parameter compared to Rivest et al.'s puzzle. In a puzzle with fixed difficulty, the busy server cannot adjust the difficulty levels of the puzzle based on its load. The second scheme of Karame and Čapkun supports various difficulty levels but it doubles the verification cost of their first scheme. Though Karame and Čapkun's puzzle is superior in efficiency compared to Rivestet al. puzzle, it still requires modular exponentiation for puzzle verification. To avoid mdoular exponentiation in the Karame-Čapkun puzzle verification, an alternative construction, namely RSApuz was proposed in [18], wherein the verification requires only few modular multiplications. However, the approach in [18] works only for the fixed difficulty level. In real-world attacks such as denial-of-service attacks, the target server is kept very busy in performing varoius computational tasks. Thus puzzles can be an effective countermeasure to DoS attacks when they support variable difficulty levels and avoid modular exponentiation cost on their side. The state-of-art puzzles, namely [12] and [18] fail to meet at least one of the above desirable properties as seen in Table 1.

1.2 Contributions

1. We give an efficient modular exponentiation-based puzzle which achieves non-parallelizability, deterministic solving time, and finer granularity. Our puzzle is superior in efficiency to Karame and Čapkun's variable puzzle difficulty level puzzle. Though our scheme is similar to [12] and [18], our puzzle does not involve any modular exponentiation during puzzle verification unlike [12] and does not require to repeat the pre-computation procedure to change the puzzle difficulty level unlike [18]. Our construction requires only a few modular multiplications

Table 1 Comparision of modular exponentiation-based puzzles

Puzzle	Difficulty level	Verification
RSWpuz [19]	variable	$\|n\|$-bit mod. exp.
KCpuz [12]	fixed	k-bit mod. exp.
KCpuz [12]	variable	$2k$-bit mod. exp.
RSApuz [18]	fixed	3 mod. mul.
Ours	variable	3 mod. mul.

Legend n is an RSA modulus, $k \ll n$ is a security parameter to verify puzzle solutions. Table 1 compares our puzzle with other puzzles of the same kind.

2. We show that our puzzle is unforgeable and difficult in the puzzle security model proposed by Chen et al. [7]

Outline: We organize the rest of the paper as follows: Sect. 2 breifly presents the related work. Design and security analysis of our puzzle construction is described in Sect. 3. Finally Sect. 4 concludes our work.

2 Background on Modular Exponentiation-Based Puzzles

This section discusses the state-of-art puzzle schemes and identifies their limitations. Throughout the paper we use the following notations: Let n be an integer and $|n|$ be the length of the interger in bits; let $\phi(n)$ be the Euler phi function of n; the set of all integers $\{a, \dots, b\}$ between and including a and b be denoted by $[a, b]$; denote by $x \leftarrow_r S$ to choose an element x uniformly at random from s set S; for an algorithm A to run on input y and produce an output x, we denote it by $x \leftarrow A(y)$; let $\mathrm{negl}(k)$ denote a function which is neglible in k, where k is a security parameter; We denote p.p.t for a probabilistic polynomial time algorithm.

2.1 *RSWpuz*

Rivest et al. [19] proposed a puzzle scheme based on repeated squarings, which we call RSWPuz. In their puzzle construction, the puzzle generating server first chooses an RSA modulus $n = pq$ using two large primes p and q and then computes the Euler totient function $\phi(n) = (p-1)\cdot(q-1)$. Now the server sends a tuple (a, Q, n) as a puzzle instance to the client after selecting the difficulty level Q and an integer $a \leftarrow_r \mathbb{Z}_n^*$. Observe that the difficulty level determines the amount of work a client has to do. Now, the client performs Q repeated squarings to compute $b \leftarrow a^{2^Q} \bmod n$ and returns b to the server as a puzzle solution. After receivng the puzzle solution, the server checks whether $a^c \stackrel{?}{\equiv} b \bmod n$ where $c = 2^Q \bmod \phi(n)$. The server can reuse the computation of c as long as the puzzle difficutly value Q is fixed. Since the server knows the trapdoor information $\phi(n)$ the server can verify the solution in one $|n|$-bit exponentiation, whereas the client is forced to do Q repeated squarings for $Q \gg |n|$.

Note that the puzzle verification step is expensive in RSWPuz scheme as it involves the computation of full $|n|$-bit modular exponentiation on the server side. A malicious client can exploit this weakness to send a large number of fake puzzle solutions. The busy server now needs to engage in computationally expensive operation to verify all of them. Hence, the client puzzle construction itself introduces a new vulnerability to a resource exhaustion-based DoS attack.

2.2 *KCpuz*

Rivest et al.'s puzzle construction was improved by Karame and Čapkun [12]. The improvement in terms of computational efficiency is the significant reduction of puzzle verification cost from $|n|$-bit exponentiation (Rivest et al. puzzle verification cost) to $2k$-bit exponentiation modulo n for a security parameter k. That is, the burden for the server is reduced by a factor of $\frac{|n|}{2k}$. Their scheme with variable difficulty level, which we call KCPuz is illustrated in Fig. 1. Unlike [19], Karame and Čapkun analyzed their puzzle scheme under the security notions from [7] and showed that the puzzle satisfies both the unforgeability and difficulty notions.

Though KCPuz scheme requires less computation cost to verify each puzzle solution compared to RSWPuz, it still needs a $2k-$bit modular exponentiation. This could still be a burdensome computation for DoS defending servers. Also, KCPuz does not provide the property of finer granularity. That is, the gap between the two adjacent difficulty levels must be large for security reasons. In particular, the next difficulty level R' must satisfy $\frac{R'}{R} \geq n^2$ where R is the current difficulty level. This reduces the number of possible and acceptable difficulty levels to be chosen by the puzzle generator.

Client		**Server**
		Puzzle Setup 1. $(n, p, q) \leftarrow \mathsf{GenMod}(1^k)$ 2. $d \leftarrow_r [2^k, 2^{k+1}]$ such that $e = d^{-1} \bmod \phi(n)$ exists and $e \geq n^2$. 3. $s \leftarrow (e, d, \phi(n))$ 4. Select $R \geq e,\ \|R\| = Q$.
	request $\longrightarrow$	Puzzle Generation 1. $Z \leftarrow_r \{0,1\}^k$ 2. $X \leftarrow \mathrm{HMAC}(d, Z)$ 3. $v \leftarrow \mathrm{HMAC}(d, X), \|v\| \geq k, \gcd(v, d) = 1$ 4. $K \leftarrow e \cdot v - (R \bmod \phi(n))$.
Finding Solution 1. $Y_1 \leftarrow X^R \bmod n$ 2. $Y_2 \leftarrow X^K \bmod n$	$\longleftarrow$ puz	5. $puz \leftarrow (n, R, K, Z, X)$
	Y_1, Y_2, puz $\longrightarrow$	Puzzle Authentication 1. $X \leftarrow \mathrm{HMAC}(d, Z)$ 2. $v \leftarrow \mathrm{HMAC}(d, X)$ Solution verification 3. Verify $(Y_1 \cdot Y_2)^d \stackrel{?}{\equiv} X^v \bmod n$

Fig. 1 The KCPuz Scheme [12]

The difficulty level for client puzzles employed in DoS scenarios is typically set between 0 to 2^{25} operations. Hence, the possible successive difficulty levels for KCPuz scheme are $R = 2^{512}$, $R' = 2^{1536}$ and $R'' = 2^{2560}$ for 512-bit moduli.

2.3 RSApuz

An alternative and more efficient vewrsion version of KCPuz was proposed in [18], which we call RSAPuz. In RSAPuz the puzzle issuing server does the most computation work offline so that it does not perform any modular exponentiation online during puzzle generation and solution verification. In fact, the solution verification requires only three bit modular multiplications and thus its efficiency is comparable with that of hash function-based puzzles [18]. RSAPuz is shown to meet the security notions of Chen et al. and additional desirable properties such as finer granularity, non-parallelizability, and deterministic solving time. The RSAPuz scheme is depicted in Fig. 2.

RSAPuz uses the (BPV) technique due to Boyko et al. [5] which reduces the online computation cost for the pair (x, X). The technique has two phases: BPV pre-processing phase, namely BPVPre and the BPV pair generation phase BPVGen. The pre-processing phase computes N pairs of the form (α_i, β_i) where $\alpha_i \leftarrow_r \mathbb{Z}_n^*$ and $\beta_i \leftarrow {\alpha_i}^u \bmod n$ for $i = 1, \ldots, N$ and stores them in a table. Whenever a new pair (x, X) is required to be computed online, the pair generation phase *randomly* chooses ℓ out of N pairs and computes the new pair as follows: $(x, X) \leftarrow (\prod_{j=1}^{\ell} \alpha_j, \prod_{j=1}^{\ell} \beta_j)$. Thus RSAPuz does not perform any computationally intensive operation online.

Though the puzzle verification requires only few modular multiplications, it works only for the fixed difficulty level. For changing one difficulty level to the other, the puzzle scheme needs to run the computationally demanding pre-computation again. That is, in the pre-computation phase (as seen in Fig. 2), the server first selects the difficulty parameter R of length Q, computes $u \leftarrow d - (2^Q \bmod \phi(n))$ and then runs the BPV pre-processing step with inputs (u, n, N) to obtain N pairs (α_i, β_i). Hence the server has to run the pre-computation phase every time the difficulty needs to be changed.

All the modular exponentiation-based puzzles in the related literature add computational burden, either offline (e.g., precomputation in Fig. 2) or online (e.g., solution verification in Fig. 1), to support change of difficulty. In the following section, we overcome the limitations in the above puzzle schemes by proposing a new modular exponentiation-based client puzzle which is as fast as RSAPuz and adds no cost to support variable difficulty. We then analyze its security properties.

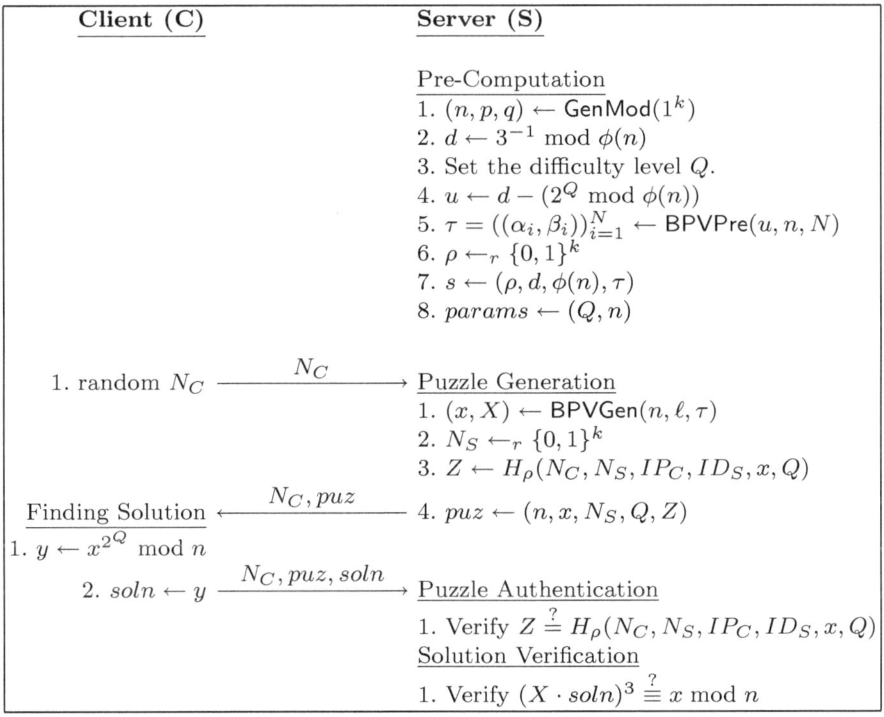

Fig. 2 The RSAPuz Scheme [18]

3 The Proposed Puzzle Scheme

Now we propose an efficient puzzle scheme requiring only few modular multiplications for puzzle generation and solution verification. Unlike in the existing puzzle schemes, our puzzle achieves both efficiency and the security properties such as unforgeability, puzzle difficulty, deterministic, non-parallelizability, and finer granularity. Our scheme uses all the algorithms used by RSAPuz in Fig. 2.

3.1 Definitions

We begin by defining an algorithm to generate a modulus $n = pq$ similar to the generation of RSA modulus as below:

Definition 1 (*Generating Modulus n*) For a security parameter k, the algorithm to generate a modulus n is a probabilistic polynomial time algorithm GenMod which accepts the input 1^k and produces (n, p, q) as output such that $n = pq$ where p and q are k-bit primes.

Like [18], our puzzle requires a server to generate a pair (x, X) for each puzzle which involves modular exponentiation. To avoid this exponentiation cost, we use the (BPV)technique proposed by Boyko et al. [5] which requires few modular multiplications and pre-computed values to generate the pairs of the form (x_i, X_i) where $X_i = x_i^u \bmod n$ for some predefined exponent u.

Definition 2 (*BPV Technique*) Suppose that $N \geq \ell \geq 1$ for the parameters N and ℓ. Let $n \leftarrow \mathsf{GenMod}(1^k)$ be an RSA modulus and u be an element in $\mathbb{Z}_{\phi(n)}$ of length m. The BPV technique has the following two phases:

- $\mathsf{BPVPre}(u, n, N)$: This pre-processing algorithm run once, generates N random integers $\alpha_1, \alpha_2, \ldots, \alpha_N \leftarrow_r \mathbb{Z}_n^*$ and computes $\beta_i \leftarrow {\alpha_i}^u \bmod n$ for each i. A table $\tau \leftarrow ((\alpha_i, \beta_i))_{i=1}^N$ consisting of pairs (α_i, β_i) is finally returned.
- $\mathsf{BPVGen}(n, \ell, \tau)$: Whenever a pair $(x, X \bmod n)$ is needed, the algorithm chooses a random set $S \subseteq_r \{1, \ldots, N\}$ of size ℓ and computes $x \leftarrow \prod_{j \in S} \alpha_j \bmod n$. If $x = 0$, then the algorithm stops and generates S again. Else, it computes $X \leftarrow \prod_{j \in S} \beta_j \bmod n$ and return (x, X). The indices S and the corresponding pairs $((\alpha_j, \beta_j))_{j \in S}$ are kept secret.

Security Analysis of BPV Technique. The results by Boyko and Goldwasser [4] and Shparlinski [20] show that the value x generated using the BPV technique are statistically close to the uniform distribution. In particular, the following theorem shows that with overwhelming probability on the choice of α_i's, the distribution of x is statistically close to the uniform distribution of a randomly chosen $x' \in \mathbb{Z}_n^*$.

Theorem 1 ([4], Chap. 2) *If* $\alpha_1, \ldots, \alpha_N$ *are chosen independently and uniformly from* $\mathbb{Z}_n^*$ *and if* $x = \prod_{j \in S} \alpha_j \bmod n$ *is computed from a random set* $S \subseteq \{1, \ldots N\}$ *of* ℓ *elements, then the statistical distance between the computed* x *and a randomly chosen* $x' \in \mathbb{Z}_n^*$ *is bounded by* $2^{-\frac{1}{2}\left(\log \binom{N}{\ell}+1\right)}$. *That is,*

$$\left| \Pr\left(\prod_{j \in S} \alpha_j = x \bmod n \right) - \frac{1}{\phi(n)} \right| \leq 2^{-\frac{1}{2}\left(\log \binom{N}{\ell}+1\right)} .$$

BPV Metrics. For defending against DoS attacks, the difficulty level Q can be set between 0 and 2^{25} operations. In [18] it is recommended to select N and ℓ such that $\binom{N}{\ell} > 2^{40}$. Instead of choosing $N = 512$ and $\ell = 6$ for the BPV generator as per Boyko et al. [4, 5], we can choose $N = 2500$ and $\ell = 4$ so as to reduce the number of online modular multiplications performed during the BPV pair generation process. We refer to [18] for more details about the choices of N and ℓ in a DoS scenario.

Making BPV pair generation process offline. At ESORICS 2014, Wang et al. [23] proposed that the BPV pair generation process can be executed offline. That is, the pairs generated during the BPV pre process $\mathsf{BPVPre}(u, n, N)$ are stored in the static table (ST) and the pairs generated during the BPV pair generation process $\mathsf{BPVGen}(n, \ell, \tau)$ are stored in the dynamic table (DT). Whenever a BPV pair is required during puzzle generation, an entry from DT is selected and the table is updated with another BPV pair in idle time. The use of dynamic table allows us to completely avoid the number of modular multiplication computations required for the BPV pair generation process and thus make our puzzle more efficient.

3.2 The Construction

Our client puzzle illustrated in Fig. 3 is executed as a series of message exchanges between a client and a DoS defending server. The server generates a puzzle instance using BPV pairs computed offline and verifies the puzzle solution sent by the client as follows:

Client (C)		Server (S)
	Off-line Phase	
		$\mathsf{Setup}(1^k)$ 1. $(n, p, q) \leftarrow \mathsf{GenMod}(1^k)$ 2. $d \leftarrow 3^{-1} \bmod \phi(n)$ 3. Select u such that $\|u\| \geq k$. 4. $\tau = ((\alpha_i, \beta_i))_{i=1}^{N} \leftarrow \mathsf{BPVPre}(u, n, N)$ 5. Set the difficulty level Q. 6. $K \leftarrow d - (\frac{2^Q}{u} \bmod \phi(n))$ 7. $\rho \leftarrow_r \{0,1\}^k$ 8. $s \leftarrow (\rho, d, \phi(n), \tau)$ 9. $params \leftarrow (Q, n)$
	On-line Phase	
1. random N_C	$\xrightarrow{N_C}$	$\mathsf{GenPuz}(s, Q, N_C)$ 1. $(x, X) \leftarrow \mathsf{BPVGen}(n, \ell, \tau)$ 2. $N_S \leftarrow_r \{0,1\}^k$ 3. $Z \leftarrow H_\rho(N_C, N_S, IP_C, ID_S, x, X, Q)$
$\mathsf{FindSoln}(puz, t)$	$\xleftarrow{puz}$	4. $puz \leftarrow (n, x, X, N_S, Q, Z)$
1. $y \leftarrow x^{2^Q} \bmod n$ 2. $Y \leftarrow X^K \bmod n$		
3. $soln \leftarrow (y, Y)$	$\xrightarrow{N_C, puz, soln}$	$\mathsf{VerAuth}(s, puz)$ 1. Verify $Z \stackrel{?}{=} H_\rho(N_C, N_S, IP_C, ID_S, x, X, Q)$ $\mathsf{VerSoln}(s, puz, soln)$ 1. Verify $(y \cdot Y)^3 \stackrel{?}{\equiv} X \bmod n$

Fig. 3 Our modular exponentiation-based puzzle with variable difficulty

- PRE- COMPUTATION. In the pre-computation phase, the server generates (n, p, Q) using the modulus generation algorithm and generates d as the inverse of 3 mod $\phi(n)$. Then the server runs the BPVPre phase by selecting N and u such that $|u| \geq k$. Unlike [18], our algorithm requires to run the BPVPre phase (to obtain N pairs of (α_i, β_i)) only once as it does not depend on the difficulty level Q.

For a client puzzle to be effective against resource exhaustion-based DoS attacks, generation of puzzles and verification of their solutions should be very efficient for the busy server as can be seen in our scheme described below:

- PUZZLE GENERATION (GenPuz). The server has to spend its significant computational resource for generating the puzzle through BPV pair generation BPVGen process which requires $2(\ell - 1)$ modular multiplications. That is, it requires $\ell - 1$ modular multiplications to compute x and another $\ell - 1$ modular multiplications to compute X. The server runs pseudo-random function H_ρ to compute the puzzle-authentication tag Z after generating a nonce N_s at random. Note that ℓ could be set between 4 and 16 so that the puzzle requires only 8 modular multiplications for $\ell = 4$ [4, 18].
- PUZZLE AUTHENTICITY VERIFICATION (VerAuth). Verifying that the puzzle is originated from the server can be done using the pseudo-random function H_ρ again and comparing the result with the received Z.
- PUZZLE VERIFICATION (VerSoln). The puzzle solution is verified by performing only 3 modular multiplications.

Observe that our puzzle scheme does not require the server perform any modular exponentiation either to generate the puzzle or to verify its solution. On the other hand, the client has to perform modular exponentiations to find the solution to the puzzle as follows:

- PUZZLE SOLUTION (FindSoln). After receiving the puzzle from the server, the client computes the puzzle solution in the form of two modular exponentiations $x^{2^Q} \bmod n$ and $x^K \bmod n$. The client can opt either to factor n or to perform repeated squarings to solve the puzzle. Since factoring is hard, the best knowm method for the client to find the solution is to implement the square and multiply algorithm and perform repeated squarings, which is believed to be a highly sequential process [10, 12, 19]. Hence the client will be performing exactly Q sequential modular multiplications to find $x^{2^Q} \bmod n$ and $O(\log K)$ sequential modular multiplications to find $x^K \bmod n$, and hence the puzzle has deterministic solving time of Q repeated squarings and non-parallelizability properties.

3.3 Security Analysis

Client puzzles were analyzed in various security models proposed in [6, 7, 21]. In this section, we analyze our puzzle scheme using difficulty notions such as unforgeability and difficulty proposed by Chen et al. [7] and prove that our puzzle is unforgeable and difficult. We refer [7] for a more formal treatment of these security difficulty notions.

Unforgeability. In our puzzle scheme, we use a pseudo-random function H_ρ in puzzle generation to generate Z. Thus showing that our puzzle meets the unforgeability notion is striaghforward following the same argument in [18] if H_ρ is a secure pseudo-random function. Hence we omit the unforgeability proof due to space constraints.

Difficulty. For proving the difficulty of our puzzle we again adapt the approach from [18] and show that our puzzle satisfies the difficulty notion of Chen et al. model as long as the KCPuz in Sect. 2 is difficult. In particular we relate the security of our puzzle to that of KCPuz with $R = 2^Q$ in Fig. 1. Note that our puzzle can be seen as the result of applying the precomputation approach in RSAPuz to KCPuz. The difficulty of our puzzle is proved in the following theorem.

Theorem 2 *Assume that k is a security parameter and Q is a difficulty parameter. If* KCPuz *with a modulus generation algorithm* GenMod *is $\epsilon_{k,Q}(t)$-difficult, then our puzzle, say puz, from Fig. 3 is $\epsilon'_{k,Q}(t)$-difficult for all probabilistic polynomial time $\mathcal{A}$ running in time at most t, where*

$$\epsilon'_{k,Q}(t) = 2 \cdot \epsilon_{k,Q}\left(t + (q_C + 1)\left(2(\ell - 1)T_{\mathrm{Mul}}\right) + c\right).$$

Here, q_C is the total number of CreatePuzSoln *queries issued in the experiment and T_{Mul} is the time complexity for computing a multiplication modulo n, and c is a constant.*

Proof We prove the theorem using the game hopping technique. Assume that $\mathcal{A}$ is a probabilistic algorithm running in time t and wins the puzzle difficulty experiment of puz. Using $\mathcal{A}$, we construct an algorithm $\mathcal{B}$ that solves KCPuz easily. Let the event E_i be such that $\mathcal{A}$ wins in game G_i.

Game G_0. Let G_0 be the original difficulty game $\mathsf{Exp}^{\mathsf{Diff}}_{\mathcal{A},puz}(k)$ defined as follows:

1. The challenger runs the Setup algorithm to generate $s \leftarrow (\rho, d, \phi(n), (\alpha_i, \beta_i)_{i=1}^N)$ and $params \leftarrow (Q, n)$. The challenger submits the parameters $params$ to $\mathcal{A}$ and keeps s.
2. Now, the challenger answers the CreatePuzSoln(N_C) query issued by $\mathcal{A}$ as follows:
   - The challenger runs the BPV pair generator BPVGen to obtain a pair (x, X) and computes Z, y and Y as per the protocol in Fig. 3.
   - The challenger submits $(puz, soln) \leftarrow ((N_S, Z, x, X), (y, Y))$ to $\mathcal{A}$.
3. At some time during the game, $\mathcal{A}$ is allowed to issue the Test(N_C^*) query to the challenger. The challenger answers the query with puz^* by generating a puzzle $puz^* = (N_S^*, Z^*, x^*, X^*)$ using GenPuz(s, Q, N_{C^*}) algorithm. The $\mathcal{A}$ may

continue to ask $\mathsf{CreatePuzSoln}(N_C)$ queries even after issuing the test query $\mathsf{Test}(N_C^*)$.

4. $\mathcal{A}$ outputs a valid solution $soln^* = (y^*, Y^*)$.
5. The challenger outputs 1 if $\mathsf{VerSoln}(puz^*, soln^*) = \mathsf{true}$, otherwise the challenger outputs 0.

Then

$$\Pr\left(\mathsf{Exp}^{\mathsf{Diff}}_{\mathcal{A},puz}(k) = 1\right) = \Pr(\mathsf{E}_0)\ . \tag{1}$$

Game G_1. The difference between Game G_1 and Game G_0 is that the KCPuz challenger is used to answer the CreatePuzSoln queries issued by $\mathcal{A}$ and the KCPuz challenge is inserted in response to the Test query. Note that we assume that $R = 2^Q$ in KCPuz shown in Fig. 1 in order to be compatible with our puzzle scheme puz. The game is defined as folows:

1. The parameters $params \leftarrow (Q, n)$ are obtained from the KCPuz challenger.
2. Initiate the adversary $\mathcal{A}$ with $params$ as input.
   The adversary is allowed oracle access to $\mathsf{CreatePuzSoln}(\cdot)$ and $\mathsf{Test}(\cdot)$ oracles. That is, $\mathcal{B}$ interacts with KCPuz challenger and the adversary $\mathcal{A}$ individually. $\mathcal{B}$ acts as a puz challenger for $\mathcal{A}$. Whenever $\mathcal{A}$ issues CreatePuzSoln queries, $\mathcal{B}$ simply forwards the queries to KCPuz challenger and returns whatever it receives from KCPuz challenger with minor modifications to $\mathcal{A}$. We explain the interaction between $\mathcal{B}$ and KCPuz and between $\mathcal{B}$ and $\mathcal{A}$ in detail below:
   - $\mathsf{CreatePuzSoln}(str)$: Whenever $\mathcal{A}$ issues $\mathsf{CreatePuzSoln}(\mathtt{str})$ query, our challenger $\mathcal{B}$ forwards the same CreatePuzSoln query to the KCPuz challenger. The KCPuz challenger sends $(\mathsf{puz} = (X, R = 2^Q, K, Z), \mathsf{soln} = (X^{2^Q}, X^K))$ to $\mathcal{B}$. Upon receiving a pair of the form $(\mathsf{puz}, \mathsf{soln})$ our challenger $\mathcal{B}$ acts as follows:
     – Assigns the puzzle values $x \leftarrow X$, $X_1 \leftarrow X^u$, for a fixed u of its choice and the solution values $y \leftarrow X^{2^Q}$ and $Y \leftarrow (X^K)^u$. Note that the value X received each time from KCPuz challenger is an output of the HMAC function, whereas in puz, (x, X) is an output of the BPV generator.
     – Return $(\mathsf{puz}, \mathsf{soln}) = ((x, X_1), (y, Y))$ to $\mathcal{A}$.
   - $\mathsf{Test}(str^*)$: When $\mathcal{A}$ issues a $\mathsf{Test}(\mathtt{str}^*)$ query, $\mathcal{B}$ simply passes the same query as its Test query to the KCPuz challenger that returns the challenge puzzle $\mathsf{puz}^* = (X^*, R^* = 2^Q, K^*, Z^*)$, where X^* is an output of HMAC. Then $\mathcal{B}$ sets $x^* \leftarrow X^*$, $X_1^* \leftarrow (X^*)^u$ and sends the target puzzle $\mathsf{puz}^* = (x^*, X_1^*, R^* = 2^Q, K^*, Z^*)$ to $\mathcal{A}$.
3. $\mathcal{A}$ may continue its CreatePuzSoln queries and $\mathcal{B}$ answers them as explained above.
4. When $\mathcal{A}$ outputs a potential solution $\mathsf{soln}^* = (y^* = (X^*)^{2^Q}, Y^* = ((X^*)^u)^K)$, $\mathcal{B}$ omits Y^*, computes $(X^*)^K$ and outputs its soln^* as $(y^*, (X^*)^K)$.

We say that if the $\mathcal{A}$ wins game G_1, then the challenger $\mathcal{B}$ wins the puzzle difficulty experiment of KCPuz. Hence,

$$\Pr(\mathsf{E}_2) \leq \mathsf{Adv}^{\mathsf{Diff}}_{\mathcal{B},\mathsf{KCPuz},Q}(k) \leq \epsilon_{k,Q}(t) \ . \tag{2}$$

where $\mathcal{B}$ runs in time $t(B) = t(A) + (q_{\mathsf{C}} + 1)\left(T_{\mathrm{Exp}}\right)$ where q_{C} is the total number of CreatePuzSoln queries issued by $\mathcal{A}$ in G_0, and T_{Exp} is the total time needed to compute an exponentiation modulo n.

In the game G_0, a puzzle is of the form $(N_S, Z, x, X, R = 2^Q, K)$ where (x, X) is an output from the BPV generator BPVGen whereas in G_1, x is an output of HMAC run by the KCPuz challenger and $X = x^u$ is uniform at random.

Hence by Theorem 1, we get

$$|\Pr(\mathsf{E}_0) - \Pr(\mathsf{E}_1)| \leq 2^{-\frac{1}{2}\left(\log\binom{N}{\ell}+1\right)} \leq \epsilon_{k,Q}(t), \tag{3}$$

where the second inequality is due to the appropriate choices of N and ℓ.

Game G_2. The messages generated by the challenger in G_2 are identical to those in G_1 except for the following modification: The value X which is returned during the Test query: in G_1 it is a random integer from $[1, n]$ generated by the challenger whereas in G_2 it is the output of KCPuz challenger. This change is indistinguishable as we basically replace one random x with another. Hence

$$|\Pr(\mathsf{E}_1) - \Pr(\mathsf{E}_2)| = 0 \ . \tag{4}$$

Combining equations (1) through (3) yields the desired result. □

4 Conclusion

In this paper, we presented an efficient non-parallelizable puzzle based on modular exponentiation. Our puzzle can be viewed as a combination of two previously known puzzles, namely KCPuz and RSAPuz. However, our puzzle inherits all the advantages of these two puzzles and eludes their disadvantages. Our puzzle is the first modular exponentiation-based puzzle without computational burden, either offline (e.g., precomputation in RSAPuz) or online (e.g., solution verification in KCPuz), to support change of difficulty. Thus our puzzle supports scalability in an efficient manner, making it more practical in detering attacks like denial-of-service attacks.

References

1. Aura, T., Nikander, P.: Stateless connections. In: Han, Y., Okamoto, T., Qing, S. (eds.) ICICS 1997, vol. 1334 of LNCS, pp. 87–97. Springer (1997)
2. Aura, T., Nikander, P., Leiwo, J.: DoS-resistant authentication with client puzzles. In: Christianson, B., Crispo, B., Malcolm, J.A., Roe, M. (eds.) Security Protocols: 8th International Workshop, vol. 2133 of LNCS, pp. 170–177. Springer (2000)
3. Back, A.: Hashcash: a denial-of-service countermeasure. Available as http://www.hashcash.org/papers/hashcash.pdf (2002)
4. Boyko, V.: A pre-computation scheme for speeding up public-key cryptosystems. Master's thesis, Massachusetts Institute of Technology. Available as http://hdl.handle.net/1721.1/47493 (1998)
5. Boyko, V., Peinado, M., Venkatesan, R.: Speeding up discrete log and factoring based schemes via precomputations. In: Nyberg, K. (ed.) EUROCRYPT '98, vol. 1403 of LNCS, pp. 221–235. Springer (1998)
6. Canetti, R., Halevi, S., Steiner, M.: Hardness amplification of weakly verifiable puzzles. In: Kilian, J. (ed.) Theory of Cryptography Conference (TCC) 2005, vol. 3378 of LNCS, pp. 17–33. Springer (2005)
7. Chen, L., Morrissey, P., Smart, N.P., Warinschi, B.: Security notions and generic constructions for client puzzles. In: Matsui, M. (ed.) ASIACRYPT 2009, vol. 5912 of LNCS, pp. 505–523. Springer (2009)
8. Dwork, C., Naor, M.: Pricing via processing or combatting junk mail. In: Brickell, E.F. (ed.) CRYPTO '92, vol. 740 of LNCS, pp. 139–147. Springer (1992)
9. Feng, W., Kaiser, E., Luu, A.: Design and implementation of network puzzles. In: INFOCOM 2005, vol. 4, pp. 2372–2382. IEEE (2005)
10. Hofheinz, D., Unruh, D.: Comparing two notions of simulatability. In: Kilian, J. (ed.) TCC 2005, vol. 3378 of LNCS, pp. 86–103. Springer (2005)
11. Juels, A., Brainard, J.: Client puzzles: a cryptographic countermeasure against connection depletion attacks. In: NDSS 1999, pp. 151–165. Internet Society (1999)
12. Karame, G., Čapkun, S.: Low-cost client puzzles based on modular exponentiation. In: Gritzalis, D., Preneel, B., Theoharidou, M. (eds.) ESORICS 2010, vol. 6345 of LNCS, pp. 679–697. Springer (2010)
13. Kilian, J. (ed.) TCC 2005, vol. 3378 of LNCS. Springer (2005)
14. Lenstra, A., Verheul, E.: Selecting cryptographic key sizes. J. Cryptology **14**(4), 255–293 (2001)
15. Mao, W.: Timed-release cryptography. In: Vaudenay, S., Youssef, A. M. (eds.) SAC 2001, vol. 2259 of LNCS, pp. 342–358. Springer (2001)
16. Miller, G.L.: Riemann's hypothesis and tests for primality. In: STOC, 1975, pp. 234–239. ACM (1975)
17. Moore, D., Shannon, C., Brown, D.J., Voelker, G.M., Savage, S.: Inferring internet denial-of-service activity. ACM Trans. Comput. Syst. (TOCS) **24**(2), 115–139 (2006)
18. Rangasamy, J., Stebila, D., Kuppusamy, L., Boyd, C., González Nieto, J.M.: Efficient modular exponentiation-based puzzles for denial-of-service protection, The 14th International Conference on Information Security and Cryptology - ICISC 2011, pp. 319–331 (2011)
19. Rivest, R.L., Shamir, A., Wagner, D.A.: Time-lock puzzles and timed-release crypto. Technical Report TR-684, MIT Laboratory for Computer Science, March 1996
20. Shparlinski, I.: On the uniformity of distribution of the RSA pairs. Math. Comput. **70**(234), 801–808 (2001)
21. Stebila, D., Kuppusamy, L., Rangasamy, J., Boyd, C., González Nieto, J.M.: Stronger difficulty notions for client puzzles and Denial-of-Service-Resistant protocols. In: Kiayias, A. (ed.) Topics in Cryptology The Cryptographers' Track at the RSA Conference (CT-RSA) 2011, vol. 6558 of LNCS, pp. 284–301. Springer (2011)

22. Wang, X., Reiter, M.K.: Defending against denial-of-service attacks with puzzle auctions. In: Proceedings 2003 IEEE Symposium on Security and Privacy (SP'03), pp. 78–92. IEEE Press (2003)
23. Wang, Y., Wu, Q., Wong, D.S., Qin, B., Chow, S.M., Liu, Z., Tan, X.: Securely Outsourcing Exponentiations with Single Untrusted Program for Cloud Storage. In: Proceedings of Computer Secuirty - ESORCICS 2014, vol. 8712 of LNCS, pp. 326–343. Springer (2014)

Computationally Secure Robust Multi-secret Sharing for General Access Structure

Angsuman Das, Partha Sarathi Roy and Avishek Adhikari

Abstract Secret sharing scheme plays a crucial role in distributed cryptosystems. Due to its extensive use in numerous applications, an important goal in this area is to minimize trust among the participants. To remove this bottleneck, robust secret sharing, which allows the correct secret to be recovered even when some of the shares presented during an attempted reconstruction are incorrect, can be an efficient tool. However, as unconditional security demands honest majority and share size to be at least equal to the size of the secret, the need for computational security of such schemes has been felt over the years, specially in case of multi-secret sharing schemes. In this paper, we provide a notion of computationally robust multi-secret sharing scheme for general access structure. We also propose a robust multi-secret sharing scheme for general access structure and prove its computational security under the proposed notation.

Keywords Robust secret sharing · General access structure · Computational security

Research supported in part by National Board for Higher Mathematics, Department of Atomic Energy, Government of India (No 2/48(10)/2013/NBHM(R.P.)/R&D II/695).

A. Das
Department of Mathematics, St. Xavier's College, Kolkata, India
e-mail: angsumandas054@gmail.com

P.S. Roy · A. Adhikari (✉)
Department of Pure Mathematics, University of Calcutta, Kolkata, India
e-mail: avishek.adh@gmail.com

P.S. Roy
e-mail: psrpm_s@caluniv.ac.in; royparthasarathi0@gmail.com

R.N. Mohapatra et al. (eds.), *Mathematics and Computing*,
Springer Proceedings in Mathematics & Statistics 139,
DOI 10.1007/978-81-322-2452-5_9

1 Introduction

Secret sharing scheme is one of the key components in various cryptographic protocols and in particular distributed systems. It is not only a building block for interactive cryptographic primitives, it is also an important tool to build non-interactive cryptosystems (specially in public-key scenario). Informally, a *secret sharing scheme* (SSS) allows a dealer $\mathcal{D}$ to split a secret s into different pieces, called *shares*, which are given to a set of players $\mathcal{P}$, such that only certain qualified subsets of players can recover the secret using their respective shares. The collection of those qualified sets of players is called *access structure* Γ_s corresponding to the secret s.

Blakley [2] and Shamir [19], in 1979, independently, came out with a scheme known as (t, n) threshold secret sharing scheme. Later on, with increasing interest in this area, secret sharing schemes with features like general access structures (where qualified subsets are not all of same size t), multiple secrets (when number of secrets to be shared is more than one), verifiability, multi-usability (reconstruction of one secret does not endanger the security of the other secrets) [5, 13, 17] came into existence.

In the basic form of (multi) secret sharing schemes, it was assumed that the dishonest players involved with the protocol is *semi-honest* i.e., honest but curious. But for the real life scenario, this assumption may not hold good due to the presence of *malicious* players. This idea led to the development of secret sharing under various adversarial models. It may happen that some players behave maliciously during the execution of the protocol. Malicious players may also submit incorrect shares resulting in incorrect secret reconstruction. This observation led to *robust secret sharing schemes* [14, 16]. Informally, robust secret sharing schemes allow the correct secret to be recovered even when some of the shares presented during an attempted reconstruction are incorrect.

Most of the robust secret sharing schemes proposed and analysed so far enjoy unconditional (or information-theoretic) security [16, 18], which means that the value of the shared secret is hidden to a computationally unbounded adversary who controls a subset of users. However, when secret sharing schemes are used in the design of distributed public key cryptosystems (that can enjoy computational security, at most), one could argue that requiring unconditional security for the underlying secret sharing schemes may be unnecessarily restrictive. Moreover, in order to achieve unconditional security, we need to have honest majority and share size at least equal to the size of all the secrets. This may be considered as system resources consuming and sometimes impracticable.

An alternative solution can be relying on computational security that serves well in practical purposes and to have a lower share size along with the tolerance of arbitrary number of dishonest participants. The trade-off to be made is how small can we make the share-size without compromising *much* security.

1.1 Related Work

Keeping these issues in mind, the idea of computationally secure secret sharing schemes and robust secret sharing schemes came into existence with various proposals [1, 4, 5, 7, 9, 12, 15, 17, 20, 21]. In 1994, He-Dawson [9] proposed a multi-stage (t, n) threshold secret sharing scheme. In 2007, Geng et al. [7] proposed a multi-use threshold secret sharing scheme using one-way hash function and pointed out that the He-Dawson scheme was actually an one-time-use scheme and can not endure conspiring attacks. A SSS is said to be *multi-use* if even after a secret is reconstructed by some players, the reconstructor cannot misuse their submitted information to reconstruct other secrets. Generally, to make a scheme multi-use, the players do not provide the reconstructor with the original share but a shadow or image of that share, which is actually an entity that depends on the original share. This image or shadow is known as the *pseudo-share*. In 2006, Pang et al. [15] proposed a multi-secret sharing scheme for general access structure in which all the secrets are revealed at a time. In 2008, Wei et al. [21] proposed a renewable secret sharing scheme for general access structure, but the secrets are to be revealed in a pre-determined order. Recently, in [20], Shao proposed a threshold multi-secret sharing scheme using hash function which also suffers the drawback of revealing all the secrets at a time as in [15].

Multi-secret sharing schemes lacked a formal computational security notion and analysis, until Herranz et al. [10], [11] came up with a proper computational notion of security for multi-secret sharing schemes and provided some concrete constructions secure in that model. In [1, 4], authors discussed formal security notion for computationally secure robust threshold (single) secret sharing scheme. But, up to the best of our knowledge, there does not exist any formal security notion for computationally secure robust multi secret sharing scheme for general access structure.

1.2 Our Contribution

In this paper, we introduce a formal notion of security for computationally secure robust multi secret sharing scheme for general access structure and propose a multi secret sharing scheme which is secure under the proposed notion in random oracle model.

2 Model and Definitions

In this section, we specify the adversarial and communication model used in the rest of the paper. We also provide formal definitions of construction and security notion of robust multi-secret sharing scheme for general access structure.

Adversarial Model: The dealer $\mathcal{D}$ and the designated reconstructor are assumed to be honest. The dealer delivers the shares to respective players over point-to-point private channels. We assume that the adversary $\mathcal{A}$ is computationally bounded and malicious. Once a player P is corrupted, the adversary learns his share and internal state. Moreover from that point onwards, $\mathcal{A}$ has full control over P. By being *malicious*, we mean that $\mathcal{A}$ can deviate from the protocol in an arbitrary manner.
Communication Model: We assume synchronous network model. There are point to point secure channels among the dealer and the players. Moreover, all of them have an access to a common broadcast channel.

Definition 1 A Robust Multi Secret Sharing Scheme (Robust MSSS) Ω consists of three probabilistic polynomial time algorithms (Setup, Dist, $\mathsf{Reconst}$) as follows:

1. The setup protocol, Setup, takes as input a security parameter $\lambda \in \mathbb{N}$, the set of players $\mathcal{P}$ and the k access structures $\Gamma_1, \Gamma_2, \ldots, \Gamma_k$, where $\Gamma_i = \{A_{i1}, A_{i2}, \ldots, A_{it_i}\}$ is the access structure for the ith secret and A_{ij} is the jth qualified subset of the access structure for the ith secret s_i, and outputs some public and common parameters pms for the scheme (such as the access structures and set of players, mathematical groups, hash functions, etc.). We implicitly assume that pms also contains the descriptions of $\mathcal{P}$ and the access structures.
2. The share distribution protocol, Dist, (run by the dealer $\mathcal{D}$) takes as input pms and the global secret $\mathbf{s} = (s_1, s_2, \ldots, s_k)$ to be distributed, and produces the set of shares $\{x_\alpha\}_{P_\alpha \in \mathcal{P}}$, possibly some public output out_{pub} and a set of public verification values $\mathcal{V} = \{V_{\varphi(x_\alpha, A_{ij})} : P_\alpha \in A_{ij} \in \Gamma_i\}$. (Note: $\varphi(x_\alpha, A_{ij})$ is a public function used to generate pseudo-shares from the share x_α and the qualified set A_{ij}.)
3. The secret reconstruction protocol, $\mathsf{Reconst}$, takes as input $\mathsf{pms}, \mathsf{out}_{pub}$, an index $i \in \{1, 2, \ldots, k\}$, $\mathcal{V}$ and the possible pseudo-shares $\{\varphi_\alpha^*\}_{P_\alpha \in A_{ij}}$ for all qualified sets $A_{ij} \in \Gamma_i = \{A_{i1}, A_{i2}, \ldots, A_{it_i}\}$ and outputs a possible value of the secret s_i^* for the i-th secret.

For correctness, we require that, for any index $i \in \{1, 2, \ldots, k\}$ and any $A_{ij} \in \Gamma_i$, it holds

$$\mathsf{Reconst}(\mathsf{pms}, \mathsf{out}_{pub}, \mathcal{V}, \{\varphi(x_\alpha, A_{ij})\}_{P_\alpha \in A_{ij}}) = s_i$$

if $\{x_\alpha\}_{P_\alpha \in A_{ij}} \subset \{x_\alpha\}_{P_\alpha \in \mathcal{P}}$ and $(\mathsf{out}_{pub}, \mathcal{V}, \{x_\alpha\}_{P_\alpha \in \mathcal{P}}) \leftarrow \mathsf{Dist}(\mathsf{pms}, \mathbf{s})$ is a distribution of the secret $\mathbf{s} = (s_1, \ldots, s_i, \ldots s_k)$ and the setup protocol has produced $\mathsf{pms} \leftarrow \mathsf{Setup}(1^\lambda, \mathcal{P}, \{\Gamma_i\}_{1 \le i \le k})$.

The computational security and *robustness* of Robust-MSSS Ω is defined by the games described in Definition 2 and Definition 3 respectively.

Definition 2 (*Indistinguishability of Shares against Chosen Secret Attack*): Indistinguishability of shares of a Robust MSSS under chosen secret attack **(IND-CSA)** is defined by the following game $\mathcal{G}$ between a challenger $\mathcal{C}$ and an adversary $\mathcal{A}$ as follows:

1. The adversary $\mathcal{A}$ publishes the set of players $\mathcal{P}$ and the k access structures $\Gamma_1, \Gamma_2, \ldots, \Gamma_k \subset 2^{\mathcal{P}}$.
2. The challenger $\mathcal{C}$ runs $\mathsf{pms} \leftarrow \mathsf{Setup}(1^\lambda, \mathcal{P}, \{\Gamma_i\}_{1\leq i\leq k})$ and sends pms to $\mathcal{A}$.
3. $\mathcal{A}$ outputs a subset $\tilde{B} \subset \mathcal{P}$ of *unqualified* players (unqualified means $\exists i \in \{1, 2, \ldots, k\}$ such that $\tilde{B} \notin \Gamma_i$) and two different global secrets $\mathbf{s}^{(0)} \neq \mathbf{s}^{(1)}$ with the restriction:

$$s_i^{(0)} = s_i^{(1)}, \forall i \in \{1, 2, \ldots, k\}, \text{ such that } \tilde{B} \in \Gamma_i.$$

4. The challenger $\mathcal{C}$ chooses at random a bit $b \in_R \{0, 1\}$, runs $\mathsf{Dist}(\mathsf{pms}, \mathbf{s}^{(b)}) \rightarrow \left(\mathsf{out}_{pub}, \mathcal{V}, \{x_\alpha\}_{P_\alpha\in\mathcal{P}}\right)$ and sends $\left(\mathsf{out}_{pub}, \mathcal{V}, \{x_\alpha\}_{P_\alpha\in\tilde{B}}\right)$ to $\mathcal{A}$.
5. Finally, $\mathcal{A}$ outputs a bit b'.

The advantage of $\mathcal{A}$ in breaking the MSSS Ω is defined as $\mathsf{Adv}_{\mathcal{A}}(\lambda) = |\mathsf{Pr}[b' = b] - \frac{1}{2}|$.

The scheme Ω is said to be computationally IND-CSA secure if $\mathsf{Adv}_{\mathcal{A}}(\lambda)$ is negligible for all polynomial-time adversaries $\mathcal{A}$.

Definition 3 (*Robustness*): Robustness of a MSSS Ω is defined by the following game $\mathcal{G}$ between a challenger $\mathcal{C}$ and an adversary $\mathcal{A}$ as follows:

1. The adversary $\mathcal{A}$ chooses the set of players $\mathcal{P}$, a secret vector $\mathbf{s} = (s_1, s_2, \ldots, s_k)$ and the corresponding k access structures $\Gamma_1, \Gamma_2, \ldots, \Gamma_k \subset 2^{\mathcal{P}}$. Then $\mathcal{A}$ runs $\mathsf{pms} \leftarrow \mathsf{Setup}(1^\lambda, \mathcal{P}, \{\Gamma_i\}_{1\leq i\leq k})$ and sends $(pms, \mathbf{s})$ to $\mathcal{C}$.
2. The challenger $\mathcal{C}$ runs $\mathsf{Dist}(\mathsf{pms}, \mathbf{s}) \rightarrow \left(\mathsf{out}_{pub}, \mathcal{V}, \{x_\alpha\}_{P_\alpha\in\mathcal{P}}\right)$ and sends $(\mathsf{out}_{pub}, \mathcal{V}, \{x_\alpha\}_{P_\alpha\in\mathcal{P}})$ to $\mathcal{A}$.
3. $\mathcal{A}$ outputs $\{x_\alpha^*\}_{P_\alpha\in\mathcal{P}}$ with restriction:

$$\forall i \in \{1, 2, \ldots, k\}, \exists B_i \in \Gamma_i \text{ such that } x_\alpha = x_\alpha^*, \forall \alpha \in B_i.$$

4. The challenger $\mathcal{C}$ runs $\forall i \in \{1, 2, \ldots, k\}$

$$\mathsf{Reconst}(\mathsf{pms}, \mathsf{out}_{pub}, \mathcal{V}, \{\varphi(x_\alpha^*, A_{ij})\}_{P_\alpha\in A_{ij}}, \forall A_{ij} \in \Gamma_i) \rightarrow s_i^*$$

to output $\mathbf{s}^* = (s_1^*, s_2^*, \ldots, s_k^*)$.
5. If $\mathbf{s}^* = \mathbf{s}$, the challenger $\mathcal{C}$ sets $b = 0$, else sets $b = 1$. Finally, $\mathcal{C}$ outputs the bit b.

The scheme Ω is said to be computationally robust if $\mathsf{Pr}[b = 1]$ is negligible for all polynomial-time adversaries $\mathcal{A}$.

3 A Robust Multi-secret Sharing Scheme

In this section, we modify the MSSS for general access structure proposed by [17] and analyse its security in the computational model of IND-CSA and robustness. (It is worth mentioning that the scheme in [17] lacked formal security analysis.) The scheme $\Omega = (\mathsf{Setup}, \mathsf{Dist}, \mathsf{Reconst})$ consists of three basic phases,

1. **Setup:** On an input security parameter λ, the set of n players or participants $\mathcal{P} = \{P_\alpha : \alpha \in \{1, 2, \ldots, n\}\}$ and k-access structures $\Gamma_1, \Gamma_2, \ldots, \Gamma_k$ for k secrets, where $\Gamma_i = \{A_{i1}, A_{i2}, \ldots, A_{it_i}\}$ is the access structure for the i-th secret and A_{ij} is the jth qualified subset of the access structure of ith secret s_i and $|A_{ij}| = r_{ij}$,
   a. Choose a $q = q(\lambda)$-bit prime p.
   b. Choose a hash function $H : \{0,1\}^{q+l+m} \rightarrow \mathbb{Z}_p \subseteq \{0,1\}^q$, where $l = [log_2 k] + 1, m = [log_2 t] + 1$ such that $t = max\{t_1, t_2, \ldots, t_k\}$.
   c. Choose distinct identifier $ID_\alpha \in_R \mathbb{Z}_p^*$ corresponding to each of the participant $P_\alpha, \alpha \in \{1, 2, \ldots, n\}$
   d. Choose a hash function $G : \{0,1\}^q \rightarrow \{0,1\}^{u(\lambda)}$.
   e. Set as $\mathsf{pms} = (p, q, k, l, m, H, G, ID_\alpha, \mathcal{P}, \Gamma_1, \Gamma_2, \ldots, \Gamma_k)$.
2. **Dist:** On input $\mathsf{pms} = (p, q, k, l, m, H, ID_\alpha, \mathcal{P}, \Gamma_1, \Gamma_2, \ldots, \Gamma_k)$, where $\alpha \in \{1, \ldots, n\}$, and k secrets $s_1, s_2, \ldots, s_k \in \mathbb{Z}_p \subseteq \{0,1\}^q$,
   a. Choose $x_\alpha \in_R \{0,1\}^q$, $\alpha = 1, 2, \ldots, n$.
   b. For A_{ij} where $i = 1, 2, \ldots, k$; $j = 1, 2, \ldots, t_i$, choose $d_1^{ij}, d_2^{ij}, \ldots, d_{r_{ij}-1}^{ij}$ $\in_R \mathbb{Z}_p \subseteq \{0,1\}^q$ and set
   $$f_{ij}(x) = s_i + d_1^{ij} x + d_2^{ij} x^2 + \cdots + d_{r_{ij}-1}^{ij} x^{r_{ij}-1}.$$
   c. For each $P_\alpha \in A_{ij}$, compute
      - $\varphi(x_\alpha, A_{ij}) = H(x_\alpha || i_l || j_m)$ where i_l denotes the l-bit binary representation of i, j_m denotes the m-bit binary representation of j and '$||$' denotes the concatenation of two binary strings.
      - $\mathcal{B}_{ij}^\alpha = f_{ij}(ID_\alpha)$ and $\mathcal{M}_{ij}^\alpha = \mathcal{B}_{ij}^\alpha - \varphi(x_\alpha, A_{ij})$.
      - the public verification values $V_{\varphi(x_\alpha, A_{ij})} = G(\varphi(x_\alpha, A_{ij}))$.
   d. Output $\{x_\alpha\}_{1 \le \alpha \le n}$ as shares, $\mathsf{out}_{pub} = \{\mathcal{M}_{ij}^\alpha : P_\alpha \in A_{ij}, 1 \le i \le k; 1 \le j \le t_i\}$ as public output.
   e. Output $\mathcal{V} = \{V_{\varphi(x_\alpha, A_{ij})} : P_\alpha \in A_{ij}, 1 \le i \le k; 1 \le j \le t_i\}$ as public verification value.
3. **Reconst:**
   - **Participant Phase:** On input pms, out_{pub}, an index $i \in \{1, 2, \ldots, k\}$,
   a. Each participant $P_\alpha \in A_{ij}$ computes $\varphi(x_\alpha, A_{ij}) = H(x_\alpha || i_l || j_m)$, $\forall A_{ij} \in \Gamma_i$.

b. $\forall A_{ij} \in \Gamma_i$, each participant $P_\alpha \in A_{ij}$ sends $\varphi(x_\alpha, A_{ij}) = H(x_\alpha || i_l || j_m)$ to the reconstructor.

- **Verification Phase:** On input $\mathcal{V} = \{V_{\varphi(x_\alpha, A_{ij})} : P_\alpha \in A_{ij}, 1 \leq i \leq k; 1 \leq j \leq t_i\}$ and $\{\varphi(x_\alpha, A_{ij}) : \forall P_\alpha \in A_{ij}, \forall A_{ij} \in \Gamma_i\}$,
  a. $\forall A_{ij} \in \Gamma_i$ and $\forall P_\alpha \in A_{ij}$, the reconstructor checks $G(\varphi(x_\alpha, A_{ij}) \overset{?}{=} V_{\varphi(x_\alpha, A_{ij})}$.
  b. If equality holds for some $A_{ij} \in \Gamma_i$, that A_{ij} is considered as an honest qualified set. Whereas, if $G(\varphi(x_\alpha, A_{ij}) \neq V_{\varphi(x_\alpha, A_{ij})}$ for some $A_{ij} \in \Gamma_i$, that A_{ij} is considered as a corrupted qualified set.
- **Secret Reconstruction Phase:** For an honest qualified set $A_{ij} \in \Gamma_i$, the reconstructor
  a. computes $f_{ij}(ID_\alpha) = \mathcal{B}_{ij}^\alpha = \mathcal{M}_{ij}^\alpha + \varphi(x_\alpha, A_{ij}), \forall P_\alpha \in A_{ij}$.
  b. computes s_i from $\{f_{ij}(ID_\alpha) : P_\alpha \in A_{ij}\}$ using Lagrange's Interpolation.

3.1 Analysis of the Scheme Ω

Theorem 1 *Ω satisfies correctness condition.*

Proof As correctness is considerable only when all the participants are honest, it is obvious that, using Lagrange's Interpolation, every qualified set of honest participants can reconstruct corresponding secret. □

Theorem 2 *Ω is IND-CSA secure MSSS in random oracle model.*

Proof Let $\mathcal{A}_\Omega$ be an adversary against IND-CSA security of Ω. Let $\mathcal{C}$ be the challenger of the security game. $\mathcal{A}_\Omega$ starts the game by choosing a set of participants $\mathcal{P} = \{P_1, P_2, \ldots, P_n\}$ and k access structures $\Gamma_1, \Gamma_2, \ldots, \Gamma_k$. $\mathcal{C}$ runs Setup of Ω to generate pms and sends everything in pms except the hash functions G, H to $\mathcal{A}_\Omega$.

$\mathcal{A}_\Omega$ outputs a set $\tilde{B} \subset \mathcal{P}$ of corrupted players and two different global secrets $\mathbf{s}^{(0)} \neq \mathbf{s}^{(1)}$ with the restriction:

$$s_i^{(0)} = s_i^{(1)}, \forall i \in \{1, 2, \ldots, k\}, \text{ such that } \tilde{B} \in \Gamma_i.$$

$\mathcal{C}$ chooses pairwise distinct $x_\alpha \in_R \{0, 1\}^q$, $\alpha = 1, 2, \ldots, n$.

Simulation of H-queries: $\mathcal{C}$ starts with two empty lists namely H-list and R-list. When $\mathcal{A}_\Omega$ submits a hash query of the form $x||i||j$ (in this proof, for simplicity, we write i_l, j_m as i, j only.), $\mathcal{C}$ checks whether $x = x_\alpha$ for some $P_\alpha \in \mathcal{P}$.

- If $x \neq x_\alpha, \forall \alpha \in \{1, 2, \ldots, n\}$ do

$$\begin{cases} \text{Choose } \gamma \in_R \{0, 1\}^q \\ \text{Add } (x||i||j, \gamma) \text{ to the R-list} \\ \text{Return } \gamma. \end{cases}$$

- If $x = x_\alpha$ for some α,

If $x = x_\alpha$ & $P_\alpha \in \tilde{B}$,	If $x = x_\alpha$ & $P_\alpha \notin \tilde{B}$,
do $\begin{cases} \text{If } P_\alpha \in A_{ij} \in \Gamma_i \\ \text{Choose } h_{\alpha,i,j} \in_R \{0,1\}^q. \\ \text{Add } (x_\alpha \| i \| j, h_{\alpha,i,j}) \text{ to H-list} \\ \text{Return } h_{\alpha,i,j}. \\ \text{If } P_\alpha \notin A_{ij} \in \Gamma_i \\ \text{Choose } \gamma \in_R \{0,1\}^q. \\ \text{Add } (x_\alpha \| i \| j, \gamma) \text{ to R-list} \\ \text{Return } \gamma. \end{cases}$	do $\begin{cases} \text{If } P_\alpha \in A_{ij} \in \Gamma_i \\ \text{Choose } h_{\alpha,i,j} \in_R \{0,1\}^q. \\ \text{Add } (x_\alpha \| i \| j, h_{\alpha,i,j}) \text{ to H-list} \\ \text{Return } h_{\alpha,i,j}. \\ \text{If } P_\alpha \notin A_{ij} \in \Gamma_i \\ \text{Choose } \gamma \in_R \{0,1\}^q. \\ \text{Add } (x_\alpha \| i \| j, \gamma) \text{ to R-list} \\ \text{Return } \gamma. \end{cases}$

If a hash query $x||i||j$ by $\mathcal{A}_\Omega$ is already in H or R-list, the stored value is sent back to $\mathcal{A}_\Omega$. It is to be noted that the entries in R-list are not required in the actual execution of the MSSS, whereas H-list will be used by the challenger $\mathcal{C}$ to simulate the out_{pub}.

Simulation of G-queries: $\mathcal{C}$ starts with two empty lists namely G-list and G'-list. When $\mathcal{A}_\Omega$ submits a hash query of the form h^*, $\mathcal{C}$ checks whether $h^* = h_{\alpha,i,j}$ for some $h^* \in H$-list.

If $h^* = h_{\alpha,i,j} \in H$-list,	If $h^* \notin H$-list,
do $\begin{cases} \text{Choose } V_{\alpha,i,j} \in_R \{0,1\}^u. \\ \text{Add } (h_{\alpha,i,j}, V_{\alpha,i,j}) \text{ to G-list} \\ \text{Return } V_{\alpha,i,j}. \end{cases}$	do $\begin{cases} \text{Choose } \eta \in_R \{0,1\}^u. \\ \text{Add } (h^*, \eta) \text{ to G'-list} \\ \text{Return } \eta. \end{cases}$

If a hash query h^* by $\mathcal{A}_\Omega$ is already in G or G'-list, the stored value is sent back to $\mathcal{A}_\Omega$. It is to be noted that it may happen that $\mathcal{A}_\Omega$ queries the hash function G with h^* such that at that stage $h^* \notin H$-list, but h^* was latter added to the H-list as some $h_{\alpha,i,j}$. In that case, the entry (h^*, η) is shifted from G'-list to G-list and renamed as $(h_{\alpha,i,j}, V_{\alpha,i,j})$. Observe that the entries in the final G'-list are not required in the actual execution of Dist algorithm. Only the entries in G-list are used by the challenger $\mathcal{C}$ to simulate the $\mathcal{V}$.

$\mathcal{C}$ chooses a bit $b \in_R \{0, 1\}$ and does the following:

- $\forall A_{ij} \in \Gamma_i$ where $i = 1, 2, \ldots, k$; $j = 1, 2, \ldots, t_i$, chooses $d_1^{ij}, d_2^{ij}, \ldots, d_{r_{ij}-1}^{ij} \in_R \mathbb{Z}_p \subseteq \{0, 1\}^q$ and sets

$$f_{ij}(x) = s_i + d_1^{ij} x + d_2^{ij} x^2 + \cdots + d_{r_{ij}-1}^{ij} x^{r_{ij}-1}$$

- For each $P_\alpha \in A_{ij}$, computes $\mathcal{B}_{ij}^\alpha = f_{ij}(ID_\alpha)$, $\mathcal{M}_{ij}^\alpha = \mathcal{B}_{ij}^\alpha - h_{\alpha,i,j}$.

The values of $h_{\alpha,i,j}$ are either recollected from H-list, if they exist, or they are chosen randomly from $\{0, 1\}^q$. In the latter case, the entry is added to the H-list for answering further hash queries. Moreover, $\mathcal{C}$ generates a simulated set $\mathcal{V} = \{V_{\alpha,i,j} :$

$P_\alpha \in A_{ij} \in \Gamma_i\}$ where $V_{\alpha,i,j}$'s are either collected from G-list, if they exist, or randomly chosen from $\{0, 1\}^u$ and added in the G-list.

$\mathcal{C}$ returns the public output $\mathsf{out}_{pub} = \{\mathcal{M}_{ij}^{\alpha} : P_\alpha \in A_{ij}, 1 \le i \le k; 1 \le j \le t_i\}$, $\mathcal{V} = \{V_{\alpha,i,j} : P_\alpha \in A_{ij} \in \Gamma_i\}$ and the shares $\{x_\alpha : P_\alpha \in \tilde{B}\}$ of the corrupted participants to $\mathcal{A}_\Omega$. Finally, $\mathcal{A}_\Omega$ outputs its guess b' for b.

Therefore, to compute the probability that $\mathcal{A}_\Omega$ outputs the correct bit, we distinguish between two cases, depending on whether $\mathcal{A}_\Omega$ somehow manages to get the pseudo-share $h_{\alpha,i,j}$ for some non-corrupted participant $P_\alpha \notin \tilde{B}$ and $P_\alpha \in A_{ij} \in \Gamma_i$ or not. If $\mathcal{A}_\Omega$ gets $h_{\alpha,i,j}$ for some $P_\alpha \notin \tilde{B}$, say with probability δ, this is the best case for $\mathcal{A}_\Omega$ and he can correctly guess the secret bit. On the other hand, if $\mathcal{A}_\Omega$ is not able to output any pseudo-share corresponding to a non-corrupted participant, which happens with probability $1 - \delta$, then the probability of $\mathcal{A}_\Omega$ guessing the correct bit is exactly $1/2$. Hence, in any case, the probability of $\mathcal{A}_\Omega$ guessing the correct bit is $\delta + \frac{1}{2}(1 - \delta) = \frac{\delta}{2} + \frac{1}{2}$ i.e., $\mathsf{Adv}_{\mathcal{A}_\Omega}(\lambda) = |(\frac{\delta}{2} + \frac{1}{2}) - \frac{1}{2}| = \frac{1}{2}\delta$.

Now, let E_1 be the event that $\mathcal{A}_\Omega$ make a hash query $x_\alpha||i||j$, where x_α is the share of $P_\alpha \in \mathcal{P} \setminus \tilde{B}$ and $P_\alpha \in A_{ij} \in \Gamma_i$ and $|\tilde{B}| = \tilde{t}$. The probability that a single H query leads to E_1 is $\dfrac{n - \tilde{t}}{2^q - \tilde{t}}$. Now, taking Q_H to be the total number of H-queries, we get

$$\begin{aligned}\mathsf{Pr}[E_1] &= 1 - \left(1 - \frac{n - \tilde{t}}{2^q - \tilde{t}}\right)\left(1 - \frac{n - \tilde{t}}{2^q - \tilde{t} - 1}\right)\cdots\left(1 - \frac{n - \tilde{t}}{2^q - \tilde{t} - Q_H + 1}\right) \\ &\le 1 - \left(1 - \frac{n - \tilde{t}}{2^q - \tilde{t}}\right)^{Q_H} \approx \frac{Q_H(n - \tilde{t})}{2^q - \tilde{t}} \le \frac{n \cdot Q_H}{2^q - \tilde{t}} \approx \frac{n \cdot Q_H}{2^q}\end{aligned}$$

as $\tilde{t}$, Q_H are negligible compared to 2^q. Let E_2 be the event that $\mathcal{A}_\Omega$ guesses the $h_{\alpha,i,j}$ for some $P_\alpha \notin \tilde{B}$ and $P_\alpha \in A_{ij} \in \Gamma_i$ from the publicly available $V_{\alpha,i,j}$. Since, $V_{\alpha,i,j}$ is randomly chosen and letting Q_G to be the total number of G-queries, we get,

$$\mathsf{Pr}[E_2] = 1 - \left(1 - \frac{1}{2^q}\right)^{Q_G} \approx \frac{Q_G}{2^q}$$

Now, $\delta = \mathsf{Pr}[E_1 \cup E_2] \le \mathsf{Pr}[E_1] + \mathsf{Pr}[E_2] \approx \dfrac{n \cdot Q_H + Q_G}{2^q}$. Thus,

$$\mathsf{Adv}_{\mathcal{A}_\Omega}(\lambda) \approx \frac{n \cdot Q_H + Q_G}{2^{q+1}}.$$

□

Theorem 3 *Ω is robust, if H and G are collision resistant.*

Proof The adversary $\mathcal{A}$ chooses the set of players $\mathcal{P}$, a secret vector $\mathbf{s} = (s_1, s_2, \ldots, s_k)$ and the corresponding k access structures $\Gamma_1, \Gamma_2, \ldots, \Gamma_k \subset 2^{\mathcal{P}}$. Then $\mathcal{A}$ runs

$\mathsf{Setup}(1^\lambda, \mathcal{P}, \{\Gamma_i\}_{1\le i\le k}) \rightarrow \mathsf{pms} = (p, q, k, l, m, H, G, ID_\alpha)$ and sends $(\mathsf{pms}, \mathbf{s})$ to $\mathcal{C}$. The challenger $\mathcal{C}$ runs $\mathsf{Dist}(\mathsf{pms}, \mathbf{s})$ to output the shares $\{x_\alpha\}_{P_\alpha\in\mathcal{P}}$, public outputs $\mathsf{out}_{pub} = \{\mathcal{M}_{ij}^\alpha : P_\alpha \in A_{ij}, 1 \le i \le k; 1 \le j \le t_i\}$ and public verification value $\mathcal{V} = \{V_{\varphi(x_\alpha, A_{ij})} : P_\alpha \in A_{ij}, 1 \le i \le k; 1 \le j \le t_i\}$ and sends $(\mathsf{out}_{pub}, \mathcal{V}, \{x_\alpha\}_{P_\alpha\in\mathcal{P}})$ to $\mathcal{A}$. $\mathcal{A}$ outputs $\{x_\alpha^*\}_{P_\alpha\in\mathcal{P}}$ with restriction: $\forall i \in \{1, 2, \ldots, k\}, \exists B_i \in \Gamma_i$ such that $x_\alpha = x_\alpha^*, \forall \alpha \in B_i$. Finally, the challenger $\mathcal{C}$ runs

$$\mathsf{Reconst}(\mathsf{pms}, \mathsf{out}_{pub}, \mathcal{V}, \{\varphi(x_\alpha^*, A_{ij})\}_{P_\alpha\in A_{ij}}, \forall A_{ij} \in \Gamma_i) \rightarrow s_i^*, \quad \forall i \in \{1, 2, \ldots, k\}$$

to output $\mathbf{s}^* = (s_1^*, s_2^*, \ldots, s_k^*)$.

Now, let us consider the case when $\mathcal{A}$ wins the game i.e., when $\mathcal{C}$ outputs $b = 1$. Note that $b = 1 \Rightarrow \mathbf{s}^* \ne \mathbf{s} \Rightarrow \exists$ at least one $i \in \{1, 2, \ldots, k\}$ such that $s_i \ne s_i^*$. However, as $s_i^* = \mathsf{Reconst}(\mathsf{pms}, \mathsf{out}_{pub}, \mathcal{V}, \{\varphi(x_\alpha^*, A_{ij})\}_{P_\alpha\in A_{ij}}, \forall A_{ij} \in \Gamma_i)$, there exists one $A_{ij} \in \Gamma_i$ such that $\{\varphi(x_\alpha^*, A_{ij})\}_{P_\alpha\in A_{ij}}$ passed the verification phase of $\mathsf{Reconst}$ algorithm, i.e.,

$$G(\varphi(x_\alpha^*, A_{ij})) = V_{\varphi(x_\alpha, A_{ij})}, \ \forall P_\alpha \in A_{ij}$$

However, from the public verification values computed by $\mathcal{C}$, we have

$$G(\varphi(x_\alpha, A_{ij})) = V_{\varphi(x_\alpha, A_{ij})}, \ \forall P_\alpha \in A_{ij}$$

$$\Rightarrow G(\varphi(x_\alpha^*, A_{ij})) = G(\varphi(x_\alpha^*, A_{ij})), \ \forall P_\alpha \in A_{ij}.$$

If $\varphi(x_\alpha^*, A_{ij}) \ne \varphi(x_\alpha, A_{ij})$ for some $P_\alpha \in A_{ij}$, we get a collision of G. On the other hand,

$$\varphi(x_\alpha^*, A_{ij}) = \varphi(x_\alpha, A_{ij}), \ \forall P_\alpha \in A_{ij}$$

$$\Rightarrow H(x_\alpha^*||i||j) = H(x_\alpha||i||j), \ \forall P_\alpha \in A_{ij}.$$

Now, if $x_\alpha^* = x_\alpha, \ \forall P_\alpha \in A_{ij}$, then A_{ij} is an honest qualified set in Γ_i, which in turn implies $s_i = s_i^*$, a contradiction. Hence, $\exists$ at least one $P_\alpha \in A_{ij}$ with $x_\alpha^* \ne x_\alpha$, thereby finding a collision $(x_\alpha^*||i||j, x_\alpha||i||j)$ for H.

Let us denote the events of finding collision for H and G by Col_H and Col_G respectively and let $\mathsf{Pr}[\mathsf{Col}_H] = \delta_H$ and $\mathsf{Pr}[\mathsf{Col}_G] = \delta_G$. Thus, the adversary wins the game if Col_H or Col_G occurs, i.e.,

$$\mathsf{Pr}[b = 1] \le \mathsf{Pr}[\mathsf{Col}_H \cup \mathsf{Col}_G] = \delta_H + \delta_G - \delta_H \cdot \delta_G < \delta_H + \delta_G$$

Since, G and H are collision resistant, δ_G and δ_H are negligible and as a result, $\mathsf{Pr}[b = 1]$ is negligible. Hence Ω is robust. □

4 Conclusion

In this paper, the notion of computational robustness for multi-secret sharing schemes for general access structure is established. We also provide construction and proofs of security of a robust MSSS for general access structure. As a topic of future research, one can think of more efficient construction of robust multi-secret sharing schemes for general access structure.

References

1. Bellare, M., Rogaway, P.: Robust computational secret sharing and a unified account of classical secret-sharing goals. In: Proceedings of the 14th ACM Conference on Computer and Communications Security, pp. 172–184 (2007)
2. Blakley, G.R.: Safeguarding cryptographic keys. In: The National Computer Conference 1979 AFIPS, vol. 48, pp. 313–317 (1979)
3. Chang, T.Y., Hwang, M.S., Yang, W.P.: An improvement on the Lin-Wu (t, n) threshold verifiable multi-secret sharing scheme. Appl. Math. Comput. **163**, 169–178 (2005)
4. Damgard, I., Jakobsen, T.P., Nielsen, J.B., Pagter, J.I.: Secure key management in the cloud. In: Cryptography and Coding, pp. 270–289. Springer, Heidelberg (2013)
5. Das, A., Adhikari, A.: An efficient multi-use multi-secret sharing scheme based on hash function. Appl. Math. Lett. **23**(9), 993–996 (2010)
6. Dehkordi, M.H., Mashhadi, S.: An efficient threshold verifiable multi-secret sharing. Comp. Stand. Inter. **30**, 187–190 (2008)
7. Geng, Y.J., Fan, X.H., Hong, F.: A new multi-secret sharing scheme with multi-policy. In: The 9th International Conference on Advanced Communication Technology, Vol. 3, pp. 1515–1517 (2007)
8. He, J., Dawson, E.: Multisecret-sharing scheme based on one-way function. Electron. Lett. **31**(2), 93–95 (1994)
9. He, J., Dawson, E.: Multi-stage secret sharing based on one-way function. Electron. Lett. **30**(19), 1591–1592 (1994)
10. Herranz, J., Ruiz, A., Saez, G.: New results and applications for multi-secret sharing schemes. In: Design, Codes and Cryptography, pp. 1–24, Springer, Heidelberg (2013)
11. Herranz, J., Ruiz, A., Saez, G.: Sharing many secrets with computational provable security. Inf. Process. Lett. **113**, 572–579 (2013)
12. Krawczyk, H.: Distributed ngerprints and secure information dispersal. In: 12th Annual ACM Symposium on Principles of Distributed Computing (PODC 1993), ACM Press, Austria, pp. 207–218 (1993)
13. Martin, K.M.: Challenging the adversary model in secret sharing schemes. In: Coding and Cryptography II, Proceedings of the Royal Flemish Academy of Belgium for Science and the Arts, pp. 45–63 (2008)
14. McEliece, R., Sarwate, D.: On sharing secrets and reed-solomon codes. Commun. ACM **24**(9), 583–584 (1981)
15. Pang, L.J., Li, H., Wang, Y.: An efficient and secure multi-secret sharing scheme scheme with general access structure. J. Wuhan Univ. Natur. Sci. **11**(6) (2006)
16. Rabin, T., Ben-Or, M.: Verifiable secret sharing and multiparty protocols with honest majority (extended abstract). STOC **1989**, 73–85 (1989)
17. Roy, P.S., Adhikari, A.: Multi-use multi-secret sharing scheme for general access structure. Annals of the University of Craiova, Math. Comput. Sci. Ser. **37**(4), 50–57 (2010)
18. Roy, P.S., Adhikari, A., Xu, R., Kirill, M., Sakurai, K.: An efficient robust secret sharing scheme with optimal cheater resiliency. SPACE. LNCS **8804**, 47–58 (2014)

19. Shamir, A.: How to share a secret. Commun. ACM **22**, 612–613 (1979)
20. Shao, J.: Efficient verifiable multi-secret sharing scheme based on hash function. Inf. Sci. **278**, 104–109 (2014)
21. Wei, Y., Zhong, P., Xiong, G.: A Multi-stage secret sharing scheme with general access structures. In: Wireless Communications, Networking and Mobile Computing, Beijing, pp. 1–4 (2008)

Key Chain-Based Key Predistribution Protocols for Securing Wireless Sensor Networks

Prasun Hazra, Debasis Giri and Ashok Kumar Das

Abstract Since the conception of the seminal work proposed by Eschenauer and Gligor in 2002, several key predistribution mechanisms have been proposed in the literature in order to establish symmetric secret keys between any two neighbor sensor nodes in wireless sensor networks (WSNs) for secure communication. However, due to lack of prior deployment knowledge, limited resources of sensor nodes and security threats posed in the unattended environment of WSNs, it is always a challenging task to propose a better secure key predistribution scheme apart from existing schemes. In this paper, we aim to propose two new key predistribution schemes based on hashed key chains, which provide secure communication between the sensor nodes with the desired storage and communication overheads. The proposed schemes provide better tradeoff among security, network connectivity, and overheads as compared to those for other existing schemes.

Keywords Wireless sensor networks · Key predistribution · Key establishment · Security

1 Introduction

Usage of sensor nodes in several fields is growing rapidly as it is very economical [1]. A sensor is composed of some digital logic unit to perform basic computations such as addition–multiplication comparison, memory chip (ranging from 8 to 128 MB)

P. Hazra (✉)
Paladion Networks, Bangalore 560078, India
e-mail: prasun.hazra@paladion.net

D. Giri
Department of Computer Science and Engineering, Haldia Institute of Technology, Haldia 721657, India
e-mail: debasis_giri@hotmail.com

A.K. Das
Center for Security, Theory and Algorithmic Research, International Institute of Information Technology, Hyderabad 500032, India
e-mail: iitkgp.akdas@gmail.com; ashok.das@iiit.ac.in

R.N. Mohapatra et al. (eds.), *Mathematics and Computing*,
Springer Proceedings in Mathematics & Statistics 139,
DOI 10.1007/978-81-322-2452-5_10

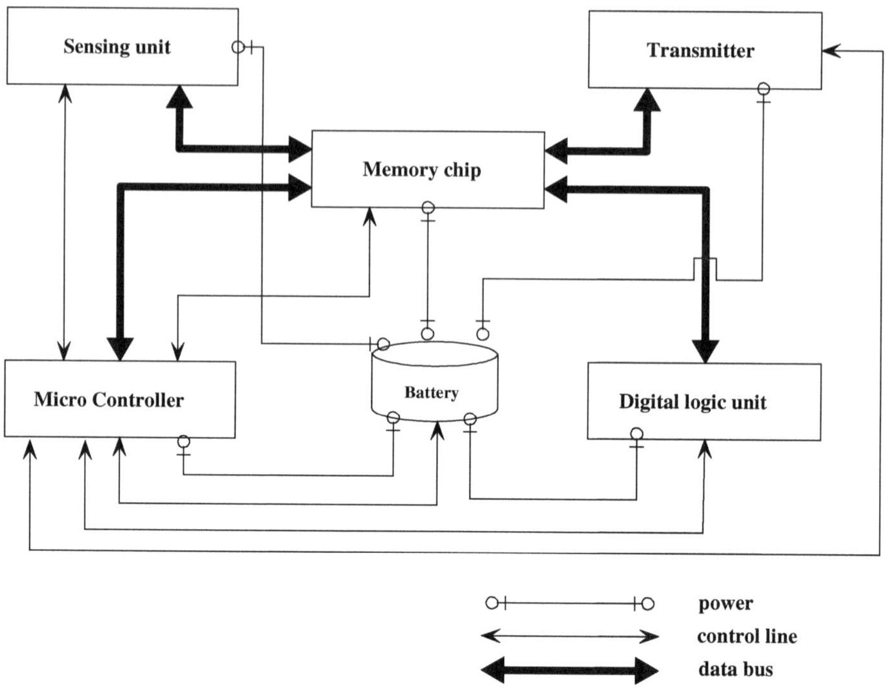

Fig. 1 Sensor node architecture (*Source* [2])

with microcontroller, where we burn the key establishment logic, sensing unit to retrieve raw information, radio unit with some communication range and energy source. Figure 1 explains how a sensor node works. First, the sensing unit digitizes sensed data using ADC (Analog to Digital Converter). Then, these data are stored in the memory location for further processing by digital logic under the instructions of microcontroller. After that the transmitter sends these data to other sensor nodes and also receives the processed data from other sensor nodes within the communication range. All these components are directly connected with a constant power supply, which is battery powered.

In a wireless sensor network, thousands of tiny sensor nodes are distributed over an area and these nodes sense surrounding data from the area. These data need to be transmitted to other neighbor nodes or to a base station via some intermediate nodes securely. As sensor nodes are resource constrained, heavy computation is not desirable. The symmetric key cryptography is feasible due to efficiency. We have no prior information about sensor node's deployment area and we have limited storage capacity of sensor nodes and also have the possibility to node capture along with node fabrication attack. All these make wireless sensor network communication very sensitive.

1.1 Related Work

In 2002, Eschenauer and Gligor proposed the seminal basic random key predistribution scheme [3] (which is also known as the EG scheme), where the key ring of each sensor node is generated by randomly picking a fixed amount of keys from a key pool without any replacement. The key pool size and key ring size are selected in a manner that any two key rings will share at least one key with some predefined probability. However, it has low connectivity when the key pool size is large, and its resilience to node capture becomes poor when the key pool size is small. In 2003, Chan et al. proposed the q-composite scheme [4], which is an improvement of the EG scheme, where a number of shared keys is greater than one. When the number of node capture is less, the resilience against node capture is high in this scheme. The random pairwise key predistribution is also proposed by Chan et al. [4], where its resiliency is perfect, but it does not support a large network. Also, its network connectivity is quite poor.

Castelluccia and Spognardi proposed a key management scheme, called RoK (a robust key predistribution protocol for multiphase wireless sensor networks) [5]. In this scheme, the sensors are deployed at different times in order to establish secure channels. They showed that its resiliency against node capture is much better than [4, 6]. Zo-RoK [7] is an improvement of RoK. This scheme is based on zone-based deployment. Later, Unlu et al. [8] presented key predistribution schemes for continuous deployment scenario. Their scheme performs better than other location-aware schemes in terms of connectivity, resiliency, memory usage, and communication cost. Other schemes such as [9–11] proposed in the literature provide significantly better performance than [4, 6].

The main drawback of the basic random key predistribution scheme is that if a node is captured, all keys stored in the memory are known to an adversary, and these keys might have been chosen from the key pool by other noncaptured nodes also. In the basic scheme, there is no concept of threshold cryptography such that by knowing all keys of captured nodes, an adversary could not guess the keys of other noncaptured nodes. The advantage of using the threshold cryptography is that when the number of captured sensor nodes is less than a certain number of nodes, it is not sufficient for the adversary to know the keys of other nodes. It is possible, if the key is used to compose to several other keys. In order to compromise the keys of noncaptured nodes, the number of captured nodes must exceed a threshold. For details, one can refer to [12, 13].

1.2 Motivation

Random selection of the keys from a key pool has few drawbacks. The drawbacks of random selection of keys can be overcome up to certain level using the threshold cryptography concept. That is, we should not use a key, rather the shares of a key are used. Nodes with different shares of a common key can communicate securely.

We generate key chain based on two keyed hash chains, where a keyed hash chain is generated by the help of the repeated hashing (up to some level) on a random secret information. It has drawback as anybody can compute next hashed value if he/she knows previous hash value of that key chain. In μTESLA protocol [14], it is shown that asymmetry is provided by delaying the disclosure of symmetric keys, where key chain is used to provide authentication. If we restrict the link (here, the link means the computation of next higher hash values from a known hash value) between a hash value of a key chain and its all next hash values, it can be used as a key. We know that the deterministic approaches enhance network scalability with low resiliency, whereas probabilistic approaches provide high resiliency with low connectivity. By using key chain with some reasonable chain length, we can achieve high connectivity along with high resiliency compared to the existing schemes.

1.3 Threat Model

In this paper, we use the Dolev–Yao threat model [2] in which two communicating parties communicate over an insecure channel. Any adversary (attacker or intruder) can eavesdrop the transmitted messages over the public insecure channel and he/she has the ability to modify, delete, or change the contents of the transmitted messages. We further assume that the sensor nodes are not equipped with tamper-resistant hardware. If an adversary captures or compromises some sensor nodes from the unattended target or deployment field of the sensor network, we assume that the adversary will know all the sensitive information including the keying materials, data, and codes from the captured sensor nodes' memory.

1.4 Our Contributions

Our contributions are listed below:

- In this paper, we propose two schemes: Key predistribution using linear key chain, and key predistribution using multilevel key chain for wireless sensor networks.
- These schemes provide better network connectivity.
- They also provide perfect resilience against node capture attacks.
- Moreover, the proposed schemes support large-scale sensor network, that is, our schemes are scalable.
- In addition, our schemes are also efficient in terms of computation and communication overheads required for the resource-constrained sensor nodes.

1.5 Roadmap of the Paper

The rest of the paper is organized as follows: In Sect. 2, we provide some preliminaries on sensor networks and their challenges, which are useful for describing and analyzing the proposed schemes. In Sect. 3, we proposed two new key predistribution schemes. In Sect. 4, we derive the probability for establishing direct keys between neighbor nodes and then analyze the security against node capture attack. In Sect. 5, we compare the performance of our proposed schemes with other existing schemes. Finally, we conclude the paper with some concluding remarks in the last section.

2 Sensor Network Preliminaries and Challenges

Sensor network is build up with hundreds to thousands of sensor nodes. Sensor nodes are densely deployed over an area or monitoring object such that they can communicate with each other efficiently. Generally, tiny nodes are distributed from aircrafts or trucks over an area. After deployment of nodes, they establish their secret keys. Each node transmits the collected data via other nodes or directs them to base station node securely. Base node is a gateway to other network or human interface of a sensor network. Information are gathered here to make a decision.

Sensor networks are challenging to implement due to following reasons.

- Low memory, low computation power and low life time: Sensor nodes are very small in size with small storage capacity, battery backup, and computation power. Thus symmetric key cryptography is used where secret key sharing between nodes is a big issue. Though by using symmetric key cryptography we reduce computation power and enhance sensor node's life time, it is still impractical regarding storage capacity to store a massive number of secret keys for large network.
- No prior knowledge before deployment: Nodes are preinitialized with secret keys and randomly distributed over an area which implies nodes are uncertain about their location and also about their neighbors. Due to short communication range of a sensor node, it will always have less number of neighbors. So the probability to direct key establishment is less.
- Node capture and replication: Adversary can physically capture sensor nodes and may know all secret information stored in that node. The adversary can also replicate that node in the network for his benefit.
- Dynamic nature of network topology: Sensor nodes may change their location as well as new nodes can be added later.

3 Proposed Schemes

In this section, we propose two schemes which can provide perfect resiliency with high connectivity between sensor nodes. Hash function is used to establish secret shared keys. We generate secret key chain using two hash chains and predistribute it to sensor nodes by offline. Generation of secret key chain of our first scheme(Scheme 1) is described below.

3.1 Scheme 1: Key Predistribution Using Linear Key chain

In this section, we describe linear key chain generation and the key predistribution of our Scheme 1 using this linear key chain in sequel.

Linear Key Chain Generation

Key chain is formed from hash chains where hash function [15] is used repeatedly on a randomly seed value, which can also be considered as a secret. We generate a key chain by merging two hash chains of equal length. Suppose two hash chains are Ch_{1a} and Ch_{1b}, where $Ch_{1a} => h^1(s_{1a}) \longrightarrow h^2(s_{1a}) \longrightarrow \cdots \longrightarrow h^l(s_{1a})$ such that $h^l(s_{ia}) = \underbrace{h(h...h(s_{ia}))}_{l \text{ times}}$ and $Ch_{1b} => h^l(s_{1b}) \longleftarrow h^{l-1}(s_{1b}) \longleftarrow \cdots \longleftarrow h^1(s_{1b})$, s_{1a} and s_{1b} are two different seed values for chain Ch_{1a} and Ch_{1b}, respectively. A key chain Ch_1 is produced by merging these two hash chains, $Ch_1 => (Ch^1_{1a}\|Ch^1_{1b}) \rightleftarrows (Ch^2_{1a}\|Ch^2_{1b}) \rightleftarrows \cdots \rightleftarrows (Ch^l_{1a}\|Ch^l_{1b})$, where $Ch^i_{1a} = h^i(s_{1a})$ and $Ch^i_{1b} = h^{l-i+1}(s_{1b})$ for $i = 1, 2, 3, \ldots, l$. Since hash chain length is l, so key chain length is also l denoted as $Ch^1_1, Ch^2_1, \ldots, Ch^l_1$ for key chain Ch_1 and we can distribute these l secret keys $(Ch^1_i, Ch^2_i, ...)$ to l sensor nodes. In Fig. 2, we show the keychain formation.

Each key chain is generated by a pair of hash chain of two different seeds. For each key chain, we use two different seeds randomly. Therefore, for c key chains, we use $2c$ random seed values. If two key values are known of a key chain, one can compute the intermediate keys corresponding to the key chain where the nodes contain keys from the same key chain. Therefore, one can easily compute secret keys of other nodes if secret keys of any two nodes are compromised and the secret keys are generated from the same key chain.

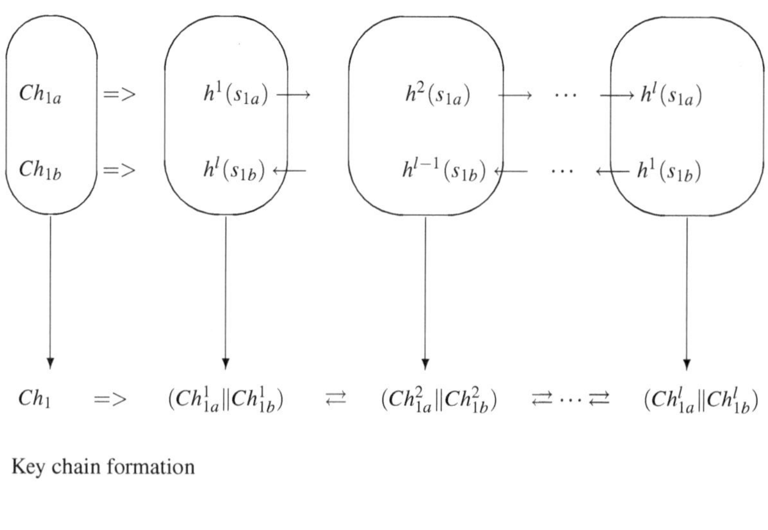

Fig. 2 Key chain formation

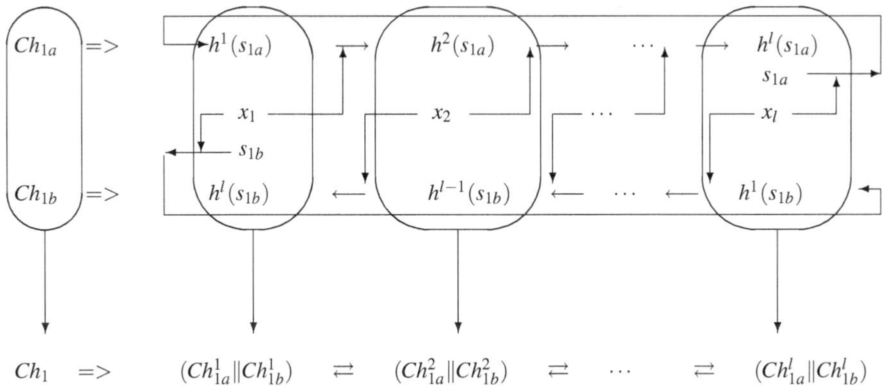

Fig. 3 Linear key chain formation using keyed hash function

To remedy the weakness, we can design it in such a way that key chain's secrecy not only depends upon the first and last hash values, but also depends upon each hash value. Then, resiliency against node capture becomes perfectly unconditional. We achieve this by using keyed hash function. In each hash value, we use different keys. If we do not know the actual key, we cannot get higher level hash values of a linear key chain by hashing only. In Fig. 3, we show key generation where each node stores a secret key along with corresponding secret information which is hashed to get next higher level secret key(hash value).

Each node in a network consists of some secret information, where the information can be a random key of length 64–256 bits and two keyed hash values of two hash chain. For an example, a node can consists of a triple of information $k^i_j = (Ch^i_{ja}, Ch^i_{jb}, x_i)$ as secret information, where $i(1 \leq i \leq l)$ represents the number of hashing and $j(1 \leq j \leq c$ where c is the number of linear key chain) can be c, represents the index of key chain. The nodes which contain the first and last key values of a linear key chain need to store one more secret seed value of the corresponding hash chain. The generalized hashing technique to form linear key chain is as:

$$h^i_{1a}(s_{1a}) = h(h^{i-1}(s_{1a})\|x_{i-1}),\, x_0 = x_l$$

and

$$h^i_{1b}(s_{1b}) = h(h^{i-1}(s_{1b}\|x_{l+2\text{-}i \bmod l}),$$

where $\|$ represents concatenation operator.

Key Predistribution Using Linear Key Chain

In linear key chain, key chain length l depends upon the total number of key chains (c) and key ring size m. We discuss it briefly in below:

1. Let the total number of secret key chains is c and each secret key chain length is l.
2. Assume that key ring size is m. Now m distinct linear key chain is chosen randomly l times from c precomputed linear key chains with replacement. A secret key of a key ring is taken randomly from each chosen linear key chain of length l with replacement. Total number of possible key rings is $\binom{c}{m}$. Each linear key chain is chosen $l = \binom{c-1}{m-1}$ times which is very large. Therefore, the linear key chain length (l) is equal to the number of uses of a linear key chain that is $l = \binom{c-1}{m-1}$.

Remark 1 Each secret key of linear key chains needs to identify uniquely by calculating a offset value of length $\lceil \log_2 c \rceil + \lceil \log_2 l \rceil$ in terms of bits. First, $\lceil \log_2 c \rceil$ bits identify linear key chain number. Next, $\lceil \log_2 l \rceil$ bits identify linear key chain's repeated hash value number. A sensor node's key ring is filled with previously unused secret keys of length $2|h| + |r|$ (where, h represents hash length and r represents length of random number) from each m linear key chain along with offset value used as identifier to identify each secret key uniquely and a unique node id provided by key setup server. So each node is stored m secret keys with m identifiers and a unique id.

3.2 Scheme 2: Key Predistribution Using Multilevel Key Chain

In this section, we first discuss about multilevel key chain formation and next, we discuss predistribution of the keys using multilevel key chain.

Multilevel Key Chain Generation

Using different secret information in each node, we provide perfect security, but the connectivity of a network pretends to be very poor, if the key chain length is very long. To overcome the weakness, we propose a design where we generate multilevel key chain based on the number of secret information stored in a node. Similar to the previous design, we use keyed hash function. Unlike previous design, seed information is hashed r times with r different secret information to connect with r different nodes directly. That means, if a node contains r secret information, it can communicate to maximum n nodes depending on building block key chain length (x) which is described below in same level and $n \times (r-1)$ nodes in next $r-1$ levels. Now lengthy hash chain is composed of multilevel interrelated hashed values. Higher length implies higher number of level, that is, higher number of direct link from a seed. The number of hashed values which are used as secret keys distinctly depends upon previous level's total number of hashed values and the number of secret

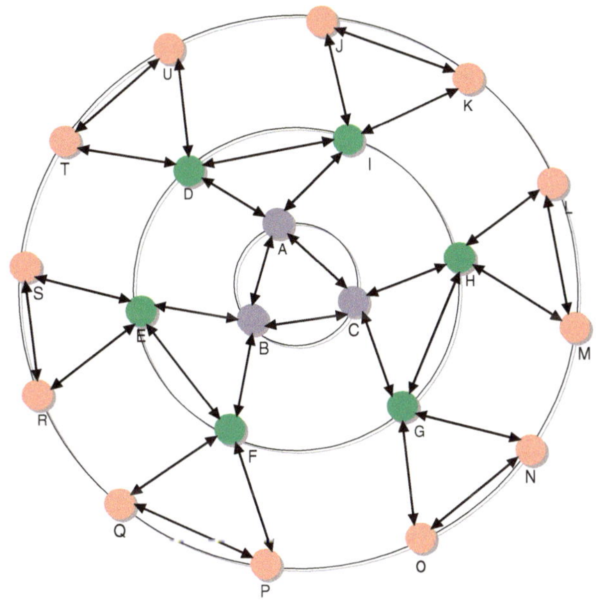

Fig. 4 Multilevel key chain formation using keyed hash function and multiple secret information

information used in that particular level. In Fig. 4, there are three nodes A, B, and C in first most inner level(circle). Next level consists of six nodes, namely, D, E, F, G, H, and I and the outermost level consists of 12 nodes, namely J, K, L, M, N, O, P, Q, R, S, T, and U. There are minimum two secret information in a node, if it is in the first level. A node contains one secret information if it is in the last level(outer most level) which is described in Fig. 4.

In Fig. 4, A, B, and C are in a same chain and in same level. Each of them containing two secret information. Each secret information is used in two different key chain which are used in same level and immediate next level. By using these two secret information, they are directly connected to four other nodes. In the first circle, that is in first level, there are two secret information and in the immediate next level there are also two secret information. In the last level, there is only one secret information. Multilevel key chain is composed of small length interrelated key chains. First level is composed of one key chain with certain length x. This chain is called the *building block key chain* of multilevel key chain. Here, we consider $x = 3$. All small interrelated key chains' length is x except last level's key chain block length which we discuss in the next Sect. 3.2.2.

Key Predistribution Using Multilevel Key Chain

After multilevel key chain generation discussed in Sect. 3.2.1, we form key ring and predistribute each secret key to appropriate key ring with multiple secret information. This procedure is done offline by key server. In multilevel key chain formation, key chain length, l_c is calculated exactly to desired key chain length, $l = \binom{c-1}{m-1}$. If not exact then it is calculated to be very nearer, but less value ($l_c < l$) of desired chain length by manipulating level number and as well as number of secret information in each level. Let total number of levels be L and in each level, $r_i (1 \leq i \leq L)$ secret information are there. Then always $r_L = 1$ as it is the last level of multilevel key chain and r_1 is greater than one unless first level cannot link next outer levels. Except Lth level, other inner level's secret information $r_1, r_2, r_3, ..., r_{L-1}$, respectively, are chosen smartly to match with desired chain length l. If $l_c < l$, then we calculate $l_r = l - l_c$. These l_r keys belong to last level that is in Lth level with one secret information, that is, $r_L = 1$ and will be stored in the last level of multilevel nodes. If multilevel key chain's *building block key chain* known as unit of multilevel key chain, is x then some of last level's key chain blocks length will be higher than x to hold total l_r keys in last level. l_c is calculated as,

$$l_c = x + x[2^1(r_1 - 1) + 2^2(r_1 - 1)(r_2 - 1) + 2^3(r_1 - 1)(r_2 - 1)(r_3 - 1) + \cdots + 2^{L-1}(r_1 - 1)(r_2 - 1)(r_3 - 1) \cdots (r_{L-1} - 1)].$$

The above expression can be generalized as:

$$l_c = x\left[1 + \sum_{i=1}^{r-1}\left\{2^i \prod_{j=1}^{i}(r_j - 1)\right\}\right].$$

when, $r = r_1 = r_2 + 1 = r_3 + 2 = \cdots = r_{L-1} + L - 2$.

Remark 2 Each key is given a unique id same as Remark 1. After calculation of levels (L), we can get different number of consecutive keys to different level secret keys(at least two). These keys are now predistributed as described in earlier subsections.

3.3 Master Shared Secret Key Establishment

In this section, we describe master shared secret key establishment for both Scheme 1 and Scheme 2. After deployment of sensor nodes over an area, first of all each node detects its neighbor nodes. Suppose a node B is in the communication range of a node A. They interchange their m identifiers corresponding of m secret keys which are stored before deployment. To communicate securely, A and B establish master shared secret key using common identifiers. Of course they establish secret key if they share the keys from the same secret key chain. But as secret key chain length varies with network size, for large network a node may have to compute large number (maximum $l - 1$) of hashing upon same secret key chain match. To reduce possibility of huge computation, we establish master shared secret key by fixing up the number of hashing. We want maximum number of one time hashing on each matched secret key chain share for each pair of communicating nodes. That means to communicate, both nodes not only share same secret key chain, but also have to share any two consecutive secret keys of same secret key chain. In a secret key chain, first secret key which is generated by one time hashing on a secret seed has next hash value known as right consecutive key and similarly last secret key has previous hash value known as left consecutive key. Except these two positions, all other secret keys have both side hash value known as right and left consecutive keys. If we store seeds as described in Fig. 3, the first secret key may have circularly next left secret key and last secret key may have circularly next right secret key. In the following steps, we describe how master secret key is shared between two communicating nodes.

Step 1: Both A and B nodes can establish a key, if they share two consecutive secret keys of same secret key chain. Node A checks whether the first $\lceil \log_2 c \rceil$ bits of $\lceil \log_2 c \rceil + \lceil \log_2 l \rceil$ bits sent by the node B is appear in the memory. If it is true, node A then compares matched identifier's second part $\log_2 l$ to check whether they are consecutive or not. If consecutive, it retrieves matched key chain's secret key locations(hash position) j, k in node A and node B, respectively, where $j, k \in \{1, 2, \ldots, l\}$.

Node A may have several matched key chains's secret key, $n(1 \leq n < m)$ which is consecutive with node B's secret key. For n matched secret key chains, node A computes secret key chain number(i_p) where $p \in \{1, 2, \ldots, n\}$ and $i_p \in \{1, 2, \ldots, c\}$ by comparing identifier's first part. Node A also computes j_p and k_p, the contribution position of secret key chain i_p to node A and node B, for each p where $j_p, k_p \in \{1, 2, \ldots, l\}$ and they are consecutive by comparing i_pth secret key chain identifier's second part. For each secret key chain i_p, node A computes shared secret key $ssk_{AB}^{i_p}$ and node B also computes $ssk_{BA}^{i_p}$ where $ssk_{AB}^{i_p}$=$ssk_{BA}^{i_p}$. The shared secret key $ssk_{AB}^{i_p}$ is calculated as follows.

Case I: if $j_p < k_p$, node A computes $ssk_{AB}^{i_p} = h^{k_p - j_p}(Ch_{i_{pa}}^{j_p}) \oplus Ch_{i_{pb}}^{l_{i_p}-(j_p-1)}$, where $\oplus$ denotes bitwise exclusive OR operation. Node B computes $ssk_{BA}^{i_p} = Ch_{i_{pa}}^{k_p} \oplus h^{k_p - j_p}(Ch_{i_{pb}}^{l_{i_p}-(k_p-1)})$.

Case II: if $j_p > k_p$, node A computes $ssk_{AB}^{i_p} = Ch_{i_{pa}}^{j_p} \oplus h^{j_p - k_p} (Ch_{i_{pb}}^{l_{i_p}-(j_p-1)})$. Node B computes $ssk_{BA}^{i_p} = h^{j_p - k_p}(Ch_{i_{pa}}^{k_p}) \oplus Ch_{i_{pb}}^{l_{i_p}-(k_p-1)}$.

Now node A and B compute master shared secret key, $ssk_{AB}^{M} = ssk_{AB}^{i_1} \oplus ssk_{AB}^{i_2} \oplus \cdots \oplus ssk_{AB}^{i_n} = ssk_{BA}^{M}$. This master shared secret key establishment is called direct key establishment. After direct key establishment, if network connectivity is still poor, we then further need to compute indirect key establishment as described in the next step(Step2).

Step 2: Nodes A and B also can establish direct key even if they do not share consecutive secret keys of same secret key chain or they do not have any common secret key chain. Two cases can be figured out when nodes are establishing master shared secret key with the help of path key.

case 1: Suppose nodes A and B hold the keys generated from the same key chains, but they do not share consecutive keys. In this case, both nodes interchange their identifier's ids, nonces. And then they calculate hash value positions j, k of a shared key chain as described previously. As j and k are not consecutive to each other, so $l - 2 \geq |j - k| \geq 2$, where l, j, k are the length of key chain, hash value position in node A, and hash value position in node B. Now nodes A and B both transmit following identifiers to its immediate neighbors with whom it has direct key. This transmission is done up to h number of hops. Generally to reduce computational overhead, h is chosen between 2 and 3.

- *For $k = 1$:* If k is the first hash value of the shared secret key chain, then node A needs $(k+1)$th and $(k+l-1)$th contributions of shared secret key chain to establish a direct key with node B. Node A transmits desired secret key's identifiers to its direct neighbors with whom it has direct key. Neighbor nodes again retransmit to its direct neighbors until at least one identifier gets matched. Transmissions take place as follows:

  Step i: Let a identifier of A is matched with node C via h intermediate nodes $A_1, A_2, \ldots, A_h$. So the secret key is found in node C through a path $< A = A_0, A_1, A_2, \ldots, A_{h+1} = C >$. During node to node transmission, the path is saved which is followed reversely to go back to initial node A after identifier match. $A_{i-1} \rightarrow A_i$:$(\sum_{j=1}^{i} id_{A_{j-1}} \| \{\text{list of identifiers}\}$ $\| MAC_{ssk^M_{A_{i-1}A_i}} (\sum_{j=1}^{i} id_{A_{j-1}} \| \{\text{list of identifiers}\}))$ for $i = 1, 2, 3, \ldots, h+1$.

  Step ii: Now node C securely transmits secret key to node A after identifiers are matched. $A_i \rightarrow A_{i-1}$:$(\sum_{j=1}^{i-1} id_{A_{j-1}} \| E_{ssk^M_{A_iA_{i-1}}} \{\text{secret key from node C}\} \| MAC_{ssk^M_{A_iA_{i-1}}} (\sum_{j=1}^{i-1} id_{A_{j-1}} \| E_{ssk^M_{A_iA_{i-1}}} \{\text{secret key from node C}\}))$ for $i = h+1, h, h-1, \ldots, 1$. Node A establishes direct key with node B using currently received secret key which is right consecutive of kth hash value in node B. Step 1 is followed to establish direct key.

- *For $k = l$:* If k is the last contribution of shared key chain, it needs 1st and $(k-1)$th contributions to establish a direct key. The transmissions are same as previous two steps, Step i and Step ii when $k = 1$.
- *For $1 < k < l$:* In that case, it searches $k + 1$th and $k - 1$th contributions of same secret key chain and transmissions are same as previous Step i and Step ii.

case 2: Suppose nodes A and B do not hold the keys generated from same key chain. First, nodes A and B interchange their identifier ids, nonces. After that they calculate hash value positions j, k to find out a common key chain and then consecutive keys of that key chain. Nodes A and B both transmit these identifiers to their neighbor nodes with whom they have direct keys up to h(2–3) number of hops. Unlike case 1, here node A and B both first check desired identifier's first part, that is, $\lceil \log_2 c \rceil$ bits of $\lceil \log_2 c \rceil + \lceil \log_2 l \rceil$. After successful matching of identifer's first part, second part of that identifier that is, $\lceil \log_2 l \rceil$ is checked. Upon successful matching of identifier's both parts, transmissions are as Step i and Step ii. In case 2, communication overhead increases than case 1 and initially nodes prefer not to follow. If, after case 1, probability of connectivity remains poor, only then case 2 is followed.

4 Analysis and Simulation Results

We derive the probability for establishing direct keys between neighbor nodes and analyze the security issues when nodes are vulnerable to node compromise attack. We find the connectivity of our two proposed schemes.

4.1 Probability of Establishing Direct Keys in Scheme 1

We calculate the probability, p, that any two nodes can establish direct key. We can see that two neighbor nodes A and B can establish direct keys when they have secret keys from the same secret key chain and secret keys are consecutive. As in each key ring have m keys from m key chains out of total c key chains and m is more than half of total keychains, c. So we have minimum $m - (c - m)$ common key chains in any two nodes. Now, we choose m key chains from total c key chains with replacement. So any two nodes may have maximum $m - 1$ common key chains.

So we have, $p = \sum_{i=m-(c-m)}^{m-1}$(probability of having i common key chain(s) $\times$ (1 $-$ probability of not being consecutive key for each i key chain)). We derive and calculate the formula as follows : probability of having i common key chain(s) is derived as if both nodes have i common key chain(s) that is they have $(m - i)$ uncommon key chain(s). If one node chooses m key chains from c then to have i key chain(s) $((m - (c - m)) \leq i < (m - 1))$ in common other node will choose i from $c - (m - i)$ keychains, which is already used key chain(s) by any node previously. Mathematically,

$$p = \sum_{i=m-(c-m)}^{m-1} \left[\frac{\binom{c-(m-i)}{i}}{\binom{c}{m}} \times \left\{ 1 - \left(\frac{l-3}{l} \right)^i \right\} \right],$$

where c, m, l are total secret key chains, key ring size, and secret key chain length, respectively. The above expression can be written as,

$$p = \sum_{i=m-(c-m)}^{m-1} \left[\frac{\prod_{j=0}^{c-m-1} \frac{c-(m-i)-1}{c-m-j}}{\prod_{j=0}^{c-m-1} \frac{c-j}{c-m-j}} \times \left\{ 1 - \left(\frac{l-3}{l} \right)^i \right\} \right].$$

We plot the simulation results in Fig. 5.

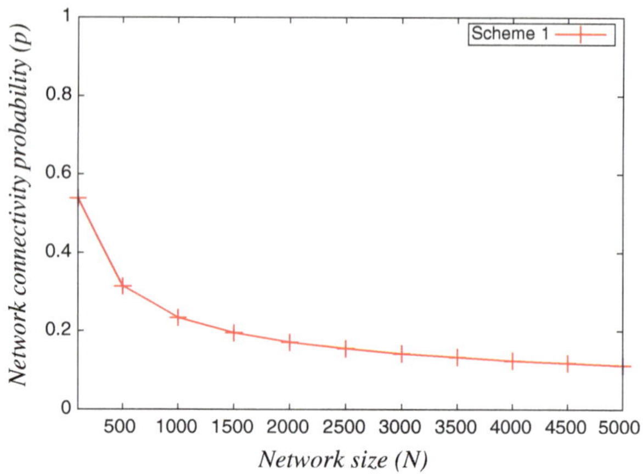

Fig. 5 Probability of direct key establish between any two nodes using linear-level key chain

Here we increase the network size by increasing total number key chains and key ring size, respectively. We simulate with $m = c - 2$ in each time to get better connectivity. From the simulation result, it is clear that if network size increases, the probability, p, decrease. We also notice that fall of probability depends upon the network size increase and it is comparatively less than other schemes like basic scheme [3], EPKDSN scheme [12] and IBPRF [6]. We describe it elaborately in Sect. 5.

4.2 Probability of Establishing Direct Keys in Scheme 2

We further improve the probability of direct key establishment between any two nodes using multilevel hierarchical secret key chain. A secret key from a key chain can communicate directly with other two consecutive secret keys of same key chain. As key chain length increases, the probability of any two nodes having consecutive keys of same key chain decreases. But in multilevel key chain, a secret key can link directly to maximum $2r$ secret keys and minimum two secret keys, where r is the number of secret information in that level. We have shown that with key chain length increase, either total level number (L) will increase or value of r will increase in particular levels, or both will increase. The number of consecutive secret keys of a secret key also increases dynamically. So obviously the probability of connectivity will increase.

Probability of having consecutive keys $= 1 -$ probability of having nonconsecutive keys. Now total number of consecutive keys of a key depends upon the number of secret information r_i in the level L_i. If a key has r_i secret information then it has $2r_i$ consecutive keys that is $l - (2r_i + 1)$ nonconsecutive keys, where l is desired key chain length. We easily calculate that the probability of having nonconsecutive keys is $\frac{l-(2r_i+1)}{l}$. This probability of having nonconsecutive keys depends upon the probability of the key having the secret information r_i. Now the probability of having

r_i is calculated as total number of secret keys of that multilevel key chain having r_i over total number of secret keys in that multilevel key chain. Let l_{ci} is the number of secret keys having r_i secret information, then the probability of having r_i is $\frac{l_{ci}}{l}$.

So, $\frac{l_{ci}}{l} \times \frac{l-(2r_i+1)}{l}$ is the probability of having nonconsecutive keys of a key if the key has r_i secret information. As r_i changes with $L_i, i \in \{1, 2, 3, \ldots, L\}$ (where L represents total number of levels of network), the number of nonconsecutive keys, $l - (2r_i + 1)$ also changes. Now total probability (p_{nc}) of having nonconsecutive keys is calculated as:

$$p_{nc} = \sum_{i=1}^{L} \left\{ \frac{l_{ci}}{l} \times \frac{l - (2r_i + 1)}{l} \right\},$$

where L represents total number of levels, $r_i \in \{r_1, r_2, r_3, \ldots., r_L\}$ and $l_{ci} \in \{l_{c1}, l_{c2}, l_{c3}, \ldots, l_{cL}\}$. We design the multilevel key chain in such a way that the last level keys have one secret information that is $r_L = 1$ and $l_{c1} = 3$ as we assume the multilevel key chain building block unit is 3. The $l_r = l - l_c$ is added with l_{cL} that is in last level. These l_r secret keys have at most two consecutive keys. So the total probability of establishing direct key between any two nodes, p_m, is calculated as $p_m = \sum_{i=m-(c-m)}^{m-1}$(probability of having i common secret key chains $\times$ (1 − probability of not being consecutive key for each i key chain)). The formula is after derivation,

$$p_m = \sum_{i=m-(c-m)}^{m-1} \left[\frac{\prod_{j=0}^{c-m-1} \frac{c-(m-i)-1}{c-m-j}}{\prod_{j=0}^{c-m-1} \frac{c-j}{c-m-j}} \times \left\{ 1 - \left(p_{nc}\right)^i \right\} \right],$$

where c, m, p_{nc} are total secret key chains, key ring size, and probability of having nonconsecutive keys, respectively. We plot the simulation results for multilevel key chain in Fig. 6 and show that multilevel key chain has better connectivity over linear key chain. During simulation, we assume parameter values given in following Table 1.

If network still remains disconnected with high probability initially, we can obtain high network connectivity after applying few hops of LAKES [16]. In LAKES [16], it is shown that the network becomes connected with high probability if the number of hops as well as the average number of neighbors of each sensor node are increased.

4.3 Security Analysis

The security of our schemes depend on one-way hash functions and a node's secret information stored in nodes. We achieve perfect resiliency by using distinct secret key in each node unlike basic scheme [3] where a key is used repeatedly in several nodes. By compromising a node, an adversary can only know the half portion of previous and next secret key of that secret key chain. Each secret key is formed by

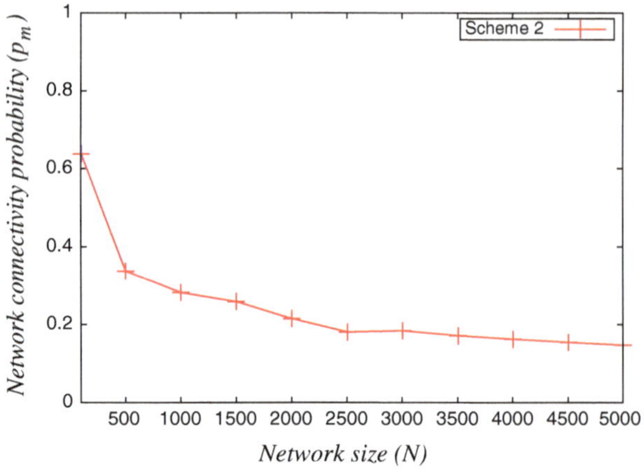

Fig. 6 Probability, p_m of direct key establish between any two nodes using multilevel secret key chain

Table 1 Simulation parameters

Total key chain (c)	Key ring size (m)	Level no (L)	r_i in level $1, 2, \ldots, L$
5	3	1	1
7	5	2	3, 1
10	8	3	2, 3, 1
15	13	3	4, 3, 1
20	18	3	6, 3, 1
40	38	3	12, 6, 1
60	58	3	18, 9, 1
80	78	3	23, 12, 1
100	98	3	27, 16, 1
101	99	3	29, 15, 1

merging two equal length hash chains reversely. Now it is easy to capture all secret keys if the adversary compromise two nodes containing first and last hash value of same key chain. But using keyed hash function, we restrict computations from first to last hash values of a key chain. Nodes store a secret information which is used as a key to hash function. Each key is generated by merging two hash over two previous hash values of two key chains with two secret information stored along with that hash values in two different nodes. Now the adversary cannot compute any secret key of a noncompromised node without knowing the two secret information stored along with immediate after and previous secret key of same secret key chain. The resiliency is perfect that is adversary cannot compromise links between nodes which are still noncompromised. We achieve this by merging two hash chains reversely and restricting the link from each hash value to next higher hash value of a key chain by hashing each time with a unique secret information which is different to different nodes. Thus no matter how many sensor nodes are compromised, the direct pairwise keys between noncompromised nodes remain secure. So our scheme provides perfect resiliency in this way.

5 Performance Comparison with Existing Related Schemes

We compare our schemes with EG scheme [3], q-composite scheme [4], polynomial pool-based key scheme [12] and with IBPRF [6]. It is shown that our schemes provide better connectivity and resiliency compare to them.

5.1 Security Issues

The EG scheme [3] and the q-composite scheme [4] may reveal a large fraction of pairwise keys shared between noncompromised nodes even if the number of nodes capture is small. Polynomial pool-based scheme [12] is unconditionally secure with t-collusion resistant. For large network t is not dynamic; that is, when network size increases t can be increased to a certain limit considering huge computation. Our schemes provide better resiliency than EG and q-composite schemes and for large network, our schemes are more resilient than polynomial pool-based scheme.

In IBPRF [6] if a master key Mk_u of a node u is compromised, then all shares (symmetric keys) with x nodes that is $PRF_{MK_u}(v_i)$ $(1 \leq i \leq x)$ is compromised. During key predistribution phase, node u selects m randomly nodes and stores m symmetric keys corresponding these m nodes. Now MK_u is known only when an adversary captures node u and m symmetric keys is known to adversary. It replies total $(x + m)$ symmetric keys is known to the adversary by compromising one node u. But in our scheme, each key of a secret key chain is treated as a share; that is, if a secret key chain length is l then it has total l shares and this l shares are derived by l times hashing over previous hash value with unique secret information each time. So total l secret information is used. Now an adversary needs to know all l secret information to know all l shares unlike one secret information in IBPRF. In our schemes by knowing one secret information, an adversary can know only half portion of two shares, not a full single share, whereas in IBPRF, an adversary knows all x shares of Mk_u to x nodes. In our schemes, if a node stores m shares, then an adversary can know only half portion of $2m$ different shares (because each key is composed of two hash values) that is not even a single share. Though we choose these m shares from m different secret key chains, the adversary knows $2m$ half shares from m distinct key chain rather than of one secret key chain. This phenomenon improves security against node compromise along with better connectivity.

5.2 Network Connectivity

From EG and q-composite schemes, we can see that the connectivity depends upon key pool size and key ring size. Polynomial pool-based key predistribution scheme's connectivity depends upon polynomial pool size and shares of each polynomial. Our schemes depend upon secret key chain length(l), total number of key chain(c), and key ring size, $m(= l)$. Our schemes support large network with better connectivity than polynomial pool-based scheme shown in Fig. 7.

From Fig. 7, we conclude that when network size is very large our schemes provide better connectivity with perfect resiliency. Initially, it has less probability comparing with polynomial pool scheme and IBPRF, but high enough that network model is remain connected. In our schemes, probability of connectivity falls slowly unlike polynomial pool scheme. In polynomial pool scheme, for small increases in network size, probability of connectivity fall comparatively near to double.

In IBPRF, a node u can be picked up by other x nodes randomly during key predistribution phase and node u also randomly chooses m nodes. The probability of having common nodes in x and m is calculated as: 1 − probability of having distinct nodes. Mathematically, it is equal to $1 - \binom{M-m}{x}/\binom{M}{m}$, where M is pool of ids of n sensor nodes. From this formula, we notice that there is very less probability of having common values when $x = m$ even this probability tends to zero for larger M and m values, and higher probability of having common values when $(x < m)$. As there is a tradeoff between network size and values of m, x to get higher network connectivity, so x is always less than m. It replies, $m + x < 2m$, that is a node u has direct connectivity to nodes where the number of nodes is less than $2m$. In the proposed schemes each m secret keys is chosen from m distinct secret key chains and each of secret key can be used to communicate with two nodes where the nodes contain the consecutive secret keys from same secret key chain. We compare the connectivity of our schemes with IBPRF and showed the simulation results in following Fig. 8. From

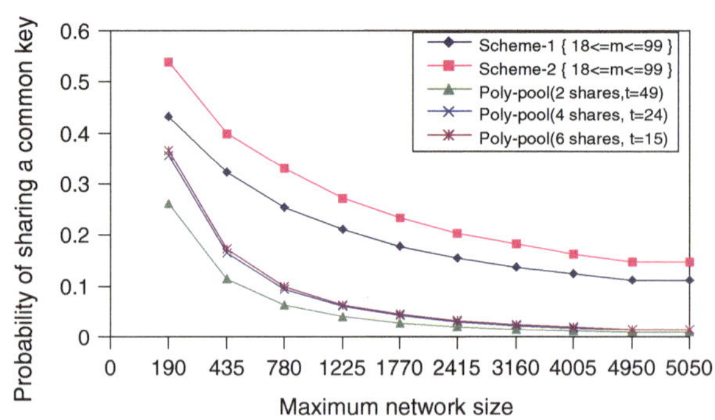

Fig. 7 Comparison with polynomial pool, assuming a node is capable of holding 100 keys

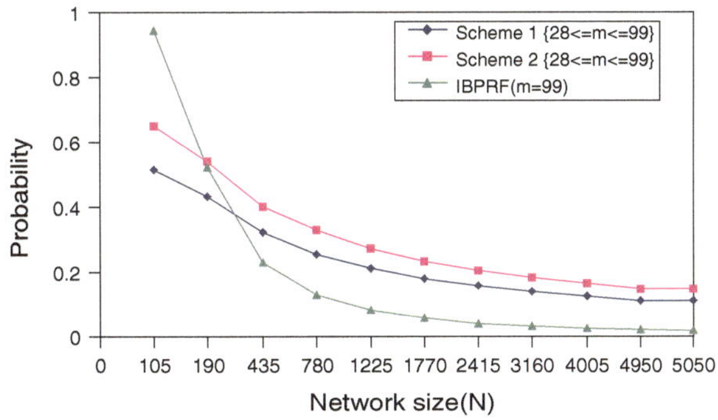

Fig. 8 Comparison with IBPRF with $m = 99$. Our schemes achieve more probability with m, ranging in $28 <= m <= 99$

Fig. 8, we conclude that when network size is very large, our schemes provide better connectivity. Initially, both schemes have less probability of connectivity comparing with IBPRF [6] and after that, as network size (N) increases our schemes provide better connectivity.

6 Conclusion

In this paper, we have proposed two protocols which are improvement of bootstrapping protocols with the help of hashchain for direct key establishment. We have designed a way to restrict the drawbacks of direct use of hash chain. We have shown our schemes provide better network connectivity for large networks with perfect resilience against node capture than EG scheme, polynomial pool-based scheme and IBPRF scheme. We have also shown diligently that uses of our secret key chain design in scheme 2, that is multilevel secret key chain, will support more large networks with better network connectivity and resiliency.

Acknowledgments The authors would like to acknowledge the many helpful suggestions of the anonymous reviewers which have improved the content and the presentation of this paper.

References

1. Akyildiz, I.F., Su, W., Sankarasubramaniam, Y., Cayirci, E.: Wireless sensor networks: a survey. Comput. Netw. **38**(4), 393–422 (2002)
2. Dolev, D., Yao, A.: On the security of public key protocols. IEEE Trans. Inf. Theory **29**(2), 198–208 (1983)
3. Eschenauer, L., Gligor, V.D.: A key management scheme for distributed sensor networks. In Proceedings of the 9th ACM Conference on Computer and Communication Security, pp. 41–47, Nov 2002
4. Chan, H., Perrig, A., Song, D.: Random key predistribution schemes for sensor networks. In: IEEE Symposium on Security and Privacy, Berkeley, California (2003)

5. Castelluccia, C., Spognardi, A.R.: A robust key pre-distribution protocol for multi-phase wireless sensor networks. In: Proceedings of Third International Conference on Security and Privacy in Communication Networks (SecureComm'07), pp. 351–360 (2007)
6. Das, A.K., Giri, D.: An identity based key management scheme in wireless sensor networks. In: Proceedings of 4th Asian International Mobile Computing Conference (AMOC 2006), Kolkata, India, pp. 70–76. Tata McGrawHill Press, Kolkata (2006)
7. Kalkan, K., Yilmaz, S., Yilmaz, O.Z., Levi, A.: A highly resilient and zone-based key predistribution protocol for multiphase wireless sensor networks. In: Proceedings of the 5th ACM Symposium on QoS and Security for Wireless And Mobile Networks (Q2SWinet'09), pp. 29–36 (2009)
8. Unlu, A., Armagan, O., Levi, A., Savas, E., Ercetin O.: Key predistribution schemes for sensor networks for continuous deployment scenario. In: Proceedings of IFIP International Conferences on Networking (Networking 2007), Lecture Notes in Computer Science (LNCS), vol. 479, pp. 239–250 (2007)
9. Das, A.K.: An efficient random key distribution scheme for large-scale distributed sensor networks. Secur. Commun. Netw. **4**(2), 162–180 (2011)
10. Das, A.K.: A random key establishment scheme for multi-phase deployment in large-scale distributed sensor networks. Int. J. Inf. Secur. **11**(3), 189–211 (2012)
11. Das, A.K.: Improving Identity-based random key establishment scheme for large-scale hierarchical wireless sensor networks. Int. J. Netw. Secur. **14**(1), 1–21 (2012)
12. Liu, D., Ning, P.: Establishing pairwise keys in distributed sensor networks. In: Proceedings of 10th ACM Conference on Computer and Communications Security (CCS), pp. 52–61. Washington DC, 27–31 Oct 2003
13. Du, W., Deng, J., Han, Y.S., Varshney, P.K.: A pairwise key pre-distribution scheme for wireless sensor networks. In: Proceedings of the 9th ACM Conference on Computer and Communications Security, Washington, DC, USA, Nov 2003
14. Perrig, A., Szewczyk, R., Wen, V., Culler, D., Tygar, D.: SPINS: security protocols for sensor networks. In: Proceedings of Seventh Annual International Conference on Mobile Computing and Networks, July 2001
15. Forouzan, B.A.: Cryptography and Network Security. Tata McGraw-Hill, New York (2007)
16. Das, A.K.: A location-adaptive key establishment scheme for large-scale distributed wireless sensor networks. J. Comput. **4**(9), 896–904 (2009)

IMSmining: A Tool for Imaging Mass Spectrometry Data Biomarker Selection and Classification

Jingsai Liang, Don Hong, Fengqing (Zoe) Zhang and Jiancheng Zou

Abstract We developed IMSmining, a free software tool combining functions of intuitive visualization of imaging mass spectrometry (IMS) data with advanced analysis algorithms in a single package which is easy to operate. Main functions of IMSmining include data visualization, biomarker selection, and classification using advanced multivariate analysis methods such as elastic net, sparse PCA, as well as wavelets. It can be used to study the correlation and distribution of the IMS data by incorporating the spatial information in the entire image cube and to help finding the distinction of the possible features caused by the biological structure and the potential biomarkers. This software package can be downloaded from http://capone.mtsu.edu/dhong/IMSmining.htm.

Keywords IMS data processing · Statistical computing · Wavelet application · Biomarker selection and Classification · Software package

---

J. Liang · D. Hong (✉) · J. Zou
Computational Science Program, Middle Tennessee State University, Murfreesboro, TN, USA
e-mail: Don.Hong@mtsu.edu

J. Liang
e-mail: JL4Z@mtmail.mtsu.edu

J. Zou
e-mail: zjc@ncut.edu.cn

D. Hong · J. Zou
College of Sciences, North China University of Technology, Beijing, China

F.Z. Zhang
Department of Psychology, Drexel University, Philadelphia, PA, USA
e-mail: fengqingzoezhang@gmail.com

R.N. Mohapatra et al. (eds.), *Mathematics and Computing*,
Springer Proceedings in Mathematics & Statistics 139,
DOI 10.1007/978-81-322-2452-5_11

1 Introduction

Mass spectrometry (MS) and imaging mass spectrometry (IMS) are both important techniques in proteomics. IMS is a novel technology that is able to incorporate spatial biochemical information in full molecular range [1]. However, there are still many challenges in data processing due to high dimensionality, huge differences between the number of predictors and the sample size, and the incorporation of both spectral and spatial information. All these challenges pose great difficulties in model selection and data processing.

Several software tools are commonly used for IMS/MS data analysis. Biomap and Tissue View are mainly for data visualization. These software tools lack advanced data analysis functionality such as multivariate analysis methods for biomarker selection and classification. MarkerView and ClinProTools are packages for MS data analysis. Technically, IMS data after using Biomap or Tissue View based on visualization can be exported and then imported to MarkerView or ClinProTools for further data analysis. However, this is not feasible for IMS data processing, especially for those in high resolution. PCA and clustering are most commonly used for IMS data analysis [2]. LDA and multivariate analysis of variance [3] and PCA combined with support vector machine (SVM) [4] were used to process IMS data. However, these methods have their limitations of handling high-dimensional IMS cubes and incorporating spatial information.

It is essential to extract the complex/hidden patterns from the IMS data. Modern statistical methods should be used to complete a series of operations for biomarker selection and classification in potential application to disease and cancer diagnosis.

IMSmining software package is mainly for IMS data visualization, biomarker selection, model validation, and classification. Visualization functions include the spectrum of a single pixel, the average spectrum of an area, and intensity distribution matrix at a fixed m/z value. The analysis functions include not only PCA, SVM, and LDA methods, but also the most recently developed models SPCA [5, 6], Wavelet4IMS [7], EN4IMS (Elastic Net) [8], and WEN (Weighted Elastic Net) [9] using the spatial information. The motivation is to provide a convenient and automatic way to analyze and extract useful information from the high-dimensional and complex IMS data by not only utilizing the spectrum information within individual pixels, but also studying the correlation and distribution using the spatial information.

The remainder of the paper is organized as follows: In Sect. 2, the main algorithms such as EN4IMS, WEN, Wavelet4IMS are briefly introduced. In Sect. 3, we give the detail of the implementation of the software. A summary of the pipeline of this software is given in Sect. 4. Finally, remarks and a brief discussion are presented in Sect. 5.

2 Algorithm Content

2.1 EN4IMS

Let us consider the multiple linear regression model with n observations. Suppose that $x_j = (x_{1j}, \ldots, x_{nj})^T, j = 1, \ldots, p$ are linear independent predictors and $y = (y_1, \ldots, y_n)^T$ is the response vector. If we use $X = [x_1, \ldots, x_p]$ represent the predictor matrix, the linear regression model can be expressed as

$$y = X\beta + \varepsilon \tag{1}$$

where $\beta = (\beta_1, \ldots, \beta_p)^T$ and the noise term $\varepsilon \sim N(0, \sigma^2 I_n)$. The naive EN criterion is to minimize the following function [10]:

$$L(\lambda_1, \lambda_2, \beta) = \|y - X\beta\|_2^2 + \lambda_1\|\beta\|_1 + \lambda_2\|\beta\|_2^2. \tag{2}$$

There are totally two penalty parts in Eq. 2. The ℓ_1 term enforces the model to generate sparse solution and the quadratic term can achieve the group effect. Zou et al. [10] mentioned that the naive EN has some weakness that will result in double amount of shrinkage. Therefore, the EN algorithm modified the naive elastic net as

$$\hat{\beta}_1 = (1 + \lambda_2)\hat{\beta}_0. \tag{3}$$

where β_1 is named elastic net and β_0 is the naive elastic net. Also, the EN estimates $\hat{\beta}$ is given in [10] by

$$\hat{\beta} = \arg\min_{\beta} \beta^T((X^TX + \lambda_2 I)/(1 + \lambda_2))\beta - 2y^TX\beta + \lambda_1\|\beta\|_1. \tag{4}$$

In the IMSmining software, we apply EN4IMS based on the above EN algorithm to estimate the biomarkers. EN4IMS algorithm incorporates a spatial penalty term into the EN model. IMS information provides huge spatial information located in each individual pixel. One important fact is that pixels in different locations of the same disease should have similar ion intensity values, which means the standard deviation of the intensities at the true biomarkers should be small. Conversely, the standard deviation would be very large among the complex tissue structure like bones.

So in EN4IMS, we use a parameter τ to balance two items together. One is the RSS of the linear model and another is the average of spatial standard deviations of the selected ion intensities. In detail, we use tenfold CV to minimizing the following formula:

$$(1 - \tau)\|y - \hat{y}\|_2^2 + \frac{\tau}{M}\sum_{j=1}^{M}\sqrt{\frac{\sum_{i=1}^{N}(x_{ij} - \mu_j)^2}{N - 1}}, \quad 0 < \tau < 1. \tag{5}$$

where N is the number of all cancer pixel, x_{ij} is the intensity of a fixed jth m/z value at pixel i, μ_j is the average intensity of all cancer pixels at this fixed jth m/z value, and M is the cardinality of active set as defined in [8].

2.2 WEN

In order to consider more precise biomarker selection, Hong and Zhang [9] proposed the following model named WEN:

$$\arg\min_{\beta} \frac{1}{2}\|y - \sum_{j=1}^{p} x_j \beta_j\|_2^2 + n\lambda_1 \sum_{j=1}^{p} \omega_j |\beta_j| + \frac{n}{2}\lambda_2 \sum_{j=1}^{p} |\omega_j \beta_j|^2. \quad (6)$$

where $\omega_j > 0, j = 1, \ldots, p$ are weighted penalty coefficients. In [9], the LARS-WEN algorithm is provided to solve the above WEN model. Experiments show that WEN not only reduces the number of side features but also helps new biomarkers discovery.

2.3 Wavelet4IMS

To meet challenges in IMS data processing, an effective and efficient algorithm for IMS data biomarker selection and classification using methods of multiresolution analysis are proposed. In [7], the authors proposed Wavelet4IMS algorithm. In addition to apply wavelet transform for IMS data denoising, measurement for the similarity of wavelet coefficients is introduced, and the idea of wavelet pyramid method for image matching is applied for biomarker selection and the Naive Bayes classifier is used for classification in the wavelet coefficient space. Performance of the algorithm is evaluated with real data and the results of our experiments show that the multiresolution method has higher accuracy in classification.

3 Software Description

IMSmining allows users to visualize IMS data, to discover biomarkers, and to perform a pixel level classification for different IMS data sections. This software package is designed to give users a maximum level of convenience together with high flexibility.

3.1 Interface

Figure 1 shows the interface of the software based on MATLAB GUI. The first menu is for the data-type options. We can import the data from .mat file or .txt folder or export the biomarker. The next menu contains seven algorithmic options: EN4IMS, WEN, PCA+SVM/LDA, SPCA+SVM/LDA, and Wavelet4IMS. We can also use "view menu" to view the spectrum of a single pixel or the average spectrum of selected area. Toolbar icons can be used to zoom in, zoom out, drag, or rotate the data cube. There are also four figure windows including training, spectrum, testing, and result. We can use the mouse to drag the squares to select the cancer and noncancer area for training and testing.

3.2 Data Visualization

IMSmining provides different methods of visualization for IMS data. Users can see intensity distribution images of different m/z values by clicking on different m/z

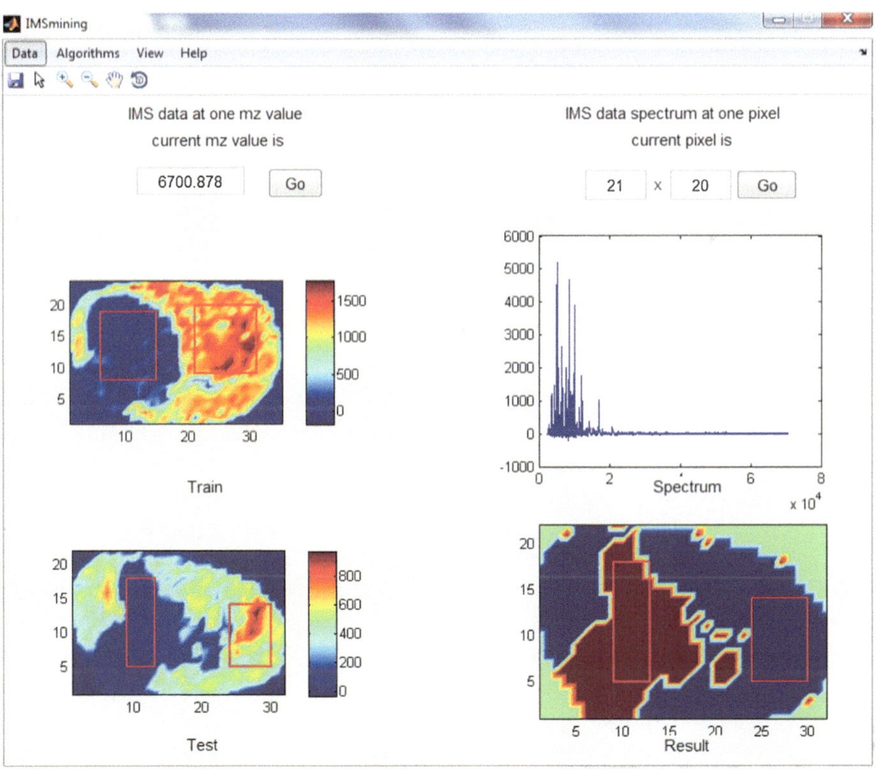

Fig. 1 Interface of GUI

values on the spectrum image. Users can also see spectrum of different pixels just by clicking on different pixel positions on the distribution images. Users can enlarge the spectrum to see whether the m/z value is corresponding to a true peak. The interactive responses between the intensity images (Left Upper Window) and the spectra (Right Upper Window) are extremely convenient and provide a better understanding of the spatial distribution information for a selected m/z peak. Furthermore, users can directly select an area of pixels from the left upper window to see the mean spectrum of these selected pixels.

3.3 Biomarker Selection

IMSmining provides a series of algorithms, which include very recently developed EN4IMS, WEN models, and Wavelet4IMS for IMS data analysis, and other methods such as PCA, SPCA, and SVM popularly used in IMS community. Here, m/z values selected by the model are considered as potential biomarkers.

In EN4IMS algorithm, a spatial penalty term is incorporated into the cross validation step of the EN model [10] for IMS data processing [8]. The WEN model associates the weighted coefficients of EN model with ion intensity spreading information, and thus provides a systematic consideration for the spatial information of the IMS data for biomarker selection and classification. Both models inherit good properties from the EN method which produces a sparse model with admirable prediction accuracy. By taking the spatial information into consideration, these two models help distinguish the IMS feature peaks caused by biological structure differences from those truly associated with diseases. In Wavelet4IMS algorithm, IMSmining transforms each mass spectrometry to wavelet space and select biomarkers based on multiresolution analysis.

3.4 Classification

IMSmining provides model validation and classifies testing samples. Users can select the training data region directly from the training data figure. After analyzing the training data sets to create the predictive model, validation of models can be done on the selected cancer and noncancer square area of the testing data sets. To enhance the chance of finding the best model, the tuning parameter λ of EN4IMS and WEN algorithms can be changed accordingly by users. As a result, we can obtain the classification rates of the selected testing area. Besides implementing EN4IMS or WEN algorithm, IMSmining has one method named Wavelet4IMS which uses feature vectors selected from wavelet domain combining with a naive Bayes classifier for classification. IMSmining can also use PCA or SPCA to reduce the dimension of the data and then continue to use SVM or LDA for classification.

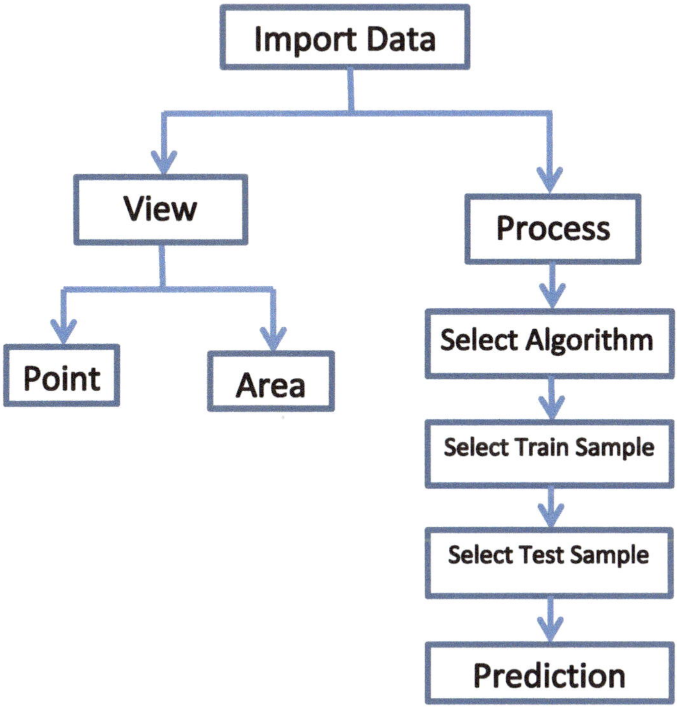

Fig. 2 Pipeline of GUI

4 Pipeline

Figure 2 shows the pipeline of IMSmining. After importing the data, we can either view the image of the data or process the data-based variety of algorithms. If you only want to view the image, you have two choices: point or area. Then you can import a single pixel or just simply click on the data image. Or you can drag the mouse to select an area to calculate the major statistical value of this specifical area. In another branch, you have three steps to complete the model prediction: algorithm selecting, training image selecting, and testing image selecting. You can stop the algorithm at each step and start over in another algorithm. And after you select the images, you need to use the mouse to drag both of the cancer and noncancer area. After the calculation, IMSmining will show the comparative cancer and noncancer result.

5 Discussion

We developed a software package called IMSmining based on algorithms of EN4IMS, WEN, SPCA, and Wavelet4IMS. We have applied this software tool to real IMS data [8, 9]. Compared with other current popular methods, the models of EN4IMS, WEN, and Wavelet4IMS work more efficiently and effectively for IMS data processing in terms of confirming new biomarkers, producing a more accurate feature list including significant peaks, and providing more accurate classification results.

Acknowledgments The authors would like to thank Shannon Cornett, Sara Frappier, and Richard M. Caprioli from the VUMSRC for valuable discussions and providing IMS data sets for the study. DH is grateful for the support from the program of Beijing Overseas High Caliber Talents.

References

1. Trede, D., Kobarg, J. H., Oetjen, J., Thiele, H., Maass, P., Alexandrov, T.: On the importance of mathematical methods for analysis of maldi imaging mass spectrometry data. J Integr Bioinform **9**(1), 189 (2012)
2. de Plas, R.V., Ojeda, F., Dewil, M., Bosch, L.V.D., Moor, B.D., Waelkens, E.: Prospective exploration of biochemical tissue composition via imaging mass spectrometry guided by principal component analysis. In: Pacific Symposium on Biocomputing, World Scientific, pp. 458–469 (2007)
3. Muir, E.R., Ndiour, I., Le Goasduff, N.A., Moffitt, R., Liu, Y., Sullards, M.C., Merrill, A., Chen, Y., Wang, M.: Multivariate analysis of imaging mass spectrometry data. Bioinform. Bioeng. 472–479 (2007)
4. Gerhard, M., Deininger, S.-O., Schleif, F.: Statistical classification and visualization of maldi-imaging data. Comput. Based Med Syst 403–405 (2007)
5. Zou, H.T., Tibshirani, H.R.: Sparse principal component analysis. J. Comput. Graph. Stat. **15**(2), 265–286 (2006)
6. Wang, Y., Wu, Q.: Sparse PCA by iterative elimination algorithm. Adv. Comput. Math. **36**(1), 137–151 (2012)
7. Xiong, L., Hong, D.: Multi-resolution analysis method for ims data biomarker selection and classification. British J. Math. Comp. Sci. **5**(1), 64–80 (2015)
8. Zhang, F., Hong, D.: Elastic net based framework for imaging mass spectrometry data biomarker selection and classification. Stat. Med. **30**, 753–768 (2010)
9. Hong, D., Zhang, F.: Weighted elastic net model for mass spectrometry imaging processing. Math. Model. Nat. Phenom. **5**(3), 115–133 (2010)
10. Zou, H., Hastie, T.: Regularization and variable selection via the elastic net. J. R. Stat. Soc. B **67**, 301–320 (2005)

Pal Interpolation of Integral Types

Gayatri Ramesh

Abstract In this paper, the author(s) discuss existence and uniqueness results of three so-called integral types of Pal interpolation schemes which are interesting extensions/generalizations of classical Hermite-Fejer Interpolation problem. The results are of interest to approximation theory.

Keywords Pal interpolation of integral types · Hermite-Fejer interpolation · Approximation theory

1 Introduction

Let $X := \{x_1, \ldots, x_n\}$ contain n distinct nodes $x_1 < x_2 < \cdots < x_n$ on the real line. Then the roots $x_1, \ldots, x_n$ of the polynomial

$$\omega_X(x) := (x - x_1) \cdots (x - x_n) \tag{1}$$

and the roots $x_1^*, \ldots, x_{n-1}^*$ of the derivative

$$\omega'_X(x) = n(x - x_1^*) \cdots (x - x_{n-1}^*) \tag{2}$$

have the following interlacing property:

$$x_1 < x_1^* < x_2 < x_2^* < \cdots < x_{n-1}^* < x_n.$$

Pál considered the following Hermite-Fejer interpolation problem in [4]: *Find a polynomial P of lowest degree such that*

$$P(x_k) = y_k \text{ for all } 1 \le k \le n \text{ and } P'(x_l^*) = y_l^* \text{ for all } 1 \le l \le n-1 \tag{3}$$

G. Ramesh (✉)
University of Central Florida, 4000 Central Florida Blvd, Orlando, FL 32816, USA
e-mail: gayatriramesh@me.com

R.N. Mohapatra et al. (eds.), *Mathematics and Computing*,
Springer Proceedings in Mathematics & Statistics 139,
DOI 10.1007/978-81-322-2452-5_12

for any given interpolation data $\{y_k\}_{k=1}^{n}$*and* $\{y_l^*\}_{l=1}^{n-1}$.

The above interpolation is now known as Pál interpolation. In [4], Pál established the following for its existence and uniqueness.

Theorem 1.1 *Given any interpolation data* $\{y_k\}_{k=1}^{n}$ *and* $\{y_l^*\}_{l=1}^{n-1}$, *there exists a polynomial* P *of degree* $2n-1$ *that satisfies* (3). *Moreover,*

$$\begin{aligned}P(x) = &-\sum_{k=1}^{n} y_k \frac{\omega_X(x)}{(\omega'_X(x_k))^2 \prod_{i\neq k}(x_k - x_i)}\\ &\times \int \frac{\omega'_X(x)}{(x-x_k)^2}\big(\omega'_X(x_k) - \omega''_X(x_k)(x-x_k)\big)\mathrm{d}x\\ &+\sum_{l=1}^{n-1} y_l^* \frac{\omega_X(x)}{\omega_X(x_l^*)} \int \frac{\prod_{j\neq l}(x-x_j^*)}{\prod_{j\neq l}(x_l^*-x_j^*)}\mathrm{d}x.\end{aligned}$$

For any polynomial P satisfying (3), $P(x) + C\omega_X(x)$ has the same interpolation property for any constants C. The uniqueness of polynomials satisfying (3) was discussed in [4] when an additional interpolation condition is imposed.

Theorem 1.2 *Let* $\{x_k\}_{k=1}^{n}$, $\{x_l^*\}_{l=1}^{n-1}$, $\{y_k\}_{k=1}^{n}$ *and* $\{y_l^*\}_{l=1}^{n-1}$ *be as in* Theorem (1.1), *and let* $a \neq x_k$ *for all* $k = 1, 2, ..., n$. *Then the polynomial*

$$\begin{aligned}R(x) := &-\sum_{k=1}^{n} y_k \frac{\omega_X(x)}{(\omega'_X(x_k))^2 \prod_{i\neq k}(x_k - x_i)}\\ &\times \int_a^x \frac{\omega'_X(t)}{(t-x_k)^2}\big(\omega'_X(x_k) - \omega''_X(x_k)(t-x_k)\big)\mathrm{d}t\\ &+\sum_{l=1}^{n-1} y_l^* \frac{\omega_X(x)}{\omega_X(x_l^*)} \int_a^x \frac{\prod_{j\neq l}(t-x_j^*)}{\prod_{j\neq l}(x_l^*-x_j^*)}\mathrm{d}t \quad \text{for } x \in (a-\delta, a+\delta),\end{aligned}$$

is the unique polynomial of degree at most $2n-1$ *that satisfies* (3) *and* $R(a) = 0$, *where* $\delta = \min_{1\leq k\leq n}|x_k - a|$.

In the last 40 years, various extensions of Pál interpolation have been made [1–10]. In this paper, we consider Pál interpolation of integral types.

2 Pál Interpolation of Integral Types I

In this section, we consider the existence and uniqueness of polynomials $P(x)$ of lowest degree for any given interpolation data $\{y_k\}_{k=1}^{n}$ and $\{y_l^*\}_{l=1}^{n-1}$ such that

$$P(x_l^*) = y_l^*,\ 1 \le l \le n-1, \ \text{and} \ \int_{x_k}^{x_{k+1}} P(x)\,\mathrm{d}x = y_{k+1}, \quad 1 \le k \le n-1. \tag{4}$$

Theorem 2.1 *Given interpolation data* $\{y_k\}_{k=2}^n$ *and* $\{y_l^*\}_{l=1}^{n-1}$, *the polynomial* P *of degree* $2n-2$ *defined by*

$$\begin{aligned} P(x) := \frac{\mathrm{d}}{\mathrm{d}x}\Bigg\{ & -\sum_{k=2}^{n} z_k \frac{\omega_X(x)}{(\omega'_X(x_k))^2 \prod_{i\neq k}(x_k - x_i)} \\ & \times \int \frac{\omega'_X(x)}{(x-x_k)^2}\big(\omega'_X(x_k) - \omega''_X(x_k)(x-x_k)\big)\mathrm{d}x \\ & + \sum_{l=1}^{n-1} y_l^* \frac{\omega_X(x)}{\omega_X(x_l^*)} \int \frac{\prod_{j\neq l}(x - x_j^*)}{\prod_{j\neq l}(x_l^* - x_j^*)}\mathrm{d}x \Bigg\}, \end{aligned} \tag{5}$$

satisfies

$$P(x_l^*) = y_l^*, \quad 1 \le l \le n-1, \tag{6}$$

and

$$\int_{x_k}^{x_{k+1}} P(x)\,\mathrm{d}x = y_{k+1}, \quad 1 \le k \le n-1, \tag{7}$$

where $z_k = \sum\limits_{q=2}^{k} y_q,\ 2 \le k \le n.$

Theorem 2.2 *Given data* $\{y_k\}_{k=1}^n$ *and* $\{y_l^*\}_{l=1}^{n-1}$, *define a polynomial* P *of degree* $2n-2$ *as in* (5). *Then a polynomial* R *of degree at most* $2n-1$ *satisfies* (6) *and* (7) *if and only if*

$$R(x) = P(x) + \omega'_X(x)(\alpha + \beta\omega_X(x))$$

for some constants α *and* β.

2.1 Proof of Theorem 2.1

First we construct polynomials $B_l(x)$, $1 \le l \le n-1$, of degree at most $2n-1$ satisfying

$$\begin{cases} \text{(a)}\ B_l(x_i) = 0 \ \text{ for all } 1 \le l \le n-1 \text{ and } 1 \le i \le n \\ \text{(b)}\ B_l'(x_j^*) = \delta_{lj} \ \text{ for all } 1 \le l \le n-1 \text{ and } 1 \le j \le n-1. \end{cases} \tag{8}$$

Here δ_{ij} stands for the Kronecker symbol defined by $\delta_{ij} = 1$ if $i = j$ and $\delta_{ij} = 0$ otherwise. Take $1 \le l \le n-1$. From the requirement (a) in (8),

$$B_l(x) = \omega_X(x)V_l(x) \tag{9}$$

for some polynomial $V_l(x)$ of degree at most $n-1$. Consequently,

$$B_l'(x) = \omega_X'(x)V_l(x) + \omega_X(x)V_l'(x) = \frac{\omega_X'(x)}{(x-x_l^*)}W_l(x) \tag{10}$$

for some polynomial $W_l(x)$ of degree at most n, where the last equality follows from the requirement (b) in (8). Multiplying $x - x_l^*$ at both sides of the above equation leads to

$$[\omega_X'(x)V_l(x) + \omega_X(x)V_l'(x)](x-x_l^*) = \omega_X'(x)W_l(x).$$

Rearranging above equation yields

$$\omega_X'(x)\big((x-x_l^*)V_l(x) - W_l(x)\big) = -(x-x_l^*)\omega_X(x)V_l'(x). \tag{11}$$

Recall that ω_X and its derivative ω_X' do not have common roots. Then it follows from (11) that

$$(x-x_l^*)V_l'(x) = \omega_X'(x)M_l(x) \tag{12}$$

for some polynomial $M_l(x)$. Comparing the degree of both sides of equation (12) shows that $M_l(x)$ has degree zero, i.e., $M(x) = M$ for some constant M.

Evaluating (10) at $x = x_l^*$ and recalling the requirement (b) in (8) gives

$$1 = \omega_X'(x_l^*)V_l(x_l^*) + \omega_X(x_l^*)V_l'(x_l^*) = \omega_X(x_l^*)V_l'(x_l^*), \tag{13}$$

and hence

$$V_l(x_l^*) = (\omega_X(x_l^*))^{-1}.$$

Substituting this in (12) and recalling that M_l is a constant function, we obtain,

$$V_l'(x) = \frac{1}{\omega_X(x_l^*)}\frac{\prod_{j\neq l}(x-x_j^*)}{\prod_{j\neq l}(x_l^*-x_j^*)}. \tag{14}$$

Therefore

$$V_l(x) = \int \frac{1}{\omega_X(x_l^*)}\frac{\prod_{j\neq l}(x-x_j^*)}{\prod_{j\neq l}(x_l^*-x_j^*)}\mathrm{d}x. \tag{15}$$

Substituting the above expression about $V_l(x)$ into (9) yields

$$B_l(x) = \frac{\omega_X(x)}{\omega_X(x_l^*)}\int \frac{\prod_{j\neq l}(x-x_j^*)}{\prod_{j\neq l}(x_l^*-x_j^*)}\mathrm{d}x, \quad 1 \le l \le n-1.$$

The polynomials B_l, $1 \le l \le n-1$, just defined have degree at most $2n-1$. It satisfies the requirement (a) in (8), and also the requirement (b) in (8), because

$$B_l'(x) = \frac{\omega_X'(x)}{\omega_X(x_l^*)} \int \frac{\prod_{j\neq l}(x - x_j^*)}{\prod_{j\neq l}(x_l^* - x_j^*)} \mathrm{d}x + \frac{\omega_X(x)}{\omega_X(x_l^*)} \frac{\prod_{j\neq l}(x - x_j^*)}{\prod_{j\neq l}(x_l^* - x_j^*)}$$

and hence

$$B_l'(x_{j'}^*) = \frac{\omega_X(x_{j'}^*) \prod_{j\neq l}(x_{j'}^* - x_j^*)}{\omega_X(x_l^*) \prod_{j\neq l}(x_l^* - x_j^*)} = \begin{cases} 1 & \text{if } j' = l \\ 0 & \text{if } j' \neq l. \end{cases}$$

Next we find A_k, $2 \le k \le n$, of degree at most $2n-1$ that satisfies

$$\begin{cases} \text{(c) } A_k(x_i) = \delta_{ki} \;\text{ for all } 2 \le k \le n \text{ and } 1 \le i \le n \\ \text{(d) } A_k'(x_j^*) = 0 \;\text{ for all } 2 \le k \le n \text{ and } 1 \le j \le n-1. \end{cases} \tag{16}$$

From the requirement (c) in (16), it follows that

$$A_k(x) = \frac{\omega_X(x)}{x - x_k} S_k(x), \quad 1 \le k \le n, \tag{17}$$

for some polynomial $S_k(x)$ of degree at most n that satisfies

$$S_k(x_k) \neq 0.$$

Taking derivative of both sides of (17) and applying the requirement (d) in (16), we have

$$A_k'(x) = \left(\frac{\omega_X'(x)}{(x - x_k)} - \frac{\omega_X(x)}{(x - x_k)^2}\right) S_k(x) + \frac{\omega_X(x)}{(x - x_k)} S_k'(x) = \omega_X'(x) T_k(x)$$

for some polynomial $T_k(x)$ of degree at most $n-1$. Thus

$$\omega_X'(x)(x - x_k)\big(S_k(x) - T_k(x)(x - x_k)\big) = \omega_X(x)\big(S_k(x) - (x - x_k)S_k'(x)\big). \tag{18}$$

Again, recall that $\omega_X(x)$ and $\omega_X'(x)$ do not share any root. Then

$$S_k(x) - (x - x_k)S_k'(x) = \omega_X'(x) U_k(x) \tag{19}$$

and

$$S_k(x) - (x - x_k)T_k(x) = \frac{\omega_X(x)}{x - x_k} U_k(x) \tag{20}$$

for some polynomial $U_k(x)$ of degree at most one. Substituting x by x_k in (20) and recalling that $A_k(x_k) = 1$ by the requirement (c) in (16), we obtain

$$U_k(x_k) = \frac{1}{\omega'_X(x_k)\prod_{i\neq k}(x_k - x_i)}. \tag{21}$$

Taking derivative of both sides of (19) yields

$$(\omega'_X(x)U_k(x))' = -(x - x_k)S''_k(x),$$

which implies that

$$\omega''_X(x_k)U_k(x_k) + \omega'_X(x_k)U'_l(x_k) = 0. \tag{22}$$

Thus $\omega'_X U_k$ has the following Taylor expansion at $x = x_k$:

$$\omega'_X(x)U_k(x) = \omega'_X(x_k)U_k(x_k) + c_2(x - x_k)^2 + c_3(x - x_k)^2 + \cdots + c_n(x - x_k)^n. \tag{23}$$

Dividing both sides of (19) by $(x - x_k)^2$ gives

$$\frac{\omega'(x)U_k(x)}{(x - x_k)^2} = \frac{S_k(x)}{(x - x_k)^2} - \frac{S'_k(x)}{x - x_k} = -\left(\frac{S_k(x)}{x - x_k}\right)'.$$

This together with (23) implies that

$$\frac{S_k(x)}{x - x_k} = -\int \frac{\omega'_X(x)U_k(x)}{(x - x_k)^2}\mathrm{d}x.$$

Hence

$$A_k(x) = -\omega_X(x)\int \frac{\omega'_X(x)U_k(x)}{(x - x_k)^2}\mathrm{d}x. \tag{24}$$

Now it remains to figure out the polynomial U_k of degree at most one. Write

$$U_k(x) = r_0 + r_1(x - x_k). \tag{25}$$

Then

$$r_0 = U_k(x_k) = \frac{1}{\omega'_X(x_k)\prod_{i\neq k}(x_k - x_i)} \tag{26}$$

by (21). From (22) and (25) it follows that

$$r_1 = -\frac{\omega''_X(x_k)}{(\omega'_X(x_k))^2\prod_{i\neq k}(x_k - x_i)}. \tag{27}$$

Therefore

$$U_k(x) = \frac{1}{(\omega_X'(x_k))^2 \prod_{i\neq k}(x_k - x_i)} \big(\omega_X'(x_k) - \omega_X''(x_k)(x - x_k)\big).$$

Substituting this into (24), we obtain that

$$A_k(x) = -\frac{\omega_X(x)}{(\omega_X'(x_k))^2 \prod_{i\neq k}(x_k - x_i)} \times \int \frac{\omega_X'(x)}{(x - x_k)^2} \big(\omega_X'(x_k) - \omega_X''(x_k)(x - x_k)\big) \mathrm{d}x,\ 1 \le k \le n. \quad (28)$$

Finally let us verify that the functions A_k, $1 \le k \le n$, satisfy (16). Notice that

$$A_k'(x) = -\frac{\omega_X'(x)}{\prod_{i\neq k}(x_k - x_i)} \int \frac{\omega_X'(x)}{(x - x_k)^2} \Big(1 - \frac{\omega_X''(x_k)}{\omega_X'(x_k)}(x - x_k)\Big) \mathrm{d}x \quad (29)$$

$$-\frac{\omega_X(x)}{\prod_{i\neq k}(x_k - x_i)} \frac{\omega_X'(x)}{(x - x_k)^2} \Big(1 - \frac{\omega_X''(x_k)}{\omega_X'(x_k)}(x - x_k)\Big), \quad (30)$$

which implies that $A_k'(x_l^*) = 0$ for all $1 \le l \le n-1$. On the other hand, $A_k(x_{k'}) = 0$ for all $k' \neq k$ as $\omega_X(x_{k'}) = 0$, and

$$A_k(x_k) = -\lim_{x\to x_k} \frac{\omega_X(x)}{(\omega_X'(x_k))^2 \prod_{i\neq k}(x_k - x_i)} \times \int \frac{1}{(x - x_k)^2} \big((\omega_X'(x_k))^2 + Q(x - x_k)\big) \mathrm{d}x \quad (31)$$

$$= \lim_{x\to x_k} \frac{\omega_X(x)}{\prod_{i\neq k}(x_k - x_i)(x - x_k)} = 1 \quad (32)$$

where Q is a polynomial such that $Q(0) = 0$. This proves that polynomials A_k, $2 \le k \le n$, in (28) satisfies (16).

Finally, we show that the polynomial

$$P(x) := \frac{\mathrm{d}}{\mathrm{d}x} \left[\sum_{k=2}^{n} z_k A_k(x) + \sum_{l=1}^{n-1} y_l^* B_l(x) \right] \quad (33)$$

has the interpolation properties (6) and (7). Set $z_1 = 0$. By (8), (16), and (33),

$$\int_{x_i}^{x_{i+1}} P(x)\mathrm{d}x = \Big(\sum_{k=2}^{n} z_k A_k(x) + \sum_{l=1}^{n-1} y_l^* B_l(x)\Big)\Big|_{x_i}^{x_{i+1}} = z_{i+1} - z_i = y_{i+1} \quad (34)$$

for all $1 \le i \le n-1$, and

$$P(x_j^*) = \sum_{k=2}^{n} z_k A_k'(x_j^*) + \sum_{l=1}^{n-1} y_l^* B_l'(x_j^*) = y_j^* \tag{35}$$

for all $1 \le j \le n-1$. This proves that the polynomial P in (33) satisfies the interpolation requirements (6) and (7).

2.2 *Proof of Theorem 2.2*

($\Longleftarrow$) Consider a polynomial $\tilde{P}$ of the following form:

$$\tilde{P}(x) = P(x) + \omega_X'(x)(\alpha + \beta\omega_X(x)) \tag{36}$$

where $\alpha, \beta \in \mathbf{R}$. Then

$$\int_{x_k}^{x_{k+1}} \tilde{P}(x)\mathrm{d}x = \int_{x_k}^{x_{k+1}} P(x)\mathrm{d}x + \int_{x_k}^{x_{k+1}} \omega_X'(x)(\alpha + \beta\omega_X(x))\,\mathrm{d}x$$

$$= y_{k+1} + \left(\alpha\omega_X(x) + (\beta/2)(\omega_X(x))^2\right)\Big]_{x_k}^{x_{k+1}} = y_{k+1}, \quad 1 \le k \le n-1.$$

Also, observe that

$$\tilde{P}(x_l^*) = P(x_l^*) + \omega_X'(x_l^*)(\alpha + \beta\omega_X(x_l^*)) = y_l^*, \quad 1 \le l \le n-1. \tag{37}$$

Therefore a polynomial P of the form of (36) satisfies (6) and (7).

($\Longrightarrow$) Let Q be a polynomial of degree at most $2n-1$ that satisfies (6) and (7). Then $R(x) := Q(x) - P(x)$ satisfies

$$R(x_l^*) = 0,\ 1 \le l \le n-1, \text{ and } \int_{x_k}^{x_{k+1}} R(x)\mathrm{d}x = 0, \quad 1 \le k \le n-1. \tag{38}$$

From the above requirement, the antiderivative of the polynomial

$$\int R(x)\mathrm{d}x = c + \omega_X(x)S(x) \tag{39}$$

for some polynomial S of degree at most n, and

$$R(x) = \omega_X'(x)M(x) \tag{40}$$

for some polynomial M of degree at most n. Therefore

$$\omega'_X(x)S(x) + \omega_X(x)S'(x) = \omega'_X(x)M(x). \tag{41}$$

Rearranging the above equation gives

$$\omega'_X(x)(S(x) - M(x)) = -\omega_X(x)S'(x). \tag{42}$$

Recall that $\omega_X(x)$ and $\omega'_X(x)$ do not have common roots, and that $S'(x)$ has degree at most $n-1$. Therefore $S'(x) = \frac{\beta}{2}\omega'_X(x)$ for some constants β. This implies that

$$M(x) = \alpha + \beta\omega_X(x),$$

or equivalently the desired conclusion that $R(x) = P(x) + \omega'_X(x)(a + \beta\omega_X(x))$ for some constant α, β.

3 Pál Interpolation of Integral Types II

Let a, b, and c be real numbers and let $x_k^*, k = 1, 2, ..., n^*$, be the real roots of $\tilde{\omega}_X(x) := a\omega_X(x) + (bx + c)\omega'_X(x)$. Szabó and Joó [6] and Szabó [7–9] generalized Pál interpolation problem to the following: *Let a, b, c be real numbers, and let $x_l^*, l = 1, 2, ..., n^*$ be the real roots of $\tilde{\omega}_X(x) := a\omega_X(x) + (bx + c)\omega'_X(x)$. Determine a polynomial $R(x)$ of the lowest possible degree that has the properties $R(x_k) = y_k, 1 \leq k \leq n$, and $R'(x_l^*) = y_l^*, 1 \leq l \leq n^*$.* They found general polynomials for the following cases: (1) $b = 0$; and (2) $a < 0, b = 1$. If $a = b = 0$ and $c = 1$, the above interpolation becomes Pál interpolation. In this section, we modify the work done by Szabó and Joó [6] to fit the following conditions:

$$R(x_l^*) = y_l^*, \ 1 \leq l \leq n, \text{ and } \int_{x_k}^{x_{k+1}} R(x)\,dx = y_{k+1}, \quad 1 \leq k \leq n-1. \tag{43}$$

under the assumption that $a \neq 0$ and $b = 0$. In this case $n^* = n$. Moreover, $\omega_X(x)$ and $\tilde{\omega}_X(x)$ have the following interlacing property:

$$x_1 < x_1^* < x_2 < \cdots < x_n < x_n^* \ \text{ if } a/c < 0; \tag{44}$$

and

$$x_1^* < x_1 < x_2^* < \cdots < x_n^* < x_n \ \text{ if } a/c > 0. \tag{45}$$

Theorem 3.1 *Let $a, c \neq 0$, $X := \{x_1, \ldots, x_n\}$ contain n distinct nodes on the real line ordered by $x_1 < x_2 < \cdots < x_n$, and denote by $X^* := \{x_1^*, x_2^*, \ldots, x_n^*\}$ the set of the real roots of the polynomial $\tilde{\omega}_X(x) := a\omega_X(x) + c\omega'_X(x)$, and let*

$$\Omega(x_l^*) = \frac{a\omega_X(x) + c\omega_X'(x)}{x - x_l^*}\Big|_{x=x_l^*}. \tag{46}$$

Given the interpolation data $\{y_k\}_{k=2}^n$ *and* $\{y_l^*\}_{l=1}^n$, *set* $z_k = \sum_{q=2}^{k} y_q$, $2 \le k \le n$, *and define the polynomial* $R(x)$ *of degree* $2n-2$ *by*

$$\begin{aligned} R(x) := \frac{\mathrm{d}}{\mathrm{d}x}\Big[\sum_{k=2}^{n} z_k \frac{\omega_X(x)}{c\omega_X'(x_k)\prod_{i\neq k}(x_k - x_i)} e^{\frac{a}{c}x} \\ \times \int_x^{\infty} \frac{a\omega_X(t) + c\omega_X'(t)}{(t-x_k)^2}\Big(1 - \frac{\omega_X''(x_k)}{\omega_X'(x_k)}(t - x_k)\Big)e^{-\frac{a}{c}t}\mathrm{d}t \\ - \sum_{l=1}^{n} y_l^* \frac{\omega_X(x)e^{\frac{a}{c}x}}{\omega_X(x_l^*)\Omega(x_l^*)} \int_x^{\infty} \frac{a\omega_X(t) + c\omega_X'(t)}{(t - x_l^*)} e^{-\frac{a}{c}t}\mathrm{d}t\Big], \quad x > x_n \end{aligned} \tag{47}$$

if $\frac{a}{c} > 0$, *and*

$$\begin{aligned} R(x) := \frac{\mathrm{d}}{\mathrm{d}x}\Big[-\sum_{k=2}^{n} z_k \frac{\omega_X(x)}{c\omega_X'(x_k)\prod_{i\neq k}(x_k - x_i)} e^{\frac{a}{c}x} \\ \times \int_{-\infty}^{x} \frac{a\omega_X(t) + c\omega_X'(t)}{(t-x_k)^2}\Big(1 - \frac{\omega_X''(x_k)}{\omega_X'(x_k)}(t - x_k)\Big)e^{-\frac{a}{c}t}\mathrm{d}t \\ + \sum_{l=1}^{n-1} y_l^* \frac{\omega_X(x)e^{\frac{a}{c}x}}{\omega_X(x_l^*)\Omega(x_l^*)} \int_{-\infty}^{x} \frac{a\omega_X(t) + c\omega_X'(t)}{(t - x_l^*)} e^{-\frac{a}{c}t}\mathrm{d}t\Big] \text{ for } x < x_1 \end{aligned} \tag{48}$$

if $\frac{a}{c} < 0$. *Then* $R(x)$ *satisfies (43).*

Proof We start by decomposing $R(x)$ into a sum of two functions, as in the previous section, where

$$R(x) = \frac{\mathrm{d}}{\mathrm{d}x}\left[\sum_{k=2}^{n} z_k A_k(x) + \sum_{l=1}^{n} y_l^* B_l(x)\right], \tag{49}$$

and polynomials $\{A_k(x)\}_{k=2}^n$ and $\{B_l(x)\}_{l=1}^n$ of degree at most $2n-1$ satisfy

$$\begin{cases} \text{(a) } A_k(x_i) = \delta_{ki} \text{ for all } 2 \le k \le n \text{ and } 1 \le i \le n \\ \text{(b) } A_k'(x_j^*) = 0 \text{ for all } 2 \le k \le n \text{ and } 1 \le j \le n, \end{cases} \tag{50}$$

and

$$\begin{cases} \text{(c) } B_l(x_i) = 0 \text{ for all } 1 \le l \le n \text{ and } 1 \le i \le n \\ \text{(d) } B_l'(x_j^*) = \delta_{lj} \text{ for all } 1 \le l \le n \text{ and } 1 \le j \le n. \end{cases} \tag{51}$$

Similar to the previous section let us first construct the polynomials $B_l(x)$, $1 \leq l \leq n$. From the requirement (c) in (51), we know that

$$B_l(x) = \omega_X(x) V_l(x) \tag{52}$$

for a polynomial $V_l(x)$ of degree at most $n-1$. Recall that roots of $a\omega_X(x)+c\omega'_X(x)$ are real and have a multiplicity of one. Consequently,

$$B'_l(x) = \omega'_X(x)V_l(x) + \omega_X(x)V'_l(x) = \frac{a\omega_X(x) + c\omega'_X(x)}{(x - x_l^*)} W_l(x) \tag{53}$$

for some polynomial $W_l(x)$ of degree at most $n - 1$, where the last equality follows from the requirement (d) in (51). Multiplying $x - x_l^*$ at both sides of the above equation leads to

$$\omega'_X(x)[(x - x_l^*)V_l(x) - cW_l(x)] = \omega_X(x)[-(x - x_l^*)V'_l(x) + aW_l(x)].$$

Recall that ω_X and its derivative ω'_X do not have common roots. Then

$$M\omega'_X(x) = -(x - x_l^*)V'_l(x) + aW_l(x) \tag{54}$$

and

$$M\omega_X(x) = (x - x_l^*)V_l(x) - cW_l(x) \tag{55}$$

for a constant M. Multiplying (54) with c and (55) with a, and then adding them together, we obtain

$$(x - x_l^*)[aV_l(x) - cV'_l(x)] = M[a\omega_X(x) + c\omega'_X(x)]. \tag{56}$$

Multiplying both sides by $-\frac{e^{-\frac{a}{c}x}}{c(x-x_l^*)}$ gives

$$\frac{\mathrm{d}}{\mathrm{d}x}\left(e^{-\frac{a}{c}x}V'_l(x)\right) = -\frac{Me^{-\frac{a}{c}x}}{c}\frac{a\omega_X(x) + c\omega'_X(x)}{(x - x_l^*)}. \tag{57}$$

Integrating both sides leads to

$$V_l(x) = \frac{Me^{\frac{a}{c}x}}{c}\int\limits_x^{\infty}\frac{a\omega_X(t) + c\omega'_X(t)}{(t - x_l^*)}e^{-\frac{a}{c}t}\mathrm{d}t \ \text{ if } \frac{a}{c} > 0, \tag{58}$$

and

$$V_l(x) = -\frac{Me^{\frac{a}{c}x}}{c}\int\limits_{-\infty}^{x}\frac{a\omega_X(t) + c\omega'_X(t)}{(t - x_l^*)}e^{-\frac{a}{c}t}\mathrm{d}t \ \text{ if } \frac{a}{c} < 0. \tag{59}$$

The next step is to determine the constant M. Note from (53) and the condition (d) in (51) that

$$B_l'(x_l^*) = \omega_X(x_l^*)V_l'(x_l^*) + V_l(x_l^*)\omega_X'(x_l^*) = 1. \tag{60}$$

Multiplying both sides of (56) by $-c$ and replacing x with x_l^* gives

$$-cV_l'(x_l^*) + aV_l(x_l^*) = \left.\frac{a\omega_X(x) + c\omega_X'(x)}{x - x_l^*}\right|_{x=x_l^*} M. \tag{61}$$

Note that the right-hand side of the above equation is nonzero because x_l^* is a simple root of the polynomial $a\omega_X(x) + c\omega_X'(x)$. Multiplying both sides of (60) by $-\frac{c}{\omega_X(x_l^*)}$ and recalling that $a\omega_X(x_l^*) + c\omega_X'(x_l^*) = 0$, we get

$$-\frac{c}{\omega_X(x_l^*)} = -cV_l'(x_l^*) + aV_l(x_l^*). \tag{62}$$

Let $\Omega(x_l^*) = \left.\frac{a\omega_X(x)+c\omega_X'(x)}{x-x_l^*}\right|_{x=x_l^*}$. Thus combining (61) and (62) determines the constant

$$M = \frac{-c}{\omega_X(x_l^*)\Omega(x_l^*)}. \tag{63}$$

Therefore,

$$B_l(x) = -\frac{\omega_X(x)e^{\frac{a}{c}x}}{\omega_X(x_l^*)\Omega(x_l^*)} \int_x^{\infty} \frac{a\omega_X(t) + c\omega_X'(t)}{(t - x_l^*)} e^{-\frac{a}{c}t} \mathrm{d}x \quad \text{if } \frac{a}{c} > 0 \tag{64}$$

and

$$B_l(x) = \frac{\omega_X(x)e^{\frac{a}{c}x}}{\omega_X(x_l^*)\Omega(x_l^*)} \int_{-\infty}^{x} \frac{a\omega_X(t) + c\omega_X'(t)}{(t - x_l^*)} e^{-\frac{a}{c}t} \mathrm{d}t \quad \text{if } \frac{a}{c} < 0. \tag{65}$$

The polynomials B_l, $1 \leq l \leq n$, just defined have degree at most $2n - 1$, satisfy the requirement (c) in (51) as they have the factor ω_X, and also the requirement (d) in (51) as

$$\begin{aligned} B_l'(x) &= \frac{\omega_X(x)e^{\frac{a}{c}x}}{\omega_X(x_l^*)\Omega(x_l^*)} \frac{a\omega_X(x) + c\omega_X'(x)}{x - x_l^*} e^{-\frac{a}{c}x} \\ &\quad - \frac{e^{\frac{a}{c}x}(a\omega_X(x) + c\omega_X'(x))}{c\omega_X(x_l^*)\Omega(x_l^*)} \int_x^{\infty} \frac{a\omega_X(t) + c\omega_X'(t)}{(t - x_l^*)} e^{-\frac{a}{c}t} \mathrm{d}t \text{ if } \frac{a}{c} > 0 \end{aligned}$$

and

$$B_l'(x) = \frac{\omega_X(x)e^{\frac{a}{c}x}}{\omega_X(x_l^*)\Omega(x_l^*)}\frac{a\omega_X(x)+c\omega_X'(x)}{x-x_l^*}e^{-\frac{a}{c}x}$$
$$+\frac{e^{\frac{a}{c}x}(a\omega_X(x)+c\omega_X'(x))}{c\omega_X(x_l^*)\Omega(x_l^*)}\int_{-\infty}^{x}\frac{a\omega_X(t)+c\omega_X'(t)}{(t-x_l^*)}e^{-\frac{a}{c}t}\mathrm{d}t \text{ if } \frac{a}{c}<0.$$

Thus

$$B_l'(x_{j'}^*) = \frac{\omega_X(x_{j'}^*)}{\omega_X(x_l^*)}\frac{\Omega(x_{j'}^*)}{\Omega(x_l^*)} = \begin{cases} 1 \text{ if } & j'=l \\ 0 \text{ if } & j'\neq l. \end{cases}$$

This completes the construction of polynomials B_l, $1\le l\le n$, of degree at most $2n-1$ satisfying (51).

Polynomial $A_k(x)$, $1\le k\le n$, that satisfies (50) can be constructed in a similar way. Condition (a) implies that

$$A_k(x) = \frac{\omega_X(x)}{x-x_k}S(x), \tag{66}$$

where $S(x)$ is a nonzero polynomial of degree at most n. Taking derivative on both sides of (66) gives

$$A_k'(x) = \frac{\omega_X(x)}{x-x_k}S'(x) + \left[\frac{\omega_X'(x)}{x-x_k} - \frac{\omega_X(x)}{(x-x_k)^2}\right]S(x). \tag{67}$$

Recall that $a\omega_X(x)+c\omega_X'(x)$ has all roots being real and simple, and we obtained from Condition (b) that $A_k'(x) = (a\omega_X(x)+c\omega_X'(x))T(x)$ for some polynomial T of degree at most $n-2$. Thus

$$(a\omega_X(x)+c\omega_X'(x))T(x) = \frac{\omega_X(x)}{x-x_k}S'(x) + \left[\frac{\omega_X'(x)}{x-x_k} - \frac{\omega_X(x)}{(x-x_k)^2}\right]S(x). \tag{68}$$

Multiplying both sides by $(x-x_k)^2$ and then moving all terms with the factor $\omega_X(x)$ to the right-hand side, we obtain

$$\begin{aligned}&\omega_X'(x)(x-x_k)\big[S(x)-c(x-x_k)T(x)\big]\\ &= \omega_X(x)\big[S(x)-(x-x_k)S'(x)+a(x-x_k)^2T(x)\big].\end{aligned} \tag{69}$$

Since ω_X and ω_X' are relatively prime (i.e., they do not have any zeros in common),

$$\omega_X'(x)U_k(x) = S(x)-(x-x_k)S'(x)+a(x-x_k)^2T(x) \tag{70}$$

and

$$\frac{\omega_X(x)}{x-x_k}U_k(x) = S(x) - c(x-x_k)T(x) \tag{71}$$

for a polynomial $U_k(x)$ of degree at most 1. From (70) and (71), we have that

$$\frac{\omega'_X(x)U_k(x)}{(x-x_k)^2} = \frac{S(x)}{(x-x_k)^2} - \frac{S'(x)}{x-x_k} + aT(x) = -\left(\frac{S(x)}{x-x_k}\right)' + aT(x). \tag{72}$$

and

$$\frac{\omega_X(x)U_k(x)}{(x-x_k)^2} = \frac{S(x)}{x-x_k} - cT(x). \tag{73}$$

Multiplying (73) with a/c, and then adding it with (72) gives

$$-\left(\frac{S(x)}{x-x_k}\right)' + \frac{a}{c}\frac{S(x)}{x-x_k} = \frac{1}{c}\frac{a\omega_X(x) + c\omega'_X(x)}{(x-x_k)^2}U_k(x). \tag{74}$$

Multiplying both sides with $e^{-\frac{a}{c}x}$ gives

$$\frac{\mathrm{d}}{\mathrm{d}x}\left(e^{-\frac{a}{c}x}\frac{S(x)}{x-x_k}\right) = -\frac{e^{-\frac{a}{c}x}}{c}\frac{a\omega_X(x) + c\omega'_X(x)}{(x-x_k)^2}U_k(x). \tag{75}$$

Integrating both sides leads to

$$\frac{S(x)}{x-x_k} = -\frac{e^{\frac{a}{c}x}}{c}\int\frac{a\omega_X(x) + c\omega'_X(x)}{(x-x_k)^2}U_k(x)e^{-\frac{a}{c}x}\mathrm{d}x. \tag{76}$$

The above equation combined with equation (66) gives us

$$A_k(x) = -\frac{1}{c}e^{\frac{a}{c}x}\omega_X(x)\int\frac{a\omega_X(x) + c\omega'_X(x)}{(x-x_k)^2}U_k(x)e^{-\frac{a}{c}x}\mathrm{d}x. \tag{77}$$

Now it remains to figure out the polynomial U_k of degree at most one. Write

$$U_k(x) = r_0 + r_1(x-x_k). \tag{78}$$

Then by (21)

$$r_0 = U_k(x_k) = \frac{1}{\omega'_X(x_k)\prod_{i\neq k}(x_k - x_i)}, \tag{79}$$

and by (70),

$$\omega''_X(x_k)U_k(x_k) + \omega'_X(x_k)U'_k(x_k) = 0. \tag{80}$$

The above equation implies that

$$r_1 = -\frac{\omega_X''(x_k)}{(\omega_X'(x_k))^2 \prod_{i\neq k}(x_k - x_i)}. \tag{81}$$

Therefore

$$U_k(x) = \frac{1}{(\omega_X'(x_k))^2 \prod_{i\neq k}(x_k - x_i)}\big(\omega_X'(x_k) - \omega_X''(x_k)(x - x_k)\big).$$

Substituting this into (77), we obtain that

$$\begin{aligned} A_k(x) &= \frac{1}{c\omega_X'(x_k)\prod_{i\neq k}(x_k - x_i)} e^{\frac{a}{c}x}\omega_X(x) \\ &\times \int\limits_x^{\infty} \frac{a\omega_X(t) + c\omega_X'(t)}{(t - x_k)^2}\big(1 - \frac{\omega_X''(x_k)}{\omega_X'(x_k)}(t - x_k)\big)e^{-\frac{a}{c}t}\mathrm{d}t \text{ for } x > x_k, \end{aligned} \tag{82}$$

if $\frac{a}{c} > 0$, and

$$\begin{aligned} A_k(x) &= -\frac{1}{c\omega_X'(x_k)\prod_{i\neq k}(x_k - x_i)} e^{\frac{a}{c}x}\omega_X(x) \\ &\int\limits_{-\infty}^{x} \frac{a\omega_X(t) + c\omega_X'(t)}{(t - x_k)^2}\big(1 - \frac{\omega_X''(x_k)}{\omega_X'(x_k)}(t - x_k)\big)e^{-\frac{a}{c}t}\mathrm{d}t \text{ for } x < x_k, \end{aligned} \tag{83}$$

if $\frac{a}{c} < 0$. Note that $A_k(x)$ satisfies condition (b) because

$$\begin{aligned} A_k'(x) &= -\frac{e^{\frac{a}{c}x}\omega_X(x)}{c\omega_X'(x_k)\prod_{i\neq k}(x_k - x_i)} \frac{a\omega_X(x) + c\omega_X'(x)}{(x - x_k)^2} \\ &\times \left(1 - \frac{\omega_X''(x_k)}{\omega_X'(x_k)}(x - x_k)\right)e^{-\frac{a}{c}x} - \frac{1}{c^2}e^{\frac{a}{c}x}(a\omega_X(x) + c\omega_X'(x)) \\ &\times \int \frac{a\omega_X(x) + c\omega_X'(x)}{\omega_X'(x_k)(x - x_k)^2\prod_{i\neq k}(x_k - x_i)}\big(1 - \frac{\omega_X''(x_k)}{\omega_X'(x_k)}(x - x_k)\big)e^{-\frac{a}{c}x}\mathrm{d}x. \end{aligned} \tag{84}$$

Thus $A_k'(x_j^*) = 0$ because both terms have the factor $a\omega_X(x) + c\omega_X'(x)$ which takes zero value when x replaced by x_j^*. Note that

$$
\begin{aligned}
&(a\omega_X(x)+c\omega_X'(x))U_k(x)\\
&=(a\omega_X(x_k)+c\omega_X'(x_k))U_k(x_k)+((a\omega_X(x)+c\omega_X'(x)U_k(x))'|_{x=x_k}(x-x_k)\\
&\quad+c_2(x-x_k)^2+\cdots+c_N(x-x_k)^N\\
&=\frac{c}{\prod\limits_{i\neq k}(x_k-x_i)}-\frac{a}{\prod\limits_{i\neq k}(x_k-x_i)}(x-x_k)+c_2(x-x_k)^2+\cdots+c_N(x-x_k)^N,
\end{aligned}
$$

by (79) and (81). Therefore,

$$
\int\frac{a\omega_X(x)+c\omega_X'(x)}{(x-x_k)^2}U_k(x)e^{-\frac{a}{c}x}\mathrm{d}x=-\frac{ce^{-\frac{a}{c}x}}{\prod\limits_{i\neq j}(x_k-x_j)}\frac{1}{x-x_k}+Q(x)e^{-\frac{a}{c}x} \quad (85)
$$

for some polynomial Q of degree at most n. Therefore A_k is a polynomial of degree at most $2n-1$ and

$$
A_k(x_j)=0 \quad \text{for all} \quad j\neq k \quad (86)
$$

and

$$
A_k(x_k)=-\frac{1}{c}e^{\frac{a}{c}x_k}\lim_{x\to x_k}\frac{-ce^{-\frac{a}{c}x}}{\prod\limits_{i\neq j}(x_k-x_j)}\frac{\omega_X(x)}{x-x_k}=1. \quad (87)
$$

This completes the proof.

4 Pál Interpolation of Integral Type III

In this section, we consider Pál interpolation associated with $\tilde{\omega}_X(x):=a\omega_X(x)+(bx+c)\omega_X'(x)$ with $a<0$ and $b=1$.

Theorem 4.1 *Let $X:=\{x_1,\ldots,x_n\}$ contain n distinct nodes on the real line ordered by $x_1<x_2<\cdots<x_n$, $0>a\notin\{-1,-n\}$ and $0\neq c\notin -X$. Assume that the polynomial $\tilde{\omega}_X(x):=a\omega_X(x)+(x+c)\omega_X'(x)$ has n simple roots, which is denoted by $X^*:=\{x_1^*,x_2^*,\ldots,x_n^*\}$. Then the polynomial $R(x)$ of degree $2n-2$ defined by*

$$
\begin{aligned}
R(x):=\frac{\mathrm{d}}{\mathrm{d}x}\Big[&-\sum_{k=2}^{n}z_k\frac{\omega_X(x)}{|x+c|^{-a}}\int_{-c}^{x}\frac{|t+c|^{-a}}{t+c}(1-\alpha_k(t))\\
&\frac{\tilde{\omega}_X(t)}{\omega_X'(x_k)(t-x_k)^2\prod\limits_{i\neq k}(x_k-x_i)}\mathrm{d}t+\sum_{l=1}^{n-1}y_l^*\frac{\beta_l}{\Omega(x_l^*)}\omega_X(x)\\
&\times\int_{-c}^{x}\left|\frac{t+c}{x+c}\right|^{-a}\frac{1}{t+c}\frac{\tilde{\omega}_X(t)}{t-x_l^*}\mathrm{d}t\Big]
\end{aligned}
$$

satisfies

$$R(x_l^*) = y_l^*,\ 1 \le l \le n, \text{ and } \int_{x_k}^{x_{k+1}} R(x)\,\mathrm{d}x = y_{k+1},\quad 1 \le k \le n-1,$$

where

$$z_k = \sum_{q=2}^{k} y_q,\quad \alpha_k(t) = \frac{\omega_X''(x_k)}{\omega_X'(x_k)}(t - x_k),\quad 2 \le k \le n,$$

and

$$\beta_l = \frac{x_l^* + c}{\omega_X(x_l^*)},\quad \Omega(x_l^*) = \frac{\tilde{\omega}_X(x)}{x - x_l^*}\Big|_{x=x_l^*},\quad 1 \le l \le n-1.$$

Proof We begin the proof the same way as in the previous sections, by decomposing $R(x)$ into a sum of polynomials $A_k(x)$ and $B_l(x)$ which satisfy the conditions:

$$\begin{cases} \text{(a) } A_k(x_i) = \delta_{ki} \text{ for all } 2 \le k \le n \text{ and } 1 \le i \le n \\ \text{(b) } A_k'(x_j^*) = 0 \text{ for all } 2 \le k \le n \text{ and } 1 \le j \le n, \end{cases} \tag{88}$$

and

$$\begin{cases} \text{(c) } B_l(x_i) = 0 \text{ for all } 1 \le l \le n \text{ and } 1 \le i \le n \\ \text{(d) } B_l'(x_j^*) = \delta_{lj} \text{ for all } 1 \le l \le n \text{ and } 1 \le j \le n. \end{cases} \tag{89}$$

Again, we start by obtaining the polynomial $B_l(x)$, $1 \le l \le n$ first. By (89),

$$B_l(x) = \omega_X(x) V_l(x) \tag{90}$$

where $V_l(x)$ is a polynomial of degree at most $n-1$. Taking derivative of the above equality leads to

$$B_l'(x) = \omega_X(x) V_l'(x) + \omega_X'(x) V_l'(x) = \frac{a\omega_X(x) + (x + c)\omega_X'(x)}{x - x_l^*} W_l(x) \tag{91}$$

where $W_l(x)$ is a polynomial of degree at most $n-1$. Note that in the above expression equality holds by (89) and the assumption that all roots of $a\omega_X(x) + (x + c)\omega_X'(x)$ are real and simple. Recall that ω_X and ω_X' have no common roots. Hence we obtain from (91) that

$$-(x - x_l^*) V_l'(x) + a W_l(x) = M \omega_X'(x) \tag{92}$$

and

$$(x - x_l^*) V_l(x) - (x + c) W_l(x) = M \omega_X(x) \tag{93}$$

for some constant M. Dividing (92) by $x - x_l^*$, (93) by $(x+c)(x-x_l^*)/a$, and then taking their sum, we have

$$V_l'(x) - \frac{a}{x+c}V_l(x) = -\frac{M}{x+c}\frac{a\omega_X(x) + (x+c)\omega_X'(x)}{x - x_l^*}. \tag{94}$$

Multiplying both sides by $|x+c|^{-a}$ and then integrating both sides yields

$$V_l(x) = -M|x+c|^a \int_{-c}^{x} |t+c|^{-a}\frac{a\omega_X(t) + (t+c)\omega_X'(t)}{(t-x_l^*)(t+c)}\mathrm{d}t. \tag{95}$$

To find the constant M, we note that from condition (d) and (91)

$$B_l'(x_k^*) = 1 = \omega_X(x_l^*)V_l'(x_l^*) + \omega_X'(x_l^*)V_l(x_l^*). \tag{96}$$

Replacing x with x_l^* in (94) we get

$$-(x_l^*+c)V_l'(x_l^*) + aV_l(x_l^*) = M\frac{a\omega_X(x) + (x+c)\omega_X'(x)}{x - x_l^*}\Big|_{x=x_l^*} =: M\Omega(x_l^*). \tag{97}$$

We remark that $\Omega(x_l^*)$ is nonzero because roots of $a\omega_X(x) + (x+c)\omega_X'(x)$ are simple. Multiplying both sides of (96) with x_l^*+c gives

$$x_l^* + c = (x_l^*+c)\omega_X'(x_l^*)V_l(x_l^*) + (x_l^*+c)V_l'(x_l^*)\omega_X(x_l^*). \tag{98}$$

Recalling that x_l^* is a root of the polynomial $a\omega_X(x) + (x+c)\omega_X'(x)$, i.e.,

$$a\omega_X(x_l^*) + (x_l^*+c)\omega_X'(x_l^*) = 0. \tag{99}$$

This together with (96) implies that

$$x_l^* + c = -a\omega_X(x_l^*)V_l(x_l^*) + (x_l^*+c)\omega_X(x_l^*)V_l'(x_l^*). \tag{100}$$

Observe that $\omega_X(x_l^*) \neq 0$, as otherwise $(x_l^*+c)\omega_X'(x_l^*) = 0$, which contradicts to the assumptions on c and the simple root property for $\omega_X(x)$. Therefore,

$$-\frac{x_l^*+c}{\omega(x_l^*)} = -V_l'(x_l^*)(x_l^*+c) + aV_l(x_l^*). \tag{101}$$

Thus

$$M = -\frac{x_l^*+c}{\omega_X(x_l^*)\Omega(x_l^*)}, \tag{102}$$

from which we conclude that

$$B_l(x) = \frac{x_l^* + c}{\omega_X(x_l^*)\Omega(x_l^*)}\omega_X(x)\int_{-c}^{x}\left|\frac{t+c}{x+c}\right|^{-a}\frac{1}{t+c}\frac{a\omega_X(t)+(t+c)\omega_X'(t)}{t-x_l^*}\mathrm{d}t. \tag{103}$$

The polynomials $B_l, 1 \leq l \leq n-1$, satisfy the requirement (c) in (89) as they have the factor ω_X by (90), and also the requirement (d) in (89) as

$$\begin{aligned} B_l'(x) &= M\omega_X(x)\frac{1}{x+c}\frac{a\omega_X(x)+(x+c)\omega_X'(x)}{x-x_l^*} \\ &\quad + M(\omega_X(x)(x+c)^a)'\int_{-c}^{x}|t+c|^{-a}\frac{a\omega_X(t)+(t+c)\omega_X'(t)}{(t+c)(t-x_l^*)}\mathrm{d}t \end{aligned}$$

and hence

$$B_l'(x_{j'}^*) = \begin{cases} 1 & \text{if} \quad j' = l \\ 0 & \text{if} \quad j' \neq l. \end{cases}$$

This completes the construction of polynomials $B_l, 1 \leq l \leq n-1$.

We finish this section by the construction of polynomials $A_k, 2 \leq k \leq n$ that satisfies (88). Condition (a) in (88) implies that

$$A_k(x) = \frac{\omega_X(x)}{x-x_k}S(x) \tag{104}$$

where $S(x)$ is a nonzero polynomial of degree at most n. The above equation (104) together with condition (b) in (88) implies that

$$\begin{aligned} A_k'(x) &= \left(\frac{(x-x_k)\omega_X'(x)-\omega_X(x)}{(x-x_k)^2}\right)S(x) + \frac{\omega_X(x)}{x-x_k}S'(x) \\ &= \big(a\omega_X(x)+(x+c)\omega_X'(x)\big)T(x) \end{aligned} \tag{105}$$

for a polynomial $T(x)$ of degree at most $n-2$. Multiplying (105) with $(x-x_k)^2$ and rearranging the equation yields

$$\begin{aligned} &\omega_X'(x)(x-x_k)\big[S(x)-(x+c)(x-x_k)T(x)\big] \\ &= \omega_X(x)\big[S(x)-(x-x_k)S'(x)+a(x-x_k)T(x)\big]. \end{aligned}$$

Recalling that $\omega_X(x)$ and $\omega_X'(x)$ have no roots in common, we have

$$\omega_X'(x)U_k(x) = S(x)-(x-x_k)S'(x)+a(x-x_k)^2T(x) \tag{106}$$

and

$$\frac{\omega_X(x)}{x-x_k}U_k(x) = S(x) - (x+c)(x-x_k)T(x) \tag{107}$$

for some polynomial U_k of degree at most one. Rearranging equations (106) and (107) yields

$$-\Big(\frac{S(x)}{x-x_k}\Big)' + aT(x) = \frac{\omega_X'(x)U_k(x)}{(x-x_k)^2} \tag{108}$$

and

$$\frac{S(x)}{x-x_k} - (x+c)T(x) = \frac{\omega_X(x)U_k(x)}{(x-x_k)^2}. \tag{109}$$

Multiplying (109) by $a/(x+c)$ and adding it to (108) gives

$$-\Big(\frac{S(x)}{x-x_k}\Big)' + \frac{a}{x+c}\Big(\frac{S(x)}{x-x_k}\Big) = -\frac{U_k(x)}{x+c}\frac{a\omega_X(x)+(x+c)\omega_X'(x)}{(x-x_k)^2}. \tag{110}$$

Multiplying both sides of the above equation by $|x+c|^{-a}$ leads to

$$\frac{\mathrm{d}}{\mathrm{d}x}\Big(|x+c|^{-a}\frac{S(x)}{x-x_k}\Big) = -\frac{U_k(x)}{x+c}|x+c|^{-a}\big(a\omega_X(x)+(x+c)\omega_X'(x)\big). \tag{111}$$

Hence

$$\frac{S(x)}{x-x_k} = -|x+c|^a \int \frac{|x+c|^{-a}}{x+c}\frac{a\omega_X(x)+(x+c)\omega_X'(x)}{(x-x_k)^2}U_k(x)\mathrm{d}x. \tag{112}$$

Comparing (112) to (104) yields

$$A_k(x) = -\omega_X(x)|x+c|^a \int \frac{|x+c|^{-a}}{x+c}\frac{a\omega_X(x)+(x+c)\omega_X'(x)}{(x-x_k)^2}U_k(x)\mathrm{d}x. \tag{113}$$

From (106)

$$S(x_k) = \omega_X'(x_k)U_k(x_k), \tag{114}$$

and from (104) and condition (a) it follows that

$$A_k(x_k) = \omega_X'(x_k)S(x_k) = 1. \tag{115}$$

Therefore,

$$S(x_k) = \frac{1}{\omega_X'(x_k)} \tag{116}$$

and

$$U_k(x_k) = \frac{S(x_k)}{\omega'_X(x)} = \frac{1}{\left(\omega'_X(x_k)\right)^2}. \tag{117}$$

Multiplying (106) by $x + c$ and (107) by a and adding the two gives

$$(a\omega_X(x) + (x+c)\omega'_X(x))U_k(x) = S(x)[(x+c) + a(x - x_k)] - (x+c)(x - x_k)S'(x). \tag{118}$$

Replacing x with x_k yields

$$\tilde{\omega}(x)U_k(x)|_{x=x_k} = (x+c)S(x)|_{x=x_k} = \left.\frac{x+c}{\omega'_X(x)}\right|_{x=x_k}. \tag{119}$$

Now taking the derivative on both sides of (118) yields

$$\left((a\omega_X(x) + (x+c)\omega'_X(x))U_k(x)\right)' \tag{120}$$

$$= -S''(x)(x+c)(x-x_k) + S'(x)(x-x_k)(a-1) + (a+1)S(x). \tag{121}$$

Replacing x with x_k in the above equation and then applying (116) gives

$$\left((a\omega_X(x) + (x+c)\omega'_X(x))U_k(x)\right)'|_{x=x_k} = \frac{a+1}{\omega'_X(x_k)}. \tag{122}$$

Therefore, the Taylor series expansion of $\tilde{\omega}(x)U_k(x)$ about the point x_k is

$$\begin{aligned}(a\omega_X(x) + (x+c)\omega'_X(x))U_k(x) &= \frac{x_k + c}{\omega'_X(x_k)} + \frac{a+1}{\omega'_X(x_k)}(x - x_k) \\ &\quad + C_2(x-x_k)^2 + \cdots + C_N(x-x_k)^N\end{aligned} \tag{123}$$

for some constants c_i, $2 \le i \le N$, where $N = n + \deg U_k \le n+1$. Hence by (113)

$$\begin{aligned}A_k(x) = -\frac{\omega_X(x)}{|x+c|^{-a}} \int \frac{|x+c|^{-a}}{x+c} &\left(\frac{x_k+c}{\omega'_X(x_k)} \cdot \frac{1}{(x-x_k)^2} + \frac{a+1}{\omega'_X(x_k)} \cdot \frac{1}{x-x_k}\right. \\ &\left. + C_2 + \cdots + C_N(x-x_k)^{N-2}\right) \mathrm{d}x.\end{aligned} \tag{124}$$

Note that

$$\int \frac{(x+c)^{-a-1}}{(x-x_k)^2} \mathrm{d}x = -\frac{(x+c)^{-a-1}}{x-x_k} - (a+1) \int \frac{(x+c)^{-a-2}}{x-x_k} \mathrm{d}x. \tag{125}$$

Therefore for $x > c$, we have

$$\begin{aligned}
&\int \frac{|x+c|^{-a}}{x+c}\Big(\frac{x_k+c}{\omega_X'(x_k)}\cdot\frac{1}{(x-x_k)^2}+\frac{a+1}{\omega_X'(x_k)}\cdot\frac{1}{x-x_k}\Big)\mathrm{d}x\\
&=-\frac{x_k+c}{\omega_X'(x_k)}\frac{(x+c)^{-a-1}}{x-x_k}-\frac{(a+1)(x_k+c)}{\omega_X'(x_k)}\int\frac{(x+c)^{-a-2}}{x-x_k}\mathrm{d}x\\
&\quad+\frac{a+1}{\omega_X'(x_k)}\int\frac{(x+c)^{-a-1}}{x-x_k}\mathrm{d}x\\
&=-\frac{x_k+c}{\omega_X'(x_k)}\frac{(x+c)^{-a-1}}{x-x_k}-\frac{1}{\omega_X'(x_k)}(x+c)^{-a-1}+C\\
&=-\frac{(x+c)^{-a}}{\omega_X'(x_k)(x-x_k)}+C.
\end{aligned}\tag{126}$$

By (113), (124) and (126), we then obtain

$$\begin{aligned}
A_k(x)&=\frac{\omega_X(x)}{\omega_X'(x_k)(x-x_k)}-\omega_X(x)|x+c|^a\\
&\quad\times\int(x+c)^{-a-1}\big(C_2+\cdots+C_N(x-x_k)^{N-2}\big)\mathrm{d}x.
\end{aligned}$$

which implies that $A_k(x)$ is a polynomial and

$$A_k(x)=-\omega_X(x)|x+c|^a\int_{-c}^{x}\frac{|t+c|^{-a}}{t+c}\frac{\tilde{\omega}(x)}{\omega_X'(x_k)(t-x_k)^2\prod\limits_{i\neq k}(x_k-x_i)}U_k(x)\mathrm{d}x.\tag{127}$$

Recall that $U_k(x)$ is a linear function, and so we may write

$$U_k(x)=r_0+r_1(x-x_k).\tag{128}$$

From (122) and (117),

$$\big((a\omega_X(x)+(x+c)\omega_X'(x)U_k(x)\big)'|_{x=x_k}=\frac{a+1}{\omega_X'(x_k)}+\frac{(x_k+c)\omega_X''(x_k)}{\big(\omega_X'(x_k)\big)^2}\tag{129}$$

$$+(x_k+c)r_1\omega_X'(x_k)=\frac{a+1}{\omega_X'(x_k)}.\tag{130}$$

This together with (117) implies that

$$r_0=\frac{1}{(\omega_X'(x_k))^2}\text{ and }r_1=-\frac{\omega''(x_k)}{\big(\omega_X'(x_k)\big)^3}.\tag{131}$$

Finally,

$$A_k(x) = -\frac{\omega_X(x)}{|x+c|^{-a}} \int_{-c}^{x} \frac{|t+c|^{-a}}{t+c} \frac{a\omega_X(t) + (t+c)\omega_X'(t)}{\omega_X'(x_k)(t-x_k)^2 \prod_{i\neq k}(x_k - x_i)} \times \left(1 - \frac{\omega_X''(x_k)}{\omega_X'(x_k)}(t-x_k)\right) \mathrm{d}t. \tag{132}$$

Using (124) and (126), we can verify that $A_k, 2 \leq k \leq n$, just defined satisfy the requirement (a) and (b) in (88).

Acknowledgments The author would like to sincerely thank Dr. Qiyu Sun and Dr. Ram Mohapatra of the University of Central Florida, whose help and guidance made this paper possible.

References

1. Joó, I.: Weighted (0, 2)-interpolation on the roots of Jacobi polynomials. Acta Mathematica Hungarica **66**, 25–50 (1995)
2. Krebsz, A.: Weighted (0, 1, 3) interpolation on the roots of classical orthogonal polynomials. Mathematica Pannonica **15**, 21–35 (2004)
3. Mathur, P., Dutta, S.: On Pál type weighted lacunary (0, 2; 0) interpolation on infinite interval $(-\infty, +\infty)$. Approximation Theor. Appl. **17**, 1–10 (2001)
4. Pál, L.G.: A new modification of the Hermite-Fejér interpolation. Anal. Mathematica **1**, 197–205 (1975)
5. Srivastava, R.: Weighted (0;0,2)-interpolation on the roots of Hermite polynomials. Acta Mathematica Hungarica **70**, 57–73 (1996)
6. Szabó, V.E.S., Joó, I.: A generalization of Pál interpolation process. Acta Scientarum Mathematicarum (Szeged) **60**, 429–438 (1995)
7. Szabó, V.E.S.: A generalization of Pál interpolation process II. Acta Mathematica Hungarica **74**, 19–29 (1997)
8. Szabó, V.E.S.: A generalization of Pál interpolation process III: the Hermite case. Acta Mathematica Hungarica **74**, 191–201 (1997)
9. Szabó, V.E.S.: A generalization of Pál interpolation Process IV: the Jacobi case. Acta Mathematica Hungarica **74**, 287–300 (1997)
10. Szil, L.: Weighted (0,2)-interpolation on the roots of Hermite polynomials. Annales Universitatis Scientarium Budapestinensis **70**, 153–166 (1985)

Positivity Preserving Rational Cubic Trigonometric Fractal Interpolation Functions

A.K.B. Chand and K.R. Tyada

Abstract In this paper, we propose a family of $\mathscr{C}^1$-rational cubic trigonometric fractal interpolation function (RCTFIF) to preserve positivity inherent in a set of data. The proposed RCTFIF is a generalized fractal version of the classical rational cubic trigonometric polynomial spline of the form $\frac{p_i(\theta)}{q_i(\theta)}$, where $p_i(\theta)$ and $q_i(\theta)$ are cubic trigonometric polynomials. The RCTFIF involves a scaling factor and four shape parameters in each subinterval. The convergence of the RCTFIF towards the original function is studied. We deduce the simple data dependent sufficient conditions on the scaling factors and shape parameters associated with the $\mathscr{C}^1$-RCTFIF so that the proposed RCTFIF preserves the positivity property of the given positive data set. The first derivative of the proposed RCTFIF is irregular in a finite or dense subset of the interpolation interval, and matches with the first derivative of the classical rational trigonometric cubic interpolation function whenever all scaling factors are zero. The effects of the scaling factors and shape parameters on the RCTFIF and its first derivative are illustrated graphically.

Keywords Iterated function systems · Fractal interpolation · Rational cubic trigonometric interpolation · Positivity

Mathematics Subject Classification (2000): 28A80 · 41A30 · 42A15 · 41A55 · 37C25

1 Introduction

The development of interpolating schemes for shape preservation of discrete data has a great deal of significance in applied mathematics, industry and engineering. Particularly, when the data is obtained from some complex natural and scientific

A.K.B. Chand · K.R. Tyada (✉)
Department of Mathematics, Indian Institute of Technology Madras, Chennai 600036, India
e-mail: kurmaths86@gmail.com

A.K.B. Chand
e-mail: chand@iitm.ac.in

R.N. Mohapatra et al. (eds.), *Mathematics and Computing*,
Springer Proceedings in Mathematics & Statistics 139,
DOI 10.1007/978-81-322-2452-5_13

phenomena, it becomes vital to incorporate the inherited geometric features of given data. In general a user demands an interpolant which preserves the hidden geometric properties of the given data. In the literature a wide range of shape preserving classical spline interpolation techniques (see [2, 5, 11, 14, 15, 18–20]) have been discussed. The classical splines interpolate the data smoothly but, certain derivatives of the classical interpolants are either piecewise smooth or globally smooth in nature. Therefore the classical interpolants are not suitable to approximate functions that have irregular nature or fractality in their first order derivatives. On the other hand fractal interpolation can be applied in such scenario as well.

Fractal interpolation is a better technique to analyze various scientific data. Barnsley [3] introduced the concept of fractal interpolation functions (FIFs) based on the structure of iterated functions system (IFS). An IFS provides the attractor which is the graph of an approximated function that interpolates given data points. FIFs are the fixed points of Read-Bajraktaverić operator [4], which is defined on suitable function spaces. The functional relation involved in the definition of a FIF gives self similarity on small scales. Barnsley and Harrington [4] introduced the construction of k-times differentiable spline FIF with a fixed type of boundary conditions. The spline FIF with general boundary conditions is studied in a constructive manner in [6, 10]. The α-fractal polynomial spline with general Hermite boundary conditions is studied by Chand and Navascués [7]. A specific feature of spline FIF is that its certain derivative can be used to capture the irregularity associated with interpolation data. The graph of the derivative of a spline FIF possesses a fractal dimension which provides a geometric characterization of the measured variable.

Since the classical polynomial spline interpolant representation available in the literature is unique for given data, and it simply depends on the data points, it is difficult to preserve all the hidden shape properties of the given data, for example data over a straight line, positivity, monotonicity or convexity. For this reason the user need a capable smooth curve representation of shape preserving interpolating schemes which preserve the shape of the data. In this case, rational interpolation functions provide the proficient shape preserving interpolating techniques. Splines cannot represent the transcendental curves like circular arc, elliptical arc, cylinder, sphere, hyperbola, etc. To overcome this issue, many bases are presented using trigonometric functions or the blending of polynomial and trigonometric functions.

A data set obtained by scientific phenomena or by a complex function can be categorized as positive, monotonic or convex based on its distribution. Out of all the geometric properties, positivity plays important role at several places. There are many physical circumstances where the variables have meaning only when they are non negative. For example, in a probability distribution, the presentation is always positive. Similarly, when dealing with the samples of populations, the data are always in positive figures. Another area of application is in the observation of gas discharge when certain chemical experiments are in process. Therefore, it is important to discuss positive interpolation to provide a computationally economical and visually pleasing solution to the problems of different scientific phenomena.

A considerable amount of research is available in the literature on positive data interpolation (see [2, 5, 14, 15, 18–20]). Abbas [2] constructed a $\mathscr{C}^2$ piecewise rational cubic interpolation scheme that preserves the shape of the positive data. Butt [5] and Schmidt [20] generated algorithms through $\mathscr{C}^1$ cubic Hermite interpolant for positivity preservation. Hussain and Sarfraz developed a $\mathscr{C}^1$ rational cubic interpolant to preserve positivity [14] including other shape preserving properties in [15, 18].

Abbas [1] developed a $\mathscr{C}^1$ cubic trigonometric spline with four shape parameters for the data visualizations of positive data. Han developed a new kind of $\mathscr{C}^1$ rational quadratic [12] and cubic [13] rational trigonometric interpolation functions to preserve the shape of positive data. Hussain [16] used the quadratic trigonometric polynomial interpolation functions for shape preserving data visualization. Ibraheem [17] proposed a rational cubic trigonometric interpolant to preserve 2D and 3D positive data. Bashir [21] developed the rational quadratic trigonometric interpolation scheme to visualize positive, monotonic and convex data.

The shape preservation of scientific data through different types of smooth rational FIF are studied in [8–10, 22, 23]. In this paper, we have presented the smooth rational cubic trigonometric fractal interpolation function for the first time in the literature. Since our proposed RCTFIF contains scaling factors and four shape parameters, it is more efficient for preserving the shape of the data. In order to study the shape preserving aspects by the RCTFIF, we have studied the positive interpolation. In particular, when the interpolation data set is positive, the parameters of the proposed RCTFIF are restricted so that the corresponding fractal trigonometric FIF itself is positive.

The paper is organized as follows: In Sect. 2, the general frame work of FIF based on the IFS theory is reviewed. The construction of $\mathscr{C}^1$ RCTFIFs passing through a set of data points is discussed in Sect. 3. In Sect. 4, the error estimation of the RCTFIF to an original function is proven. Section 5 establishes the theory of rational cubic trigonometric fractal interpolation functions. Sufficient conditions for positivity preserving interpolation by the RCTFIF is developed for which the range of scaling factors and shape parameters are restricted in Sect. 5.1 and the examples of positive RCTFIF are given in Sect. 5.2 followed by conclusions in Sect. 6.

2 Review of Fractal Interpolation Functions

Let (X, d_X) be a complete metric space. For $\Lambda := \{1, 2, \ldots, n-1\}$, let $\omega_i : X \to X, i \in \Lambda$ be continuous functions. Then the set $\mathscr{I} = \{X; \omega_i, i \in \Lambda\}$ is called an IFS. If each ω_i, $i \in \Lambda$ is contraction with contractive factor s_i then $\mathscr{I}$ is known as a *hyperbolic* IFS. Let $\mathscr{H}(X)$ be the set of all non empty compact subsets of X. Then there exists a natural metric called Hausdorff metric which completes $\mathscr{H}(X)$. The Hausdorff metric $\mathscr{H}(X)$ is defined by $d_{\mathscr{H}(X)}(A, B) = \max\{\mathscr{D}_B(A), \mathscr{D}_A(B)\}$, where $\mathscr{D}_B(A) = \max_{a \in A} \min_{b \in B} d_X(a, b)$. Associated with the IFS $\mathscr{I}$, there is a set valued Hutchinson map W on $\mathscr{H}(X)$ defined by $W(A) = \bigcup_{i=1}^{n-1} \omega_i(A)$ for all $A \in \mathscr{H}(X)$. If IFS $\mathscr{I}$ is *hyperbolic*, then it is easy to verify that W is a contraction map on $\mathscr{H}(X)$ with the contractive factor $s = \max\{s_i : i = 1, 2, \ldots, n-1\}$. Then by the Banach fixed point theorem, W has a unique fixed point (say) G such that for any initiator $A \in \mathscr{H}(X)$, $\lim_{m\to\infty} W^{o(m)}(A) = G$, and the limit is taken with respect to the Hausdorff metric. The fixed point G is called the attractor or deterministic fractal corresponding to the IFS $\mathscr{I}$.

Let $\mathscr{P} : \{t_1, t_2, \ldots, t_n\}$ be a partition of the real compact interval $I = [t_1, t_n]$, where $t_1 < t_2 < \cdots < t_n$. For $\Lambda^* := \{1, 2, \ldots, n\}$, let a set of data points $\{(t_j, f_j) \in I \times \mathbb{K} : j \in \Lambda^*\}$ be given, where $\mathbb{K}$ is a compact set in $\mathbb{R}$. Let $I_i = [t_i, t_{i+1}]$ and $L_i : I \to I_i,\ i \in \Lambda$ be contractive homeomorphisms such that

$$L_i(t_1) = t_i,\ L_i(t_n) = t_{i+1} \text{ for } i \in \Lambda. \tag{1}$$

$$|L_i(t) - L_i(t^*)| \le l_i\ |t - t^*| \quad \forall\, t, t^* \in I \text{ for some } 0 < l_i < 1.$$

Let $C = I \times \mathbb{K}$, and consider $n - 1$ continuous mappings $F_i : C \to \mathbb{K}$ satisfying

$$F_i(t_1, f_1) = f_i,\quad F_i(t_n, f_n) = f_{i+1},\quad i \in \Lambda, \tag{2}$$

$$|F_i(t, x) - F_i(t, y)| \le |\lambda_i||x - y| \quad \forall\ t \in I, \quad \forall\, x, y \in \mathbb{K} \text{ and for some } 0 \le |\lambda_i| < 1. \tag{3}$$

Now, define functions $\omega_i : C \to I_i \times \mathbb{K}$ such that $\omega_i(t, f) = (L_i(t), F_i(t, f))$ $\forall\, i \in \Lambda$.

Proposition 1 (Barnsley [3]) *The IFS $\{C;\ \omega_i, i = 1, 2, \ldots, n-1\}$ defined above admits a unique attractor G such that G is the graph of a continuous function $f^* : I \to \mathbb{K}$ which obeys $f^*(t_j) = f_j$ for $j \in \Lambda^*$.*

The above function f^* is called a FIF corresponding to the IFS $\{I \times \mathbb{K}; \omega_i(t, f) = (L_i(t), F_i(t, f)), i = 1, \ldots, n-1\}$. The functional representation of f^*, which is the fixed point of the Read-Bajraktarević operator is based on the following results: Let $\mathscr{G} = \{g : I \to \mathbb{R} \mid g \text{ is continuous},\ g(t_1) = f_1 \text{ and } g(t_n) = f_n\}$. Then $(\mathscr{G}, d_v)$ is a complete metric space, where the metric d_v is induced from the supremum norm on $\mathscr{C}(I)$. Define the Read-Bajraktarević operator T on $(\mathscr{G}, d_v)$ as

$$Tg(t) = F_i(L_i^{-1}(t), g(L_i^{-1}(t))),\ t \in I_i,\ i \in \Lambda. \tag{4}$$

According to (1) and (2), Tg is continuous on $I_i = [t_i, t_{i+1}]$, $i \in \Lambda$ and at each of the internal grids $t_2, \ldots, t_{n-1}$. Further, T is a contraction map on the complete metric space $(\mathscr{G}, d_v)$, i.e.,

$$d_v(Tf, Tg) = \|Tf - Tg\|_\infty \le |\lambda|_\infty \|f - g\|_\infty, \tag{5}$$

where $|\lambda|_\infty = \max\{|\lambda_i| : i \in \Lambda\}$. By the Banach fixed point theorem T possesses a unique fixed point (say) f^* on $\mathscr{G}$, i.e., $f^* \in \mathscr{G}$ such that $(Tf^*)(t) = f^*(t)\ \forall\, t \in I$. According to (4), the FIF f^* satisfies the following functional equation:

$$f^*(t) = F_i(L_i^{-1}(t), f^* \circ L_i^{-1}(t)), \quad t \in I_i, \quad i \in \Lambda. \tag{6}$$

The following IFS is popular in FIF theory:

$$\{C; \omega_i(t, f) = (L_i(t), F_i(t, f)), i = 1, 2, \ldots, n-1\}, \tag{7}$$

where $L_i(t) = a_i t + b_i$, $F_i(t, f) = \lambda_i f + M_i(t)$ with $M_i : I \to \mathbb{R}$ are suitable continuous functions such that (2–3) are satisfied. The multiplier λ_i is called a scaling factor of the transformation ω_i, and $\lambda = (\lambda_1, \lambda_2, \ldots, \lambda_{n-1})$ is the scale vector associated with the IFS. The scaling factors give an additional degree of freedom to FIFs over its counter parts in classical interpolation and allow us to modify its geometric properties. In this paper, we take $M_i(t)$ as a rational function, whose numerator and denominators are cubic trigonometric polynomials involving four shape parameters. The existence of a spline FIF is given by Barnsley and Harrington [4] and that result has been extended for the existence of rational spline FIF in the following theorem [8].

Theorem 1 *Let $\{(t_j, f_j) : j \in \Lambda^*\}$ be the given data set such that $t_1 < t_2 < \cdots < t_n$. Suppose that $L_i(t) = a_i t + b_i$, where $a_i = \frac{t_{i+1}-t_i}{t_n-t_1}$, $b_i = \frac{t_n t_i - t_1 t_{i+1}}{t_n - t_1}$ and $F_i(t, f) = \alpha_i f + M_i(t)$, $M_i(t) = \frac{p_i(t)}{q_i(t)}$, $p_i(t)$ and $q_i(t)$ are chosen polynomials of degree r and s respectively, and $q_i(t) \neq 0\ \forall\, t \in [t_1, t_n]$ for $i \in \Lambda$. Suppose for some integer $p \geq 0$, $|\alpha_i| < a_i^p$, $i \in \Lambda$. Let $F_{i,m}(t, f) = \frac{\lambda_i f + M_i^{(m)}(t)}{a_i^m}$, $f_{1,m} = \frac{M_1^{(m)}(t_1)}{a_1^m - \lambda_1}$, $f_{n,m} = \frac{M_{n-1}^{(m)}(t_n)}{a_{n-1}^m - \lambda_{n-1}}$, $m = 1, 2, \ldots, p$, where $M_i^{(m)}(t)$ represents the mth derivative of $M_i(t)$ with respect to t. If $F_{i,m}(t_n, f_{n,m}) = F_{i+1,m}(t_1, f_{1,m})$, $i = 1, 2, \ldots, n-2$, $m = 1, 2, \ldots, p$, then the IFS $\{I \times \mathbb{K}; \omega_i(t, f) = (L_i(t), F_i(t, f)), i = 1, 2, \ldots, n-1\}$ determines a rational FIF $\phi \in \mathscr{C}^p[t_1, t_n]$ such that $\phi(L_i(t)) = \alpha_i \phi(t) + M_i(t)$, and $\phi^{(m)}$ is the FIF determined by $\{I \times \mathbb{K}; w_{i,m}(t, f) = (L_i(t), F_{i,m}(t, f)), i = 1, \ldots, n-1\}$ for $m = 1, 2, \ldots, p$.*

Remark 1 The above theorem not only valid for algebraic polynomials but also works even if the algebraic polynomials $p_i(t)$ and $q_i(t)$ are replaced with trigonometric polynomials.

3 Construction of Rational Cubic Trigonometric Fractal Interpolation Functions

In this section, we construct the RCTFIF ϕ with four shape parameters in each subinterval with the help of Theorem 1. Let $\{(t_j, f_j), j \in \Lambda^*\}$ be a given set of interpolation data for an original function ψ such that $t_1 < t_2 < \cdots < t_n$. Consider the IFS $\{I \times \mathbb{K}; \omega_i(t, f) = (L_i(t), F_i(t, f)), i \in \Lambda\}$, where $L_i(t) = a_i t + b_i$ and $F_i(t, f) = \lambda_i f(t) + M_i(t)$, $M_i(t) = \frac{p_i(t)}{q_i(t)}$, where $p_i(t)$ and $q_i(t)$ are cubic trigonometric polynomials, $q_i(t) \neq 0 \; \forall \; t \in [t_1, t_n]$, and $|\lambda_i| < a_i, i \in \Lambda$. Let $F_i^{(1)}(t,d) = \frac{\lambda_i d + M_i^{(1)}(t)}{a_i}$, where $M_i^{(1)}(t)$ is the first order derivative of $M_i(t), t \in [t_1, t_n], i \in \Lambda$. $F_i(t, f)$ satisfying the following $\mathscr{C}^1$-interpolatory conditions:

$$F_i(t_1, f_1) = f_i, \; F_i(t_n, f_n) = f_{i+1}, \; F_i^{(1)}(t_1, d_1) = d_i, \; F_i^{(1)}(t_n, d_n) = d_{i+1}, \quad (8)$$

where d_i denote the first order derivative of ψ with respect to t at knot t_i. The attractor of the above IFS will be the graph of a $\mathscr{C}^1$-rational cubic trigonometric FIF. From (7) one can observe that our RCTFIF ϕ can be written as:

$$\phi(L_i(t)) = \lambda_i \phi(t) + M_i(t), \quad (9)$$

where $M_i(t) = \frac{p_i(\theta)}{q_i(\theta)}$ with $p_i(\theta) = (1-\sin\theta)^3 U_i + \sin\theta(1-\sin\theta)^2 V_i + \cos\theta(1-\cos\theta)^2 W_i + (1-\cos\theta)^3 X_i$, $q_i(\theta) = (1-\sin\theta)^3\alpha_i + \sin\theta(1-\sin\theta)^2\beta_i + \cos\theta(1-\cos\theta)^2\gamma_i + (1-\cos\theta)^3\delta_i$, and $\theta = \frac{\pi}{2}(\frac{t-t_1}{l}), l = t_n - t_1, t \in I$ and $\alpha_i, \beta_i, \gamma_i$ and δ_i are positive real shape parameters. To ensure that the rational cubic trigonometric FIF is $\mathscr{C}^1$-continuous, the following interpolation properties are imposed:

$$\phi(L_i(t_1)) = f_i, \phi(L_i(t_n)) = f_{i+1}, \phi'(L_i(t_1)) = d_i, \phi'(L_i(t_n)) = d_{i+1}, i \in \Lambda. \quad (10)$$

From (9), at $t = t_n$, we observe that $\phi(L_i(t_1)) = f_i \implies \lambda_i f_1 + U_i/\alpha_i = f_i$ with $f_i^* := f_i - \lambda_i f_1$ yields $U_i = \alpha_i(f_i - \lambda_i f_1) = \alpha_i f_i^*$.
Similarly, at $t = t_1$, it is clear that $\phi(L_i(t_n)) = f_{i+1}$ with $f_{i+1}^* := f_{i+1} - \lambda_i f_n$ provides us $f_{i+1} = \lambda_i f_n + X_i/\delta_i \implies X_i = \delta_i(f_{i+1} - \lambda_i f_n) = \delta_i f_{i+1}^*$.
Again $\phi'(L_i(t_1)) = d_i$ with $d_i^* := a_i d_i - \lambda_i d_1$ gives us

$$V_i = \beta_i(f_i - \lambda_i f_1) + \frac{2l\alpha_i(a_i d_i - \lambda_i d_1)}{\pi} = \beta_i f_i^* + \frac{2l\alpha_i d_i^*}{\pi}.$$

Similarly, $\phi'(L_i(t_n)) = d_{i+1}$ with $d_{i+1}^* := a_i d_{i+1} - \lambda_i d_n$ provides

$$W_i = \gamma_i(f_{i+1} - \lambda_i f_n) - \frac{2l\delta_i(a_i d_{i+1} - \lambda_i d_n)}{\pi} = \gamma_i f_{i+1}^* - \frac{2l\delta_i d_{i+1}^*}{\pi}.$$

Substituting the values of U_i, V_i, W_i and X_i in (9), we get the required $\mathscr{C}^1$-RCTFIF with the numerator,

$$p_i(\theta) = \alpha_i f_i^*(1-\sin\theta)^3 + \left\{\beta_i f_i^* + \frac{2\ell\alpha_i d_i^*}{\pi}\right\}\sin\theta(1-\sin\theta)^2$$
$$+ \left\{\gamma_i f_{i+1}^* - \frac{2\ell\delta_i d_{i+1}^*}{\pi}\right\}\cos\theta(1-\cos\theta)^2 + \delta_i f_{i+1}^*(1-\cos\theta)^3.$$

In most applications, the derivatives $d_j (j \in \Lambda^*)$ are not given, and hence must be calculated either from the given data or by some numerical methods. In this paper we have calculated $d_j,\ j \in \Lambda^*$ from the given data using the arithmetic mean method [11].

Remark 2 If $\lambda_i = 0$ for all $i \in \Lambda$, the RCTFIF 'ϕ' becomes the classical rational cubic trigonometric interpolation function $P(t)$(say) that is defined in [17] on each sub interval $[t_i, t_{i+1}]$ as

$$P(t) = p_i(z)/q_i(z), \tag{11}$$

where

$$p_i(z) = (1-\sin z)^3\overline{U}_i + \sin z(1-\sin z)^2\overline{V}_i + \cos z(1-\cos z)^2\overline{W}_i + (1-\cos z)^3\overline{X}_i,$$
$$q_i(z) = (1-\sin z)^3\alpha_i + \sin z(1-\sin z)^2\beta_i + \cos z(1-\cos z)^2\gamma_i + (1-\cos z)^3\delta_i,\ t \in [t_i, t_{i+1}]$$

with $z = \frac{\pi}{2}\left(\frac{t-t_i}{h_i}\right)$, $h_i = t_{i+1} - t_i$ and $\overline{U}_i = \alpha_i f_i, \overline{V}_i = \beta_i f_i + \frac{2h_i\alpha_i d_i}{\pi}, \overline{W}_i = \gamma_i f_{i+1} - \frac{2h_i\delta_i d_{i+1}}{\pi}, \overline{X}_i = \delta_i f_{i+1}$.

4 Convergence Analysis

In order to show that the convergence of the $\mathscr{C}^1$-RCTFIF 'ϕ' towards a data generating function $\psi \in \mathscr{C}^3[t_1, t_n]$, we need an upper bound for the uniform distance between them. Since 'ϕ' has an implicit expression, it is difficult to compute the uniform error $\|\phi - \psi\|_\infty$ by using any standard technique in numerical analysis. Hence we derive an upper bound of the error by using the classical counterpart P of 'ϕ' with the help of

$$\|\phi - \psi\|_\infty \le \|\phi - P\|_\infty + \|P - \psi\|_\infty, \tag{12}$$

where P is given by (11).

Now the error estimation between the original function ψ and the classical rational cubic trigonometric function P in an arbitrary subinterval $I_i = [t_i, t_{i+1}]$ can be found by using the Peano-Kernel theorem since P is exact for any quadratic polynomial and the details are given in [17].

Proposition 2 *The error between the classical rational cubic trigonometric function defined in* (11) *and the original function* $\psi \in C^3[t_1, t_n]$ *is*

$$|\psi(t) - P(t)| \le \frac{1}{2}\|\psi^{(3)}\|h_i^3 c_i,\ t \in [t_i, t_{i+1}], \tag{13}$$

$c_i = \max_{0\le z\le 1} \Theta(\alpha_i, \beta_i, \gamma_i, \delta_i, z)$,

$$\Theta(\alpha_i, \beta_i, \gamma_i, \delta_i, z) = \begin{cases} \max \Theta_1(\alpha_i, \beta_i, \gamma_i, \delta_i, z) & \text{for } 0 \le \gamma_i \le 1, 0 \le z \le 1, \\ \max \Theta_2(\alpha_i, \beta_i, \gamma_i, \delta_i, z) & \text{for } \gamma_i > 1 + \frac{2\delta_i}{\pi}, 0 \le z \le z^*, \\ \max \Theta_3(\alpha_i, \beta_i, \gamma_i, \delta_i, z) & \text{for } z^* \le z \le 1 \end{cases}$$

where $z^* = 1 - \frac{2\delta_i}{\pi(\gamma_i - \delta_i)}$ *and* $\Theta_1(\alpha_i, \beta_i, \gamma_i, \delta_i, z)$, $\Theta_2(\alpha_i, \beta_i, \gamma_i, \delta_i, z)$ *and* $\Theta_3(\alpha_i, \beta_i, \gamma_i, \delta_i, z)$ *are obtained from the proof of Theorem 3.1 in* [17].

Theorem 2 *Let 'ϕ' is the $\mathscr{C}^1$ continuous RCTFIF and $\psi \in C^3[t_1, t_n]$ is the data generating function with respect to the given data $\{(t_j, f_j), j \in \Lambda^*\}$. Let d_j, $j \in \Lambda^*$ be the bounded first order derivative at the knot t_j. Let $|\lambda|_\infty = \max\{|\lambda_i|, i \in \Lambda\}$ and the shape parameters α_i, β_i, γ_i and δ_i for $i \in \Lambda$ are non negative and $\beta_i \ge \alpha_i$, $\gamma_i \ge \delta_i$. Then*

$$\|\psi - \phi\|_\infty \le \frac{1}{2}\|\psi^{(3)}\|_\infty h^3 c + \frac{|\lambda|_\infty}{1 - |\lambda|_\infty}(E(h) + E^*(h)), \tag{14}$$

where $E(h) = \|\psi\|_\infty + \frac{4h}{\pi} E_1$, $E^*(h) = F + \frac{4h}{\pi} E_2$, $E_1 = \max_{1 \le j \le n-1}\{|d_j|\}$, $F = \max\{|f_1|, |f_n|\}$, $E_2 = \max\{|d_1|, |d_n|\}$, *and c is defined as in* Proposition 2.

Proof We have $\phi(t) = \lambda_i \phi(L_i^{-1}(t)) + P(L_i^{-1}(t), \lambda_i)\ \forall\, i = 1, 2, \ldots, n-1$. From (7), the Read-Bajraktarević operator ($T = T_\lambda^*$) with respect to the scaling vector $\lambda \ne 0$ on $\mathscr{G}$ can be written as:

$$T_\lambda^* f^*(t) = \lambda_i f^*(L_i^{-1}(t)) + M_i(L_i^{-1}(t), \lambda_i) \text{ for } t \in I_i,\ i \in \Lambda. \tag{15}$$

It is clear that RCTFIF 'ϕ' is the fixed point of T_λ^* and the classical rational cubic trigonometric function is the fixed point of $T_{\mathbf{0}}^*$. For $\lambda \ne \mathbf{0}$, T_λ^* is a contraction map with contraction factor $|\lambda|_\infty$. Thus

$$\|T_\lambda^* \phi - T_\lambda^* P\|_\infty \le |\lambda|_\infty \|\phi - P\|_\infty. \tag{16}$$

Also

$$\begin{aligned} |T_\lambda^* P(t) - T_{\mathbf{0}}^* P(t)| &= |\lambda_i P(L_i^{-1}(t)) + M_i(L_i^{-1}(t), \lambda_i) - M_i(L_i^{-1}(t), 0)|, \\ &= \left| \lambda_i P(L_i^{-1}(t)) + \frac{p_i(L_i^{-1}(t), \lambda_i)}{q_i(L_i^{-1}(t))} - \frac{p_i(L_i^{-1}(t), 0)}{q_i(L_i^{-1}(t))} \right|, \\ &\le |\lambda|_\infty \left(\|P\|_\infty + \left| \frac{\partial \left\{ \frac{p_i(L_i^{-1}(t), \tau_i)}{q_i(L_i^{-1}(t))} \right\}}{\partial \lambda_i} \right| \right), \ |\tau_i| \in (0, \lambda_i). \end{aligned} \tag{17}$$

Now we wish to find out the error bounds of the terms on the right-hand side of (12). From the classical rational cubic trigonometric function (11), it is easy to observe that

$$P(t) = \sigma_1(\alpha_i, \beta_i, \gamma_i, \delta_i, z) f_i + \sigma_2(\alpha_i, \beta_i, \gamma_i, \delta_i, z) f_{i+1} + \sigma_3(\alpha_i, \beta_i, \gamma_i, \delta_i, z) d_i \\ - \sigma_4(\alpha_i, \beta_i, \gamma_i, \delta_i, z) d_{i+1}, \tag{18}$$

where

$$\sigma_1(\alpha_i, \beta_i, \gamma_i, \delta_i, z) = \frac{1}{q_i(z)}\{\alpha_i(1-\sin z)^3 + \beta_i \sin z(1-\sin z)^2\} \geq 0,$$
$$\sigma_2(\alpha_i, \beta_i, \gamma_i, \delta_i, z) = \frac{1}{q_i(z)}\{\gamma_i \cos z(1-\cos z)^3 + \delta_i(1-\cos z)^2\} \geq 0,$$
$$\sigma_3(\alpha_i, \beta_i, \gamma_i, \delta_i, z) = \frac{2h_i}{\pi q_i(z)}\{\alpha_i \sin z(1-\sin z)^2\} \geq 0,$$
$$\sigma_4(\alpha_i, \beta_i, \gamma_i, \delta_i, z) = \frac{2h_i}{\pi q_i(z)}\{\delta_i \cos z(1-\cos z)^3\} \geq 0.$$

It is easy to observe that $\sigma_1(\alpha_i, \beta_i, \gamma_i, \delta_i, z) + \sigma_2(\alpha_i, \beta_i, \gamma_i, \delta_i, z) = 1$.
Also, for $\alpha_i > 0, \beta_i, \gamma_i > 0$ and $\delta_i > 0$ and choosing $\beta_i \geq \alpha_i$ and $\gamma_i \geq \delta_i$ we obtain the following inequality,

$$\begin{aligned} &\sigma_3(\alpha_i, \beta_i, \gamma_i, \delta_i, z) + \sigma_4(\alpha_i, \beta_i, \gamma_i, \delta_i, z) \\ &= \frac{2h_i}{\pi q_i(z)}\{\alpha_i \sin z(1-\sin z)^2 + \delta_i \cos z(1-\cos z)^2\}, \\ &\leq \frac{2h_i}{\pi}\left\{\frac{\alpha_i \sin z(1-\sin z)^2}{\beta_i \sin z(1-\sin z)^2} + \frac{\delta_i \cos z(1-\cos z)^2}{\gamma_i \cos z(1-\cos z)^2}\right\}, \\ &= \frac{2h_i}{\pi}\left\{\frac{\alpha_i}{\beta_i} + \frac{\delta_i}{\gamma_i}\right\} \leq \frac{4h_i}{\pi}. \end{aligned}$$

Thus, $|P(t)| \leq \max_{j=i,i+1}\{|f_j|\} + \frac{4h_i}{\pi}\max_{j=i,i+1}\{|d_j|\} \leq \|\psi\|_\infty + \frac{4h_i}{\pi}E_1$. Since the above estimation is true for $i \in \Lambda$, we get the following estimation:

$$\|P\|_\infty \leq E(h) := \|\psi\|_\infty + \frac{4h_i}{\pi}E_1, \tag{19}$$

Since $q_i(t)$ is independent of λ_i, from the first term in the right side of (17),

$$\frac{\partial\left\{\frac{p_i(L_i^{-1}(t),\tau_i)}{q_i(L_i^{-1}(t))}\right\}}{\partial \alpha_i} = \sigma_1(\alpha_i, \beta_i, \gamma_i, \delta_i, z) f_1 + \sigma_2(\alpha_i, \beta_i, \gamma_i, \delta_i, z) f_n + \sigma_3(\alpha_i, \beta_i, \gamma_i, \delta_i, z) d_1 \\ - \sigma_4(\alpha_i, \beta_i, \gamma_i, \delta_i, z) d_n.$$

Now by applying a similar argument, the following estimate can be obtained:

$$\left| \frac{\partial \left\{ \frac{p_i(L_i^{-1}(t),\tau_i)}{q_i(L_i^{-1}(t))} \right\}}{\partial \lambda_i} \right| \leq E^*(h) := F + \frac{4h_i}{\pi} E_2. \tag{20}$$

Substituting (19) and (20) in (17) , we have

$$|T_\lambda^* P(t) - T_{\mathbf{0}}^* P(t)| \leq |\lambda|_\infty (E(h) + E^*(h)),\ t \in [t_i, t_{i+1}].$$

Consequently, we obtain

$$\|T_\lambda^* P - T_{\mathbf{0}}^* P\|_\infty \leq |\lambda|_\infty (E(h) + E^*(h)). \tag{21}$$

Using (16) and (21)

$$\begin{aligned} \|\phi - P\|_\infty = \|T_\lambda^* \phi - T_{\mathbf{0}}^* P\|_\infty &\leq \|T_\lambda^* \phi - T_\lambda^* P\|_\infty + \|T_\lambda^* P - T_{\mathbf{0}}^* P\|_\infty, \\ &\leq |\lambda|_\infty \|\phi - P\|_\infty + |\lambda|_\infty (E(h) + E^*(h)), \end{aligned}$$

$$\Rightarrow \|\phi - P\|_\infty \leq \frac{|\lambda|_\infty (E(h) + E^*(h))}{1 - |\lambda|_\infty}. \tag{22}$$

From Theorem 2, we have $|\psi(t) - P(t)| \leq \frac{1}{2}\|\psi^{(3)}\|_\infty h^3 c$. Using this inequality with (22) in (12), we obtain the desired upper bound in (14). □

Convergence Result Assume that $\max_{1\leq j\leq n}\{|d_j|\}$ are bounded for every partition of the domain I. Since $|\lambda_i| < a_i, i \in \Lambda \Rightarrow |\lambda|_\infty < \frac{h}{\ell}$, and hence $\|\psi - \phi\|_\infty = O(h^2)$. Therefore Theorem 2 proves that the rational cubic trigonometric fractal interpolation function ϕ converges uniformly to the original function ψ as $h \to 0$. Further, if we select scaling factors such that $|\lambda_i| < a_i^3 = \frac{h_i^3}{\ell^3}$, then we get $\|\psi - \phi\|_\infty = O(h^3)$.

5 Theory of Positivity Preserving Rational Cubic Trigonometric Interpolation

In this section, we discuss the theory and construction of positive rational cubic trigonometric interpolation functions. Section 5.1 illustrates the construction of positive RCTFIF and deriving suitable restrictions on the scaling factors and shape parameters such that the RCTFIF is positive. The positive nature of the RCTFIF is demonstrated visually through an example in Sect. 5.2.

5.1 Positivity Preserving RCTFIF

The classical polynomial spline interpolation methods usually do not guarantee the shape preservation of the given data and interactive adjustment to the shape. But rational spline FIFs with shape parameters can be used to preserve the shape of the data effectively over the classical interpolants due to presence of scaling factors and shape parameters in its structure.

In this section, we discuss about the construction of a positive RCTFIF whose graph preserve the positivity nature of the data. In general, a RCTFIF may not preserve positivity with arbitrary choice of IFS parameters. In order to avoid this, it is required to deduce sufficient data dependent restrictions on the scaling factor λ_i and on the shape parameters $\alpha_i, \beta_i, \gamma_i$ and δ_i so that the RCTFIF preserves the shape of the positive data.

Theorem 3 *Let ϕ be the RCTFIF* (9) *defined over the interval $[t_1, t_n]$ for given data $\{(t_j, f_j); j \in \Lambda^*\}$. Assume that the data points are positive, i.e., $f_j > 0$ for all $j \in \Lambda^*$. Then the RCTFIF ϕ preserves the positive nature of the data if the following conditions are satisfied for all $i \in \Lambda$:*

(i) the scaling factors are chosen such that:

$$0 \leq \lambda_i < \min\left\{a_i, \frac{f_i}{f_1}, \frac{f_{i+1}}{f_n}\right\}, \tag{23}$$

(ii) the shape parameters are chosen such that:
$\alpha_i > 0$ and $\delta_i > 0$ and

$$\beta_i > \max\left\{0, \frac{-2\ell\alpha_i d_i^*}{\pi f_i^*}\right\}, \quad and \tag{24}$$

$$\gamma_i > \max\left\{0, \frac{2\ell\delta_i d_{i+1}^*}{\pi f_{i+1}^*}\right\}, \tag{25}$$

where $f_i^* = f_i - \lambda_i f_1$, $f_{i+1}^* = f_{i+1} - \lambda_i f_n$, $d_i^* = a_i d_i - \lambda_i d_1$, and $d_{i+1}^* = a_i d_{i+1} - \lambda_i d_n$.

Proof Let $\{(t_j, f_j); j \in \Lambda^*\}$ be the given set of positive data points i.e., $f_j > 0 \; \forall \; j \in \Lambda^*$. Thus the curve is positive if the $\mathscr{C}^1$-RCTFIF 'ϕ' satisfies the following condition:

$$\phi(L_i(t)) > 0 \; \forall \; t \in [t_1, t_n], \; i \in \Lambda. \tag{26}$$

It is clear that (26) is true at all node points. From (9), the above relation can be expressed as

$$\alpha_i \phi(t) + \frac{p_i(\theta)}{q_i(\theta)} > 0, \tag{27}$$

where $p_i(\theta)$ and $q_i(\theta)$ are defined Sect. 3.

In order to show that (27) holds for all $t \in [t_1, t_n]$ and to deduce the sufficient data dependent conditions for the strictly positive RCTFIF, we assume that $\lambda_i \geq 0\, \forall i \in \Lambda$. As per our assumptions $\phi(t_j) = f_j \geq 0\, \forall\, j \in \Lambda^*$. In order to show that the RCTFIF is strictly positive over $[t_1, t_n]$, it is sufficient to verify that $\frac{p_i(\theta)}{q_i(\theta)} > 0,\ i \in \Lambda,\ t \in [t_1, t_n]$. The shape parameters $\alpha_i > 0,\ \beta_i > 0,\ \gamma_i > 0$ and $\delta_i > 0$, guarantee that the denominator $q_i(\theta)$ is strictly positive. Hence the positivity of the RCTFIF depends upon the positivity of the cubic trigonometric polynomial $p_i(\theta)$. It is clear that $p_i(\theta) > 0$ if each of the terms $U_i > 0,\ V_i > 0,\ W_i > 0$, and $X_i > 0$ holds for all $i \in \Lambda$.

Since $U_i = \alpha_i f_i^* = \alpha_i(f_i - \lambda_i f_1)$, it is clear that $U_i > 0$ if $\lambda_i < \frac{f_i}{f_1}$.

Similarly, $X_i = \delta_i f_{i+1}^* = \delta_i(f_{i+1} - \lambda_i f_n) > 0$ if $\lambda_i < \frac{f_{i+1}}{f_n}$. Thus, we obtain that (23) is true in this case.

Now consider $V_i = \beta_i f_i^* + \frac{2\ell\alpha_i d_i^*}{\pi} = \beta_i(f_i - \lambda_i f_1) + \frac{2\ell\alpha_i d_i^*}{\pi}$.

If $d_i^* \geq 0$, then arbitrary $\alpha_i \geq 0,\ \beta_i > 0$ and $(f_i - \lambda_i f_1) > 0$ for $i \in \Lambda$ provides $V_i > 0$, i.e., we need $\lambda_i < \frac{f_i}{f_1}$ for $i \in \Lambda$. Otherwise we can choose $\beta_i > \frac{-2\ell\alpha_i d_i^*}{\pi(f_i - \lambda_i f_1)}$, for ensuring $V_i > 0$ as (23) gives $(f_i - \lambda_i f_1) \geq 0$. Hence we choose β_i according to (24).

At the end, we have $W_i = \delta_i f_{i+1}^* - \frac{2\ell\delta_i d_{i+1}^*}{\pi} = \delta_i(f_{i+1} - \lambda_i f_n) - \frac{2\ell\delta_i d_{i+1}^*}{\pi}$.

If $d_{i+1}^* < 0$ then arbitrary $\delta_i \geq 0, \gamma_i > 0$ and $(f_{i+1} - \lambda_i f_n) > 0$ for $i \in \Lambda$ provides $W_i > 0$, i.e., we need $\lambda_i < \frac{f_{i+1}}{f_n}$ for $i \in \Lambda$. Otherwise we can choose $\gamma_i > \frac{2\ell\delta_i d_{i+1}^*}{\pi(f_{i+1} - \lambda_i f_n)}$, so that $Y_i > 0$ as (23) gives $(f_{i+1} - \lambda_i f_n) > 0$. Hence we choose γ_i according to (25). □

5.2 Numerical Example

In this section we illustrate the numerical demonstration of the positivity preserving aspect of the given positive data by the proposed RCTFIF. We consider the positive data set as in Table 1. Figure 1h represents the graph of the classical rational cubic trigonometric interpolant. Since the classical interpolant is unique for the predefined shape parameters, rational trigonometric fractal interpolation functions are more

Table 1 2D data set

t	1	3	8	10	11	12	16
f	14	2	0.8	0.65	0.75	0.7	0.69

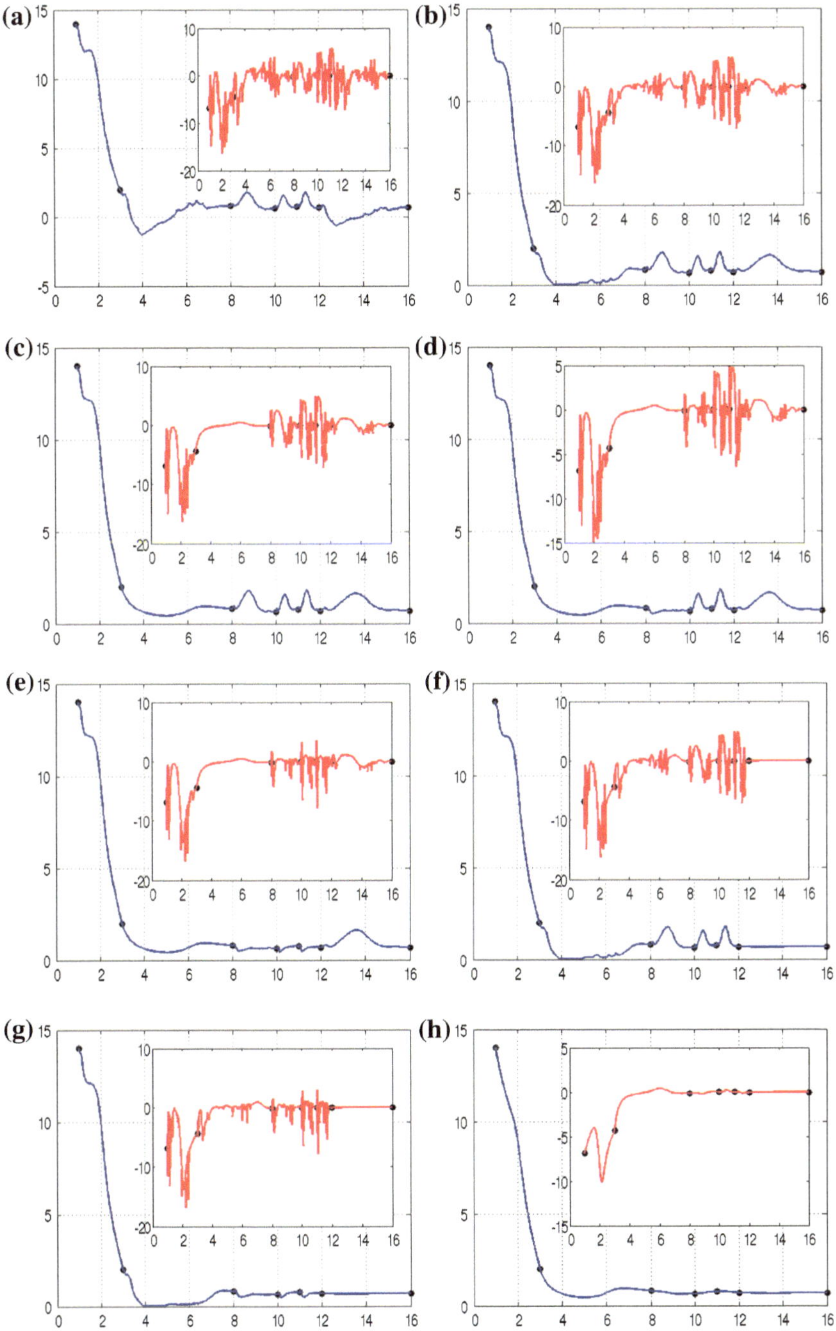

Fig. 1 Positive RTCFIF curves and their first order derivatives curves (*inset figures*) with respect to the free parameters in the Table 2. **a** Non positive RCTFIF, **b** standard positive RCTFIF ϕ_1, **c** ϕ_2 effect of λ_2 and β_2 on ϕ_1, **d** ϕ_3 effect of β_3 and γ_3 on ϕ_2, **e** ϕ_4 effect of β_4, γ_4, β_5 and γ_5 on ϕ_3, **f** ϕ_5 effect of λ_6 and β_6 on ϕ_1, **g** ϕ_6 effect of β_4, γ_4, β_5 and γ_5 on ϕ_3, **h** Positive classical interpolant

Table 2 Scaling factors and shape parameters used in the construction of positive RCTFIFs (Fig. 1)

Figure	Scaling factors	Shape parameters
Figure 1a	$\lambda =$ (0.1323, 0.2419, 0.0561, 0.0454, 0.0526, 0.149)	$\beta = (0.5028, 1.1853, 0.5, 0.5, 0.5, 3.9649)$, $\gamma = (0.5, 0.5, 0.5868, 0.5221, 0.5, 0.5)$
Figure 1b	$\lambda =$ (0.1323, 0.1419, 0.0561, 0.0454, 0.0526, 0.049)	$\beta = (0.5028, 172.6956, 0.5, 0.5, 0.5, 0.5)$, $\gamma = (0.5, 0.5, 0.5868, 0.5221, 0.5, 0.5)$
Figure 1c	$\lambda =$ (0.1323, **0.001**, 0.0561, 0.0454, 0.0526, 0.049)	$\beta = (0.5028, \mathbf{3.9731}, 0.5, 0.5, 0.5, 0.5)$, $\gamma = (0.5, 0.5, 0.5868, 0.5221, 0.5, 0.5)$
Figure 1d	$\lambda =$ (0.1323, **0.001**, 0.0561, 0.0454, 0.0526, 0.049)	$\beta = (0.5028, \mathbf{3.9731}, \mathbf{2.575}, 0.5, 0.5, 0.5)$, $\gamma = (0.5, 0.5, \mathbf{2.817}, 0.5221, 0.5, 0.5)$
Figure 1e	$\lambda =$ (0.1323, **0.001**, 0.0561, 0.0454, 0.0526, 0.049)	$\beta = (0.5028, \mathbf{3.9731}, \mathbf{2.575}, \mathbf{2.124}, \mathbf{2.515}, 0.5)$, $\gamma = (0.5, 0.5, \mathbf{2.817}, \mathbf{2.868}, \mathbf{2.221}, 0.5)$
Figure 1f	$\lambda =$ (0.1323, 0.2419, 0.0561, 0.0454, 0.0526, **0.001**)	$\beta = (0.5028, 172.6956, 0.5, 0.5, 0.5, \mathbf{0.5277})$, $\gamma = (0.5, 0.5, 0.5868, 0.5221, 0.5, 0.5)$
Figure 1g	$\lambda =$ (0.1323, 0.2419, 0.0561, 0.0454, 0.0526, **0.001**)	$\beta = (0.5028, 172.6956, \mathbf{2.575}, \mathbf{2.124}, \mathbf{2.515}, \mathbf{0.5277})$, $\gamma = (0.5, 0.5, \mathbf{0.5817}, \mathbf{2.868}, \mathbf{2.221}, 0.5)$
Figure 1h	$\lambda = (0, 0, 0, 0, 0, 0)$	$\beta = (0.8103, 3.9650, 0.5972, 0.5, 0.5, 0.5737)$, $\gamma = (0.5, 0.5, 0.5816, 0.5212, 0.5, 0.5)$

Here $\alpha = 0.5$ and $\delta = 1$

convenient due to the presence scaling factors and shape parameters. With a random choice of scaling factors and shape parameters we obtain Fig. 1a illustrates that the proposed RCTFIF may not preserve the shape of the data for the random choice of free parameters. Therefore for positivity preserving RCTFIF, we choose scaling factors and shape parameters according to the Theorem 3. The deduced scaling factors and shape parameters used in our construction are shown in Table 2. Due to availability the free parameters we get different positivity preserving RCTFIFs and we denote them by ϕ_j, $j = 1, 2, \ldots, 6.$, see Fig. 1. Out of the four shape parameters involved in the RCTFIF, two of them, $\alpha_i = 0.5$ and $\delta_i = 1, i \in \Lambda$, are fixed while the rest will participate in interactive curve design. The derivative values (d_j, $j \in \Lambda^*$) are calculated by arithmetic mean method. The scaling factors are restricted as $\lambda_1 \in [0, 0.1333]$, $\lambda_2 \in [0, 0.1429]$, $\lambda_3 \in [0, 0.0571]$, $\lambda_4 \in [0, 0.0464]$, $\lambda_5 \in [0, 0.0536]$, and $\lambda_6 \in [0, 0.05]$ to preserve the positivity feature of the given positive data.

The positivity preserving RCTFIF curves $\phi_2 - \phi_6$ are generated using the modified free parameters from the Table 2 and are shown in Fig. 1. The modified scaling factors and shape parameters are shown in bold in the Table 2. The graphs of the first derivative of the positive RCTFIFs are also inserted as inset figures of the corresponding positive RCTFIFs respectively. Figure 1b illustrates the positivity preserving RCT-FIF ϕ_1 and its first derivative and is generated by the suitable restricted scaling factors and the shape parameters. We take Fig. 1b as the reference positive RCTFIF curve. The perturbation in the scaling factor λ_2 and shape parameter β_2 effects the shape of ϕ_2 and its first order derivative in the subinterval $[t_2, t_3]$ and variations in the remaining intervals are negligible, see Fig. 1c. Figure 1d is generated by changing the shape parameters β_3, γ_3 in Fig. 1c, it is clear that the shape parameters effects the shape of ϕ_3 in the subinterval $[t_3, t_4]$. By perturbing shape parameters β_4, γ_4 and β_5, γ_5, we generate Fig. 1e and modify the shape of ϕ_4 in the subintervals $[t_4, t_5]$ and $[t_5, t_6]$ respectively. A small disruption in the scaling factor λ_6 and the shape parameters β_6 generate Fig. 1f and modifies the shape of ϕ_5 and its first derivative in the subinterval $[t_6, t_7]$. Similarly the perturbed shape parameters β_4, γ_4 and β_5, γ_5 provide Fig. 1g and modify the shape of ϕ_6 and their derivative in the subinterval $[t_4, t_5]$ and $[t_5, t_6]$ respectively. Finally a positive classical rational trigonometric interpolant, see Fig. 1h, can be extracted by setting all the scaling factors to zero in (9). One can observe that all the scaling factors and shape parameters used in this example produces local effects. The optimal scaling factors and shape parameters can be determined by the genetic algorithm to obtain desired accuracy with the original function.

6 Conclusions

In this paper, a smooth rational cubic trigonometric fractal interpolation function is introduced for the first time in the literature. With a zero scaling vector, the developed RCTFIF reduces to the classical rational cubic trigonometric interpolant with four shape parameters. A uniform error bound has been determined between the original function and RCTFIF. The developed RCTFIF converges uniformly to the original function as $h \to 0$. We have deduced the range of the scaling factors and shape parameters so that the RCTFIF preserves the positive aspect of the given positive data. The effects of the rational IFS parameters on the shape of the curves are illustrated. The irregularity nature of the first order RCTFIF curves is studied and demonstrated through suitable example. The developed RCTFIF can be used for the visualization of both data with the slopes and the data without slopes at the knots. Applications of the proposed RCTFIF in geometric modeling problems are under investigation.

Acknowledgments The partial support of the Department of Science and Technology of Govt. of India (SERC DST Project No. SR/S4/MS: 694/10) is gratefully acknowledged.

References

1. Abbas, M., Majid, A.A., Ali, J. Md.: Positivity preserving interpolation of positive data by cubic trigonometric spline. Mathematika **27**(1), 41–50 (2011)
2. Abbas, M., Majid, A.A. Md.: Positivity-preserving $\mathscr{C}^2$ rational cubic spline interpolation. Sci. Asia **39**, 208–213 (2013)
3. Barnsley, M.F.: Fractals everywhere. Constr. Approx. **2**, 205–222 (1986)
4. Barnsley, M.F., Harrington, A.N.: The calculus of fractal interpolation functions. J. Approx. Theory. **57**(1), 14–34 (1989)
5. Butt, S., Brodlie, K.W.: Preserving positivity using piecewise cubic interpolation. Comput. Graph. **17**(1), 55–64 (1993)
6. Chand, A.K.B., Kapoor, G.P.: Generalized cubic spline fractal interpolation functions. SIAM J. Numer. Anal. **44**(2), 655–676 (2006)
7. Chand, A.K.B., Navascués, M.A.: Generalized Hermite fractal interpolation. Rev. R. Acad. Cienc. Zaragoza **64**(2), 107–120 (2009)
8. Chand, A.K.B., Vijender, N., Navascués, M.A.: Shape preservation of scientific data through rational fractal splines. Calcolo **51**(2), 329–362 (2014)
9. Chand, A.K.B., Vijender, N., Agarwal, R.P.: Rational iterated function system for positive/monotonic shape preservation. Adv. Diff. Equ. **2014**(30), 1–19 (2014)
10. Chand, A.K.B., Viswanathan, P.: A constructive approach to cubic Hermite fractal interpolation function and its constrained aspects. BIT Numer. Math. **53**(4), 841–865 (2013)
11. Gregory, J.A., Delbourgo, R.: Shape preserving piecewise rational interpolation. SIAM J. Stat. Comput. **6**(4), 967–976 (1985)
12. Han, X.: Quadratic trigonometric polynomial curves with a shape parameter. Comput. Aided Geom. Des. **19**(7), 503–512 (2002)
13. Han, X.: Cubic trigonometric polynomial curves with a shape parameter. Comput. Aided Geom. Des. **21**, 479–502 (2004)
14. Hussain, M.Z., Sarfraz, M.: Positivity-preserving interpolation of positive data by rational cubics. J. Comp. Appl. Math **218**(2), 446–458 (2008)
15. Hussain, M.Z., Sarfraz, M.: Scientific data visualization with shape preserving $\mathscr{C}^1$-rational cubic interpolation. Eur. J. Pure Appl. Math. **3**(2), 194–212 (2010)
16. Hussain, M.Z., Hussain, M., Waseem, A.: Shape preserving trigonometric finctions. Comput. Appl. Math. **33**(2), 411–431 (2014)
17. Ibraheem, F., Hussain, M., Hussain, M.Z., Bhatti, A.A.: Positive data visualization using trigonometric function. J. Appl. Math. 1–19 (2012) doi:10.1155/2012/247120
18. Sarfraz, M., Hussain, M.Z.: Data visualization using rational spline interpolation. J. Comput. Appl. Math **189**, 513–525 (2006)
19. Sarfraz, M., Hussain, M.Z., Hussain, M.: Shape-preserving curve interpolation. Mathematika **89**(1), 35–53 (2012)
20. Schmidt, J.W., Heβ, W.: Positivity of cubic polynomials on intervals and positive spline interpolation. BIT Numer. Math. **28**(2), 340–352 (1988)
21. Bashir, U., Ali, J. Md.: Data visualization using rational trigonometric spline. J. Appl. Math. 1–10 (2013). doi:10.1155/2013/531497
22. Viswanathan, P., Chand, A.K.B.: Fractal rational functions and their approximation properties. J. Approx. Theory. **185**, 31–50 (2014)
23. Viswanathan, P., Chand, A.K.B.: A $\mathscr{C}^1$-rational cubic fractal interpolation function: convergence and associated parameter identification problem. Acta. Appl. Math. **136**(1), 19–41 (2015)

A Monotonic Rational Fractal Interpolation Surface and Its Analytical Properties

A.K.B. Chand and N. Vijender

Abstract A $\mathscr{C}^1$-continuous rational cubic fractal interpolation function was introduced and its monotonicity aspect was investigated in [Adv. Difference Eq. (30) 2014]. Using this univariate interpolant and a blending technique, in this article, we develop a monotonic rational fractal interpolation surface (FIS) for given monotonic surface data arranged on the rectangular grid. The analytical properties like convergence and stability of the rational cubic FIS are studied. Under some suitable hypotheses on the original function, the convergence of the rational cubic FIS is studied by calculating an upper bound for the uniform error of the surface interpolation. The stability results are studied when there is a small perturbation in the corresponding scaling factors. We also provide numerical examples to corroborate our theoretical results.

Keywords Fractals · Fractal interpolation functions · Fractal interpolation surfaces · Monotonicity · Blending functions

1 Introduction

Fractal interpolation is a modern interpolation technique developed to represent a prescribed data set with a smooth or nonsmooth continuous function. This is in contrast to the classical interpolants like polynomial splines, exponential splines, trigonometric splines, etc., which produce functions that are differentiable infinite number of times except perhaps at a finite number of points in the interpolation interval. Barnsley [1] proposed the concept of a fractal interpolation function (FIF)

A.K.B. Chand
Department of Mathematics, Indian Institute of Technology Madras,
Chennai 600036, Tamil Nadu, India
e-mail: chand@iitm.ac.in

N. Vijender (✉)
Department of Mathematics, VIT University, Chennai 600127, India
e-mail: vijendernallapu@gmail.com

R.N. Mohapatra et al. (eds.), *Mathematics and Computing*,
Springer Proceedings in Mathematics & Statistics 139,
DOI 10.1007/978-81-322-2452-5_14

based on the theory of an iterated function system (IFS). Later a good number of fractal polynomial splines [2–7] dealing with univariate fractal interpolation have been developed by various authors. These spline FIFs cannot actually be fractals. However, the name fractal interpolation function is retained because of the flavor of the scaling in its definition and because of the fact that certain derivative of this function is typically a fractal. Further, the graph of a fractal spline is a union of its transformed copies, and hence possesses self-similarity, a characteristic feature of the fractal sets.

In general, both the classical polynomial spline interpolation and the fractal polynomial spline interpolation may ignore the intrinsic form implied by the given data points. Consequently, interpolants produced by these methods may have undesirable inflections or oscillations. To obtain a valid physical interpretation of the underlying process, it is important to develop interpolation schemes that honor the properties inherent in the data, particularly when the data is produced by some scientific phenomena. The problem of searching a sufficiently smooth function that preserves qualitative shape properties inherent in the input data is generally called a shape preserving interpolation problem. Various shape properties are mathematically expressed in terms of conditions such as positivity, monotonicity, and convexity. Owing to the difficulties to develop shape preserving interpolation with polynomial FIFs, recently our group has introduced rational fractal splines for shape preserving interpolation [8–12].

In a natural way, different univariate fractal interpolation functions have been extended to suitable fractal bivariate interpolation functions [13–17] to model natural surfaces such as rocks, metals, planets, terrains, and so on. Among the existing FIS schemes, many are developed by imposing suitable restriction on choice of the scaling factors, surface data, or interpolation domain. For instance, the continuity of the fractal interpolation surface is achieved in [18] with the assumption that interpolation points on the boundary are collinear. The aforementioned fractal surface schemes give self-similar, self-affine, or more generally self-referential fractal interpolation surfaces. To generate self-referential and nonself-referential surfaces simultaneously, Chand and Kapoor [19–21] developed the notion of coalescence hidden variable fractal interpolation surfaces. Including the aforementioned FISs, all the existing spline FISs lack most important shape preserving aspects of given surface data. Also from the existing FIS schemes, it is observed that (i) attractor of a bivariate IFS (ii) tensor product of FIFs may not be suitable for the construction of shape preserving FISs. Owing to these reasons, using a blending surface technique, the monotonicity preserving rational FIS is developed in the present work. In addition to having monotonicity preserving capabilities, an interpolant should also possess some other important analytical properties such as stability and convergence. To facilitate such investigations for the developed rational FIS, we have studied its stability and convergence properties.

In Sect. 2, we shall revisit the construction of $\mathscr{C}^p$-univariate rational cubic FIF detailed in [12]. In Sect. 3, for the given surface data arranged on a rectangular grid, we construct a rational cubic FIS as blending of the rational cubic FIFs defined along the grid lines of rectangular domain. Analytical properties of the rational cubic FIS are studied in Sect. 4. Monotonicity aspects of the rational cubic FIS is studied in Sect. 5. Some numerical results are provided in Sect. 6 to verify our monotonicity results.

2 Fractal Interpolation Functions

In this section, we briefly recall the fractal interpolation and rational fractal interpolation from [1, 12]. Let $x_1 < x_2 < \cdots < x_{m-1} < x_m (m > 2)$ be a partition of the closed interval $I = [x_1, x_m]$, and $f_1, f_2, \ldots, f_m$, be a collection of real numbers. Let θ_i, $i = 1, 2, \ldots, m-1$, be a set of homeomorphism mappings from I to $I_i = [x_i, x_{i+1}]$ satisfying $\theta_i(x_1) = x_i$, $\theta_i(x_m) = x_{i+1}$, and K be a compact subset of $\mathbb{R}$ such that $f_i \in K, i = 1, 2, \ldots, m$. Again, let $F_i(x, f) = a_i[\xi_i f + r_i(x)]$, $|\xi_i| < \kappa < 1$, be a continuous function from $I \times K$ to K such that $F_i(x_1, f_1) = f_i$, $F_i(x_m, f_m) = f_{i+1}$, $i = 1, 2, \ldots, m-1$. Furthermore, $r_i(x)$ is a suitable continuous real-valued function on I. Define a set of maps $w_i : I \times K \to I_i \times K$ as $w_i(x, f) = (\theta_i(x), F_i(x, f))$, $(x, f) \in I \times K$, $i = 1, 2, \ldots, m-1$. Then $\mathscr{I} = \{I \times K; w_i(x, f), i = 1, 2, \ldots, m-1\}$ is called an IFS related to a given interpolation data $\{(x_i, f_i), i = 1, 2, \ldots, m\}$. According to [1], the IFS $\mathscr{I}$ has a unique attractor G which is the graph of a continuous function $\Phi : I \to \mathbb{R}$, $\Phi(x_i) = f_i, i = 1, 2, \ldots, m$. The function Φ is called a FIF generated by the IFS $\mathscr{I}$, and it takes the form: $\Phi(L_i(x)) = \xi_i \Phi(x) + r_i(x)$, $x \in I$. The differentiable spline FIF was introduced in [2]. This result was extended to $\mathscr{C}^p$-rational spline fractal functions in the following proposition [12].

Proposition 2.1 *Let $\{(x_i, f_i), i = 1, 2, \ldots, m\}$ be given data set, where $d_i^{(k)}$ $(i = 1, 2, \ldots, m, k = 1, 2, \ldots, p)$ are the kth derivative values at knots. Consider the rational IFS $\mathscr{I}^* \equiv \{\mathscr{I} \times K; w_i(x, f) = (\theta_i(x), F_i(x, f)), i = 1, 2, \ldots, m-1\}$, where $\theta_i(x) = a_i x + b_i$ satisfies $\theta_i(x_1) = x_i$, $\theta_i(x_m) = x_{i+1}$, $F_i(x, f) = a_i^p\left(\xi_i f + r_i(x)\right)$, $r_i(x) = \frac{\Omega_{i,1}(x)}{\Omega_{i,2}(x)}$, $\Omega_{i,1}(x)$ is a polynomial containing $2p+2$ arbitrary constants, and $\Omega_{i,2}(x)$ is a nonvanishing quadratic polynomial with three shape parameters defined on I, and $|\xi_i| < \kappa < 1, i = 1, 2, \ldots, m-1$. Let $F_i^{(k)}(x, f) = a_i^{p-k}\left(\xi_i f + r_i^{(k)}(x)\right)$, where $r_i^{(k)}(x)$ represents the kth derivative of $r_i(x)$ with respect to x. With the setting $f_i = d_i^{(0)}, i = 1, 2, \ldots, m$, if*

$$F_i^{(k)}(x_1, d_1^{(k)}) = d_i^{(k)},\ F_i^{(k)}(x_m, d_m^{(k)}) = d_{i+1}^{(k)},\ i = 1, 2, \ldots, m-1,\ k = 0, 1, \ldots, p, \tag{1}$$

then the fixed point of the rational IFS $\mathscr{I}^*$ is the graph of a $\mathscr{C}^p$- rational FIF.

3 Rational Cubic Fractal Interpolation Surfaces

Based on Proposition 2.1, the rational fractal boundary curves along the grid lines in the domain of surface interpolation are constructed in Sect. 3.1. The $\mathscr{C}^1$-rational cubic fractal interpolation surface is constructed as a combination of blending functions and these fractal boundary curves in Sect. 3.2.

3.1 Construction of $\mathscr{C}^1$-Rational Cubic FIFs (Fractal Boundary Curves)

Consider the surface data $\Delta = \{(x_i, y_j, z_{i,j}, z^x_{i,j}, z^y_{i,j}) : i = 1, 2, \dots, m,\ j = 1, 2, \dots, n\}$, where $z^x_{i,j}$ and $z^y_{i,j}$ be the x-partials and y-partials at the point (x_i, y_j). Let $x_1 < x_2 < \cdots < x_{m-1} < x_m, y_1 < y_2 < \cdots < y_{n-1} < y_n$, be the grids on $D = I \times J, I = [x_1, x_m], J = [y_1, y_n]$. Suppose $D_{i,j} = I_i \times J_j, i = 1, 2, \dots, m-1$, $j = 1, 2, \dots, n-1$, forms a rectangular partition of D, where $I_i = [x_i, x_{i+1}]$, $J_j = [y_j, y_{j+1}]$. Now $T_j = \{(x_i, z_{i,j}, z^x_{i,j}) : i = 1, 2, \dots, m\}$ is the interpolation data along the jth grid line parallel to the x-axis, $j = 1, 2, \dots, n$. Similarly, $T_i^\dagger = \{(y_j, z_{i,j}, z^y_{i,j}) : j = 1, 2, \dots, n\}$ is the interpolation data along the ith grid line parallel to the y-axis, $i = 1, 2, \dots, m$. In order to construct a rational cubic FIF (fractal boundary curve) which interpolates the interpolation data set T_j for $j = 1, 2, \dots, n$, consider Proposition 2.1 with $p = 1$, interpolation data T_j, and $r_{i,j}(x) = \frac{p_{i,j}(\theta)}{q_{i,j}(\theta)}, i = 1, 2, \dots, m-1, p_{i,j}(\theta) = A_{i,j}(1-\theta)^3 + C_{i,j}\theta(1-\theta)^2 + D_{i,j}\theta^2(1-\theta) + B_{i,j}\theta^3$, $q_{i,j}(\theta) = \alpha_{i,j}(1-\theta)^2 + \gamma_{i,j}\theta(1-\theta) + \beta_{i,j}(1-\theta)^2$, where $\theta = \frac{\theta_i^{-1}(x) - x_1}{x_m - x_1}$, $x \in I_i$, $A_{i,j}, B_{i,j}, C_{i,j}, D_{i,j}$ are arbitrary constants, $\alpha_{i,j} > 0$, $\beta_{i,j} > 0$, and $\gamma_{i,j} > 0$ are the shape parameters. Then we obtain the following functional equations for $j = 1, 2, \dots, n$:

$$S(x, y_j) = a_i[\xi_{i,j} S(\theta_i^{-1}(x), y_j) + r_{i,j}(x)],\ x \in I_i, \tag{2}$$

where $|\xi_{i,j}| < \kappa < 1, i = 1, 2, \dots, m-1$. Constants $A_{i,j}, B_{i,j}, C_{i,j}, D_{i,j}$ in $r_{i,j}$ are evaluated using the following interpolatory conditions, respectively: $S(x_i, y_j) = z_{i,j}$, $S(x_{i+1}, y_j) = z_{i+1,j}$, $S^{(1)}(x_i, y_j) = z^x_{i,j}$, $S^{(1)}(x_{i+1}, y_j) = z^x_{i+1,j}$. Thus, the desired rational cubic FIF (fractal boundary curve) interpolating the data set T_j is

$$S(x, y_j) = a_i \xi_{i,j} S(\theta_i^{-1}(x), y_j) + \frac{p_{i,j}(\theta)}{q_{i,j}(\theta)},\ x \in I_i, \tag{3}$$

$$p_{i,j}(\theta) = \alpha_{i,j}\big[z_{i,j} - \xi_{i,j} z_{1,j} a_i\big](1-\theta)^3 + \big[z_{i,j}(\gamma_{i,j}+\alpha_{i,j}) + z^x_{i,j}\alpha_{i,j}h_i - \xi_{i,j}\{\alpha_{i,j}h_i z^x_{1,j} + z_{1,j}a_i(\gamma_{i,j}+\alpha_{i,j})\}\big]\theta(1-\theta)^2 + \big[z_{i+1,j}(\gamma_{i,j}+\beta_{i,j}) - z^x_{i+1,j}\beta_{i,j}h_i + \xi_{i,j}\{\beta_{i,j}h_i z^x_{m,j} - z_{m,j}a_i(\gamma_{i,j}+\beta_{i,j})\}\big]\theta^2(1-\theta) + \beta_{i,j}\big[z_{i+1,j} - \xi_{i,j}z_{m,j}a_i\big]\theta^3,$$

$$q_{i,j}(\theta) = \alpha_{i,j}(1-\theta)^2 + \gamma_{i,j}\theta(1-\theta) + \beta_{i,j}\theta^2,\ x \in I_i,\ i = 1, 2, \ldots, m-1.$$

Thus we have exactly n different x-direction fractal boundary curves. The parameters $\xi_{i,j}$, $\alpha_{i,j}$, $\beta_{i,j}$, and $\gamma_{i,j}$, $i = 1, 2, \ldots, m-1$, $j = 1, 2, \ldots, n$, involved in the x-direction fractal boundary curves are arranged in the matrix form as $\boldsymbol{\xi} = [\xi_{i,j}]_{(m-1)\times n}$, $\boldsymbol{\alpha} = [\alpha_{i,j}]_{(m-1)\times n}$, $\boldsymbol{\beta} = [\beta_{i,j}]_{(m-1)\times n}$, and $\boldsymbol{\gamma} = [\gamma_{i,j}]_{(m-1)\times n}$.

By reiterating the above procedure, the fractal boundary curve $S^\dagger(x_i, y)$ interpolating the data set $T_i^\dagger = \{(y_j, z_{i,j}, z^y_{i,j});\ j = 1, 2, \ldots, n\}$ is given by

$$S^\dagger(x_i, y) = c_j\xi^\dagger_{i,j}S^\dagger(x_i, \phi_j^{-1}(y)) + \frac{p^\dagger_{i,j}(\phi)}{q^\dagger_{i,j}(\phi)},\ y \in J_j, \tag{4}$$

$$p^\dagger_{i,j}(\phi) = \alpha^\dagger_{i,j}\big[z_{i,j} - \xi^\dagger_{i,j}z_{i,1}c_j\big](1-\phi)^3 + \big[z_{i,j}(\gamma^\dagger_{i,j}+\alpha^\dagger_{i,j}) + z^y_{i,j}\alpha^\dagger_{i,j}h^\dagger_j - \xi^\dagger_{i,j}\{\alpha^\dagger_{i,j}h^\dagger_j z^y_{i,1} + z_{i,1}c_j(\gamma^\dagger_{i,j}+\alpha^\dagger_{i,j})\}\big]\phi(1-\phi)^2 + \big[z_{i,j+1}(\gamma^\dagger_{i,j}+\beta^\dagger_{i,j}) - z^y_{i,j+1}\beta^\dagger_{i,j}h^\dagger_j + \xi^\dagger_{i,j}\{\beta^\dagger_{i,j}h^\dagger_j z^y_{i,n} - z_{i,n}c_j(\gamma^\dagger_{i,j}+\beta^\dagger_{i,j})\}\big]\phi^2(1-\phi) + \beta^\dagger_{i,j}\big[z_{i,j+1} - \xi^\dagger_{i,j}z_{i,n}c_j\big]\phi^3,$$

$$q^\dagger_{i,j}(\phi) = \alpha^\dagger_{i,j}(1-\phi)^2 + \gamma^\dagger_{i,j}\phi(1-\phi) + \beta^\dagger_{i,j}\phi^2,\ \phi = \frac{\phi_j^{-1}(y) - y_1}{y_n - y_1},\ y \in J_j,$$

$h^\dagger_j = y_{j+1} - y_j$, $\xi^\dagger_{i,j}$ is the scaling factor in the y-direction satisfies $|\xi^\dagger_{i,j}| < \kappa < 1$, $\alpha^\dagger_{i,j} > 0$, $\beta^\dagger_{i,j} > 0$, and $\gamma^\dagger_{i,j} \geq 0$ are the shape parameters for $i = 1, 2, \ldots, m$, $\phi_j(y) = c_j y + d_j = \frac{(y_{j+1}-y_j)y}{y_n-y_1} + \frac{y_n y_j - y_1 y_{j+1}}{y_n - y_1} : J \to J_j$ is a homeomorphism such that $\phi_j(y_1) = y_j$, $\phi_j(y_n) = y_{j+1}$, $j = 1, 2, \ldots, n-1$.

From (4), we have exactly m different y-direction fractal boundary curves. The parameters $\xi^\dagger_{i,j}$, $\alpha^\dagger_{i,j}$, $\beta^\dagger_{i,j}$, and $\gamma^\dagger_{i,j}$, $i = 1, 2, \ldots, m$, $j = 1, 2, \ldots, n-1$ seeded in the y-direction fractal boundary curves are put in the matrix form as $\boldsymbol{\xi}^\dagger = [\xi^\dagger_{i,j}]_{m\times(n-1)}$, $\boldsymbol{\alpha}^\dagger = [\alpha^\dagger_{i,j}]_{m\times(n-1)}$, $\boldsymbol{\beta}^\dagger = [\beta^\dagger_{i,j}]_{m\times(n-1)}$, and $\boldsymbol{\gamma}^\dagger = [\gamma^\dagger_{i,j}]_{m\times(n-1)}$.

3.2 $\mathscr{C}^1$-Rational Cubic Fractal Interpolation Surface

The union of the four straight lines $I_i \times y_j$, $I_i \times y_{j+1}$, $x_i \times J_j$, and $x_{i+1} \times J_j$ forms the boundary of sub-rectangle $D_{i,j}$. Utilizing the above fractal boundary curves (3) and (4), we define the rational cubic fractal surface patch over the sub-rectangle $D_{i,j}$, $i = 1, 2, \dots, m-1$, $j = 1, 2, \dots, n-1$, as

$$\Theta(x, y) = -A\Pi(x, y)B^T, \ (x, y) \in D_{i,j}, \tag{5}$$

$$\text{where } \Pi(x, y) = \begin{bmatrix} 0 & S(x, y_j) & S(x, y_{j+1}) \\ S^\dagger(x_i, y) & z_{i,j} & z_{i,j+1} \\ S^\dagger(x_{i+1}, y) & z_{i+1,j} & z_{i+1,j+1} \end{bmatrix},$$

$$A = [-1 \ a_{x,0}(\theta) \ a_{x,1}(\theta)], \ a_{x,0}(\theta) = (1-\theta)^2(1+2\theta), \ a_{x,1}(\theta) = \theta^2(3-2\theta),$$

$$\theta = \frac{\theta_i^{-1}(x) - x_1}{x_m - x_1}, B = [-1 \ b_{y,0}(\phi) \ b_{y,1}(\phi)], \ b_{y,0}(\phi) = (1-\phi)^2(1+2\phi),$$

$$b_{y,1}(\phi) = \phi^2(3-2\phi), \phi = \frac{\phi_j^{-1}(y) - y_1}{y_n - y_1}.$$

The functions $a_{x,0}$, $a_{x,1}$, $b_{y,0}$, and $b_{y,1}$ are called the *blending functions*. The fractal boundary curves $S(x, y_{j+1})$, $j = 1, 2, \dots, n-1$,, and $S^\dagger(x_{i+1}, y)$, $i = 1, 2, \dots, m-1$, are defined from (3) and (4) by replacing j by $j+1$ and i by $i+1$, respectively.

Theorem 3.1 *The surface Θ in (5) interpolates the surface data Δ, and possesses $\mathscr{C}^1$-continuity.*

Proof Using interpolation properties of the fractal boundary curves and the properties of the blending functions, it is straightforward to prove that Θ satisfies the Hermite type interpolation conditions $\Theta(x_i, y_j) = z_{i,j}$, $\frac{\partial \Theta}{\partial x}(x_i, y_j) = z^x_{i,j}$, and $\frac{\partial \Theta}{\partial y}(x_i, y_j) = z^y_{i,j}$ for $i = 1, 2, \dots, m$, $j = 1, 2, \dots, n$. Since the fractal boundary curves and blending functions are $\mathscr{C}^1$-continuous, we observe that Θ is $\mathscr{C}^1$-continuous in the interior of $D_{i,j}$, that is over $(x_i, x_{i+1}) \times (y_j, y_{j+1})$. If (x, y) is on the boundary $\partial(I_i \times J_j)$ of $I_i \times J_j$, then one of the following holds:

$$(i) \ (x, y) \in \partial(I_i \times J_j) \cap \partial(I_i \times J_{j-1}),$$

$$(ii) \ (x, y) \in \partial(I_i \times J_j) \cap \partial(I_{i+1} \times J_j),$$

$$(iii) \ (x, y) \in \partial(I_i \times J_j) \cap \partial(I_i \times J_{j+1}),$$

$$(iv) \ (x, y) \in \partial(I_i \times J_j) \cap \partial(I_{i-1} \times J_j).$$

Considering the first case, we have to prove that each of the expressions $\Theta(x, y)$, $\frac{\partial\Theta}{\partial x}(x, y)$, and $\frac{\partial\Theta}{\partial y}(x, y)$ has the same value irrespective of whether (x, y) is considered as a point in $I_i \times J_j$ or as a point in $I_i \times J_{j-1}$. Similarly for the other cases. This is possible with help of general calculations via the following properties of the blending functions:

$$a_{x,0}(0) = 1,\ a_{x,1}(0) = 0,\ a_{x,0}(1) = 0,\ a_{x,1}(1) = 1,$$

$$a^{(1)}_{x,0}(0) = 0,\ a^{(1)}_{x,1}(0) = 0,\ a^{(1)}_{x,0}(1) = 0,\ a^{(1)}_{x,1}(1) = 0,$$

$$b_{y,0}(0) = 1,\ b_{y,1}(0) = 0,\ b_{y,0}(1) = 0,\ b_{y,1}(1) = 1,$$

$$b^{(1)}_{y,0}(0) = 0,\ b^{(1)}_{y,1}(0) = 0,\ b^{(1)}_{y,0}(1) = 0,\ b^{(1)}_{y,1}(1) = 0.$$

This proves the continuity of Θ and its partial derivatives on D, which in turn gives $\Theta \in \mathscr{C}^1(D)$, it completes the proof.

Definition 3.1 Since Θ is blending of rational cubic FIFs (fractal boundary curves) and $\mathscr{C}^1$-continuous over D, we call Θ as *$\mathscr{C}^1$-rational cubic FIS*.

Remark 3.1 If $\boldsymbol{\xi} = [0]_{(m-1)\times n}$ and $\boldsymbol{\xi}^\dagger = [0]_{m\times(n-1)}$, then we get the classical rational cubic surface interpolant C as

$$\left.\begin{aligned} C(x, y) = {} & b_{y,0}(\phi)C(x, y_j) + b_{y,1}(\phi)C(x, y_{j+1}) + a_{x,0}(\theta)C^\dagger(x_i, y) \\ & + a_{x,1}(\theta)C^\dagger(x_{i+1}, y) - a_{x,0}(\theta)b_{y,0}(\phi)z_{i,j} - a_{x,0}(\theta)b_{y,1}(\phi)z_{i,j+1} \\ & - a_{x,1}(\theta)b_{y,0}(\phi)z_{i+1,j} - a_{x,1}(\theta)b_{y,1}(\phi)z_{i+1,j+1}, \end{aligned}\right\} \tag{6}$$

where $C(x, y_j)$, $j = 1, 2, \dots, n$ and $C^\dagger(x_i, y)$, $i = 1, 2, \dots, m$ are the classical rational cubic interpolants obtained in [22].

4 Analytical Properties of Rational FIS

We discuss analytical properties like convergence and stability in the following Theorems 4.1 and 4.2, respectively. In this section, we use the following notation: $a_\infty = \max_{1\le i\le m-1} a_i$, $c_\infty = \max_{1\le j\le n-1} c_j$, $h = \max_{1\le i\le m-1} h_i$, $h^\dagger = \max_{1\le j\le n-1} h^\dagger_j$.

Theorem 4.1 *Let Θ be the rational cubic FIS with respect to the surface data $\{(x_i, y_j, z_{i,j}), i = 1, 2, \dots, m, j = 1, 2, \dots, n\}$ generated from the original function $F \in \mathscr{C}^4(D)$. Then*

$$\|F-\Theta\|_\infty \le \frac{\kappa a_\infty (H_{x,1}(h)+H_{x,2}(h))}{1-\kappa a_\infty} + \frac{\kappa c_\infty (H_{y,1}(h^\dagger)+H_{y,2}(h^\dagger))}{1-\kappa c_\infty}$$
$$+h\Big(\Big\|\frac{\partial F}{\partial x}\Big\|_\infty + K_C\Big) + h^\dagger \max_{1\le i\le n}\Big[\max_{1\le j\le m-1}\Big\{\frac{\omega_{i,j}h_i}{2\nu_{i,j}\tau_{i,j}}\Big(\zeta^*_{i,j}+\frac{E^\dagger_{i,j}(F)}{96}\Big)\Big\}\Big], \tag{7}$$

where

$$H_{x,1}(h) = \max_{1\le j\le n} H_{x,1,j}(h),\ H_{x,1,j}(h) = \max_{1\le i\le m-1}\{3(|z_{i,j}|+|z_{i+1,j}|)+h_i(|z^x_{i,j}|+|z^x_{i+1,j}|)\},$$
$$H_{x,2}(h) = \max_{1\le j\le n} H_{x,2,j}(h),\ H_{x,2,j}(h) = \max_{1\le i\le m-1}\{3a_i(|z_{1,j}|+|z_{m,j}|)+h_i(|z^x_{1,j}|+|z^x_{m,j}|)\},$$
$$H_{y,1}(h^\dagger) = \max_{1\le i\le m} H_{y,1,i}(h^\dagger),\ H_{y,1,i}(h^\dagger) = \max_{1\le j\le n-1}\{3(|z_{i,j}|+|z_{i,j+1}|)+h^\dagger_j(|z^y_{i,j}|+|z^y_{i,j+1}|)\},$$
$$H_{y,2}(h^\dagger) = \max_{1\le i\le m} H_{y,2,i}(h^\dagger),\ H_{y,2,i}(h^\dagger) = \max_{1\le j\le n-1}\{3c_j(|z_{i,1}|+|z_{i,n}|)+h^\dagger_j(|z^y_{i,1}|+|z^y_{i,n}|)\},$$
$$E^\dagger_{i,j}(F) = h_j^{\dagger 3}\Big\|\frac{\partial^4 F(x_i,y)}{\partial^4 y}\Big\|_\infty A_{i,j}(F) + 16\zeta_{i,j}h_j^{\dagger 2}\Big\|\frac{\partial^3 F(x_i,y)}{\partial^3 y}\Big\|_\infty + 24\zeta_{i,j}h_j^{\dagger 2}\Big\|\frac{\partial^2 F(x_i,y)}{\partial^2 y}\Big\|_\infty,$$
$$A_{i,j}(F) = \Big\|\frac{\partial F(x_i,y)}{\partial y}\Big\|_\infty + \frac{\zeta_{i,j}}{2},\ \zeta_{i,j} = \max\Big\{\Big|\frac{\partial F(x_i,y_j)}{\partial y} - z^y_{i,j}\Big|, \Big|\frac{\partial F(x_i,y_{j+1})}{\partial y} - z^y_{i,j+1}\Big|\Big\},$$
$$\nu_{i,j} = \min_{y_j\le y\le y_{j+1}}\Big|\frac{\partial F(x_i,y)}{\partial y}\Big|,\ \omega_{i,j} = \max\{\alpha_{i,j},\beta_{i,j}\},\ \tau_{i,j} = \min\{\alpha_{i,j},\beta_{i,j}\}.$$

Proof Since Θ and C, respectively, are the rational cubic FIS and the classical rational cubic surface interpolant for the surface data $\{(x_i, y_j, z_{i,j}), i = 1, 2, \ldots, m, j = 1, 2, \ldots, n\}$, from (5) and Remark 3.1, we have

$$\left.\begin{aligned}|\Theta(x,y)-C(x,y)| \le\ & b_{y,0}(\phi)|S(x,y_j)-C(x,y_j)| \\ & + b_{y,1}(\phi)|S(x,y_{j+1})-C(x,y_{j+1})| \\ & + a_{x,0}(\theta)|S^\dagger(x_i,y)-C^\dagger(x_i,y)| \\ & + a_{x,1}(\theta)|S^\dagger(x_{i+1},y)-C^\dagger(x_{i+1},y)|.\end{aligned}\right\} \tag{8}$$

Using Eq. (24) in Theorem 2 of [12], we obtain

$$\left.\begin{aligned}|S(x,y_j)-C(x,y_j)| &\le \tfrac{|\xi_j|_\infty}{1-|\xi_j|_\infty} H_{x,1,j}(h),\ j=1,2,\ldots,n, \\ |S^\dagger(x_i,y)-C(x_i,y)| &\le \tfrac{|\xi_i^\dagger|_\infty}{1-|\xi_i^\dagger|_\infty} H_{y,1,i}(h^\dagger),\ i=1,2,\ldots,m.\end{aligned}\right\} \tag{9}$$

Also it follows that

$$a_{x,0}(\theta)\le 1, a_{x,1}(\theta)\le 1, \theta\in[0,1],\ b_{y,0}(\phi)\le 1, b_{y,1}(\phi)\le 1, \phi\in[0,1]. \tag{10}$$

Using (9)–(10) in (8), it is estimated that

$$
\begin{aligned}
|\Theta(x, y) - C(x, y)| \leq & \frac{|\xi_j|_\infty}{1 - |\xi_j|_\infty} H_{x,1,j}(h) + \frac{|\xi_{j+1}|_\infty}{1 - |\xi_{j+1}|_\infty} H_{x,1,j+1}(h) \\
& + \frac{|\xi_i^\dagger|_\infty}{1 - |\xi_i^\dagger|_\infty} H_{y,1,i}(h^\dagger) + \frac{|\xi_{j+1}^\dagger|_\infty}{1 - |\xi_{j+1}^\dagger|_\infty} H_{y,1,i+1}(h^\dagger), \\
\leq & \frac{|\xi|_\infty}{1 - |\xi|_\infty} H_{x,1}(h) + \frac{|\xi^\dagger|_\infty}{1 - |\xi^\dagger|_\infty} H_{y,1}(h^\dagger).
\end{aligned}
$$

Since the above inequality is true for every $(x, y) \in D_{i,j}$, $i = 1, 2, \ldots, m-1$, $j = 1, 2, \ldots, n-1$, the following estimation is obtained:

$$
\|\Theta - C\|_\infty \leq \frac{|\xi|_\infty}{1 - |\xi|_\infty} H_{x,1}(h) + \frac{|\xi^\dagger|_\infty}{1 - |\xi^\dagger|_\infty} H_{y,1}(h^\dagger). \tag{11}
$$

Expanding the function F using Taylor formula at the point $(x_i, y) \in D_{i,j}$, we have

$$
F(x, y) = F(x_i, y) + (x - x_i)\frac{\partial F(\xi, y)}{\partial x}, \quad (\xi, y) \in D_{i,j}.
$$

This implies

$$
|F(x, y) - F(x_i, y)| \leq h \left\| \frac{\partial F}{\partial x} \right\|_\infty. \tag{12}
$$

Similarly by applying the above procedure to the classical rational cubic surface C, we have

$$
|C(x, y) - C(x_i, y)| \leq h \left\| \frac{\partial C}{\partial x} \right\|_\infty. \tag{13}
$$

It is observed that

$$
|F(x, y) - C(x, y)| \leq |F(x, y) - F(x_i, y)| + |F(x_i, y) - C(x_i, y)| + |C(x_i, y) - C(x, y)|.
$$

Now using (12)–(13) in the above inequality, we obtain

$$
|F(x, y) - C(x, y)| \leq h\left(\left\| \frac{\partial F}{\partial x} \right\|_\infty + \left\| \frac{\partial C}{\partial x} \right\|_\infty\right) + |F(x_i, y) - C(x_i, y)|. \tag{14}
$$

From [22], it is known that

$$
|F(x_i, y) - C(x_i, y)| \leq \max_{1 \leq j \leq m-1} \left\{ \frac{\omega_{i,j} h_i}{2 \nu_{i,j} \tau_{i,j}} \left(\zeta_{i,j}^* + \frac{E_{i,j}^\dagger(F)}{96} \right) \right\}. \tag{15}
$$

Substituting (15) in (14) yields

$$|F(x,y)-C(x,y)| \leq h\Big(\Big\|\frac{\partial F}{\partial x}\Big\|_\infty + \Big\|\frac{\partial C}{\partial x}\Big\|_\infty\Big) + h^\dagger \max_{1\leq i\leq n}\Big[\max_{1\leq j\leq m-1}\Big\{\frac{\omega_{i,j}h_i}{2\nu_{i,j}\tau_{i,j}}\Big(\zeta^*_{i,j}+\frac{E^\dagger_{i,j}(F)}{96}\Big)\Big\}\Big]. \quad (16)$$

Since (16) is true for every $(x, y) \in D_{i,j}$, $i = 1, 2, \ldots, m-1$, $j = 1, 2, \ldots, n-1$, the uniform error bound between F and C is given by

$$\|F-C\|_\infty \leq h\Big(\Big\|\frac{\partial F}{\partial x}\Big\|_\infty + \Big\|\frac{\partial C}{\partial x}\Big\|_\infty\Big) + h^\dagger \max_{1\leq i\leq n}\Big[\max_{1\leq j\leq m-1}\Big\{\frac{\omega_{i,j}h_i}{2\nu_{i,j}\tau_{i,j}}\Big(\zeta^*_{i,j}+\frac{E^\dagger_{i,j}(F)}{96}\Big)\Big\}\Big]. \quad (17)$$

Since $\frac{\partial C}{\partial x} \in \mathscr{C}^1(D)$, there exists a positive constant K_C such that

$$\Big\|\frac{\partial C}{\partial x}\Big\|_\infty \leq K_C. \quad (18)$$

Using (11) and (17)–(18) together with inequality

$$\|\Theta - F\|_\infty \leq \|\Theta - C\|_\infty + \|C - F\|_\infty,$$

the desired bound for $\|\Theta - F\|_\infty$ is obtained.

Convergence result: Since $a_\infty = \frac{h}{x_m - x_1}$ and $c_\infty = \frac{h^\dagger}{y_n - y_1}$, Theorem 4.1 gives that the rational cubic FIS Θ uniformly converges to the original function F as $h \to 0^+$ and $h^\dagger \to 0^+$.

Theorem 4.2 *Suppose $\Theta_{\varepsilon,\varepsilon^\dagger}$ is the perturbed rational cubic FIS when the corresponding scaling factors of Θ are perturbed as $\xi_{i,j} + \varepsilon_{i,j}$ and $\xi^\dagger_{i,j} + \varepsilon^\dagger_{i,j}$, where $\varepsilon_{i,j}$ and $\varepsilon^\dagger_{i,j}$ are real numbers such that $0 < |\xi_j|_\infty + |\varepsilon_j|_\infty < 1$, $j = 1, 2, \ldots, n$ and $0 < |\xi^\dagger_i|_\infty + |\varepsilon^\dagger_i|_\infty < 1$, $i = 1, 2, \ldots, m$, $|\xi_j|_\infty = \max_{1\leq i\leq m-1} |\xi_{i,j}|$, $|\varepsilon_j|_\infty = \max_{1\leq i\leq m-1} |\varepsilon_{i,j}|$, $|\xi^\dagger_i|_\infty = \max_{1\leq j\leq n-1} |\xi^\dagger_{i,j}|$, and $|\varepsilon^\dagger_i|_\infty = \max_{1\leq j\leq n-1} |\varepsilon^\dagger_{i,j}|$. Then*

$$\|\Theta - \Theta_{\varepsilon,\varepsilon^\dagger}\|_\infty \leq 2\Big[\max_{1\leq j\leq n} M_{j,\xi,\varepsilon} + \max_{1\leq i\leq m} M_{i,\xi^\dagger,\varepsilon^\dagger}\Big],$$

where

$$M_{j,\xi,\varepsilon} = \frac{4a_\infty|\varepsilon_j|_\infty(12 + 4[x_m - x_1])\max_{1\leq i\leq m}\{|z_{i,j}|, |z^x_{i,j}|\}}{(1 - a_\infty|\xi_j|_\infty)(1 - a_\infty[|\xi_j|_\infty + |\varepsilon_j|_\infty])},$$

$$M_{i,\xi^\dagger,\varepsilon^\dagger} = \frac{4|\varepsilon_i^\dagger|_\infty(12 + 4[y_n - y_1]) \max_{1\le j\le n}\{|z_{i,j}|, |z_{i,j}^y|\}}{(1 - c_\infty|\xi_i^\dagger|_\infty)(1 - c_\infty[|\xi_i^\dagger|_\infty + |\varepsilon_i^\dagger|_\infty])}.$$

Proof Since $\Theta_{\varepsilon,\varepsilon^\dagger}$ is the perturbation of Θ with respect to the scaling factors, using the functional equation of Θ (cf. (5)) we write the functional equation of $\Theta_{\varepsilon,\varepsilon^\dagger}$ as

$$\left.\begin{aligned}\Theta_{\varepsilon,\varepsilon^\dagger}(x, y) &= b_{y,0}(\phi)\tilde{S}_\varepsilon(x, y_j) + b_{y,1}(\phi)\tilde{S}_\varepsilon(x, y_{j+1}) + a_{x,0}(\theta)\tilde{S}^\dagger_{\varepsilon^\dagger}(x_i, y)\\ &\quad + a_{x,1}(\theta)S^\dagger_{\varepsilon^\dagger}(x_{i+1}, y) - a_{x,0}(\theta)b_{y,0}(\phi)z_{i+1,j} - a_{x,0}(\theta)b_{y,1}(\phi)z_{i+1,j+1}\\ &\quad - a_{x,1}(\theta)b_{y,0}(\phi)z_{i+1,j} - a_{x,1}(\theta)b_{y,1}(\phi)z_{i+1,j+1},\end{aligned}\right\} \tag{19}$$

where $\tilde{S}_\varepsilon(x, y_j)$ and $\tilde{S}^\dagger_{\varepsilon^\dagger}(x_i, y)$ are perturbed rational cubic FIFs of $S(x, y_j)$ and $S^\dagger(x_i, y)$, respectively, with respect to the perturbed scaling factors, and they satisfy

$$\tilde{S}_\varepsilon(x, y_j) = a_i(\xi_{i,j} + \varepsilon_{i,j})\tilde{S}_\varepsilon(\theta_i^{-1}(x), y_j) + \frac{\tilde{p}_{\varepsilon,i,j}(\theta)}{q_{i,j}(\theta)}, \tag{20}$$

$$\begin{aligned}\tilde{p}_{\varepsilon,i,j}(\theta) &= \alpha_{i,j}\big[z_{i,j} - (\xi_{i,j} + \varepsilon_{i,j})z_{1,j}a_i\big](1 - \theta)^3 + \big[z_{i,j}(\gamma_{i,j} + \alpha_{i,j}) + z_{i,j}^x\alpha_{i,j}h_i\\ &\quad - (\xi_{i,j} + \varepsilon_{i,j})\{\alpha_{i,j}h_iz_{1,j}^x + z_{1,j}a_i(\gamma_{i,j} + \alpha_{i,j})\}\big]\theta(1 - \theta)^2\\ &\quad + \big[z_{i+1,j}(\gamma_{i,j} + \beta_{i,j}) - z_{i+1,j}^x\beta_{i,j}h_i + (\xi_{i,j} + \varepsilon_{i,j})\{\beta_{i,j}h_iz_{m,j}^x\\ &\quad - z_{m,j}a_i(\gamma_{i,j} + \beta_{i,j})\}\big]\theta^2(1 - \theta) + \beta_{i,j}\big[z_{i+1,j} - (\xi_{i,j} + \varepsilon_{i,j})z_{m,j}a_i\big]\theta^3,\end{aligned}$$

$$\tilde{S}^\dagger_{\varepsilon^\dagger}(x_i, y) = c_j(\xi_{i,j}^\dagger + \varepsilon_{i,j}^\dagger)\tilde{S}^\dagger_{\varepsilon^\dagger}(x_i, \phi_j^{-1}(y)) + \frac{\tilde{p}^\dagger_{\varepsilon^\dagger,i,j}(\phi)}{q^\dagger_{i,j}(\phi)}, \tag{21}$$

$$\begin{aligned}p^\dagger_{\varepsilon^\dagger,i,j}(\phi) &= \alpha_{i,j}^\dagger\big[z_{i,j} - (\xi_{i,j}^\dagger + \varepsilon_{i,j}^\dagger)z_{i,1}c_j\big](1 - \phi)^3 + \big[z_{i,j}(\gamma_{i,j}^\dagger + \alpha_{i,j}^\dagger) + z_{i,j}^y\alpha_{i,j}^\dagger h_j^\dagger\\ &\quad - (\xi_{i,j}^\dagger + \varepsilon_{i,j}^\dagger)\{\alpha_{i,j}^\dagger h_j^\dagger z_{i,1}^y + z_{i,1}c_j(\gamma_{i,j}^\dagger + \alpha_{i,j}^\dagger)\}\big]\phi(1 - \phi)^2 + \big[z_{i,j+1}(\gamma_{i,j}^\dagger + \beta_{i,j}^\dagger)\\ &\quad - z_{i,j+1}^y\beta_{i,j}^\dagger h_j^\dagger + (\xi_{i,j}^\dagger + \varepsilon_{i,j}^\dagger)\{\beta_{i,j}^\dagger h_j^\dagger z_{i,n}^y - z_{i,n}c_j(\gamma_{i,j}^\dagger + \beta_{i,j}^\dagger)\}\big]\phi^2(1 - \phi)\\ &\quad + \beta_{i,j}^\dagger\big[z_{i,j+1} - (\xi_{i,j}^\dagger + \varepsilon_{i,j}^\dagger)z_{i,n}c_j\big]\phi^3,\end{aligned}$$

From (5) and (19), we have

$$\begin{aligned}|\Theta(x,y)-\Theta_{\varepsilon,\varepsilon^{\dagger}}(x,y)| \le\ & b_{y,0}(\phi)|S(x,y_j)-\tilde{S}_{\varepsilon}(x,y_j)| \\ &+ b_{y,1}(\phi)|S(x,y_{j+1})-\tilde{S}_{\varepsilon}(x,y_{j+1})| \\ &+ a_{x,0}(\theta)|S^{\dagger}(x_i,y)-\tilde{S}^{\dagger}_{\varepsilon^{\dagger}}(x_i,y)| \\ &+ a_{x,1}(\theta)|S^{\dagger}(x_{i+1},y)-\tilde{S}^{\dagger}_{\varepsilon^{\dagger}}(x_{i+1},y)|.\end{aligned}$$

Since $a_{x,k}(\theta) \le 1$ and $b_{y,k}(\phi) \le 1$ for $k = 0, 1$, the above inequality reduces to

$$\left.\begin{aligned}|\Theta(x,y)-\Theta_{\varepsilon,\varepsilon^{\dagger}}(x,y)| \le\ & |S(x,y_j)-\tilde{S}_{\varepsilon}(x,y_j)| + |S(x,y_{j+1})-\tilde{S}_{\varepsilon}(x,y_{j+1})| \\ &+|S^{\dagger}(x_i,y)-\tilde{S}^{\dagger}_{\varepsilon^{\dagger}}(x_i,y)| + |S^{\dagger}(x_{i+1},y)-\tilde{S}^{\dagger}_{\varepsilon^{\dagger}}(x_{i+1},y)|.\end{aligned}\right\} \tag{22}$$

To study the sensitivity of the rational cubic FIS Θ with respect to a slight perturbation in the scaling factors, it is necessary to find an upper bound for each term in the right-hand side of (22). From (20), $\tilde{S}_{\varepsilon}(x, y_j)$ is rewritten as

$$\begin{aligned}\tilde{S}_{\varepsilon}(x,y_j) =\ & a_i\xi_{i,j}\tilde{S}_{\varepsilon}(\theta_i^{-1}(x),y_j) + a_i\varepsilon_{i,j}\tilde{S}_{\varepsilon}(\theta_i^{-1}(x),y_j) + g_j(\theta_i^{-1}(x)) \\ &- a_i(\xi_{i,j}+\varepsilon_{i,j})\bar{b}_j(\theta_i^{-1}(x)),\end{aligned} \tag{23}$$

where $g_j(\theta_i^{-1}(x)) = \frac{\tilde{p}^1_{\varepsilon,i,j}(\theta)}{q_{i,j}(\theta)}$, $\bar{b}_j(\theta_i^{-1}(x)) = \frac{\tilde{p}^2_{\varepsilon,i,j}(\theta)}{q_{i,j}(\theta)}$,

$$\begin{aligned}\tilde{p}^1_{\varepsilon,i,j}(\theta) =\ & \alpha_{i,j}z_{i,j}(1-\theta)^3 + \big[z_{i,j}(\gamma_{i,j}+\alpha_{i,j}) + z^x_{i,j}\alpha_{i,j}h_i\big]\theta(1-\theta)^2 \\ &+ \big[z_{i+1,j}(\gamma_{i,j}+\beta_{i,j}) - z^x_{i+1,j}\beta_{i,j}h_i\big]\theta^2(1-\theta) + \beta_{i,j}\, z_{i+1,j}\theta^3,\end{aligned}$$

$$\begin{aligned}\tilde{p}^2_{\varepsilon,i,j}(\theta) =\ & \alpha_{i,j}z_{1,j}(1-\theta)^3 + \big[\alpha_{i,j}(x_m-x_1)z^x_{1,j} + z_{1,j}(\gamma_{i,j}+\alpha_{i,j})\big]\theta(1-\theta)^2 \\ &- \big[\beta_{i,j}(x_m-x_1)z^x_{m,j} - z_{m,j}a_i(\gamma_{i,j}+\beta_{i,j})\big]\theta^2(1-\theta) + \beta_{i,j}z_{m,j}\theta^3.\end{aligned}$$

From (3) and (23), we obtain

$$\begin{aligned}|S(x,y_j)-\tilde{S}_{\varepsilon}(x,y_j)| \le\ & a_{\infty}|\xi_j|_{\infty}|S(\theta_i^{-1}(x),y_j)-\tilde{S}_{\varepsilon}(\theta_i^{-1}(x),y_j)| \\ &+ a_{\infty}|\varepsilon_j|_{\infty}|\tilde{S}_{\varepsilon}(\theta_i^{-1}(x),y_j)-\bar{b}_j(\theta_i^{-1}(x))|, \\ \le\ & a_{\infty}|\xi_j|_{\infty}||S(.,y_j)-\tilde{S}_{\varepsilon}(.,y_j)||_{\infty} + a_{\infty}|\varepsilon_j|_{\infty}||\tilde{S}_{\varepsilon}(.,y_j)-\bar{b}_j||_{\infty}.\end{aligned}$$

Since the above inequality is valid for all $x \in I_i$, $i = 1, 2, \ldots, m-1$, we have

$$||S(.,y_j)-\tilde{S}_{\varepsilon}(.,y_j)||_{\infty} \le a_{\infty}|\xi_j|_{\infty}||S(.,y_j)-\tilde{S}_{\varepsilon}(.,y_j)||_{\infty} + a_{\infty}|\varepsilon_j|_{\infty}||\tilde{S}_{\varepsilon}(.,y_j)-\bar{b}_j||_{\infty},$$

$$\Rightarrow ||S(.,y_j)-\tilde{S}_\varepsilon(.,y_j)||_\infty \le \frac{a_\infty|\varepsilon_j|_\infty}{1-a_\infty|\xi_j|_\infty}||\tilde{S}_\varepsilon(.,y_j)-\bar{b}_j||_\infty.$$

Consequently, we can write

$$||S(.,y_j)-\tilde{S}_\varepsilon(.,y_j)||_\infty \le \frac{a_\infty|\varepsilon_j|_\infty}{1-a_\infty|\xi_j|_\infty}[||\tilde{S}_\varepsilon(.,y_j)-g_j||_\infty+||g_j-\bar{b}_j||_\infty]. \quad (24)$$

(23) asserts that

$$|\tilde{S}_\varepsilon(x,y_j)-g_j(\theta_i^{-1}(x)| \le a_\infty[|\xi_j|_\infty+|\varepsilon_j|_\infty]|\tilde{S}_\varepsilon(\theta_i^{-1}(x),y_j)-\bar{b}_j(\theta_i^{-1}(x))|,$$
$$\le a_\infty[|\xi_j|_\infty+|\varepsilon_j|_\infty]||\tilde{S}_\varepsilon(.,y_j)-\bar{b}_j||_\infty.$$

Since the above inequality is valid for $x \in I_i$, $i=1,2,\ldots,m-1$, we obtain

$$||\tilde{S}_\varepsilon(.,y_j)-g_j||_\infty \le a_\infty[|\xi_j|_\infty+|\varepsilon_j|_\infty]||\tilde{S}_\varepsilon(.,y_j)-\bar{b}_j||_\infty,$$
$$\le a_\infty[|\xi_j|_\infty+|\varepsilon_j|_\infty][||\tilde{S}_\varepsilon(.,y_j)-g_j||_\infty+||g_j \quad \bar{b}_j||_\infty],$$

which implies

$$||\tilde{S}_\varepsilon(.,y_j)-g_j||_\infty \le \frac{a_\infty[|\xi_j|_\infty+|\varepsilon_j|_\infty]||g_j-\bar{b}_j||_\infty}{1-a_\infty[|\xi_j|_\infty+|\varepsilon_j|_\infty]}. \quad (25)$$

Substituting (25) in (24), we get

$$||S(.,y_j)-\tilde{S}_\varepsilon(.,y_j)||_\infty \le \frac{a_\infty|\varepsilon_j|_\infty||g_j-\bar{b}_j||_\infty}{(1-a_\infty|\xi_j|_\infty)(1-a_\infty[|\xi_j|_\infty-|\varepsilon_j|_\infty])}. \quad (26)$$

Using the expression for g_j and $\bar{b}_j$ (cf. (23)),

$$|g_j(L_i^{-1}(x))-\bar{b}_j(L_i^{-1}(x))| \le |g_j(L_i^{-1}(x))|+|\bar{b}_j(L_i^{-1}(x))|. \quad (27)$$

Next, from (23), it is calculated that

$$|g_j(L_i^{-1}(x))| \le \frac{\alpha_{i,j}|z_{i,j}|(1-\theta)^3}{\alpha_{i,j}(1-\theta)^2}+\left[|z_{i,j}|\left(\frac{\gamma_{i,j}}{\gamma_{i,j}\theta(1-\theta)}+\frac{\alpha_{i,j}}{\alpha_{i,j}(1-\theta)^2}\right)\right.$$
$$\left.+|z_{i,j}^x|\frac{\alpha_{i,j}}{\alpha_{i,j}(1-\theta)^2}h_i\right]\theta(1-\theta)^2+\left[|z_{i+1,j}|\left(\frac{\gamma_{i,j}}{\gamma_{i,j}\theta(1-\theta)}+\frac{\beta_{i,j}}{\beta_{i,j}\theta^2}\right)\right.$$
$$\left.+|z_{i+1,j}^x|\frac{\beta_{i,j}}{\beta_{i,j}\theta^2}h_i\right]\theta^2(1-\theta)+\frac{\beta_{i,j}}{\beta_{i,j}\theta^2}|z_{i+1,j}|\theta^3,$$
$$\le (6+2h)\max_{1\le i\le m}\{|z_{i,j}|,|z_{i,j}^x|\}.$$

Similarly, we can show that

$$|\bar{b}_j(L_i^{-1}(x))| \leq (6 + 2[x_m - x_1]) \max_{1 \leq i \leq m} \{|z_{i,j}|, |z_{i,j}^x|\}.$$

Substituting the upper bounds of $|g_j(L_i^{-1}(x))|$ and $|\bar{b}_j(L_i^{-1}(x))|$ in (27), we get

$$|g_j(L_i^{-1}(x)) - \bar{b}_j(L_i^{-1}(x))| \leq (12 + 4[x_m - x_1]) \max_{1 \leq i \leq m} \{|z_{i,j}|, |z_{i,j}^x|\}.$$

Since the above inequality is valid for $x \in I_i$, $i = 1, 2, \ldots, m - 1$, we get

$$||g_j - \bar{b}_j||_\infty \leq (12 + 4[x_m - x_1]) \max_{1 \leq i \leq m} \{|z_{i,j}|, |z_{i,j}^x|\}. \tag{28}$$

Using (28) in (26), we get

$$||S(., y_j) - \tilde{S}_\varepsilon(., y_j)||_\infty \leq M_{j,\xi,\varepsilon}. \tag{29}$$

By reiterating the above procedure for $S^\dagger(x_i, y)$, we obtain

$$||S^\dagger(x_i, .) - \tilde{S}^\dagger_{\varepsilon^\dagger}(x_i, .)||_\infty \leq M_{i,\xi^\dagger,\varepsilon^\dagger}. \tag{30}$$

Substituting (29)–(30) in (22), it follows that

$$|\Theta(x, y) - \Theta_{\varepsilon,\varepsilon^\dagger}(x, y)| \leq M_{j,\xi,\varepsilon} + M_{j+1,\xi,\varepsilon} + M_{i,\xi^\dagger,\varepsilon^\dagger} + M_{i+1,\xi^\dagger,\varepsilon^\dagger}. \tag{31}$$

Since the above inequality is valid for $(x, y) \in D_{i,j}$, $i = 1, 2, \ldots, m - 1$, $j = 1, 2, \ldots, n - 1$, we get the required bound for $||\Theta - \Theta_{\varepsilon,\varepsilon^\dagger}||_\infty$ in Theorem 4.2 .

Remark 4.1 When the perturbations in the scaling matrices are very small (i.e., $|\varepsilon_j|_\infty \to 0$, $|\varepsilon_i^\dagger|_\infty \to 0$), it can be verified that both $M_{j,\xi,\varepsilon} \to 0$ and $M_{i,\xi^\dagger,\varepsilon^\dagger} \to 0$. Consequently, from Theorem 4.2, it follows that $\Theta_{\varepsilon,\varepsilon^\dagger} \to \Theta$. Thus the rational cubic FIS Θ is stable with respect to perturbations in the scaling matrices.

5 Monotonic Rational Cubic FIS

Here we shall offer the final objective of this article wherein we wish to constrain the scaling factors and shape parameters of rational cubic FIS Θ so that it is monotonic in nature whenever given surface data $\triangle$ is monotonic. Let $\triangle$ be a monotonic interpolation data, i.e., either

$$\left.\begin{array}{l} z_{i+1,j} \geq z_{i,j},\ i = 1, 2, \ldots, m-1,\ \text{for every fixed } j \in \{1, 2, \ldots, n\}, \\ z_{i,j+1} \geq z_{i,j},\ j = 1, 2, \ldots, n-1,\ \text{for every fixed } i \in \{1, 2, \ldots, m\}, \end{array}\right\} \quad (32)$$

or

$$\left.\begin{array}{l} z_{i+1,j} \leq z_{i,j},\ i = 1, 2, \ldots, m-1,\ \text{for every fixed } j \in \{1, 2, \ldots, n\}, \\ z_{i,j+1} \leq z_{i,j},\ j = 1, 2, \ldots, n-1,\ \text{for every fixed } i \in \{1, 2, \ldots, m\}, \end{array}\right\} \quad (33)$$

with $sgn(z^x_{i,j}) = sgn(\Delta_{i,j})$, $\Delta_{i,j} = \frac{z_{i+1,j}-z_{i,j}}{x_{i+1}-x_i}$, and $sgn(z^y_{i,j}) = sgn(\Delta^\dagger_{i,j})$, $\Delta^\dagger_{i,j} = \frac{z_{i,j+1}-z_{i,j}}{y_{j+1}-y_j}$. We wish to furnish conditions so as to ensure that the rational cubic FIS Θ is monotonic over D, that is, for all (x^*, y^*), $(x^{**}, y^{**}) \in \mathbb{R}^2$ with $x^* \geq x^{**}$ and $y^* \geq y^{**}$,

$$\Theta(x^*, y^*) \geq \Theta(x^{**}, y^{**})(\text{increasing}) \quad \text{or} \quad \Theta(x^*, y^*) \leq \Theta(x^{**}, y^{**})(\text{decreasing}).$$

We shall recall that the surface generated by the rational cubic FIS Θ monotonic if the fractal boundary curves $S(x, y_j)$ for all $j = 1, 2, \ldots, n$ and $S^*(x_i, y)$ for all $i = 1, 2, \ldots, m$ are monotonic [23]. By using the Theorem 4 of [12], we can see that the fractal boundary curve $S(x, y_j)$ is monotonic if scaling factors $|\xi^\dagger_{i,j}| < \kappa < 1$ and the shape parameters $\alpha_{i,j} > 0$, $\beta_{i,j} > 0$, and $\gamma_{i,j} > 0$ are selected according to the conditions:

$$\xi_{i,j} \in \begin{cases} \left[0, \min\left\{\frac{z^x_{i+1,j}}{z^x_{m,j}}, \frac{z^x_{i,j}}{z^x_{1,j}}\right\}\right], & \text{if } \min\left\{\frac{z^x_{i+1,j}}{z^x_{m,j}}, \frac{z^x_{i,j}}{z^x_{1,j}}\right\} < \min\left\{\frac{\Delta_{i,j}(x_m-x_1)}{z_{m,j}-z_{1,j}}, \kappa\right\}, \\ \left[0, \min\left\{\frac{\Delta_{i,j}(x_m-x_1)}{z_{m,j}-z_{1,j}}, \kappa\right\}\right), & \text{if } \min\left\{\frac{z^x_{i+1,j}}{z^x_{m,j}}, \frac{z^x_{i,j}}{z^x_{1,j}}\right\} \geq \min\left\{\frac{\Delta_{i,j}(x_m-x_1)}{z_{m,j}-z_{1,j}}, \kappa\right\}, \end{cases} \quad (34)$$

$$sgn(\alpha_{i,j}) = sgn(\beta_{i,j}), \text{and } \gamma_{i,j} = \frac{\alpha_{i,j}(z^x_{i,j} - \xi_{i,j} z^x_{1,j}) + \beta_{i,j}(z^x_{i+1,j} - \xi_{i,j} z^x_{m,j})}{\Delta_{i,j} - \xi_{i,j}\frac{z_{m,j}-z_{1,j}}{x_m-x_1}}. \quad (35)$$

Again using the Theorem 4 of [12], we can see that the fractal boundary curve $S^\dagger(x_i, y)$ is monotonic if the scaling factors $|\xi^\dagger_{i,j}| < \kappa < 1$ and the shape parameters $\alpha^\dagger_{i,j} > 0$, $\beta^\dagger_{i,j} > 0$, and $\gamma^\dagger_{i,j} > 0$ are selected according to the conditions:

$$\xi^\dagger_{i,j} \in \begin{cases} \left[0, \min\left\{\frac{z^y_{i,j+1}}{z^y_{i,n}}, \frac{z^y_{i,j}}{z^y_{i,1}}\right\}\right], & \text{if } \min\left\{\frac{z^y_{i,j+1}}{z^y_{i,n}}, \frac{z^y_{i,j}}{z^y_{i,1}}\right\} < \min\left\{\frac{\Delta^\dagger_{i,j}(y_n-y_1)}{z_{i,n}-z_{i,1}}, \kappa\right\}, \\ \left[0, \min\left\{\frac{\Delta^\dagger_{i,j}(y_n-y_1)}{z_{i,n}-z_{i,1}}, \kappa\right\}\right), & \text{if } \min\left\{\frac{z^y_{i,j+1}}{z^y_{i,n}}, \frac{z^y_{i,j}}{z^y_{i,1}}\right\} \geq \min\left\{\frac{\Delta_{i,j}{}^\dagger(y_n-y_1)}{z_{i,n}-z_{i,1}}, \kappa\right\}, \end{cases} \quad (36)$$

$$sgn(\alpha_{i,j}^{\dagger}) = sgn(\beta_{i,j}^{\dagger}), \text{ and } \gamma_{i,j}^{\dagger} = \frac{\alpha_{i,j}^{\dagger}(z_{i,j}^{y} - \xi_{i,j}^{\dagger} z_{i,1}^{y}) + \beta_{i,j}^{\dagger}(z_{i,j+1}^{y} - \xi_{i,j}^{\dagger} z_{i,n}^{y})}{\Delta_{i,j}^{\dagger} - \xi_{i,j}^{\dagger} \frac{z_{i,n}-z_{i,1}}{y_n - y_1}}. \quad (37)$$

For a quick reference, the entire discussion can be put in the form of following theorem.

Theorem 5.1 *Let $\triangle$ be a monotonic data. Then the rational cubic FIS Θ corresponding to surface data $\triangle$ is monotonic provided the (horizontal) scaling parameters $\xi_{i,j}$ for $i = 1, 2, \ldots, m-1$, $j = 1, 2, \ldots, n$, the (vertical) scaling parameters $\xi_{i,j}^{\dagger}$ for $i = 1, 2, \ldots, n$, $j = 1, 2, \ldots, n-1$, the (horizontal) shape parameters $\alpha_{i,j} > 0$, $\beta_{i,j} > 0$, and $\gamma_{i,j} > 0$ for $i = 1, 2, \ldots, m-1$, $j = 1, 2, \ldots, n$, and the (vertical) shape parameters $\alpha_{i,j}^{\dagger} > 0$, $\beta_{i,j}^{\dagger} > 0$, and $\gamma_{i,j}^{\dagger} > 0$ for $i = 1, 2, \ldots, n$, $j = 1, 2, \ldots, n-1$, satisfy the conditions prescribed in (34)–(37).*

6 Some Graphical Examples

For the validation of the proposed fractal surface scheme for the construction of monotonic surfaces, consider the monotonically increasing surface interpolation data (Table 1) with 16 points. For the given surface data, the horizontal and vertical homeomorphisms, respectively, are $\theta_1(x) = 0.1429x + 0.0857$, $\theta_2(x) = 0.1429x + 0.1857$, $\theta_3(x) = 0.7143x + 0.2286$, $x \in [0.1, 0.8]$ and $\phi_1(y) = 0.33y + 0.6667$, $\phi_2(y) = 0.33y + 1.6667$, $\phi_3(y) = 0.33y + 2.6667$, $y \in [1, 4]$. Utilizing the prescription given Theorem 5.1, we have calculated the suitable scaling factors and shape parameters (see Table 2) to obtain fractal surfaces (see Fig. 1a–c). By observing the fractal surfaces in Fig. 1a–c, one can notice the influence of the scaling factors and shape parameters in shape of the fractal surface. To verify stability results in Theorem 4.2, perturbed rational cubic FIS $\Theta_{\varepsilon,\varepsilon^{\dagger}}$ is generated in Fig. 1d by taking small perturbation in the scaling factors of rational cubic FIS Θ in Fig. 1a. Furthermore, it is calculated that $||\Theta_{\varepsilon,\varepsilon^{\dagger}} - \Theta||_{\infty} = 0.78$. It demonstrates that the monotonic rational cubic FIS is stable with respect to any small perturbation in the corresponding scaling factors.

Table 1 Monotonically increasing surface data

$\downarrow x/y \rightarrow$	1	2	3	8
0.1	(1, 1, 2.4)	(2, 3, 8.5)	(29, 7, 13.6)	(37, 8, 11.5)
0.2	(4, 9, 4)	(6, 8, 7)	(41, 12, 9)	(52, 7.7, 1)
0.3	(18, 16.8, 9.8)	(31, 11.3, 8.2)	(149, 18.2, 7.1)	(178, 0.1, 11)
0.8	(22, 1.1, 1.4)	(38, 9.5, 1.7)	(167, 12.7, 1.8)	(189, 11.8, 9)

Table 2 Matrices of the scaling factors and shape parameters used in the construction of rational cubic FISs in Fig. 1a–d

Figure	Scaling and shape matrices
Fig. 1a	$\xi = \begin{bmatrix} 0.12 & 0.12 & 0.12 & 0.12 \\ 0.12 & 0.12 & 0.12 & 0.12 \\ 0.2 & 0.2 & 0.2 & 0.2 \end{bmatrix}, \xi^{\dagger} = \begin{bmatrix} 0.15 & 0.2 & 0.2 \\ 0.2 & 0.2 & 0.2 \\ 0.2 & 0.2 & 0.2 \\ 0.2 & 0.2 & 0.15 \end{bmatrix}, \boldsymbol{\alpha} = 5*[1]_{3\times 4}, \boldsymbol{\beta} = 5*[1]_{3\times 4},$ $\gamma = \begin{bmatrix} 22.56 & 18.55 & 12.83 & 7.46 \\ 9.7 & 3.77 & 1.35 & 0.22 \\ 33.61 & 19.89 & 15.31 & 45.8 \end{bmatrix}, \boldsymbol{\alpha}^{\dagger} = 5*[1]_{4\times 3}, \boldsymbol{\beta}^{\dagger} = 5*[1]_{4\times 3},$ $\gamma^{\dagger} = \begin{bmatrix} 19.28 & 0.37 & 19.53 \\ 7.95 & 0.22 & 5.43 \\ 0.82 & 0.04 & 5.67 \\ 0.04 & 0.005 & 5.62 \end{bmatrix}$
Fig. 1b	$\xi = \begin{bmatrix} 0.12 & 0.12 & 0.12 & 0.12 \\ 0.12 & 0.12 & 0.12 & 0.12 \\ 0.01 & 0.01 & 0.01 & 0.01 \end{bmatrix}, \xi^{\dagger} = \begin{bmatrix} 0.15 & 0.2 & 0.2 \\ 0.2 & 0.2 & 0.2 \\ 0.2 & 0.2 & 0.2 \\ 0.2 & 0.2 & 0.15 \end{bmatrix}, \boldsymbol{\alpha} = 5*[1]_{3\times 4}, \boldsymbol{\beta} = 5*[1]_{3\times 4},$ $\gamma = \begin{bmatrix} 22.56 & 18.55 & 12.83 & 7.46 \\ 9.7 & 3.77 & 1.35 & 0.22 \\ 22.74 & 15.02 & 8.75 & 5.57 \end{bmatrix}, \boldsymbol{\alpha}^{\dagger} = 5*[1]_{4\times 3}, \boldsymbol{\beta}^{\dagger} = 5*[1]_{4\times 3},$ $\gamma^{\dagger} = \begin{bmatrix} 19.28 & 0.37 & 19.53 \\ 7.95 & 0.22 & 5.43 \\ 0.82 & 0.04 & 5.67 \\ 0.04 & 0.005 & 5.62 \end{bmatrix}$
Fig. 1c	$\xi = \begin{bmatrix} 0.12 & 0.12 & 0.12 & 0.12 \\ 0.12 & 0.12 & 0.12 & 0.12 \\ 0.2 & 0.2 & 0.2 & 0.2 \end{bmatrix}, \xi^{\dagger} = \begin{bmatrix} 0.15 & 0.2 & 0.2 \\ 0.2 & 0.2 & 0.2 \\ 0.2 & 0.2 & 0.2 \\ 0.2 & 0.2 & 0.15 \end{bmatrix}, \boldsymbol{\alpha} = 5*[1]_{3\times 4}, \boldsymbol{\beta} = 5*[1]_{3\times 4},$ $\gamma = \begin{bmatrix} 22.56 & 18.55 & 12.83 & 7.46 \\ 9.7 & 3.77 & 1.35 & 0.22 \\ 33.61 & 19.89 & 15.31 & 45.8 \end{bmatrix}, \boldsymbol{\alpha}^{\dagger} = 100*[1]_{4\times 3}, \boldsymbol{\beta}^{\dagger} = 5*[1]_{4\times 3},$ $\gamma^{\dagger} = \begin{bmatrix} 104.07 & 3.3 & 237.65 \\ 56.31 & 1.97 & 99.44 \\ 9.65 & 0.57 & 45.41 \\ 0.99 & 0.005 & 5.62 \end{bmatrix}$
Fig. 1d	$\xi = \begin{bmatrix} 0.13 & 0.13 & 0.13 & 0.13 \\ 0.13 & 0.12 & 0.12 & 0.13 \\ 0.2 & 0.2 & 0.2 & 0.2 \end{bmatrix}, \xi^{\dagger} = \begin{bmatrix} 0.15 & 0.2 & 0.2 \\ 0.2 & 0.2 & 0.2 \\ 0.2 & 0.2 & 0.2 \\ 0.2 & 0.2 & 0.15 \end{bmatrix}, \boldsymbol{\alpha} = 5*[1]_{3\times 4}, \boldsymbol{\beta} = 5*[1]_{3\times 4},$

(continued)

Table 2 (continued)

Figure	Scaling and shape matrices
	$\gamma = \begin{bmatrix} 23.27 & 19.21 & 13.65 & 7.8 \\ 9.7 & 3.77 & 1.35 & 0.22 \\ 33.61 & 19.89 & 15.31 & 45.8 \end{bmatrix}, \alpha^{\dagger} = 5 * [1]_{4\times3}, \beta^{\dagger} = 5 * [1]_{4\times3},$ $\gamma^{\dagger} = \begin{bmatrix} 19.28 & 0.37 & 19.53 \\ 7.95 & 0.22 & 5.43 \\ 0.82 & 0.04 & 5.67 \\ 0.04 & 0.005 & 5.62 \end{bmatrix}$

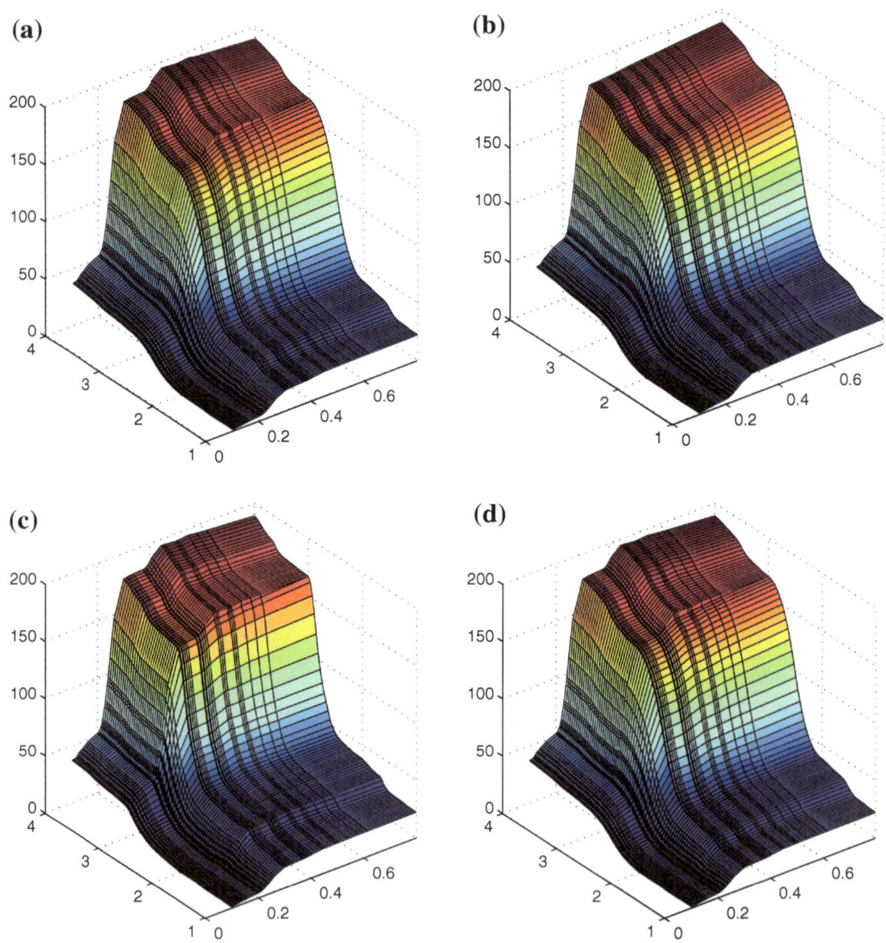

Fig. 1 Rational cubic FISs. **a** Rational cubic FIS. **b** Effects in surface Fig. 1a due to the changes in the scaling matrix ξ. **c** Effects in surface Fig. 1a due to the changes in the shape matrix $\alpha^{\dagger}$. **d** Rational cubic FIS with respect to perturbed scaling factors

7 Conclusion

In this paper, a new kind of fractal surface construction is developed over a rectangular grid. The proposed rational fractal surface interpolant not only stitch the data points arranged over the rectangular grid in a smooth way but also preserve the inherent shape feature, namely the monotonicity of the surface data. An upper bound of the interpolation error have been calculated. From this it is observed that the rational cubic FIS has linear convergence with the original function. An upper bound for the error in fractal surface interpolation is obtained when there is a slight perturbation in the corresponding scaling factors. From this it is observed that the rational cubic FIS is stable with respect to the small perturbations in the corresponding scaling factors. Data-dependent constraints on the scaling factors and shape parameters have been derived for achieving monotonic rational cubic FIS.

Acknowledgments The partial support of the Department of Science and Technology of Govt. of India (SERC DST Project No. SR/S4/MS: 694/10) is gratefully acknowledged.

References

1. Barnsley, M.F.: Fractal functions and interpolations. Constr. Approx. **2**, 303–329 (1986)
2. Barnsley, M.F., Harrington, A.N.: The calculus of fractal interpolation functions. J. Approx. Theory. **57**(1), 14–34 (1989)
3. Chand, A.K.B., Kapoor, G.P.: Generalized cubic spline fractal interpolation functions. SIAM J. Numer. Anal. **44**(2), 655–676 (2006)
4. Chand, A.K.B., Viswanathan, P.: Cubic Hermite and cubic spline fractal interpolation functions. BIT Numer. Math. **53**, 1467–1470 (2012)
5. Navascués, M.A.: Fractal polynomial interpolation. Z. Anal. Anwend. **24**(2), 1–20 (2005)
6. Navascués, M.A., Sebastián, M.V.: Generalization of Hermite functions by fractal interpolation. J. Approx. Theory **131**(1), 19–29 (2004)
7. Navascués, M.A., Sebastián, M.V.: Smooth fractal interpolation. J. Inequal. Appl. Article ID **78734**, 1–20 (2006)
8. Chand, A.K.B., Vijender, N.: Monotonicity preserving rational quadratic fractal interpolation functions. Advances in Numerical Analysis 2014, 17 p. (2014)
9. Viswanathan, P., Chand, A.K.B.: A fractal procedure for monotonicity preserving interpolation. Appl. Math. Comput. (2014)
10. Viswanathan, P., Chand, A.K.B., Navascués, M.A.: Fractal perturbation preserving fundamental shapes: bounds on the scale factors. J. Math. Anal. Appl. **419**, 804–817 (2014)
11. Chand, A.K.B., Vijender, N., Navascués, M.A.: Shape preservation of scientific data through rational fractal splines. Calcolo **51**, 329–362
12. Chand, A.K.B., Vijender, N., Agarwal, R.P.: Rational iterated function system for positive/monotonic shape preservation. Advances in Difference Equations 2014(30), 17 p (2014)
13. Chand, A.K.B., Navascués, M.A.: Natural bicubic spline fractal interpolation. Nonlinear Anal. **69**, 3679–391 (2008)
14. Massopust, P.R.: Fractal surfaces. J. Math. Anal. Appl. **151**, 275–290 (1990)
15. Geronimo, J.S., Hardin, D.P.: Fractal interpolation functions from $\mathbb{R}^n \to \mathbb{R}^m$ and their projections. Z. Anal. Anwend. **12**, 535–548 (1993)

16. Zhao, N.: Construction and application of fractal interpolation surfaces. Vis. Comput. **12**, 132–146 (1996)
17. Xie, H., Sun, H.: The study of bivariate fractal interpolation functions and creation of fractal interpolated surfaces. Fractals **5**(4), 625–634 (1997)
18. Dalla, L.: Bivariate fractal interpolation function on grids. Fractals **10**(1), 53–58 (2002)
19. Chand, A.K.B., Kapoor, G.P.: Spline coalescence hidden variable fractal interpolation functions. J. Appl. Math. Article ID **36829**, 1–17 (2006)
20. Chand, A.K.B., Kapoor, G.P.: Hidden variable bivariate fractal interpolation surfaces. Fractals **11**(3), 277–288 (2003)
21. Chand, A.K.B.: Natural cubic spline coalescence hidden variable fractal interpolation surfaces. Fractals **20**(2), 117–131 (2012)
22. Sarfraz, M., AL-Muhammed M., Ashraf, F.: Preserving monotonic shape of the data using piecewise rational cubic functions. Compt. Graph. **21**, 5–14 (1997)
23. Casciola, G., Romani, L.: Rational interpolants with tension parameters. J. Curve Surf. Des. 41–50 (2003)

Toward a Unified Methodology for Fractal Extension of Various Shape Preserving Spline Interpolants

S.K. Katiyar and A.K.B. Chand

Abstract Fractal interpolation, one in the long tradition of those involving the interpolatary theory of functions, is concerned with interpolation of a data set with a function whose graph is a fractal or a self-referential set. The novelty of fractal interpolants lies in their ability to model a data set with either a smooth or a nonsmooth function depending on the problem at hand. To broaden their horizons, some special class of fractal interpolants are introduced and their shape preserving aspects are investigated recently in the literature. In the current article, we provide a unified approach for the fractal generalization of various traditional nonrecursive polynomial and rational splines. To this end, first we shall view polynomial/rational FIFs as α-fractal functions corresponding to the traditional nonrecursive splines. The elements of the iterated function system are identified befittingly so that the class of α-fractal function f^α incorporates the geometric features such as positivity, monotonicity, and convexity in addition to the regularity inherent in the generating function f. This general theory in conjuction with shape preserving aspects of the traditional splines provides algorithms for the construction of shape preserving fractal interpolation functions. Even though the results obtained in this article are generally enough, we wish to apply it on a specific rational cubic spline with two free shape parameters.

Keywords Fractals · Iterated function system · Fractal interpolation functions · Rational cubic fractal functions · Rational cubic interpolation · Positivity

MSC 28A80 · 41A20 · 65D10 · 41A25 · 65D05 · 26A48

S.K. Katiyar (✉) · A.K.B. Chand
Department of Mathematics, Indian Institute of Technology Madras, Chennai 600036, India
e-mail: sbhkatiyar@gmail.com

A.K.B. Chand
e-mail: chand@iitm.ac.in

R.N. Mohapatra et al. (eds.), *Mathematics and Computing*,
Springer Proceedings in Mathematics & Statistics 139,
DOI 10.1007/978-81-322-2452-5_15

1 Introduction

The problem of classical interpolation is to find a continuous or a differentiable function such that the graph of the function contains a given set of data points. In most of the cases, classical interpolants are about constructing a very smooth function passing through the given data. However, in several physical experiments, the data arises from highly irregular curves and surfaces found in nature and may not be generated from smooth functions. To model such a data set, Barnsley [1] introduced the notion of Fractal Interpolation Function (FIF) based on the theory of Iterated Function System (IFS). A FIF is obtained as a fixed point of a suitable map defined on a space of continuous functions. FIFs are used to approximate naturally occurring functions which show some kind of self-similarity on magnification and the fractal dimensions of their graph are nonintegers.

FIFs provide a basis for the constructive approximation theory of nondifferentiable functions. Further, differentiable FIFs [2] constitute an alternative to the traditional nonrecursive interpolation and approximation methods (see, for instance, [3, 5, 12]). In this way, the fractal methodology provides more flexibility and versatility on the choice of an interpolant. Consequently, this function class can be useful for mathematical and engineering problems where the classical spline interpolation approach may not work satisfactorily.

By using suitable IFS, Barnsley and Navascués have provided a method to perturb a continuous function $f \in \mathscr{C}(I)$ so as to yield a class of continuous functions $f^{\alpha} \in \mathscr{C}(I)$, where α is a free parameter, called scale vector. For suitable values of the scale vector α, the fractal functions f^{α} simultaneously interpolate and approximate f. By this method, one can define fractal analogues of any continuous function. This function f^{α} retains some properties such as continuity and integrability of f, but in general does not posses differentiability. However, if the problem is of differentiable type, the parameters can be chosen in a specific way and f^{α} can be made to share the regularity of f. Thus, the parameter α can be used to modify or preserve the properties of f.

The problem of searching a sufficiently smooth function that preserves the qualitative shape property inherent in the data is generally referred to as shape preserving interpolation/approximation, which is important in practical ground. The shape properties are mathematically expressed in terms of conditions such as positivity, monotony, and convexity. As a submissive contribution to this goal, Chand and group have initiated the study on shape preserving fractal interpolation and approximation using various families of polynomial and rational IFSs (see, for instance, [5–7, 20]). These shape preserving fractal interpolation schemes possess the novelty that the interpolants inherit the shape property in question and at the same time the suitable derivatives of these interpolants own irregularity in finite or dense subsets of the interpolation interval. This attribute of shape preserving FIFs finds potential applications in various nonlinear phenomena.

Current article intends to provide a uniform approach to define fractal analogues of various polynomial/rational splines widely used in the field of shape preserving interpolation. Toward this goal, first we recognize a FIF as an α-fractal function corresponding to a traditional spline, where the IFS involves a family of base functions. Next, we propose a theorem that enables one to choose elements of the IFS appropriately so that the corresponding fractal functions f^α retain the order of the continuity and the important geometric features, namely positivity, monotonicity, and convexity of the germ f. Conditions for shape preservation of traditional spline f when coupled with this theorem provide restraints that ensure shape preservation of FIFs. For an illustrative purpose, we discuss fractal analogue of rational spline introduced in [18] and establish its positivity property. To illustrate that the proposed method indeed provides a single platform for generalizing polynomial/rational fractal splines, various shape preserving FIFs studied so far in the literature are reported, however within the current conceptual framework.

2 Basic Facts

It has not proved possible, and would perhaps not be appropriate to provide a detailed exposition of fractal interpolation theory. Neverthless, to ease the acess for the non-expert the essentials of fractal interpolation that form background for our study have been collected in this section. For a detailed study, the reader may consult [1, 2, 15].

2.1 IFS for Fractal Functions

For $r \in \mathbb{N}$, let $\mathbb{N}_r$ denote the subset $\{1, 2, \ldots, r\}$ of $\mathbb{N}$. Let a set of data points $\mathcal{D} = \{(x_i, y_i) \in \mathbb{R}^2 : i \in \mathbb{N}_N\}$ satisfying $x_1 < x_2 < \cdots < x_N$, $N > 2$, be given. Set $I = [x_1, x_N]$, $I_i = [x_i, x_{i+1}]$ for $i \in \mathbb{N}_{N-1}$. Suppose $L_i : I \to I_i$, $i \in \mathbb{N}_{N-1}$ be contraction homeomorphisms such that

$$L_i(x_1) = x_i, \;\; L_i(x_N) = x_{i+1}. \tag{1}$$

Let $0 < r_i < 1, i \in \mathbb{N}_{N-1}$, and $X := I \times \mathbb{R}$. Let $N - 1$ continuous mappings $F_i : X \to \mathbb{R}$ be given satisfying:

$$F_i(x_1, y_1) = y_i, \;\; F_i(x_N, y_N) = y_{i+1}, \;\; |F_i(x, y) - F_i(x, y^*)| \le r_i|y - y^*|, \tag{2}$$

where $(x, y), (x, y^*) \in X$. Define $w_i : X \to I_i \times \mathbb{R} \subseteq X, \;\; w_i(x, y) = \big(L_i(x), F_i(x, y)\big)\; \forall\, i \in \mathbb{N}_{N-1}$. It is known [1] that there exists a metric on $\mathbb{R}^2$, equivalent to the Euclidean metric, with respect to which $w_i, i \in \mathbb{N}_{N-1}$, are contractions. The collection $\mathcal{I} = \{X; w_i : i \in \mathbb{N}_{N-1}\}$ is called an iterated function system (IFS). Associated with the IFS $\mathcal{I}$, there is a set valued Hutchinson map $W : H(X) \to H(X)$ defined by $W(B) = \bigcup_{i=1}^{N-1} w_i(B)$ for $B \in H(X)$, where $H(X)$

is the set of all nonempty compact subsets of X endowed with the Hausdorff metric h_d. The Hausdorff metric h_d completes $H(X)$. Further, W is a contraction map on the complete metric space $(H(X), h_d)$. By the Banach Fixed Point Theorem, there exists a unique set $G \in H(X)$ such that $W(G) = G$. This set G is called the attractor or deterministic fractal corresponding to the IFS $\mathscr{I}$. For any choices of L_i and F_i satisfying the conditions prescribed in (1–2), the following result holds.

Proposition 1 (Barnsley [1]) *The IFS $\{X; w_i : i \in \mathbb{N}_{N-1}\}$ defined above admits a unique attractor G, and G is the graph of a continuous function $g : I \to \mathbb{R}$ which obeys $g(x_i) = y_i$ for $i \in \mathbb{N}_N$.*

Definition 1 The aforementioned function g whose graph is the attractor of an IFS is called a **Fractal Interpolation Function** (FIF) or a **self-referential function** corresponding to the IFS $\{X;\ w_i : i \in \mathbb{N}_{N-1}\}$.

The above fractal interpolation function g is obtained as the fixed point of the Read-Bajraktarević (RB) operator T defined on a complete metric space $(\mathscr{G}, \rho)$:

$$(Th^*)(x) = F_i\left(L_i^{-1}(x), h^* \circ L_i^{-1}(x)\right)\ \forall\, x \in I_i,\ i \in \mathbb{N}_{N-1}.$$

where $\mathscr{G} := \{h^* : I \to \mathbb{R} : h^* \text{ is continuous on } I,\ h^*(x_1) = y_1, h^*(x_N) = y_N\}$ is equipped with the uniform metric. It can be seen that T is a contraction mapping on $(\mathscr{G}, \rho)$ with a contraction factor $r^* := \max\{r_i : i \in \mathbb{N}_{N-1}\} < 1$. The fixed point of T is the FIF g corresponding to the IFS $\mathscr{I}$. Therefore, g satisfies the functional equation:

$$g(x) = F_i\left(L_i^{-1}(x), g \circ L_i^{-1}(x)\right),\ x \in I_i,\ i \in \mathbb{N}_{N-1}, \tag{3}$$

The most popular IFS for fractal interpolants are defined by the maps:

$$L_i(x) = a_i x + b_i,\ F_i(x, y) = \alpha_i y + q_i(x),\ i \in \mathbb{N}_{N-1}. \tag{4}$$

Here $-1 < \alpha_i < 1$ and $q_i : I \to \mathbb{R}$ are suitable continuous functions satisfying (2). The parameter α_i is called a scaling factor of the transformation w_i, and $\alpha = (\alpha_1, \alpha_2, \ldots, \alpha_{N-1})$ is the scale vector corresponding to the IFS. The properties such as approximation order, smoothness, differentiability, stability, senstivity with respect to perturbation in parameters, and fractal dimension of FIFs are well-studied and reported in many places in the literature (see, for instance, [8, 13, 22]).

2.2 *α-fractal Function*

Let $f \in \mathscr{C}(I)$ be a continuous function and consider the case.

$$q_i(x) = f \circ L_i(x) - \alpha_i b(x). \tag{5}$$

Here $b : I \to \mathbb{R}$ is a continuous map that fulfills the conditions $b(x_1) = f(x_1)$, $b(x_N) = f(x_N)$, and $b \neq f$. This case is proposed by Barnsley [1] and Navascués [11] as generalization of any continuous function. Here the interpolation data are $\{(x_i, f(x_i)) : i \in \mathbb{N}_N\}$. We define the α-fractal function corresponding to f in the following:

Definition 2 The continuous function $f^\alpha : I \to \mathbb{R}$ whose graph is the attractor of the IFS defined by (4–5) is referred to as **α-fractal function** associated with f, with respect to "base function"b, scale vector α, and the partition $\mathscr{D}$.

According to (3), f^α satisfies the functional equation:

$$f^\alpha(x) = f(x) + \alpha_i[(f^\alpha - b) \circ L_i^{-1}(x)] \; \forall x \in I_i, \; i \in \mathbb{N}_{N-1}. \tag{6}$$

Recently, it is observed that (see [21]) the α-fractal function f^α obtained by perturbing a given continuous function $f \in \mathscr{C}(I)$ with the help of a finite sequence of base functions $B := \{b_i \in \mathscr{C}(I), \; b_i(x_1) = f(x_1), \; b_i(x_N) = f(x_N), \; b_i \neq f\}$ instead of a single base function is more advantageous. For instance, in generalizing rational splines with different shape parameters in different subintervals determined by interpolation points. That is, consider

$$q_i(x) = f \circ L_i(x) - \alpha_i b_i(x). \tag{7}$$

According to (3) , f^α satisfies the functional equation:

$$f^\alpha(x) = f(x) + \alpha_i[(f^\alpha - b_i) \circ L_i^{-1}(x)] \; \forall x \in I_i, \; i \in \mathbb{N}_{N-1}. \tag{8}$$

Note that for $\alpha = 0$, $f^\alpha = f$. Thus aforementioned equation may be treated as an entire family of functions f^α with f as its germ. By this method, one can define fractal analogues of any continuous function.

2.3 *Differentiable FIFs (Fractal Splines)*

For a prescribed data set, a FIF with $\mathscr{C}^r$-continuity is obtained as the fixed point of IFS (4), where the scaling factors α_i and the functions q_i are chosen according to the following proposition.

Proposition 2 (Barnsley and Harrington [2]) *Let $\{(x_i, y_i) : i \in \mathbb{N}_N\}$ be given interpolation data with strictly increasing abscissae. Let $L_i(x) = a_i x + b_i$, $i \in \mathbb{N}_{N-1}$, satisfy (1) and $F_i(x, y) = \alpha_i y + q_i(x)$, $i \in \mathbb{N}_{N-1}$, satisfy (2). Suppose that for some integer $r \geq 0$, $|\alpha_i| \leq \kappa a_i^r$, $0 < \kappa < 1$, and $q_i \in \mathscr{C}^r(I)$, $i \in \mathbb{N}_{N-1}$. Let*

$$F_{i,k}(x, y) = \frac{\alpha_i y + q_i^{(k)}(x)}{a_i^k}, \quad y_{1,k} = \frac{q_1^{(k)}(x_1)}{a_1^k - \alpha_1}, \quad y_{N,k} = \frac{q_{N-1}^{(k)}(x_N)}{a_{N-1}^k - \alpha_{N-1}}, \quad k = 1, 2, \ldots, r.$$

If $F_{i-1,k}(x_N, y_{N,k}) = F_{i,k}(x_1, y_{1,k})$ *for* $i = 2, 3, \ldots, N-1$ *and* $k = 1, 2, \ldots, r$, *then the IFS* $\{X; (L_i(x), F_i(x, y)) : i \in \mathbb{N}_{N-1}\}$ *determines a FIF* $g \in \mathscr{C}^r[x_1, x_N]$, *and* $g^{(k)}$ *is the FIF determined by the IFS* $\{X; (L_i(x), F_{i,k}(x, y)) : i \in \mathbb{N}_{N-1}\}$ *for* $k = 1, 2, \ldots, r$.

2.4 Smooth α-fractal Functions

In general, it may be difficult to obtain IFS satisfying the hypotheses of Barnsley and Harrington theorem. The equality proposed in the above proposition demands the resolution of systems of equations. Sometimes the system has no solution, mainly whenever some boundary conditions are imposed on the function (see [2]). However, for the special class of IFS define through (4–7) used to construct α-fractal function f^α, the procedure can be easily carried out, which is the content of the following proposition. Details may be consulted in [14, 20, 21].

Proposition 3 *Let* $f \in \mathscr{C}^r(I)$. *Suppose* $\mathscr{D} = \{x_1, x_2, \ldots, x_N\}$ *be an arbitrary partition on* I *satisfying* $x_1 < x_2 < \cdots < x_N$. *Let* $|\alpha_i| < a_i^r$ *for all* $i \in \mathbb{N}_{N-1}$. *Further suppose that* $B = \{b_i \in \mathscr{C}^r(I) : i \in \mathbb{N}_{N-1}\}$ *fulfills* $b_i^{(k)}(x_1) = f^{(k)}(x_1)$, $b_i^{(k)}(x_N) = f^{(k)}(x_N)$ *for* $k = 0, 1, \ldots, r$. *Then the corresponding fractal function* f^α *is* r*-smooth, and* $(f^\alpha)^{(k)}(x_i) = f^{(k)}(x_i)$ *for* $i \in \mathbb{N}_N$ *and* $k = 0, 1, \ldots, r$.

This completes our preparations for the current study, and we are now ready for our main section.

3 α-fractal Rational Cubic Spline Preserving Positivity

In this section, first we shall find the strip condition for the rth derivative of α-fractal function then we construct a FIF as an α-fractal function corresponding to a traditional spline obtained through a family of base functions. We apply the present formalism to study positivity of a special class of rational FIFs.

3.1 Strip Condition for rth Derivative of α-fractal Function

In this subsection, we shall provide conditions on the parameters so as to ensure the rth derivative of a $\mathscr{C}^r$-continuous fractal function lies in a rectangle whenever its classical counterpart f has similar property. Our proof is patterned after [20]. However, we work with a slightly more general obstacle for the rth derivative, where $r \in \mathbb{N} \cup \{0\}$ is arbitrary and we consider a finite sequence of base functions $B = \{b_i \in \mathscr{C}^r(I),\ b_i(x_1) = f(x_1),\ b_i(x_N) = f_i(x_N),\ b_i \neq f\}$, perhaps with some more additional conditions. This is contrast to the analysis in [20] which deals with constraining graph of f^α (constructed with a single base function) and its first two derivatives within a rectangle $I \times [0, M]$. For a brief presentation of the theorem, let us introduce the following notation for a continuous function g defined on a compact interval J :

$$m(g; J) = min\,\{g(x) : x \in J\}, \qquad M(g; J) = max\,\{g(x) : x \in J\}.$$

Theorem 1 *Let $f \in \mathscr{C}^r(I)$ be such that $M_1 \leq f^{(r)}(x) \leq M_2$ for all $x \in I$ and for some suitable constant M_1 and M_2. The α-fractal function f^α (cf. (6)) corresponding to f satisfies $M_1 \leq (f^\alpha)^{(r)}(x) \leq M_2$ for all $x \in I$, provided the finite sequence of base functions satisfying $B = \{b_i \in \mathscr{C}^r(I),\ b_i^{(k)}(x_1) = f^{(k)}(x_1),\ b_i^{(k)}(x_N) = f_i^{(k)}(x_N),\ b_i \neq f, k = 0, 1, \ldots, r\}$ and the scaling factors $|\alpha_i| < a_i^r$ for all $i \in \mathbb{N}_{N-1}$ obeying the following additional conditions are selected.*

$$max\left\{\frac{a_i^r[M_1 - m(f^{(r)}; I_i)]}{M_2 - m(b_i^{(r)}; I)}, \frac{-a_i^r[M_2 - M(f^{(r)}; I_i)]}{M(b_i^{(r)}; I) - M_1}\right\} < \alpha_i$$
$$\leq min\left\{\frac{a_i^r[m(f^{(r)}; I_i) - M_1]}{M(b_i^{(r)}; I) - M_1}, \frac{a_i^r[M_2 - M(f^{(r)}; I_i)]}{M_2 - m(b_i^{(r)}; I)}\right\}$$

Proof With the stated conditions on the scale factors and the function b_i, we can ensure from Proposition 3 that corresponding fractal function f^α is r-smooth. Note that $(f^\alpha)^{(r)}$ is a fractal function corresponding to the IFS $\{X; (L_i(x), F_{i,r}(x, y)) : i \in \mathbb{N}_{N-1}\}$ (see Proposition 3), $(f^\alpha)^{(r)}(x_i) = f^{(r)}(x_i)$ and $(f^\alpha)^{(r)}$ is constructed iteratively using the following functional equation:

$$(f^\alpha)^{(r)}(L_i(x)) = F_{i,r}(x, (f^\alpha)^{(r)}(x)) = f^{(r)}(L_i(x)) + \frac{\alpha_i}{a_i^r}\{(f^\alpha)^{(r)} - b_i^{(r)}\}(x).$$

Therefore to prove $M_1 \leq (f^\alpha)^{(r)}(x) \leq M_2$ for all $x \in I$, by the property of the attractor of the IFS, it is enough to prove that $M_1 \leq y \leq M_2$ implies $M_1 \leq F_{i,r}(x, y) \leq M_2\ \forall i \in \mathbb{N}_{N-1}$.

Firstly, let $0 \leq \alpha_i < a_i^r$. We note that $M_1 \leq y \leq M_2$ implies $\frac{\alpha_i}{a_i^r}M_1 + f^{(r)}(L_i(x)) - \frac{\alpha_i}{a_i^r}b_i^{(r)}(x) \leq \frac{\alpha_i}{a_i^r}y + f^{(r)}(L_i(x)) - \frac{\alpha_i}{a_i^r}b_i^{(r)}(x) \leq \frac{\alpha_i}{a_i^r}M_2 + f^{(r)}(L_i(x)) - \frac{\alpha_i}{a_i^r}b_i^{(r)}(x)$. Therefore our target $M_1 \leq \frac{\alpha_i}{a_i^r}y + f^{(r)}(L_i(x)) - \frac{\alpha_i}{a_i^r}b_i^{(r)}(x) \leq M_2$ is achieved if $M_1(1 - \frac{\alpha_i}{a_i^r}) \leq f^{(r)}(L_i(x)) - \frac{\alpha_i}{a_i^r}b_i^{(r)}(x) \leq M_2(1 - \frac{\alpha_i}{a_i^r})$.

Note that $f^{(r)}(L_i(x)) \geq m(f^{(r)}; I_i)$ and $b_i^{(r)}(x) \leq M(b_i^{(r)}; I)$ is true for all $x \in I$. Therefore, $M_1(1 - \frac{\alpha_i}{a_i^r}) \leq f^{(r)}(L_i(x)) - \frac{\alpha_i}{a_i^r}b_i^{(r)}(x)$ holds if $M_1(1 - \frac{\alpha_i}{a_i^r}) \leq m(f^{(r)}; I_i) - \frac{\alpha_i}{a_i^r}M(b_i^{(r)}; I)$, i.e., if $\alpha_i \leq \dfrac{a_i^r[m(f^{(r)}; I_i) - M_1]}{M(b_i^{(r)}; I) - M_1}$. Similarly,

$f^{(r)}(L_i(x)) \leq M(f^{(r)}; I_i)$ and $b_i^{(r)}(x) \geq m(b_i^{(r)}; I)$ is true for all $x \in I$. Therefore, $f^{(r)}(L_i(x)) - \frac{\alpha_i}{a_i^r} b_i^{(r)}(x) \leq M_2(1 - \frac{\alpha_i}{a_i^r})$ holds if $M(f^{(r)}; I_i) - \frac{\alpha_i}{a_i^r} m(b_i^{(r)}; I) \leq M_2(1 - \frac{\alpha_i}{a_i^r})$, which in turn holds if

$$\alpha_i \leq \frac{a_i^r[M_2 - M(f^{(r)}; I_i)]}{M_2 - m(b_i^{(r)}; I)}.$$

Now assume $-a_i^r < \alpha_i \leq 0$. In this case, $M_1 \leq y \leq M_2$ implies that $\frac{\alpha_i}{a_i^r} M_2 + f^{(r)}(L_i(x)) - \frac{\alpha_i}{a_i^r} b_i^{(r)}(x) \leq \frac{\alpha_i}{a_i^r} y + f^{(r)}(L_i(x)) - \frac{\alpha_i}{a_i^r} b_i^{(r)}(x) \leq \frac{\alpha_i}{a_i^r} M_1 + f^{(r)}(L_i(x)) - \frac{\alpha_i}{a_i^r} b_i^{(r)}(x)$. Consequently, for $M_1 \leq F_{i,r}(x, y) \leq M_2$, it is sufficient to verify $M_1 - \frac{\alpha_i}{a_i^r} M_2 \leq f^{(r)}(L_i(x)) - \frac{\alpha_i}{a_i^r} b_i^{(r)}(x) \leq M_2 - \frac{\alpha_i}{a_i^r} M_1$. By appropriately using the definition of $m(b_i^{(r)}; I)$, $M(b_i^{(r)}; I)$, $m(f^{(r)}; I_i)$, $M(f^{(r)}; I_i)$ on lines similar to the first part, we get

$\alpha_i \geq \dfrac{a_i^r[M_1 - m(f^{(r)}; I_i)]}{M_2 - m(b_i^{(r)}; I)}$ and $\alpha_i \geq \dfrac{-a_i^r[M_2 - M(f^{(r)}; I_i)]}{M(b_i^{(r)}; I) - M_1}$. Combination of the obtained conditions on the scale factors completes the proof. □

Remark 1 A persual of the foregoing theorem reveals that if $f : I = [a, b] \to \mathbb{R}$ is $\mathscr{C}^r$-continuous r-convex function (i.e., $f^{(r)} \geq 0$), then we can select a scale vector α such that $f^\alpha \in \mathscr{C}^r(I)$ and f^α preserves r-convexity of f. Note that for $r = 0, 1$, and 2 r-convexity reduces to positivity, monotonicity, and convexity, respectively.

Remark 2 Let us confine to the positivity case (i.e., $r = 0$ and $M_1 = 0$). Then the condition for $0 \leq f^\alpha \leq M_2$ can be obtained as $|\alpha_i| < 1$ and

$$\max\left\{-\frac{m(f; I_i)}{M_2 - m(b_i; I)}, \frac{M(f; I_i) - M_2}{M(b_i; I)}\right\} \leq \alpha_i \leq \min\left\{\frac{m(f; I_i) - M_1}{M(b_i; I)}, \frac{M_2 - M(f; I_i)}{M_2 - m(b_i; I)}\right\}.$$

If it is enough to consider the nonnegative scale factors, then the following condition on the scale factors ensure the nonnegativity of f^α: $0 \leq \alpha_i \leq \frac{m(f; I_i)}{M(b_i; I)}$, where $|\alpha_i| < 1$ is assumed.

Remark 3 $f \in \mathscr{C}^r(I)$ be such that $f^{(r)}(x) \leq 0$ for all $x \in I$, then we may construct f^α satisfying $(f^\alpha)^{(r)}(x) \leq 0$ for all $x \in I$ by employing Theorem 1. Taking $M_2 = 0$ then the condition for $M_1 \leq f^\alpha \leq 0$ can be obtained as: $|\alpha_i| < 1$ and

$$\max\left\{-\frac{a_i^r[M_1 - m(f^{(r)}; I_i)]}{m(b_i^{(r)}; I)}, \frac{-a_i^r M(f^{(r)}; I_i)}{M_1 - M(b_i^{(r)}; I)}\right\} \leq \alpha_i$$
$$\leq \min\left\{\frac{a_i^r[M_1 - m(f^{(r)}; I_i)]}{M_1 - M(b_i^{(r)}; I)}, \frac{a_i^r[M(f^{(r)}; I_i)]}{m(b_i^{(r)}; I)}\right\}.$$

3.2 Construction of Rational Cubic Spline FIF with Shape Parameters

Consider a set of data points $\mathscr{D} = \{(x_i, y_i, d_i) : i \in \mathbb{N}_N\}$ where $x_1 < x_2 < \cdots < x_N$. Here y_i denote the function value and d_i denote the first derivative at the knot point x_i for each $i \in \mathbb{N}_{N-1}$. For the data set $\mathscr{D}$, a traditional nonrecursive rational cubic spline $f \in \mathscr{C}^1(I)$ is defined in a piecewise manner as follows (see [18] for details). For $\theta := \frac{x-x_1}{x_N-x_1}$ and $h_i = x_{i+1} - x_i$, $x \in I$,

$$f\big(L_i(x)\big) = \frac{A_i(1-\theta)^3 + B_i\theta(1-\theta)^2 + C_i\theta^2(1-\theta) + D_i\theta^3}{u_i(1-\theta)^2 + 2\theta(1-\theta) + v_i\theta^2}, \tag{9}$$

where $A_i = u_i y_i$, $B_i = u_i y_i + h_i u_i d_i$, $C_i = v_i y_{i+1} - h_i v_i d_{i+1}$, $D_i = v_i y_{i+1}$, and $u_i > 0$, $v_i > 0$ are free shape (tension) parameters. The rational interpolant f satisfies the Hermite interpolation conditions, $f(x_i) = y_i$ and $f^{(1)}(x_i) = d_i$, for $i \in \mathbb{N}_N$. To develop the α-fractal rational cubic spline corresponding to f, assume $|\alpha_i| \le a_i$, and select a family $B = \{b_i \in \mathscr{C}^1(I) : i \in \mathbb{N}_N\}$ satisfying the conditions $b_i(x_1) = f(x_1) = y_1$, $b_i(x_N) = f(x_N) = y_N$, $b_i^{(1)}(x_1) = f^{(1)}(x_1) = d_1$, and $b_i^{(1)}(x_N) = f^{(1)}(x_N) = d_N$ (cf. Section 2). There are variety of choices for B. For our convenience, we take b_i to be a rational function of similar form as that of the classical interpolant f. For $x \in I = [x_1, x_N]$ and $\theta := \frac{x-x_1}{x_N-x_1}$, our choice for b_i is

$$b_i(x) = \frac{A_i^*(1-\theta)^3 + B_i^*\theta(1-\theta)^2 + C_i^*\theta^2(1-\theta) + D_i^*\theta^3}{u_i(1-\theta)^2 + 2\theta(1-\theta) + v_i\theta^2}, \; i \in \mathbb{N}_{N-1}, \tag{10}$$

where the coefficients A_i^*, B_i^*, C_i^*, and D_i^* are determined through the conditions $b_i(x_1) = y_1$, $b_i(x_N) = y_N$, $b_i^{(1)}(x_1) = d_1$, $b_i^{(1)}(x_N) = d_N$. After applying these conditions on b_i, we can easily get $A_i^* = u_i y_1$, $B_i^* = u_i y_1 + (x_N - x_1)u_i d_1$, $C_i^* = v_i y_N - (x_N - x_1)v_i d_N$, $D_i^* = v_i y_N$.

Consider the α-fractal rational cubic spline corresponding to f as (see (8))

$$f^\alpha\big(L_i(x)\big) = \alpha_i f^\alpha(x) + f\big(L_i(x)\big) - \alpha_i b_i(x), \; x \in I, \; i \in \mathbb{N}_{N-1}. \tag{11}$$

In view of (9) and (10), we have

$$f^\alpha\big(L_i(x)\big) = \alpha_i f^\alpha(x) + \frac{P_i(x)}{Q_i(x)}, \tag{12}$$

$$\begin{aligned}P_i(x) =& \{y_i - \alpha_i y_1\}u_i(1-\theta)^3 + \big\{(2+u_i)y_i + h_i u_i d_i - \alpha_i[(2+u_i)y_1 + u_i(x_N - x_1)d_1]\big\}\\ &\theta(1-\theta)^2 + \big\{(2+v_i)y_{i+1} - h_i v_i d_{i+1} - \alpha_i[(2+v_i)y_N - v_i(x_N - x_1)d_N]\big\}\\ &\theta^2(1-\theta) + \{y_{i+1} - \alpha_i y_N\}v_i\theta^3,\end{aligned}$$

$$Q_i(x) = u_i(1-\theta)^2 + 2\theta(1-\theta) + v_i\theta^2,\ i \in \mathbb{N}_{N-1},\ \theta = \frac{x-x_1}{x_N-x_1}.$$

Note that the function $f^\alpha : I \to \mathbb{R}$ enjoys the interpolation conditions $f^\alpha(x_i) = y_i$, $(f^\alpha)^{(1)}(x_i) = d_i$.

Remark 4 When $\alpha_i = 0$ for all $i \in \mathbb{N}_{N-1}$ and for $x \in [x_i, x_{i+1}]$, using $\frac{L_i^{-1}(x)-x_1}{x_N-x_1} = \frac{x-x_i}{x_{i+1}-x_i}$, one can see that the above expression coincides with the classical rational cubic interpolant constructed by Sarfraz. et al. [18].

Remark 5 If $u_i = v_i = 1$ for all $i \in \mathbb{N}_{N-1}$, then the α-fractal rational cubic spline reduces to the standard $\mathscr{C}^1$-cubic Hermite FIF. For $\alpha_i = 0$ and $u_i = v_i = 1$ for all $i \in \mathbb{N}_{N-1}$, the α-fractal rational cubic spline recovers the classical cubic Hermite interpolant.

3.3 Positivity Preserving Rational Cubic FIF

Given a set of Hermite data $\mathscr{D} = \{(x_i, y_i, d_i) : i \in \mathbb{N}_N\}$, where $y_i > 0$, we would like to constraint the parameters so that the proposed rational cubic FIF itself is nonnegative.

Method 1: In view of functional equation in (12), the reader will undoubtedly discern with the fact that the positivity of f^α depends on positivity of cubic, once we assume $\alpha_i \geq 0\ \forall\ i \in \mathbb{N}_{N-1}$. Now for positivity of cubic, we shall proceed on lines similar to the traditional rational cubic, given in [18].

$$\begin{aligned} \{y_i - \alpha_i y_1\}u_i > 0,\ \{y_{i+1} - \alpha_i y_N\}v_i > 0, \\ (2+u_i)y_i + h_i u_i d_i - \alpha_i[(2+u_i)y_1 + u_i(x_N-x_1)d_1] > 0, \\ (2+v_i)y_{i+1} - h_i v_i d_{i+1} - \alpha_i[(2+v_i)y_N - v_i(x_N-x_1)d_N] > 0. \end{aligned} \tag{13}$$

The first two inequalities are satisfied if $\alpha_i < \frac{y_i}{y_1}$ and $\alpha_i < \frac{y_{i+1}}{y_N}$, respectively. Consequently, for f^α to be positive, we take the scaling factors according to $0 \leq \alpha_i < \min\{\frac{y_i}{y_1}, \frac{y_{i+1}}{y_N}\}$. The inequalities third and fourth in (13) can be rewritten as

$$u_i\{(y_i - \alpha_i y_1) + h_i d_i - \alpha_i(x_N - x_1)d_1\} > -2(y_i - \alpha_i y_1). \tag{14}$$

$$v_i\{(y_{i+1} - \alpha_i y_N) + h_i d_{i+1} + \alpha_i(x_N - x_1)d_N\} > -2(y_{i+1} - \alpha_i y_N). \tag{15}$$

First, we select α_i according to $0 \leq \alpha_i < \min\{\frac{y_i}{y_1}, \frac{y_{i+1}}{y_N}, a_i\}$, $i \in \mathbb{N}_{N-1}$. Consequently, the right-hand side expressions in the inequalities (14–15) are negative. Having selected α_i, the parameter u_i and v_i can be selected as

$0 < u_i < \frac{-2(y_i-\alpha_i y_1)}{(y_i-\alpha_i y_1)+h_i d_i-\alpha_i(x_N-x_1)d_1}$, $\quad 0 < v_i < \frac{-2(y_{i+1}-\alpha_i y_N)}{(y_{i+1}-\alpha_i y_N)+h_i d_{i+1}+\alpha_i(x_N-x_1)d_N}$.

Remark 6 When $\alpha_i = 0$ for all $i \in \mathbb{N}_{N-1}$, the conditions on IFS parameters obtained in the analysis reduce to $u_i\{y_i + h_i d_i\} > -2y_i$ and $v_i\{y_{i+1} + h_i d_{i+1}\} > -2y_{i+1}$. Thus the positivity preserving conditions developed here corrects and generalizes the positivity preserving conditions of classical rational spline studied by Sarfraz. et al. [18].

Method 2: In this method, we shall explore the idea of viewing the rational cubic FIF as a α-fractal function corresponding to rational cubic spline to obtain desired positivity of f^α, using Theorem 1. That is, we identify suitable elements of the IFS so that the fractal function f^α is positive whenever f is positive.
Step 1: Given a data set $\mathscr{D} = \{(x_i, y_i) : i \in \mathbb{N}_N\}$ wherein $y_i \geq 0$, construct the positive cubic spline f by selecting u_i and v_i according to Remark 6.
Step 2 Choose b_i as in (10) using u_i and v_i values in Step 1.
Step 3: For the f, b_i obtained in Step 1 and Step 2, respectively, compute the constants $m(b; I) = \min_{x\in I} b(x)$, $M(b; I) = \max_{x\in I} b(x)$, $m(f; I_i) = \min_{x\in I_i} f(x)$, $M(f; I_i) = \max_{x\in I_i} f(x)$. Choose $|\alpha_i| < a_i$ and

$$max\left\{-\frac{m(f; I_i)}{M_2 - m(b_i; I)}, \frac{M(f; I_i) - M_2}{M(b_i; I)}\right\} \leq \alpha_i \leq min\left\{\frac{m(f; I_i) - M_1]}{M(b_i; I)}, \frac{M_2 - M(f; I_i)}{M_2 - m(b_i; I)}\right\}.$$

for positivity as Remark 2.
Step 4: Input the scaling parameters as prescribed by Step 3 in the functional equation represented by (12) whereupon the points of the graph of f^α are computed.

Remark 7 The reader is bound to have noticed that in proving Theorem 1, to avoid dependence of condition on point x, we have worked with minimum and maximum of f, b. In some sense this provides a "worst case scenario". As a consequence, we conjecture that in most of the numerical examples, Method 1 may provide a wider range for the scaling factors in comparison with Method 2, if we are intrested in nonnegative scaling. However, Method 2 has advantage that it allows negative values of scaling for preserving positivity (see Sect. 4).

Remark 8 One more advantage of viewing the cubic rational FIF as α-fractal function corresponding to the traditional rational spline rests in the study of its convergence analysis. Let us skim through this in the following. Note that from (8), with a little algebra one can show that $\|f^\alpha - f\|_\infty \leq \frac{|\alpha|_\infty}{1 - |\alpha|_\infty} \max_{i\in\mathbb{N}_{N-1}} \|f - b_i\|_\infty$, where $|\alpha|_\infty = max\{|\alpha_i| : i \in \mathbb{N}_{N-1}\}$. Thus if ϕ is an original function corresponding to data $\mathscr{D}$, using the triangle inequality $\|\phi - f^\alpha\|_\infty \leq \|\phi - f\|_\infty + \|f^\alpha - f\|_\infty$ and convergence analysis of traditional spline f, one may easily obtain convergence analysis of f^α. If f has order of convergence $O(h^m)$ as $h \to 0$, then f^α will also have the same order of convergence provided, $|\alpha_i| < a_i^m = (\frac{h_i}{x_N - x_1})^m \; \forall\, i \in \mathbb{N}_{N-1}$.

3.4 Examples

Let a data set $\mathscr{D} = \{(x_i, y_i, d_i) : i \in \mathbb{N}_N\}$ with $x_1 < x_2 < \cdots < x_N$ be given. Here y_i and d_i are the function value and the first derivative value at the knot x_i, respectively. Though in Sect. 3.2 we have considered a special type of cubic rational FIF, the present approach can be applied to obtain fractal version of various polynomial/rational splines available in the literature. We illustrate this claim with certain examples. We would like to remark that in all these examples polynomials appearing in denominators are preassigned and free parameters occuring in the denominators are assumed to be positive.

Example 1: Consider a traditional cubic Hermite interpolant f defined in a piecewise manner by

$$f_1\big(L_i(x)\big) = A_{1i}(1-\theta)^3 + A_{2i}\theta(1-\theta)^2 + A_{3i}(1-\theta)\theta^2 + A_{4i}\theta^3,$$

where $A_{1i} = y_i,\ A_{2i} = 3y_i + h_i d_i,\ A_{3i} = 3y_{i+1} - h_i d_{i+1},\ A_{4i} = y_{i+1}$. Consider the family of base functions $\{b_i^1 : i \in \mathbb{N}_{N-1}\}$ such that $b_i^1 = b \in \mathscr{C}^1(I, \mathbb{R})\ \forall\, i$ and $b(x_m) = f(x_m) = y_m$ and $b^{(1)}(x_m) = f^{(1)}(x_m) = d_m$ for $m = 1, N$. For instance, b may be the two-point Hermite interpolant $b_i^1(x) = b(x) = A_{1i}^*(1-\theta)^3 + A_{2i}^*\theta(1-\theta)^2 + A_{3i}^*(1-\theta)\theta^2 + A_{4i}^*\theta^3$, where $A_{1i}^* = y_1,\ A_{2i}^* = 3y_1 + (x_N - x_1)d_1,\ A_{3i}^* = 3y_N - (x_N - x_1)d_N,\ A_{4i}^* = y_N$. Then the α-fractal function corresponding to f provides $\mathscr{C}^1$-cubic Hermite FIF

$$f_1^\alpha(L_i(x)) = f_1(L_i(x)) + \alpha_i(f_1^\alpha - b_i^1)(x)\ \forall\, x \in I,\ i \in \mathbb{N}_{N-1}.$$

This cubic FIF is studied by adopting a constuctive approach in [5].

Example 2: Consider a piecewise defined nonrecursive $\mathscr{C}^1$-rational cubic spline discussed in [17] with two shape parameter v_i, w_i defined as follows

$$f_2\big(L_i(x)\big) = \frac{C_{1i}(1-\theta)^3 + C_{2i}\theta(1-\theta)^2 + C_{3i}\theta^2(1-\theta) + C_{4i}\theta^3}{(1-\theta)^3 + v_i\theta(1-\theta)^2 + w_i\theta^2(1-\theta) + \theta^3},$$

where $C_{1i} = y_i,\ C_{2i} = v_i y_i + h_i d_i,\ C_{3i} = w_i y_{i+1} + h_i d_{i+1},\ C_{4i} = y_{i+1}$. We choose the family $B = \{b_i^2 \in \mathscr{C}^1(I) : i \in \mathbb{N}_{N-1}\}$ of base functions, where b_i^2 are rational functions with form similar to that of the classical interpolant. Our specific choice for b_i^2 is

$$b_i^2(x) = \frac{C_{1i}^*(1-\theta)^3 + C_{2i}^*\theta(1-\theta)^2 + C_{3i}^*\theta^2(1-\theta) + C_{4i}^*\theta^3}{(1-\theta)^3 + v_i\theta(1-\theta)^2 + w_i\theta^2(1-\theta) + \theta^3},$$

where $C_{1i}^* = y_1,\ C_{2i}^* = v_i y_1 + (x_N - x_1)d_1,\ C_{3i}^* = w_i y_N + (x_N - x_1)d_N,\ C_{4i}^* = y_N$. Then the corresponding α-fractal function corresponding to f is obtained

$$f_2^\alpha(L_i(x)) = f_2(L_i(x)) + \alpha_i(f_2^\alpha - b_i^2)(x)\ \forall\, x \in I,\ i \in \mathbb{N}_{N-1}.$$

This rational FIF is studied by adopting constructive approach in Chand et al. [7].
Example 3: Consider a piecewise defined nonrecursive $\mathscr{C}^1$-rational cubic spline discussed in [19] with four shape parameter $\lambda_i, \beta_i, \gamma_i, \delta_i$ defined as follows

$$f_3\big(L_i(x)\big) = \frac{D_{1i}(1-\theta)^3 + D_{2i}\theta(1-\theta)^2 + D_{3i}\theta^2(1-\theta) + D_{4i}\theta^3}{\lambda_i(1-\theta)^2 + \beta_i(1-\theta)^2\theta + \gamma_i(1-\theta)\theta^2 + \delta_i\theta^2},$$

where $D_{1i} = \lambda_i y_i$, $D_{2i} = (\lambda_i + \beta_i)y_i + \lambda_i h_i d_i$, $D_{3i} = (\gamma_i + \delta_i)y_{i+1} - \delta_i h_i d_{i+1}$, $D_{4i} = \delta_i y_{i+1}$. We choose the family $B = \{b_i^3 \in \mathscr{C}^1(I) : i \in \mathbb{N}_{N-1}\}$ of base functions with form similar to that of the classical interpolant f as

$$b_i^3(x) = \frac{D_{1i}^*(1-\theta)^3 + D_{2i}^*\theta(1-\theta)^2 + D_{3i}^*\theta^2(1-\theta) + D_{4i}^*\theta^3}{\lambda_i(1-\theta)^2 + \beta_i(1-\theta)^2\theta + \gamma_i(1-\theta)\theta^2 + \delta_i\theta^2},$$

where $D_{1i}^* = \lambda_i y_1$, $D_{2i}^* = (\lambda_i+\beta_i)y_1 + \lambda_i(x_N - x_1)d_1$, $D_{3i}^* = (\gamma_i+\delta_i)y_N - \delta_i(x_N - x_1)d_N$, $D_{4i}^* = \delta_i y_N$. Then the α-fractal function corresponding to f provides $\mathscr{C}^1$-cubic rational FIF

$$f_3^\alpha(L_i(x)) = f_3(L_i(x)) + \alpha_i(f_3^\alpha - b_i^3)(x)\ \forall\, x \in I,\ i \in \mathbb{N}_{N-1}.$$

The aforementioned FIF is studied constructively by Chand et al. [6].
Example 4: Consider a piecewise defined nonrecursive $\mathscr{C}^2$ rational quintic spline f discussed in [10] with two shape parameter β_i, γ_i defined as follows

$$f_4\big(L_i(x)\big) = \frac{\sum\limits_{j=1}^{5} E_{ji}\theta^j(1-\theta)^{5-j}}{\beta_i(1-\theta) + \gamma_i\theta},$$

where $E_{0i} = \beta_i y_i$, $E_{1i} = (4\beta_i + \gamma_i)y_i + \beta_i h_i d_i$, $E_{2i} = (6\beta_i + 4\gamma_i)y_i + (3\beta_i + \gamma_i)h_i d_i + \beta_i \frac{h_i^2}{2} D_i$, $E_{3i} = (4\beta_i + 6\gamma_i)y_{i+1} - (\beta_i + 3\gamma_i)h_i d_{i+1} + \gamma_i \frac{h_i^2}{2} D_{i+1}$, $E_{4i} = (\beta_i + 4\gamma_i)y_{i+1} - \gamma_i h_i d_{i+1}$, $E_{5i} = \gamma_i y_{i+1}$. We choose the family $B = \{b_i^4 \in \mathscr{C}^2(I) : i \in \mathbb{N}_{N-1}\}$ of base functions with form similar to that of the classical interpolant f, i.e.,

$$b_i^4(x) = \frac{\sum\limits_{j=1}^{5} E_{ji}^*\theta^j(1-\theta)^{5-j}}{\beta_i(1-\theta) + \gamma_i\theta},$$

where $E_{0i}^* = \beta_i y_1$, $E_{1i}^* = (4\beta_i + \gamma_i)y_1 + \beta_i(x_N - x_1)d_1$, $E_{2i}^* = (6\beta_i + 4\gamma_i)y_1 + (3\beta_i + \gamma_i)(x_N - x_1)d_1 + \beta_i \frac{(x_N - x_1)^2}{2} D_1$, $E_{3i}^* = (4\beta_i + 6\gamma_i)y_N - (\beta_i + 3\gamma_i)(x_N - x_1)d_N + \gamma_i \frac{(x_N - x_1)^2}{2} D_N$, $E_{4i}^* = (\beta_i + 4\gamma_i)y_N - \gamma_i(x_N - x_1)d_N$, $E_{5i}^* = \gamma_i y_N$. Then the α-fractal function corresponding to f provides the $\mathscr{C}^2$ quintic rational FIF

$$f_4^{\alpha}(L_i(x)) = f_4(L_i(x)) + \alpha_i(f_4^{\alpha} - b_i^4)(x)\ \forall x \in I,\ i \in \mathbb{N}_{N-1}.$$

Remark 9 Our predilection to the particular choice of examples is attributable merely to the convenience. To be a little more precise, our specific choice enables to combine $f(L_i(x))$ and $b_i(x)$ and provides a simple expression for f^{α}. Since shape preserving aspects of traditional splines reported in these examples are studied already in the literature, we can obtain shape preserving aspects of the corresponding fractal generalization using Theorem 1.

Remark 10 Apart from providing a common platform for shape preserving fractal interpolation, the depth of Theorem 1 can be emphasized also by utilizing it for developing fractal analogue of some shape preserving approximation results (as in [20]). However, we shall postpone this to a future work and confine the current article to interpolation.

4 Numerical Illustration

In this section, we will illustrate Method 1 and Method 2 (see Sect. 3.3) that yield positivity preserving rational cubic spline FIFs with shape parameters with some simple examples. To this end, let us take a set of positive Hermite data $\mathscr{D} = \{(0, 0.1, 2.4), (0.4, 1, 2.7), (0.75, 2, 8.2), (1, 5, 15.8)\}$. Assuming the values of scaling factors $\alpha_i = 0.5$ and shape parameters $u_i = v_i = 0.1$ for all $i \in \{1, 2, 3\}$, the fractal curve in Fig. 1a attains negative values at certain points, which may be undesirable for some practical applications.

For constructing a positivity preserving rational cubic spline FIF, we take scaling factors that satisfy the conditions given in Method 1, our choice being $\alpha_1 = 0.18$, $\alpha_2 = 0.30$ and $\alpha_3 = 0.20$. With this choice of scaling factors, the expressions

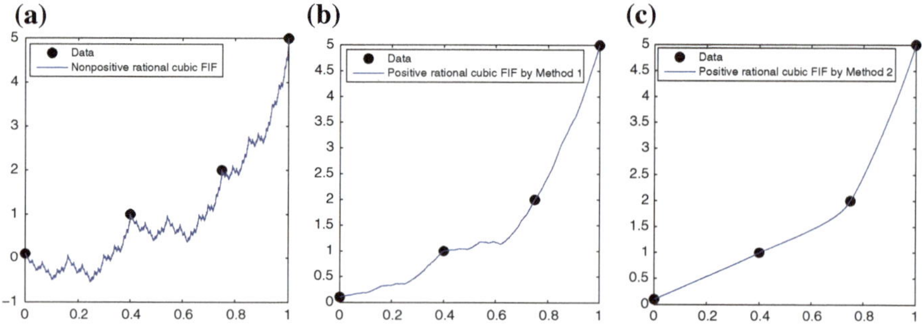

Fig. 1 Rational cubic spline FIF with shape parameters (the interpolating data points are given by the circles and the relevant rational cubic spline FIF by the *solid lines*). **a** Nonpositive rational cubic FIF, **b** Positive rational cubic spline FIF by Method 1, **c** Positive rational cubic spline FIF by Method 2

occurring in braces of the left-hand side of inequalities (14) and (15) turn out to be positive for all $i \in \mathbb{N}_{N-1}$. Consequently, $u_i > 0$ and $v_i > 0$ are free to be chosen. By taking $u_1 = 0.1, u_2 = 0.2, u_3 = 0.5$, and $v_1 = 0.1, v_2 = 0.3, v_3 = 0.7$, we generate the positive fractal spline displayed in Fig. 1b. To illustrate Method 2, we first construct the positive cubic spline f by selecting u_i and v_i according to Remark 6 and choose b_i with the same u_i, v_i. After computing the maximum and minimum value of f and b_i, we select scaling factors according to Remark 2 as mentioned in Method 2 in Step 3. In particular, we take u_i, v_i as in Fig. 1b and $\alpha_1 = 0.05, \alpha_2 = -0.001, \alpha_3 = -0.0007$. As mentioned earlier (see Remark 7), Method 2 has advantage that it also allows negative values of scaling for preserving positivity. Input the derivative values and parameters values in the functional equation represented by (12) whereupon the points of the graph of f^{α} are computed in Fig. 1c.

5 Concluding Remarks and Possible Extensions

In this article, we have constructed a unified approach for the fractal generalization of various traditional nonrecursive polynomial and rational spline. Even though the results obtained in this article are general enough, we applied it on a rational cubic spline with two shape parameters. Some more examples are reported to bring the advantage of the current study.

To keep the size of this article within reasonable limits, we have not been able to include monotonicity/convexity for the cubic rational FIF developed in Sect. 3. However, let us note that one can develop the method with the idea inherent in positivity preserving method. For instance, construct monotone cubic Hermite f using appropriate algorithm [9] then select α according to the Theorem 1 so that f^{α} preserves shape inherent in f. These details, numerical illustrations, and convergence analysis, etc., will appear elsewhere. Extension of the proposed rational fractal spline to shape preserving bivariate interpolation is also under consideration.

References

1. Barnsley, M.F.: Fractal functions and interpolation. Constr. Approx. **2**(4), 303–329 (1986)
2. Barnsley, M.F., Harrington, A.N.: The calculus of fractal functions. J. Approx. Theory **57**(1), 14–34 (1989)
3. Chand, A.K.B., Kapoor, G.P.: Generalized cubic spline fractal interpolation functions. SIAM J. Numer. Anal. **44**(2), 655–676 (2006)
4. Chand, A.K.B., Navascués, M.A.: Generalized Hermite fractal interpolation. Rev. R. Acad. de ciencias. Zaragoza **64**(2), 107–120 (2009)
5. Chand, A.K.B., Viswanathan, P.: A constructive approach to cubic Hermite fractal interpolation function and its constrained aspects. BIT Numer. Math. **53**(4), 841–865 (2013)
6. Chand, A. K. B., Katiyar, S. K., Saravaana Kumar, G.: A new class of rational fractal function for curve fitting. In: Proceeding of Computer Aided Engineering CAE 2013, pp. 78–93. ISBN No-80689-17-3

7. Chand, A.K.B., Vijender, N., Navascués, M.A.: Shape preservation of scientific data through rational fractal splines. Calcolo. **51**, 329–362 (2013)
8. Dalla, L., Drakopoulos, V., Prodromou, M.: On the box dimension for a class of nonaffine fractal interpolation functions. Anal. Theory Appl. **19**(3), 220–233 (2003)
9. Fritsch, F.N., Carlson, R.E.: Monotone piecewise cubic interpolations. SIAM J. Numer. Ana. **17**(2), 238–246 (1980)
10. Hussain, M., Hussain, M.Z., Crips, J.: Robert: $\mathscr{C}^2$ Rational qunitic function. J. Prime Res. Math. **5**, 115–123 (2009)
11. Navascués, M.A.: Fractal polynomial interpolation. Z. Anal. Anwend. **25**(2), 401–418 (2005)
12. Navascués, M.A.: Fractal approximation. Complex Anal. Oper. Theory **4**(4), 953–974 (2010)
13. Navascués, M.A., Sebastián, M.V.: Some results of convergence of cubic spline fractal interpolation functions. Fractals **11**(1), 1–7 (2003)
14. Navascués, M. A., Sebastián, M. V.: Smooth fractal interpolation, J. Inequal. Appl. Article ID 78734, p. 20 (2004)
15. Navascués, M.A., Chand, A.K.B., Viswanathan, P., Sebastián, M.V.: Fractal interpolation functions: a short survey. Appl. Math. **5**, 1834–1841 (2014)
16. Schimdt, J.W., Heß, W.: Positivity of cubic polynomial on intervals and positive spline interpolation. BIT Numer. Anal. **28**, 340–352 (1988)
17. Sarfraz, M., Hussain, M.Z.: Data visualization using rational spline interpolation. J. Comp. Appl. Math. **189**, 513–525 (2006)
18. Sarfraz, M., Hussain, M.Z., Nisar, A.: Positive data modeling using spline function. Appl. Math. Comp. **216**, 2036–2049 (2010)
19. Sarfraz, M., Hussain, M. Z., Hussain, M.: Shape-preserving curve interpolation, J. Comp. Math. **89**, 35–53 (2012)
20. Viswanathan, P., Chand, A.K.B., Navascués, M.A.: Fractal perturbation preserving fundamental shapes: bounds on the scale factors. J. Math. Anal. Appl. **419**, 804–817 (2014)
21. Viswanathan, P.: A Study on univariate shape preserving fractal interpolation and approximation. Ph. D. thesis, Indian Institute of Technology Madras, India, (2014)
22. Wang, H.Y., Yu, J.S.: Fractal interpolation functions with variable parameters and their analytical properties. J. Approx. Theory. **175**, 1–18 (2013)

Unistochastic Matrices and Related Problems

Aaron Carl Smith

Abstract A natural map sends unitary matrices to a subset of bistochastic matrices. We refer to matrices in the image of the map as being unistochastic. The map will be defined, and properties will be discussed. A necessary condition for a ray pattern to be a unitary matrix's ray pattern with be given. Observations regarding eigenvalues of unistochastic matrices and their relationship with paths of unitary matrices will also be presented.

Keywords Unistochastic · Bistochastic · Ray pattern · Circulant

1 Introduction

An $n \times n$ matrix is stochastic if all of its entries are nonnegative and all of its rows sum to one. A stochastic matrix is bistochastic if all columns sum to one also. The Euclidean norm of each row (column) from a unitary matrix is one; it follows that if the modulus of every entry of a unitary (orthonormal) matrix is squared, the resulting matrix will be bistochastic. Bistochastic matrices are also referred to as doubly stochastic matrices. The term bistochastic will be used here to avoid confusion with multistaged models [26].

Let $\mathcal{B}_n$ denote the set of $n \times n$ bistochastic matrices. By the Birkhoff-von Neumann theorem, $\mathcal{B}_n$ is the convex hull formed by the $n \times n$ permutation matrices [16]. This convex hull is referred to as Birkhoff's polytope [2]. The permutation matrices are the extreme points of Birkhoff's polytope since each one cannot be represented as an average of two distinct points in the set [21]. The volume of Birkhoff's polytope is well understood for smaller $2 \leq n \leq 10$ [1, 3, 5, 10]; Cappellini et al. [4] approximates the volume of the polytope with

A.C. Smith (✉)
Department of Mathematics, University of Central Florida, 4393 Andromeda Loop N, Orlando, FL 32816-1364, USA
e-mail: aaron.smith@ucf.edu

R.N. Mohapatra et al. (eds.), *Mathematics and Computing*,
Springer Proceedings in Mathematics & Statistics 139,
DOI 10.1007/978-81-322-2452-5_16

$$n^{n-n^2}(2\pi)^{1/2-n}\mathrm{e}^{n^2+C+\mathcal{O}(1/n)} \tag{1}$$

where C is a constant.

Let Υ be the map that sends unitary matrices, $\mathcal{U}$, to the set of bistochastic matrices, $\mathcal{B}$, by squaring the moduli of unitary entries,

$$\Upsilon : \mathcal{U} \to \mathcal{B}, \tag{2}$$

$$b_{jk} = |u_{jk}|^2. \tag{3}$$

Matrices in the image of Υ are called unistochastic. Unistochastic matrices that are the image of an orthonormal matrix are called orthostochastic [2]. Some authors use the term orthostochastic for matrices that are in the image of unitary matrices [6].

For $n = 2$, all bistochastic matrices are unistochastic. All 2×2 bistochastic matrices are of the form

$$\begin{bmatrix} p & 1-p \\ 1-p & p \end{bmatrix}. \tag{4}$$

For any arguments θ_{11}, θ_{21}, and θ_{22}

$$\begin{bmatrix} \sqrt{p}\mathrm{e}^{i\theta_{11}} & \sqrt{1-p}\mathrm{e}^{i\theta_{12}} \\ \sqrt{1-p}\mathrm{e}^{i(\theta_{11}-\theta_{21}+\theta_{22}+\pi)} & \sqrt{p}\mathrm{e}^{i\theta_{22}} \end{bmatrix} \tag{5}$$

is in the preimage. The set of 3×3 unistochastic matrices is a proper subset of $\mathcal{B}_3$. The matrices

$$\frac{1}{2}\begin{bmatrix} 1 & 0 & 1 \\ 1 & 1 & 0 \\ 0 & 1 & 1 \end{bmatrix}, \frac{1}{6}\begin{bmatrix} 3 & 3 & 0 \\ 1 & 2 & 3 \\ 2 & 1 & 3 \end{bmatrix} \tag{6}$$

are bistochastic, but not unistochastic [15, 18]. To see this, use Proposition 2 and say that the diagonals are positive, then attempt to find arguments for the off-diagonal entries. Block matrices with one of the previous 3×3 matrices show that unistochastic matrices are proper subsets of bistochastic matrices when the matrix's size is greater than 2.

For a bistochastic matrix, B, to be unistochastic it must satisfy these row and column inequalities [23, 27]:

$$\max_{m=1:n} \sqrt{B_{mj}B_{ml}} \leq \frac{1}{2}\sum_{k=1}^{n} \sqrt{B_{kj}B_{kl}} \tag{7}$$

$$\max_{m=1:n} \sqrt{B_{jm}B_{lm}} \leq \frac{1}{2}\sum_{k=1}^{n} \sqrt{B_{jk}B_{lk}} \tag{8}$$

For 3×3 bistochatic matrices, these conditions can be improved to triangle inequalities for B to be unistochastic, and equalities for B to be orthostochastic [27].

If $\left(r_{jk}e^{i\theta_{jk}}\right)$ is a unitary matrix, then

$$\sum_{k=1}^{n} r_{jk}r_{lk}e^{i(\theta_{jk}-\theta_{lk})} = \delta_{jl}. \tag{9}$$

It follows that if $j \neq l$, then the cumulative sum of these products equals zero. The previous inequalities give a method to evaluate if moduli could be the moduli from a unitary matrix. The off-diagonal sums of zero lead to a method to determine if a ray pattern matrix is a ray pattern of a unitary matrix.

Theorem 1 *Suppose that* $U = (r_{jk}e^{i\theta_{jk}})$ *is a unitary matrix, let* σ_{jl} *be a permutation such that*

$$\mathrm{Arg}\left(e^{i\left(\theta_{j\sigma_{jl}^{-1}(k)}-\theta_{l\sigma_{jl}^{-1}(k)}\right)}\right) \leq \mathrm{Arg}\left(e^{i\left(\theta_{j\sigma_{jl}^{-1}(k+1)}-\theta_{l\sigma_{jl}^{-1}(k+1)}\right)}\right) \tag{10}$$

where

$$\mathrm{Arg}\left(e^{i\left(\theta_{j\sigma_{jl}^{-1}(k)}-\theta_{l\sigma_{jl}^{-1}(k)}\right)}\right) \in (-\pi, \pi], \tag{11}$$

then the points

$$\sum_{\substack{k=1 \\ r_{j\sigma_{jl}^{-1}(k)}r_{l\sigma_{jl}^{-1}(k)}\neq 0}}^{m} r_{j\sigma_{jl}^{-1}(k)}r_{l\sigma_{jl}^{-1}(k)}e^{i\left(\theta_{j\sigma_{jl}^{-1}(k)}-\theta_{l\sigma_{jl}^{-1}(k)}\right)}, \tag{12}$$

$$m = 1, 2, \ldots, n \tag{13}$$

form a convex polygon.

Proof Since U is unitary, the total sum for each jl is zero. Thus the partial sums form a polygon. By the selection of σ_{jl}, each interior angle is less than π, it follows that the polygon is convex. □

A corollary of this result is a test to evaluate if there are no unitary matrices in a ray pattern class [19, 20].

Corollary 1 *If A is an $n \times n$ ray pattern matrix, σ_{jl} are permutations that sorts $a_{jk}\overline{a_{jl}}$ by arguments as above with zero moduli entries having zero argument, and the sum*

$$\sum_{k=1}^{n} \cos^{-1} \cos(\theta_{j\sigma_{jl}^{-1}(k)} - \theta_{j\sigma_{jl}^{-1}(k+1)} - \theta_{l\sigma_{jl}^{-1}(k)} + \theta_{l\sigma_{jl}^{-1}(k+1)} + \pi) \tag{14}$$

is not a multiple of π for some $j \neq l$, then A is not a unitary matrix's ray pattern.

For a $n \times n$ unitary matrix, if the paths formed by zero and

$$\sum_{\substack{k=1 \\ r_{j\sigma_{jl}^{-1}(k)} r_{l\sigma_{jl}^{-1}(k)} \neq 0}}^{m} r_{j\sigma_{jl}^{-1}(k)} r_{l\sigma_{jl}^{-1}(k)} \mathrm{e}^{i\left(\theta_{j\sigma_{jl}^{-1}(k)} - \theta_{l\sigma_{jl}^{-1}(k)}\right)}, \tag{15}$$

$$m = 1, 2, \ldots, n \tag{16}$$

are plotted on a $n \times n$ multiplot, the diagonal plots will be the unit interval and the off-diagonal plots will form convex polygons. Here, colinear points are included as polygons; orthonormal matrices will have colinear paths. The jl-plot is a reflection of the lj-plot.

A trend in the literature is that bistochastic and unistochastic matrices are well understood for $n = 3$ and $n = 4$, and the volume of literature for larger matrices decreases as n increases [2, 7, 11, 13, 27].

2 Properties of the Map

Since Υ acts on the entries of a matrix, maps defined by permutating rows and columns of a matrix commute with Υ.

Proposition 1 *If U is an $n \times n$ unitary matrix, P_1 and P_2 are $n \times n$ permutation matrices, then*

$$\Upsilon(P_1 U P_2) = P_1 \Upsilon(U) P_2. \tag{17}$$

Proof The permutation matrices act on the position of entries and do not change their values; Υ acts on the value of matrix entries, and do not change their position. □

Proposition 2 *If D_α and D_β are diagonal unitary matrices that are equal size as a unitary matrix U, then* [13]

$$\Upsilon(D_\alpha U D_\beta) = \Upsilon(U). \tag{18}$$

Proof Diagonal unitary matrices act on the arguments of a matrix's entries, they do not change the moduli of entries, nor do they change the location of entries. □

Definition 1 Two unitary matrices U_1 and U_2 are equivalent if there exists diagonal unitary matrices D_1 and D_2 such that [11]

$$U_1 = D_1 U_2 D_2. \tag{19}$$

Since multiplying by diagonal unitary matrices does not change the moduli in each position, Υ sends equivalent matrices to the same unistochastic matrix. A dephase representation of a unitary matrix is an equivalent matrix with nonnegative real entries in the first row and first column [8]. Fourier matrices are dephased matrices.

Definition 2 An *isolated matrix*, U, is a unitary matrix such that if $\Upsilon(U) = \Upsilon(V)$, then U and V are equivalent [25].

Definition 3 Two unitary matrices U_1 and U_2 are *Haagerup-equivalent* if there exists diagonal unitary matrices D_1 and D_2, and permutation matrices P_1 and P_2 such that [11, 14]

$$U_1 = P_1 D_1 U_2 D_2 P_2. \tag{20}$$

Haagerup equivalence is useful when using tangent maps to study unistochastic matrices [11, 25].

The map Υ sends Haagerup-equivalent matrices to unistochastic matrices with the same Shannon entropy. This is the entropy of a Markov shift (one-side or two-sided) whose measure is defined by the constant probability vector and B [17, 27].

Furthermore, Haagerup-equivalent matrices have the same squared Jarlskog invariant since multiplication by diagonal unitary and permutations do not change the area of unitarity triangles [11].

3 Proof that a Unitary Group is Connected

This section reviews the elementary proof that a unitary group, $\mathscr{U}_n$, is connected. This is done to establish notation. This proof can be found in many introductory topology texts.

The set of $n \times n$ unitary matrices is closed under multiplication and forms a group. Since Υ is continuous, and $\mathscr{U}_n$ is compact and connected under the relative topology, $\mathscr{B}_n$ is compact and connected. For all n, the set of $n \times n$ orthogonal matrices is not connected under the relative topology since there are no continuous paths from a matrix with determinant -1 to the identity matrix [22].

Proposition 3 *The set of $n \times n$ unitary matrices, $\mathscr{U}_n$ is a connected set under the relative topology.*

Proof Let U be an $n \times n$ unitary matrix. By definition

$$UU^* = I, \text{ and} \tag{21}$$
$$U^*U = I. \tag{22}$$

Thus every unitary matrix is normal and unitarily diagonalizable. Say that

$$U = SDS^* \tag{23}$$

where S is also a unitary matrix. Unitary matrices form a group and

$$S^*US = D. \tag{24}$$

Therefore D is a diagonal unitary matrix. It follows that

$$D = \text{diag}(e^{i\theta_1}, \ldots, e^{i\theta_n}) \text{ for some } \theta_1, \theta_2, \ldots, \theta_n \in \mathbb{R}. \tag{25}$$

Let f_U be the function that sends real values to unitary matrices defined by

$$f_U(t) = SD_tS^* \tag{26}$$

where D_t is the diagonal unitary matrix

$$D_t = \text{diag}(e^{i\theta_1 t}, e^{i\theta_2 t}, \ldots, e^{i\theta_n t}). \tag{27}$$

Furthermore

$$f_U(0) = I, \tag{28}$$
$$f_U(1) = U, \tag{29}$$

Thus $f_U([0, 1])$ gives a path from I to U. Since U could be any unitary matrix, every unitary matrix has a path to the identity matrix. Hence every unitary group is connected. □

4 Circulant Unistochastic Matrices

A circulant matrix with row vector $(v_1 v_2 \ldots v_n)$ is of the form

$$\begin{pmatrix} v_1 & v_2 & v_3 & \ldots & v_n \\ v_n & v_1 & v_2 & \ldots & v_{n-1} \\ v_{n-1} & v_n & v_1 & \ldots & v_{n-2} \\ \vdots & \vdots & \vdots & \ddots & \vdots \\ v_2 & v_3 & v_4 & \ldots & v_1 \end{pmatrix}. \tag{30}$$

A matrix is circulant if and only if it is diagonalizable by conjugating with a Fourier matrix [9].

For a given natural number, n, let

$$\omega = \mathrm{e}^{\frac{2\pi}{n} i}. \tag{31}$$

Let F be the $n \times n$ discrete Fourier transform matrix defined by

$$F = \left(\frac{1}{\sqrt{n}} \omega^{(j-1)(k-1)}\right). \tag{32}$$

Theorem 2 *If D is an $n \times n$ diagonal unitary matrix, then $F^* D F$ is a circulant unitary matrix and*

$$\Upsilon(F^* D F) \tag{33}$$

is a circulant unistochastic matrix. Furthermore,

$$F\Big(\Upsilon(F^* D F)\Big) F^* \tag{34}$$

gives a diagonalization of $\Upsilon(F^* D F)$.

Proof A matrix is circulant if and only if it is diagonalizable by a discrete Fourier transform matrix [9]. Discrete Fourier transform matrices are unitary, and $\mathscr{U}_n$ forms a group under matrix multiplication, hence $F^* D F$ is a circulant unitary matrix. It follows that there is a row vector $\overrightarrow{u} = (u_1, u_2, \ldots, u_n)$ such that

$$F^* D F = \begin{pmatrix} u_1 & u_2 & u_3 & \ldots & u_n \\ u_n & u_1 & u_2 & \ldots & u_{n-1} \\ u_{n-1} & u_n & u_1 & \ldots & u_{n-2} \\ \vdots & \vdots & \vdots & \ddots & \vdots \\ u_2 & u_3 & u_4 & \ldots & u_1 \end{pmatrix}. \tag{35}$$

By the definition of Υ,

$$\Upsilon(F^* D F) = \begin{pmatrix} |u_1|^2 & |u_2|^2 & |u_3|^2 & \ldots & |u_n|^2 \\ |u_n|^2 & |u_1|^2 & |u_2|^2 & \ldots & |u_{n-1}|^2 \\ |u_{n-1}|^2 & |u_n|^2 & |u_1|^2 & \ldots & |u_{n-2}|^2 \\ \vdots & \vdots & \vdots & \ddots & \vdots \\ |u_2|^2 & |u_3|^2 & |u_4|^2 & \ldots & |u_1|^2 \end{pmatrix}. \tag{36}$$

Since $F^* D F$ is unitary, $\Upsilon(F^* D F)$ is unistochastic. The matrix $\Upsilon(F^* D F)$ is circulant with row vector $(|u_1|^2, |u_2|^2, \ldots, |u_n|^2)$. Hence $\Upsilon(F^* D F)$ is circulant, and its diagonalization follows. □

5 Unistochastic Eigenpaths

For any n, the unitary group $\mathscr{U}_n$ is connected and the map Υ is continuous with respect to the Frobenius norm, thus the unistochastic matrices $\Upsilon(\mathscr{U}_n)$ are connected. Since eigenvalues depend continuously on matrix entries [12], the eigenvalues from the unistochastic image of a unitary matrix will form continuous curves. For any probability matrix, all eigenvalues are on the unit disk, and $(1, 1, \ldots, 1)^T$ is a right eigenvector with eigenvalue 1.

Definition 4 Let $g : \mathbb{R} \to \mathbb{C}^{n\times n}$ be a continuous map, I be a real interval, and $\Lambda(M)$ denote the eigenvalues of the square matrix M. The *eigenpath* of $(g, [a, b])$ is

$$\Lambda\big(g\,(I)\big). \tag{37}$$

A *unistochastic eigenpath* is the eigenpath of $\big(\Upsilon \circ f_U, I\big)$.

Unfortunately, when two unitary matrices are continuous deformation of each other, their characteristic polynomials do not identify which eigenvalues are deformations of each other. If the two matrices have the same conjugating matrix and same block form in their Jordan canonical forms, then locations on the diagonal show which eigenvalues are paired. The property of being a circulant matrix is preserved by Υ, and all circulant matrices are diagonalizable by Fourier matrices. These properties give us the ability to construct unitary matrix paths where the eigenvalues of the path's unistochastic image are easily identified as distinct curves on the closed unit disk of the complex plane (Table 1).

The figures presented here are unistochastic eigenpaths with the unitary matrices being circulant. This is done so that, computationally, it is easy to identify which eigenvalues are continuous deformations of each other. The technique used to plot eigenpaths does not use characteristic polynomial roots, it uses the discrete Fourier transform to diagonalize $\Upsilon \circ f_U(t)$. The plotted points are the diagonal entries of the diagonal matrices

$$F\big(\Upsilon \circ f_U(t)\big)F^*. \tag{38}$$

All plots were constructed with R [24]. Here is the color scheme:

From how the columns are ordered in our definition of the discrete Fourier transform matrix, conjugate eigenvalues share colors. One is always an eigenvalue of a stochastic matrix and called the stochastic eigenvalue, the figures appear to have $n - 1$ curves; the stochastic eigenvalue is a fixed point on the plots.

Table 1 The colors for each eigenpath along the diagonal

Color	Black	Red	Green	Blue	Cyan	Magenta	Yellow	Gray
Diagonal entry	1	2	3	4	5	6	7	8
	$n/2$	n	$n-1$	$n-2$	$n-3$	$n-4$	$n-5$	$n-6$

Since the change of basis matrix is determined, one needs to select the arguments of the eigenvalues of U and a real interval to construct an eigenpath. In each figure, the eigenvalues of U (rounded) are listed on the right; the interval and matrix size is in the main title (Figs. 1 and 2).

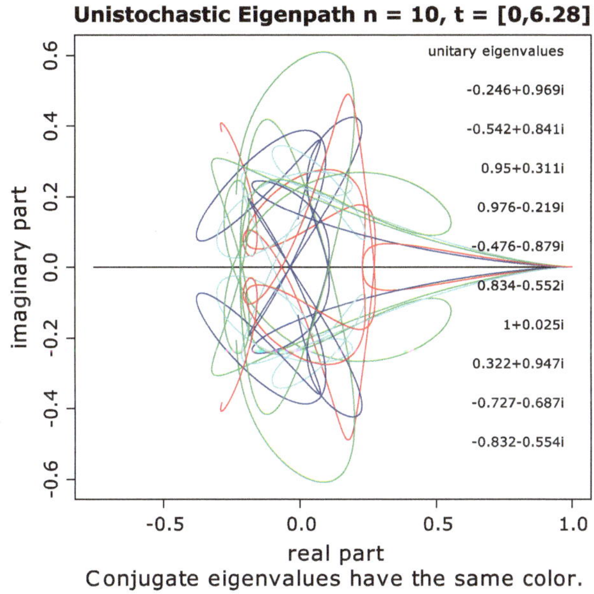

Fig. 1 A unistochastic eigenpath for a 10×10 circulant unitary matrix. The rounded off eigenvalues of the unitary matrix are listed in the figure

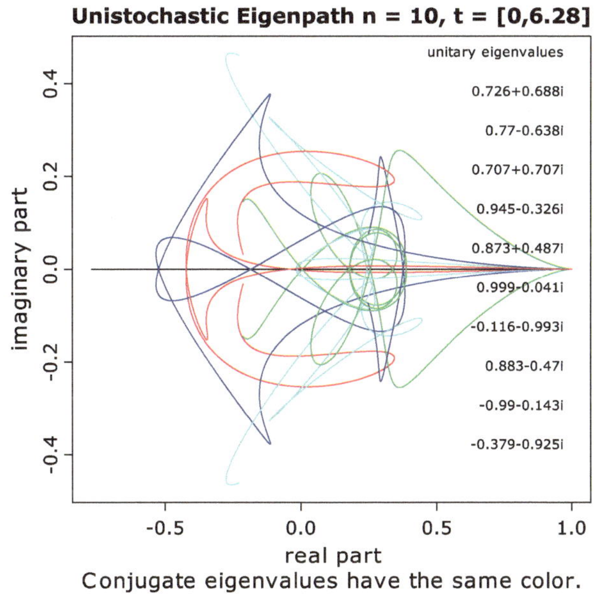

Fig. 2 A unistochastic eigenpath for a 10×10 circulant unitary matrix. The rounded off eigenvalues of the unitary matrix are listed in the figure

6 Hypocycloids and Circulant Unistochastic Eigenpaths

For eigenvalues of $1, \omega, \omega^2, \ldots, \omega^{n-1}$, and sufficiently large N, the eigenpath of $(\Upsilon \circ f_{F^*DF}, [0, N])$ forms hypocycloids [27].

Proposition 4 *If*

$$U_0 = \left(\begin{array}{c|c} 0\ldots0 & 1 \\ \hline & 0 \\ I & \vdots \\ & 0 \end{array}\right) \tag{39}$$

then $(\Upsilon \circ f_{U_0}, \mathbb{R})$'s eigenpath forms hypocycloids centered on the origin with outer radius 1 *and inner radii* $\frac{j}{n}$, $j = 1, 2, \ldots, n-1$ (Figs. 3 and 4).

Using a geometric sum and the eigenvalue formula for circulant matrices, it can be shown that the eigenvalues along $(\Upsilon \circ f_{U_0}, \mathbb{R})$ are of the form

$$\lambda_s = \frac{1}{n^2} \sin^2 \pi t \sum_{k=1}^{n} \csc^2 \tfrac{\pi}{n}[k-1+t]\omega^{(k-1)(k-s)}. \tag{40}$$

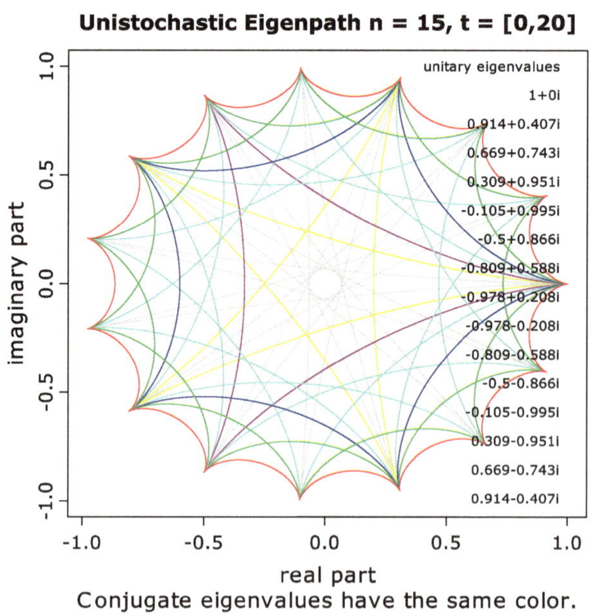

Fig. 3 The eigenpath for the 15×15 permutation matrix that sends $m \to m+1$

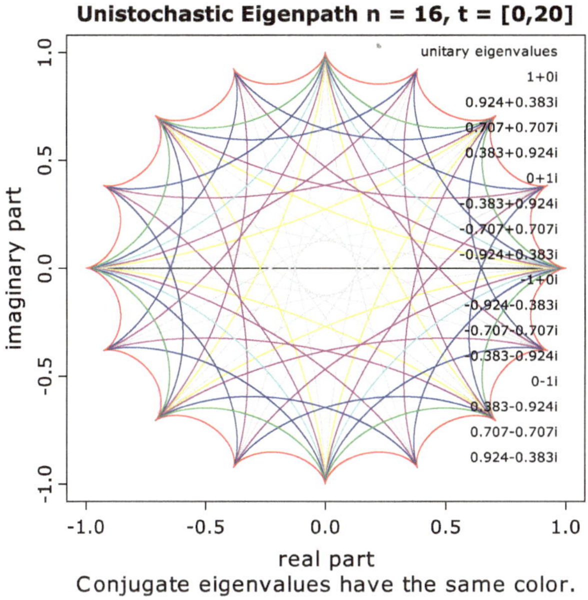

Fig. 4 The eigenpath for the 16×16 permutation matrix that sends $m \to m + 1$

References

1. Beck, M., Pixton, D.: The ehrhart polynomial of the birkhoff polytope. Discrete Comput. Geom. **30**(4), 623–637 (2003)
2. Bengtsson, I., Ericsson, A., Kus, M., Tadej, W., Zyczkowski, K.: Birkhoff's polytope and unistochastic matrices, $N = 3$ and $N = 4$. Commun. Math. Phys. **259**(2), 307–324 (2005)
3. Canfield, E.R., McKay, B.D.: The asymptotic volume of the Birkhoff polytope. Online J. Anal. Comb. **4**, 4 (2009)
4. Cappellini, V., Sommers, H.J., Bruzda, W., Życzkowski, K.: Random bistochastic matrices. J. Phys. A **42**(36), 365,209, 23 (2009)
5. Chan, C.S., Robbins, D.P.: On the volume of the polytope of doubly stochastic matrices. Exp. Math. **8**(3), 291–300 (1999)
6. Cheng, C.M.: Some results on eigenvalues, singular values and orthostochastic matrices. Ph.D. thesis, The University of Hong Kong (Pokfulam, Hong Kong) (1991)
7. Christensen, J.P.R., Fischer, P.: Positive definite doubly stochastic matrices and extreme points. Linear Algebra Appl. **82**, 123–132 (1986)
8. Chterental, O., Dokovic, D.Z.: On orthostochastic, unistochastic and qustochastic matrices. Linear Algebra Appl. **428**(4), 1178–1201 (2008)
9. Davis, P.J.: Circulant matrices. Wiley, New York (1979) (A Wiley-Interscience Publication, Pure and Applied Mathematics)
10. De Loera, J.A., Liu, F., Yoshida, R.: A generating function for all semi-magic squares and the volume of the Birkhoff polytope. J. Algebr. Combin. **30**(1), 113–139 (2009)
11. Dunkl, C., Zyczkowski, K.: Volume of the set of unistochastic matrices of order 3 and the mean jarlskog invariant. J. Math. Phys. **50**(12), 123,521, 25 (2009)
12. Franklin, J.N.: Matrix Theory. Prentice-Hall Inc., Englewood Cliffs (1968)
13. Gutkin, E.: On a multi-dimensional generalization of the notions of orthostochastic and unistochastic matrices. J. Geom. Phys. **74**, 28–35 (2013)

14. Haagerup, U.: Orthogonal maximal abelian $*$-subalgebras of the $n \times n$ matrices and cyclic n-roots. In: Operator algebras and quantum field theory (Rome, 1996), pp. 296–322. International Press, Cambridge (1997)
15. Hoffman, A.J.: A special class of doubly stochastic matrices. Aequ. Math. **2**, 319–326 (1969)
16. Horn, A.: Doubly stochastic matrices and the diagonal of a rotation matrix. Am. J. Math. **76**(3), 620–630 (1954)
17. Kitchens, B.: Countable state Markov shifts. Symbolic Dynamics. Universitext, pp. 195–240. Springer, Berlin (1998)
18. Marshall, A.W., Olkin, I., Arnold, B.C.: Inequalities: Theory of Majorization and its Applications. Springer Series in Statistics, 2nd edn. Springer, New York (2011)
19. McDonald, J.J., Stuart, J.: Spectrally arbitrary ray patterns. Linear Algebra Appl. **429**(4), 727–734 (2008)
20. Mei, Y., Gao, Y., Shao, Y., Wang, P.: The minimum number of nonzeros in a spectrally arbitrary ray pattern. Linear Algebra Appl. **453**, 99–109 (2014)
21. Mirsky, L.: Results and problems in the theory of doubly-stochastic matrices. Z. Wahrscheinlichkeitstheorie und Verw. Gebiete **1**, 319–334 (1962/1963)
22. Munkres, J.R.: Topology: a first course. Prentice-Hall Inc., Englewood Cliffs (1975)
23. Pakonski, P., Zyczkowski, K., Kus, M.: Classical 1D maps, quantum graphs and ensembles of unitary matrices. J. Phys. A **34**(43), 9303–9317 (2001)
24. R Core Team: R: A Language and Environment for Statistical Computing. R Foundation for Statistical Computing, Vienna (2014)
25. Tadej, W., Zyczkowski, K.: Defect of a unitary matrix. Linear Algebra Appl. **429**(2–3), 447–481 (2008) (With an appendix by Wojciech Slomczynski)
26. Tjstheim, D.: Some doubly stochastic time series models. J. Time Ser. Anal. **7**(1), 51–72 (1986)
27. Zyczkowski, K., Kus, M., Slomczynski, W., Sommers, H.J.: Random unistochastic matrices. J. Phys. A **36**(12), 3425–3450 (2003) (Random matrix theory)

Film Story Structure and Shot Type Analysis Using One-Way ANOVA, Kruskal–Wallis Test, and Poisson Distribution Test

Udjianna Sekteria Pasaribu and Klara Ajeng Canyarasmi

Abstract Film is a popular media of storytelling nowadays and is commonly considered as a social and art study subject. In fact, AOF can be observed mathematically such as the shot type usage as a sequence of random event, and on how a film's story adapts a particular story structure, in this case the Arch Plot Structure (APS). This study can be considered as one of the pioneering mathematical film studies. Three Indonesian children education themed movies are studied here. For the plot structure study, five ranked values were created to quantify the qualitative data. One-way ANOVA and Kruskal–Wallis test were used to study whether each film followed APS For the shot type study, four ranked values were created. Further, as a sequence of random events, the Poisson distribution is applied to fit to the shot type usage of each observed film.

Keywords Arch plot structure · Shot type · Story structure · One-way ANOVA · Kruskal–Wallis · Poisson distribution test

1 Introduction

"The moment we cry in a film is not when things are sad but when they turn out to be more beautiful than we expected them to be," [6] is an excellent quotation from Alain de Botton (a Swiss-British writer, philosopher, and television presenter) to describe that films of any sort have the power to depict stories in the most persuasive and wonderful ways to the audience. According to *Oxford Online Dictionary*, film is defined as a story or event recorded by camera as a set of moving images and shown in a cinema or on television [7].

U.S. Pasaribu (✉) · K.A. Canyarasmi
Institut Teknologi Bandung, Jalan Ganesha no. 10, Bandung 40132, Indonesia
e-mail: udjianna@math.itb.ac.id

K.A. Canyarasmi
e-mail: ajeng.canyarasmi@gmail.com; ajeng.canyarasmi@students.itb.ac.id

R.N. Mohapatra et al. (eds.), *Mathematics and Computing*,
Springer Proceedings in Mathematics & Statistics 139,
DOI 10.1007/978-81-322-2452-5_17

For all we know, film is a media to interpret story, therefore, commonly considered as a social and art study subject. This stereotype could also be what makes film to be considered as something immeasurable to some. The writers, among other preceding people with similar idea, would like to prove otherwise; there are aspects of a film that are measurable both in quantity and quality. There are many film aspects which can be considered as measurable, random events such as expense and budgeting history, its political or ethnical effect, art and audio quality or technique usage, and picture or framing quality or technique usage. Tsivian and team pioneered a statistical film study in regard to its shot called *Cinemetrics* [11]. Cinemetrics mostly study shots in the form of statistics descriptive. In this study, the writer would like to study film shot' cinematic aspects using both inference and descriptive statistics.

This paper constrained only to analyze the film story development as segmentation base and framing technique usage, the shot type usage. The reason for choosing them is because the story is considered as the essence of film, therefore it is better to analyze it before doing other analyses, and shot type usage is considered as one of the easiest and most objective cinematography techniques to be observed.

2 Theory

2.1 Arch Plot Structure (APS) in Film Storytelling

There are mainly six departments in most film productions: creative (story and scripting), art, sound, editing and special effects, production, and cinematography [5]. As mentioned before, the story is no doubt the core of most films produced; every other film aspect depends on the story the film wants to deliver. There are theories of story development structure for stories in every storytelling media including film. The most classic and still widely adapted story structure nowadays is known as the Arch Plot Story structure, abbreviated as APS. APS is a goal-oriented plot where the character will try to gain his or her goal against forces of antagonism [9]. APS comparts the film into three acts as can be seen in Fig. 1.

Figure 1 mainly shows story intensity by time progression in a typical 120-min film. The circled dots in the graphic represents highlighted events of the film which will not be explained here. As one can see, the story is getting intense from 10 to 100th minutes and declining from 100th minutes to the end of the story. This condition occurs because each act has incremental story characteristics shown in Table 1.

It is necessary to prove that the observed films adapt a particular story structure, in this case the APS, before analyzing anything else. For this purpose, the hypothesis test is applied with the null and alternative hypothesis as follows:

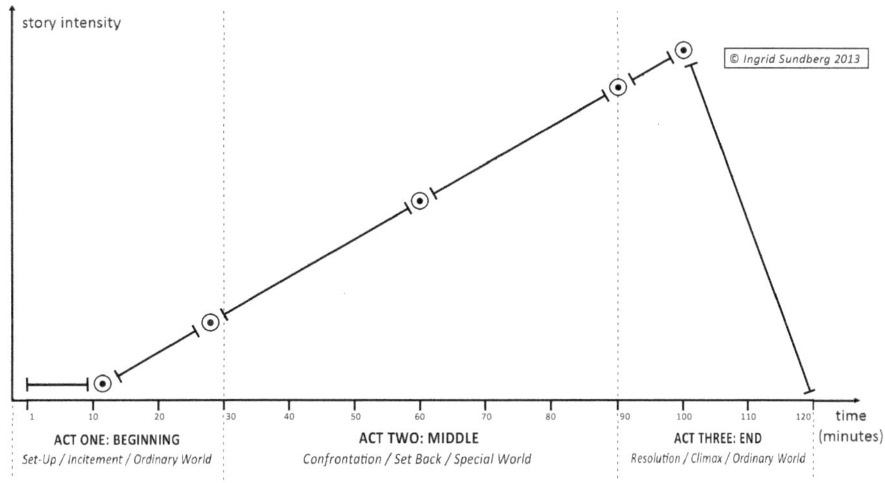

Fig. 1 The Arch plot story structure

Table 1 Story progression characteristic for acts [4]

Act 1	Act 2	Act 3
Exposition Part of a story that introduces the characters, shows some of their interrelationships, and places them within a time and place	*Obstacles* The main character encounters obstacle after obstacle that prevent him from achieving his dramatic need	*Climax (Second Culmination)* Point at which the plot reaches its maximum tension and the forces in opposition confront each other at a peak of physical or emotional action
Inciting Incident An event that sets the plot of the film in motion. It occurs approximately halfway through the first act	*First Culmination* The main character seems close to achieving his or her goal/objective, and then everything falls apart	*Denouement* Brief period of calm at the end of a film where a state of equilibrium returns
Plot Point An event that thrusts the plot in a new direction, leading into a new act of the screenplay	*Midpoint* The main character reaches his/her lowest point and seems farthest from fulfilling the dramatic need or objective	
	Plot Point	

H_0: each segment of the ith film adapt the arch plot story structure
H_1: at least one segment of the ith film does not adapt the arch plot story structure

For $i = 1, 2, \ldots, n$, as many n observed films, or in this study 3.

To perform analyses, first the observed films shall be parted into three segments: segments 1, 2, and 3 which are proportional to the proportion of Acts 1, 2, and 3, or can be written as:

$$E_{t_i}\text{: APS value for } i\text{th film, where } E_{t_i} = \{0, 1, 2, 3, 4\}$$

with $i = 1, 2, 3$ and for every $x \in t_i$, $|x| = 30$. The ranked values are defined below to analyze the observed films' segments in this paper:

0: highly not adapting, 3: greatly adapting, or
1: greatly not adapting, 4: highly adapting
2: fairly adapting.

The arch plot story progression according to the corresponding act. It will be proven that each observed film adapts the APS by calculating whether each of their segments adapts the same characteristics of each act with similar value. For each film segment, there will be randomly picked smaller subsegments. Then the ranked value will be assigned to each of the subsegments according to how adaptive the subsegments are to the corresponding act. Should the values have been assigned, two inference tests will be applied to test the hypothesis: parametric One-way ANOVA [15] and nonparametric *Kruskal–Wallis Test*. The null and alternative hypothesis for both tests are:

$$H_{0_i} : \mu_{i,1} = \mu_{i,2} = \mu_{i,j} \text{ versus } H_{1_i} : \exists\, k, l \ni \mu_{i,k} \neq \mu_{i,l}, k, l = 1, 2, 3$$

For $j = 1, 2, \ldots m$ and $i = 1, 2, 3$ [13] with m total observed segments on each film, in this case 3. One-way ANOVA can be written as

$$Y_{i_{jk}} = \mu_i + \alpha_{i_j} + \varepsilon_{i_{jk}}$$

of which μ_i is the grand mean of each Film i segment mean, that is $\mu_i = \frac{1}{k}\sum_{i=1}^{k} \mu_j$, α_{i_j} as effect of jth segment of Film i and $\varepsilon_{i_{jk}}$ is random error with $\varepsilon_{i_{jk}} \sim N(0, \sigma^2)$. To obtain valid data for this study, one must do an intensive observation on each given subsegment. For each film, Kruskal–Wallis test uses statistic

$$H_i = \frac{12}{n_i(n_i + 1)} \sum_{j=1}^{k} \frac{R_{i_j}^2}{n_{i_j}} - 3(n_i + 1)$$

with $n_i = n_{i_1} + n_{i_2} + \cdots + n_{i_k}$ for k denotes each film's observed segment which is 3 and R_{i_j} is random variable of sum of ranks corresponding to n_{i_j} and $i = 1, 2, 3$ denotes Film i [14].

Throughout this paper, many tests will prerequire independent data. Therefore, the *Chi-squared test* will be used to check data's independency, with hypothesis:

H_{0_i} : the observation on each segment is independent
H_{1_i} : there are at least two segments in which the observation is not independent

on the ith film, $i = 1, 2, 3$.

Of course, the APS is not the only story structure used worldwide. If one has seen a film and intuitively feels that it does not follow the APS, he or she can look for another story structure and apply similar analysis.

2.2 Shot Type Analysis

After proofing that each observed film adapts the arch plot story structure, the next analysis can proceed, in this research the shot type usage. Shot type is a framing technique used to interpret the points of story in a film. Shot type is a useful tool to highlight events or to show all the information the director wants the audience to notice.

No matter how thoroughly planned, the shooting process is something no one can fully predict. There are many factors contributing to this unpredictability: the actors' and actresses' mood swings, the set condition (especially if it was used a nonartificial set) such as weather or people surrounding it, even director's sudden creative decision. Due to its unpredictability, cinematography technique usage such as shot type can be considered as a sequence of random events, therefore, also considered as a stochastic process with continuous parameter value [10]. Based on each observed film following the APS, it can be written that

$$S_{t_i}\text{: shot type value for } i\text{th film, where } S_{t_i} = \{1, 2, 3, 4\}$$

With $i = 1, 2, 3$ and for every $x \in t_i$, $|x| = 30\,\text{s}$. In this study, the shot type will be categorized into four categories which also represent the ranked value of the shot type usage random event:

1: most information, least focus   3: less information, more focus
2: more information, less focus   4: least information, most focus

The same film segmentation should be done for this analysis as well. The difference is in this analysis each subsegment will have four data instead of one because there should be more than one shot in each subsegment. The data is the sum usage of shot type 1 until 4 for each segment.

Because counting usage of shot type is statistically a random events counting process, it supposedly follows phenomena in Poisson counting process. Therefore it will be proved that for each observed film, its shot type usage is a Poisson distribution. In 2002, Brown and Zhao developed a Poisson distribution test based on Anscombs statistic which will be used in this study [3]. Under the circumstances that this is a social research related to preference for films which highly vary, a relatively smaller significance level will suffice.

Assuming $\{X(t)|t \leq n, t \in \mathbb{Z}^+\}$ are independent nonnegative integer-valued random variables with $P(X = x) = f(x)$, the null and alternative hypotheses for the new Anscombe Poisson distribution test are:

$$H_{0_i} : X_{i_j} \sim \text{Poiss}(\lambda_{i_j}), \lambda_{i1} = \lambda_{i2} = \cdots = \lambda_{i_n} \text{Versus.}$$
$$H_{1_i} : X_{i_j} \sim \text{Poiss}(\lambda_i), \sum\nolimits_{j=1}^{n} (\lambda_{i_j} - \bar{\lambda}_i) > 0$$

with n indicating the number of segments in film i, which is 3, $i = 1, 2, 3$.

According to Brown and Zhao, first $Y_i = \sqrt{X_i - \frac{3}{8}}$ [1] is defined. From it the statistic

$$T_{\text{new}_i} = 4\sum_{j=1}^{n}(Y_{i_j} - \bar{Y}_i)$$

was gained to provide test for H_0. T_{new_i} has approximately a Chi-squared distribution with degree of freedom $n - 1$ [2]. Hence, if $T_{\text{new}_i} > \chi^2_{(n-1;1-\alpha)}$, H_{0_i} is rejected. Using this test, one would know whether shot type usage within the observed film follows Poisson distribution with certain rate or not, which will prove that it is a mathematically random events counting process.

3 Data Analysis

For this analysis, three Indonesian children education movies were observed: *Denias: Senandung di Atas Awan* (shortened to **DS**) by John de Rantau (2006) assigned as Film 1, *Laskar Pelangi* (**LP**) by Riri Riza (2008) as Film 2, and *Negeri 5 Menara* (**N5M**) by Affandi A. Rachman (2012) as Film 3 according to their launching year. DS tells the story of a Papuan child named Denias who seeks for proper education which barely existed in his village. LP is a movie of poor but knowledge-thirst children of Belitung, a region in southeast Sumatra, who struggle in their way to gain proper education, crusades the idea that children do not necessarily need education but must be immediately sent to work. Lastly, N5M is a movie about Alif's struggle, a West Sumatra child, to find out where he really meant to study and live: Madani Pesantren (Islam-based school) with his fellows from around Indonesia, or his dream campus in West Java, Indonesia. The reason for choosing these movies is because not only do they have the same genre, but because they have a similar story plot: a struggle to gain better education for young Indonesian individuals.

For each segment of the observed movies, random subsegments of 30 s will be picked. The following is the function of the observed films'segmentation for shot type usage.

$$E_{t_i} = \begin{cases} G1_{t_i} & \text{for } t_{i_1} \le t_i \le t_{i_2} \text{ seconds} \\ G2_{t_i} & \text{for } t_{i_2} < t_i \le t_{i_3} \text{ seconds} \\ G3_{t_i} & \text{for } t_{i_3} < t_i \le t_{i_4} \text{ seconds} \end{cases} \quad (1)$$

for $G_{p_{t_i}}$ is a stochastic random variable of APS adaptation on to-be-tested segment p of Film i, with $i, p = 1, 2, 3$ and value

$$G_{p_{t_i}} = \{0, 1, 2, 3, 4\}, \text{ for } x \in t_i, |x| = 30 \text{ s} \quad (2)$$

Table 2 Film shot segmentation function's limit (in second)

i	t_{i1}	t_{i2}	t_{i3}	t_{i4}
1	26	1131	4760	6338
2	0	1624	5247	6996
3	25	2381	5027	6694

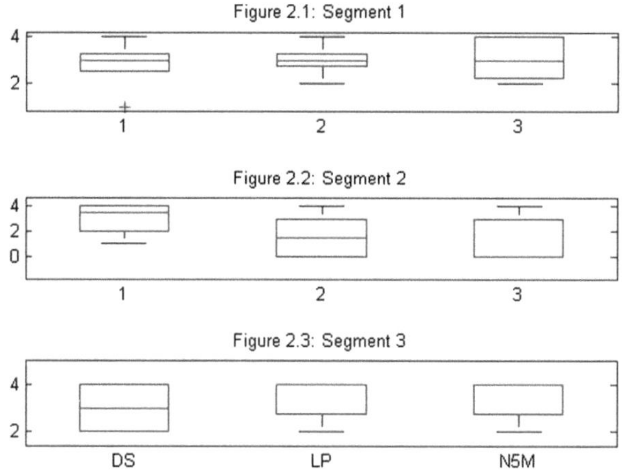

Fig. 2 Boxplot of the story structure data

The function limits and the total number of observable subsegments for each film is shown in the Table 2 and the observation result can be seen in Fig. 2, Tables 3, 4 and 5.

The writers will study the observed films's inference statistics using both parametric One-way ANOVA and nonparametric Kruskal–Wallis test, both using the following null and alternative hypotheses

$$H_{0_i} : \mu_{i,1} = \mu_{i,2} = \mu_{i,3} \text{ versus } H_{1_i} : \exists\, k, l \ni \mu_{i,k} \neq \mu_{i,l}, k, l = 1, 2, 3$$

for $i = 1, 2, 3$. First, the data must be proven independent. Chi-squared test shows that Films 1, 2, and 3 data give p-values, correspondently, $P_1 = 0.394$, $P_2 = 0.315$, and $P_3 = 0.076$. Therefore, each observed Film APS adaptation value does not reject the null hypothesis of independency under all signification levels $\alpha < 0.076$. Now one may proceed to the inference test.

As an example, Film 1 has ANOVA result table:

Table 3 ANOVA result table for Film 1's APS adaptatiton

Source of variation	SS	df	MS	F	p-value
Between groups	0.283	2	0.142	3*0.124	3*0.884
Within groups	21.717	19	1.143		
Total	22	21			

Table 4 Descriptive statistics for segment 1 data

Segment 1	Film 1	Film 2	Film 3
Mean	2.8	3	3.143
Median	3	3	3
Mode	3	3	4
Sample variance	1.2	0.5	0.809
Minimum	1	2	2
Maximum	4	4	4
Sum 14	15	22	
Count	5	5	7

Table 5 Descriptive statistics for data of segment 2

Segment 2	Film 1	Film	Film 3
Mean	3.083	1.7	1.1
Median	3.5	1.5	0
Mode	4	0	0
Sample variance	1.174	2.9	2.544
Minimum	1	0	0
Maximum	4	4	4
Sum	37	17	11
Count	12	10	10

with p-value $P_1 = 0.884$. Compared to any level of significance $\alpha < 0.884$ therefore H_{0_1} was not rejected which leads to the conclusion that DS adapts the APS. Films 2 and 3 have p-values consecutively $P_2 = 0.0680$ and $P_3 = 0.0034$. From previous results it can be concluded that using One-way ANOVA, all observed films follow APS for all significance levels $\alpha > 0.00871$.

Now the Kruskal–Wallis test will be conducted on each film. The p-values for Films 1, 2, and 3 are consecutively $P_1 = 0.351$, $P_2 = 0.119$, and $P_3 = 0.015$. Hence for each observed film, the null hypothesis will not be rejected for any significance level $\alpha < 0.015$. Comparing significance level from both tests, Kruskal–Wallis test gives a higher limit value, or in other words, showing greater chance for the observed films not rejecting the null hypothesis. Therefore for this study, one can choose Kruskal–Wallis test to show that the observed film segments, and therefore the whole film, follow APS with story progression in Fig. 1.

Under significance level limit 0.015, it shows that from 10,000 events, 150 events did not reject H_0. This value however, which is actually taken from Film 3 p-value, is much smaller than Film 1 and 2 p-values which is taken limit on 0.119. This occurred due to occasional nonincrementing intensity progression adapted in N5M especially in its Act 2: Act 2 tells Alif and his friends'life in Pesantren including their sub-stories, which have rather flat progression. The following figure shows how the APS graphic would be for N5M.

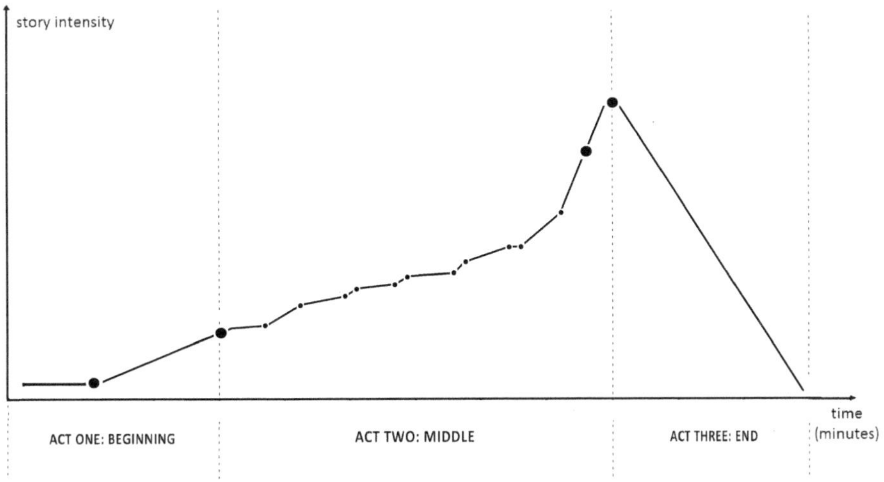

Fig. 3 The arch plot structure for Negeri 5 Menara

Table 6 Descriptive statistics for data of segment 3

Segment 3	Film 1	Film 2	Film 3
Mean	3	3.4	3.4
Median	3	4	4
Mode	2	4	4
Sample variance	1	0.8	0.8
Minimum	2	2	2
Maximum	4	4	4
Sum	15	17	17
Count	5	5	5

In Fig. 3, the circles indicate story point just like in Fig. 1 while the dots indicate the beginning of substories or transition to the main story. As can be seen, the influence of having many substories on the overall film progression is suppressed to a point. This creates a rather exponential story intensity progression instead of linear, which causes low value on APS evaluation that is rather linear as can be seen in Fig. 1. But in the end all observed films do adapt from APS with better interpretation using nonparametric test, Kruskal–Wallis test.

By segmenting each film into three acts, there should be differences in film aspects' usage from one act to another. For the shot type the films will follow story structure segmentation function with subsegment sampling shown in Table 6. The result of each shot type usage is given in Fig. 4 as well (Table 7).

Figure 4b shows that Film 1 has outlying value for more information, less focus type shot usage which is 8 times, and much bigger compared to the usual usage which is within no usage to two usage. The usage occurred on subsegment 2, with Denias and Noel, where the antagonist child, who liked to pick on Denias, fight and their

Fig. 4 Boxplot of all shot type usage

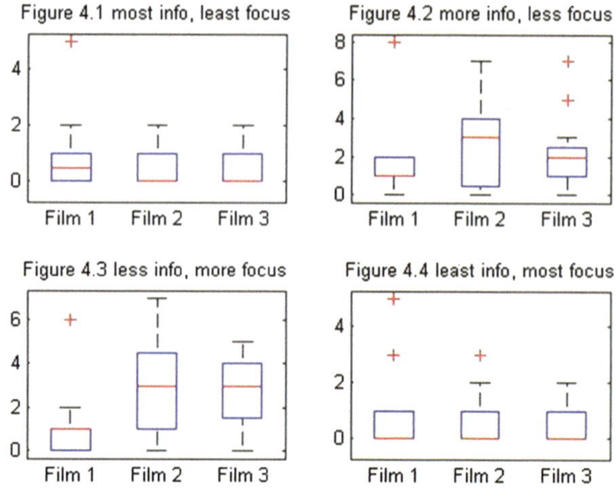

Table 7 Total number of observed films' observable subsegments

Film i	Act 1		Act 2		Act 3		Total	
	N	*n*	*N*	*n*	*N*	*N*	*N*	*n*
1	37	3	121	6	53	3	211	12
2	54	3	121	6	58	3	233	12
3	56	3	111	6	56	3	233	12

teacher came to mediate between them. It was delivered with back-to-back shots, hence it is reasonable to have up to 8 usages in the subsegment. Outliers contained in observation data can be analyzed using *Grubb's Test* if it is single outlier, or analyzed separately from the rest of data if they are more than one just as done by Pasaribu and team [8].

In this study, it will be proved that particularly the "less information, more focus" shot type usage has Poisson distribution. The reason for choosing that shot type usage as example is because the writer considers *shot type* 3 as the key shot type whose usage shows how intense a film explain its story. Similar to (1), the following is the function of the observed films's segmentation for shot type usage:

$$S_{t_i} = \begin{cases} A1_{t_i} & \text{for } t_{i_1} \leq t_1 \leq t_{i_2} \text{ seconds} \\ A2_{t_i} & \text{for } t_{i_2} < t_1 \leq t_{i_3} \text{ seconds} \\ A3_{t_i} & \text{for } t_{i_3} < t_1 \leq t_{i_4} \text{ seconds} \end{cases} \tag{3}$$

for $A_{p_{t_i}}$ is a stochastic random variable of shot type 3 usage on Act p of Film i with $i, p = 1, 2, 3$, and value similar to (2),

$$A_{p_{t_i}} = \{1, 2, 3, 4\} \text{ for } x \in t_i, |x| = 30 \tag{4}$$

and function limits following Table 2.

According to Anscombe new Poisson test on the data, shot type 3 usage for Films 1, 2, and 3 has rate, consecutively, $\lambda_{3,1} = 1.0833$, $\lambda_{3,2} = 3.0833$, and $\lambda_{3,3} = 2.75$. Assuming that the "less information, more focus" shot type usage is denoted with $X_{i,3}$, it will be proven that $X_{i,3} \sim \text{Pois}(\lambda_{3,i})$ for $i = 1, 2, 3$ denoting the ith film.

According to the Chi-squared independency test, the p-value for Films 1,2, and 3 are, consecutively: $P_1 = 0.000594$, $P_2 = 0.000549$, and $P_3 = 0.16866$. Compared with the significance level, the null hypothesis for each movie is not rejected, therefore proving the observed films' usage of shot type 3 is independent for each subsegment. Now the new Anscombe Poisson distribution test can be conducted.

According to Anscombe Poisson distribution test, the statistics values for each film are consecutively: $T_{\text{new}_1} = 13.745$, $T_{\text{new}_2} = 21.459$, $T_{\text{new}_3} = 11.293$. With $\alpha = 0.02$, gained $\chi^2_{0.98,11} = 22.618$. Because $T_{\text{new}_i} < \chi^2_{0.98,11}$, the null hypothesis for each movie is not rejected which leads to the conclusion that the "less information, more focus" shot type usage Film 1 follows Poisson distribution with rate $\lambda_{3,1} = 1.0833$, Film 2 with $\lambda_{3,2} = 3.0833$, and Film 3 with $\lambda_{3,3} = 2.75$.

Optionally, *Goodness-of-Fit test* [12] can be conducted on each result. The following is a difference table between the observed and expected value:

As can be seen in Table 8, excluding the 6 times usage, the difference for Film 1 is considerably small, less than 0.9, with the greatest difference occurring for 3 and 4 times usage. However, the greatest difference on Films 2 and 3 is increased from the greatest 3.827 occurred on one time usage in Film 2 and 2.580 for 5 times usage in Film 3. The shot type 3 with six times usage difference for Film 1 is distinctively large. This occurred due to the expected number of segments having 6 times usage from 12 segments according to $X_{1,3} \sim \text{Pois}(1.083)$ is 0.0091 (or rounded to none), while the actually observed value is 1. To avoid this bias, the p-value evaluation will be conducted instead. Consecutively, the p-values are $P_{\text{gof}_{3,1}} = 0.989$, $P_{\text{gof}_{3,2}} = 0.959$, and $P_{\text{gof}_{3,3}} = 0.995$. For level of significance $\alpha < 0.959$, it can be concluded that each movie "most information, least focus" shot type usage greatly follows Poisson distribution. These results prove that the usage of shot type 3 is a random event with occurrence probability according to their rate. One can also use the same method for other shot types (Tables 9, 10, 11 and 12).

Table 8 Descriptive statistics for shot type 1

Shot type 1	Film 1	Film 2	Film 3
Mean	0.583	0.583	0.333
Median	0.5	0	0
Mode	0	0	0
Sample variance	0.447	0.629	0.424
Minimum	0	0	0
Maximum	2	2	2
Sum	7	7	4

Table 9 Descriptive statistics for shot type 2

Shot type 2	Film 1	Film 2	Film 3
Mean	1.917	2.5	2.417
Median	1	2.5	2
Mode	1	0	1
Sample variance	4.811	4.454	4.811
Minimum	0	0	0
Maximum	8	7	7
Sum	23	30	29

Table 10 Descriptive statistics for shot type 3

Shot type 3	Film 1	Film 2	Film 3
Mean	1.333	3	3
Median	1	3	3.5
Mode	1	3	4
Sample variance	2.969	5.091	2.727
Minimum	0	0	0
Maximum	6	7	5
Sum	16	36	36

Table 11 Descriptive statistics for shot type 4

Shot type 4	Film 1	Film 2	Film 3
Mean	0.917	0.667	0.417
Median	0	0	0
Mode	2.447	1.333	0.447
Sample variance	2.447	1.333	0.447
Minimum	0	0	0
Maximum	5	3	2
Sum	11	8	5

By relating the shot type 3 usage rate and the story structure, it is proven to say that in general, Film 2 which is *Laskar Pelangi* movie has the most story intensity increment and Film 1 which is *Denias: Senandung di Atas Awan* has the least. It is practically true for both cases: *Laskar Pelangi* tells a story full of turmoil, making the story dense with emotional surprises. As for *Denias: Senandung di Atas Awan*, it is a wonderful movie full of many shots of Indonesian natural beauty (which explains the rate for the shot type 1, the "most information, least focus," usage is the highest compared to other movies) but delivered in a less dramatic, more subtle way, explaining the less usage of shot type 3.

Table 12 Difference table between the observed and expected value of shot type 3 times of usage

Times of usage	Difference between observed and expected		
	Film 1	Film 2	Film 3
0	0.25	3.827	0.0707
1	0.082	0.055	0.006
2	0.803	2.613	0.280
3	0.861	0.037	0.163
4	0.233	0.002	2.580
5	0.050	0.060	2.937×10^{-5}
6	107.673	0.180	0.461
7	0.001	1.750	0.181
Sum	109.953	8.524	3.742

After knowing how much each observed film uses shot type 3, one can consider how often more focused shots are used, supposedly implying to how much the director of each film wants to emphasize its story points. These results can be related to further study of how well the audience understands and enjoys the observed films.

4 Conclusions and Suggestions

Representing the observed film genre, *Denias: Senandung di Atas Awan*, *Laskar Pelangi*, and *Negeri 5 Menara* follow the arch plot story structure. Furthermore, with significance level $\alpha = 0.02$, the "less information, more focus" class shot type usage in the three movies was proved following Poisson distribution with rate, consecutively, $\lambda_{3,1} = 1.0833$, $\lambda_{3,2} = 3.0833$, and $\lambda_{3,3} = 2.75$. For further study, it is suggested that the conductor analyzes the observed movies in smaller and more subsegments. Subsegmenting the whole movie is recommended for greater accuracy.

References

1. Brown, L.D., Zhao, L.H.: Test for the Poisson distribution. Sankhya Indian J. Stat. **64**, 614 (2002)
2. Brown, L.D., Zhao, L.H.: Test for the Poisson distribution. Sankhya Indian J. Stat. **64**, 615 (2008)
3. Brown, L.D., Zhao, L.H.: Test for the Poisson distribution. Sankhya Indian J. Stat. **64**, 611 (2002)
4. College of DuPage: three-act structure (2014). http://www.cod.edu/people/faculty/pruter/film/threeact.htm
5. Film Victoria: crew roles and departments (2014). http://www.film.vic.gov.au

6. Goodreads Inc.: Alain de Botton: Quotes (2014). http://www.goodreads.com/author/quotes/13199.Alain_de_Botton
7. Oxford University Press: film (2014). http://www.oxforddictionaries.com/definition/english
8. Pasaribu, U.S., Hawkes, A.G., Wainwright, S.J.: Statistical assumptions underlying the fitting of the Michaelis-Menten equation. J. Appl. Stat. **26**, 329–332 (1999)
9. Sundberg, I.: What is arch plot and classic design? (2013). http://ingridsnotes.wordpress.com/2013/06/05/what-is-arch-plot-and-classic-design
10. Taylor, H.M., Karlin, S.: An introduction to stochastic modeling, 3 edn., pp. 1–6. Academic Press, New York (1998)
11. Tsivian, Y., Salt, B., Redfern, N., Baxter, M., Dukic, V.: Cinemetrics (2005). http://www.cinemetrics.lv
12. Walpole, R.E., Myers, R.H., Myers, S.L., Ye, K.: Probability and statistics for engineers and scientists, 9 edn. Chapter 9: One-factor experiments: general, p. 371. Pearson, New York (2012)
13. Walpole, R.E., Myers, R.H., Myers, S.L., Ye, K.: Probability and statistics for engineers and scientists, 9 edn. Chapter 9: One-factor experiments: general, pp. 657–658. Pearson, New York (2012)
14. Walpole, R.E., Myers, R.H., Myers, S.L., Ye, K.: Probability and statistics for engineers and scientists, 9 edn. Chapter 9: One-factor experiments: general, pp. 668–669. Pearson, New York (2012)
15. Walpole, R.E., Myers, R.H., Myers, S.L., Ye, K.: Probability and statistics for engineers and scientists, 9 edn. Chapter 9: One-factor experiments: general, pp. 509–516. Pearson, New York (2012)

Characterization of Total Very Excellent Trees

N. Sridharan and S. Amutha

Abstract Let $G = (V, E)$ be a simple graph. A subset D of V is said to be a total dominating set of G if every vertex $v \in V$ is adjacent to at least one vertex of D. The total domination number $\gamma_t(G)$ is the minimum cardinality of a total dominating set of G. A total dominating set with $\gamma_t(G)$ cardinality is said to be a γ_t-set of G. A graph G is said to be total excellent if given any vertex x of G, there is a $\gamma_t(G)$-set of G containing x. A γ_t-set D of G is said to be total very excellent γ_t-set of G if for each vertex $u \in V - D$, there is a vertex $v \in D$ such that $(D - v) \cup \{u\}$ is a γ_t-set of G. The graph G is said to be total very excellent if it has at least one total very excellent γ_t-set. Total very excellent graphs are total excellent. In this paper we characterize total very excellent caterpillars and total very excellent trees.

Keywords Dominating set · Excellent graphs · Total domination number · Total very excellent graphs

2000 Mathematics Subject Classification: 05C

1 Introduction

We consider only finite simple undirected graphs. If $G = (V, E)$ is a graph and $u \in V$, the neighborhood $N(u)$ of u is the set $\{u \in V : uv \text{ is an edge of G}\}$ and the closed neighborhood $N[u]$ of u is the set $N(u) \cup u$. A subset D of V is said to be a dominating set in G if every vertex in $V - D$ is adjacent to some vertex in D., i.e.,$V = \underset{u \in D}{\cup} N[u]$. The domination number of G is the minimum cardinality

N. Sridharan
Department of Mathematics, Alagappa University, Karaikudi, TamilNadu, India
e-mail: math.sridhar@yahoo.co.in

S. Amutha (✉)
Ramanujan Centre for Higher Mathematics, Alagappa University, Karaikudi, TamilNadu, India
e-mail: amutha.angappan@rediffmail.com

R.N. Mohapatra et al. (eds.), *Mathematics and Computing*, Springer Proceedings in Mathematics & Statistics 139,
DOI 10.1007/978-81-322-2452-5_18

of a dominating set in G and is denoted by $\gamma(G)$. A dominating set D of G with cardinality $\gamma(G)$ is called a γ-set of G. A subset D of vertex set $V(G)$ of a graph G is said to be a total dominating set of G if each vertex $u \in V(G)$ is adjacent to at least one vertex $v \in D$., i.e., $(V = \bigcup_{u \in D} N(u))$. A graph has a total dominating set only if it has no isolated vertices. The minumum cardinality of a total dominating set of G is called the total domination number and is denoted by $\gamma_t(G)$. A total dominating set D of G with cardinality $\gamma_t(G)$ is called a γ_t-set of G. An exhaustive treatment of fundamentals of domination and several advanced topics in domination are given in [3]. For graph theoretic terminologies, we refer to [1].

G. H. Fricke et. al. [2] call a vertex of a graph G to be good if it is contained in some γ-set of G and bad if it is not. They call a graph G to be γ-excellent if every vertex of G is good. Henning at al. [4] defined γ_t-good, γ_t-bad vertices and introduced the concept of γ_t- excellent graphs. They also provided a constructive characterization of γ_t-excellent trees. They proved that a path P_n is a γ_t-excellent iff either $n = 3$ or $n = 4k + 2$ for $k \geq 0$. Yamuna [6] provided a construction where a non-excellent graph G is imbedded in an excellent graph H such that $\gamma(H) \leq \gamma(G) + 2$, and also proved that if a graph G is not excellent, then there is a subdivision graph H of G which is excellent. Yamuna [5, 7, 8] introduced new classes of excellent graphs such as just excellent graphs, very excellent graphs, and rigid very excellent graphs. Yamuna also characterized very excellent trees.

In this paper we introduce the class called class of total very excellent graphs and intiate a study on this class.

2 Definition and Examples

In this section, we define and give examples for total very excellent graphs and characterize total very excellent paths, cycles, and caterpillars.

Definition 2.1 A vertex u of a graph G is said to be γ_t-good if it is contained in some γ_t-set of G, otherwise it is said to be γ_t-bad. A graph G is said to be γ_t-excellent or very excellent if every vertex of G is γ_t-good.

A total excellent graph G is said to be total very excellent(TVE), if there is a γ_t-set D of G such that for each vertex $u \in V - D$, there exist a vertex $v \in D$ such that $(D - v) \cup \{u\}$ is a γ_t-set of G. A γ_t-set D of G satisfying this property is called a total very excellent γ_t-set (TVE γ_t-set) of G.

Example 2.2

(1) The cycles $C_3, C_4, C_5, C_6, C_{10}$ are total very excellent cycles.
(2) The paths P_2, P_3, P_6, P_{10} are total very excellent paths. [for each of these paths, a TVE γ_t-set is as shown in the following figure].
(3) Complete graphs $K_n (n \geq 2)$ are total very excellent.
(4) The cycle C_7 is not total very excellent.

P_2

P_3

P_6

P_{10}

Let D be a γ_t-set of G. To each $u \in D$, the total private neighbour of u with respect to D is difined as $PN_t(D, u) = \{v \in V | N(v) \cap D = \{u\}\}$. Note that $PN_t(D, u) \neq \phi$ for all $u \in D$.

Remark 2.3 If G is a total very excellent graph and D is a total very excellent γ_t-set of G, for each $v \in V - D$, there exists $u \in D$ such that $D_v = (D - u) \cup \{v\}$ is a γ_t-set of G. Clearly $PN_t(D, u) \subset N[v]$. Otherwise D_v is not a γ_t-set of G.

The following theorem characterizes TVE paths.

Theorem 2.4 *P_2, P_3, P_6 and P_{10} are the only paths which are total very excellent*

Proof It is enough to show that P_{4k+2} for $k \geq 3$, is not total very excellent. Let P be $v_0, v_1, v_2, \ldots, v_{4k+1}$. Assume that P is TVE and D is a TVE γ_t-set for P. Each component of $\langle V - D \rangle$ is either K_1 or P_2. We can assume that $v_0, v_{4k+1} \notin D$ (and hence $v_1, v_2, v_{4k-1}, v_{4k} \in D$). Hence at least two components of $\langle V - D \rangle$ are K_1. So the number of components of $\langle V - D \rangle$ is either $k + 1$ or $k + 2$. (As $k \geq 3$, the number of components of $\langle V - D \rangle \geq k + 1 \geq 4$).

Case(i)
The number of components of $\langle V - D \rangle$ is $k + 1$. Then except the components v_0 and $\{v_{4k+1}\}$, all other components of $\langle V - D \rangle$ are P_2, and the number of components of $\langle D \rangle$ is k. Either there is a component of $\langle D \rangle$ having four vertices (all other components have exactly two vertices) or exactly two of the components of $\langle D \rangle$ having three vertices, while others are P_2. Find i such that $v_i \notin D$ but $v_j \in D$ for all $1 \leq j < i$. (Clearly $3 \leq i \leq 5$). Note that $v_{i+1} \notin D$. If $i = 3$ or 4, there is no $v \in D$ such that $PN_t(D, v) \subseteq N[v_0] = \{v_0, v_1\}$. So $i \neq 3$ or 4. If $i = 5$ then $D = \{v_1, v_2\} \cup \{v_{j+3}, v_{j+4} / 0 \leq j \leq k - 1\}$ and there is no $v \in D$ such that $PN_t(D, v) \subseteq N[v_{4k+1}]$, which is a contradiction.

Case(ii)
The number of components of $\langle V - D \rangle$ is $k + 2$. In this case $\langle D \rangle$ has $k + 1$ components and each component is P_2. As $v_0, v_3, v_4 \notin D$ and $v_1, v_2, v_5, v_6 \in D$, there is no $v \in D$ such that $PN_t(D, v) \subseteq N[v_0] = \{v_0, v_1\}$, which is a contradiction. Thus P_{4k+2} is not TVE for all $k \geq 3$. □

Similar to Theorem 2.4, we obtain the following theorem which characterizes TVE cycles.

Theorem 2.5 *Cycles C_3, C_4, C_5, C_6 and C_{10} are the only cycles which are TVE.*

Total very excellent caterpillars:
A caterpillar is a tree T such that removal of all the pendant vertices of T leaves a path P which is called the spine of T.

If T is a caterpillar and $P : u_1, u_2, \ldots, u_k$ is the spine of T, to each $i (1 \leq i \leq k)$, let a_i be the number of pendant vertices of T which are adjacent to the vertex v_i. Then the caterpillar T can be represented by the finite sequence $(a_1, a_2, \ldots, a_k)$. Note that $a_1 > 0$ and $a_k > 0$; and for all other i, $a_i \geq 0$. For example, the sequence (1, 2, 3, 0, 3, 2) represents the caterpillar shown in the following figure.

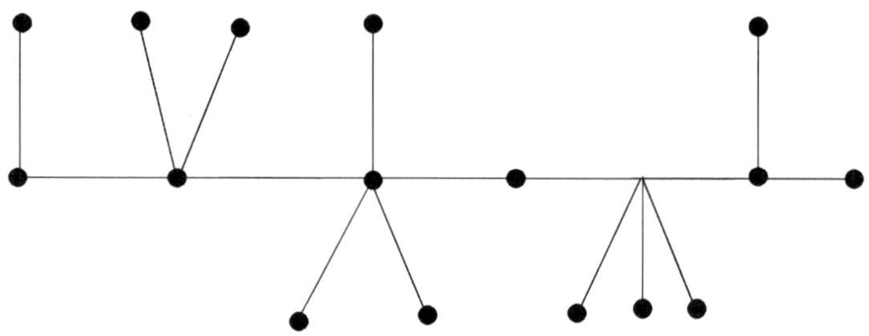

Caterpillar (1, 2, 3, 0, 3, 2)

We now characterize the caterpillars which are total very excellent.

Theorem 2.6 *Let $T = \{a_1, a_2, \cdots, a_k\}$ be a caterpillar ($T \neq K_2$). T is TVE if and only if the following condition holds:*

(i) *If $a_i \neq 0 (i < k)$, then either $a_{i+1} = a_{i+2} = 0$ and $a_{i+3} \neq 0$ or $a_{i+s} = 0$ for $1 \leq s \leq 6$ and $a_{i+7} \neq 0$.*
(ii) *If $a_i \neq 0$, $a_{i+7} \neq 0$ then $a_{i+14} = 0$.*

Proof Assume that the given caterpillar is TVE. Let $P : u_1, u_2, \cdots, u_k$ be the spine of T and D be a TVE γ_t-set of T. To each i for which $a_i \neq 0$, select a pendant vertex v_i of T adjacent to u_i. Note that every γ_t-set of T contains u_i, whenever $a_i \neq 0$. If $x \in D$ such that $D' = (D - x) \cup \{v_i\}$ is a γ_t-set of T, then $D \cap N(u_i) = \{x\}$, otherwise $D' - v_i$ is also total dominating set of T, leading to a contradiction. So the γ_t-set D' does not contain u_{i-1} and u_{i+1}, as $D' \cap N(u_i) = \{v_i\}$. As u_{i-1} and u_{i+1} are not in a γ_t-set, it follows that $a_{i-1} = 0 = a_{i+1}$. Let j be such that

(i) $i < j \leq k$
(ii) $a_s = 0$ for all $i < s < j$ and
(iii) $a_j \neq 0$.

Let $A = \{ u_{i-1}, u_i, u_{i+1}, \cdots, u_j, u_{j+1} \} \cup N(u_i) \cup N(u_j)$.

As $(D - \{u_{i-1}, u_{i+1}\}) \cup \{v_i\}$ and $(D - \{u_{j-1}, u_{j+1}\}) \cup \{v_j\}$ are γ_t-sets of T, it follows that $D - A$ is a γ_t-set of $T - A$, and hence $D \cap A$ is a γ_t-set of $\langle A \rangle$. Let $D^* = ((D \cap A) - \{u_{i-1}, u_{j+1}\}) \cup \{u_{i+1}, u_{j-1}\}$. Let D^* is a TVE γ_t-set for the path $v_i u_i u_{i+1} \cdots u_j v_j$. As this TVE path contains more than four vertices $(i + 1 < j)$, by the Theorem 2.4, this path is either P_6 or P_{10}. So either $j = i + 3$ or $i + 7$. Then $a_i \neq 0 \Rightarrow$ either $a_{i+3} \neq 0$ and $a_{i+1} = a_{i+2} = 0$ or $a_{i+7} \neq 0$ and $a_{i+1} = \cdots = a_{i+6} = 0$. This proves (i). To prove (ii), assume that for some $i, a_i, a_{i+7}, a_{i+14}$ are positive. Let $A = \{u_s / i \leq s \leq i + 14\} \cup N(u_i) \cup N(u_{i+7}) \cup N(u_{i+14})$ and $D^* = ((D \cap A) - \{u_{i-1}, u_{i+15}\}) \cup \{u_{i+1}, u_{i+13}\}$. Then as before, D^* is a TVE γ_t-set for subtree H shown in the following figure.

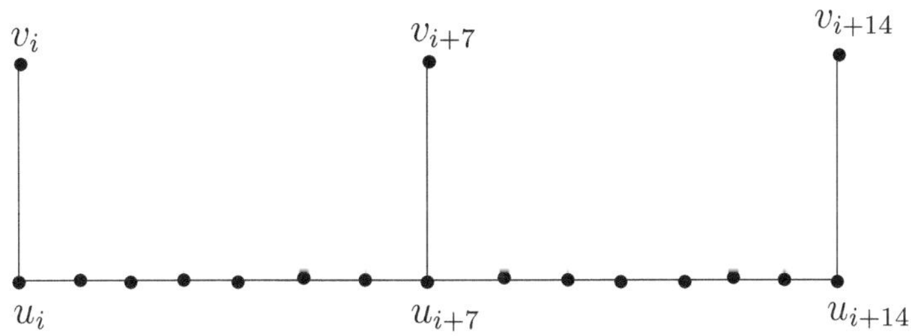

Note that $\gamma_t(H) = 10$ and $|D^* \cap \{u_{i+6}, u_{i+8}\}| = 1$. If $u_{i+6} \in D^*$, then $D^* = \{u_i, u_{i+1}, u_{i+3}, u_{i+4}, u_{i+6}, u_{i+7}, u_{i+10}, u_{i+11}, u_{i+13}, u_{i+14}\}$. In this case there is no $x \in D^*$ such that $(D^* - x) \cup \{u_{i+12}\}$ is a γ_t-set of H.

If $u_{i+8} \in D^*$, then $D^* = \{u_i, u_{i+1}, u_{i+3}, u_{i+4}, u_{i+7}, u_{i+8}, u_{i+10}, u_{i+11}, u_{i+13}, u_{i+14}\}$ and in this case there is no $x \in D^*$, such that $(D^* - x) \cup \{u_{i+2}\}$ is a γ_t-set of H. In either case we get a contradiction to the fact that D^* is a TVE γ_t-set of H. Thus $a_i \neq 0$ and $a_{i+7} \neq 0 \Rightarrow a_{i+14} = 0$ and hence we get (ii).

Now we assume that the given caterpillar $T = (a_1, \ldots, a_k)$ satisfies the conditions (i) and (ii), we give one TVE γ_t-set D for T. Let $S_0 = \{ u_i / 1 \leq i \leq k \text{ and } a_i \neq 0 \}$. In other words, S_0 is the set of support vertices of T. To each $i < k$, let

$$S_i = \begin{cases} \{u_{i+1}, u_{i+3}, u_{i+4}, u_{i+6}\} & \text{if } a_i \neq 0 \text{ and } a_{i+7} \neq 0 \\ \phi & \text{if either } a_i = 0 \text{ or } a_{i-7} \neq 0 \\ u_{i+1} & \text{if } a_{i-7} = 0;\ a_i \neq 0 \text{ and } a_{i+3} \neq 0 \end{cases}$$

and let $S_k = \{u_{k-1}\}$. One can verify that $D = \bigcup_{i=0}^{k} S_i$ is a TVE γ_t-set of T. Thus we have proved the theorem. □

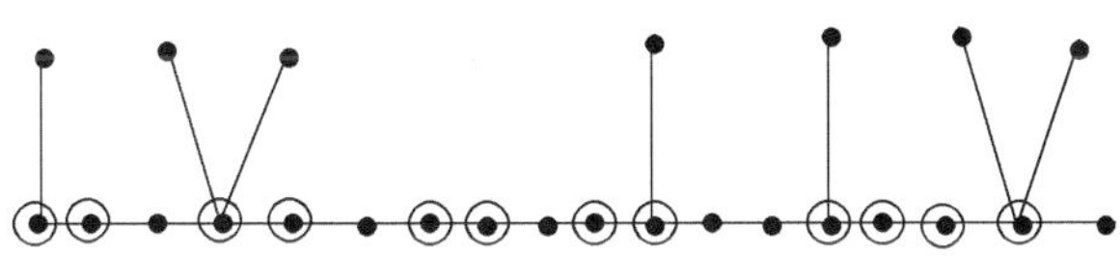

Example for a TVE - Caterpillar.

3 New TVE graphs from known TVE graphs:

The following lemmas are useful to produce graphs which are not TVE graphs.

Lemma 3.1 *Let G be a graph with $\delta(G) \geq 1$, and w_1uvw_2 be a path in G such that $degw_1 = degw_2 = 1$ in G, then G is not total very excellent.*

Proof The vertices u and v belong to every γ_t-set of G. If D is a total dominating set containing the vertex w_1, then $D - w_1$ is also a total dominating set of G. Thus w_1, w_2 are not an element of any γ_t-set of G. So w_1 and w_2 are γ_t-bad vertices of G. Thus G is not even γ_t-excellent. □

Corollary 3.2 *If G is a graph with $\delta(G) \geq 1$, then its corona $G \circ K_2$ is not γ_t-excellent.* □

Lemma 3.3 *Let u be a vertex of a graph G with $\delta(G) \geq 1$. If $w_1w_2uv_2v_1$ is a path in G such that $degw_1 = degv_1 = 1$ and $degw_2 = degv_2 = 2$, then G is not γ_t-excellent and hence not TVE.*

Proof For vertices w_2, v_2 belong to every total dominating set of G. If D is a total dominating set of G and if $u \notin D$, then both $w_1, v_1 \in D$ and $(D - \{w_1, v_1\}) \cup \{u\}$ is a total dominating set of G. It follows that D is not a γ_t-set of G. Thus u belongs to every γ_t-set of G. If D is any γ_t-set of G, then $u, w_2, v_2 \in D$ and $\{w_1, v_1\} \cap D = \phi$. In otherwords, w_1 and v_1 are γ_t-bad vertices of G. Thus G is not TVE. □

The following theorems are useful to obtain new TVE graphs from known TVE graphs.

Theorem 3.4 *Let u be a vertex of a graph G and $\delta(G) \geq 1$. A graph H is obtained from G by attaching a path P_4 at u. Then H is total very excellent iff there exist total very excellent γ_t-set D of G such that*

(i) *$u \in D$*
(ii) *there exists $v \in N(u) \cap D$ such that $PN_t(v, D) = \{u\}$.*

Proof Clearly $\gamma_t(H) = \gamma_t(G) + 2$. Let $uw_1w_2w_3w_4$ be the path attached to u to obtain H from G. Assume that H is TVE and D is a TVE γ_t-set of H. Assume that $w_2, w_3 \in D$. As D is TVE γ_t-set of H, $(D - w_2) \cup \{w_4\}$ is a γ_t-set of H. (Note that $(D - x) \cup \{w_4\}$ is a γ_t-set of H only when $PN_t(D, x) \subseteq N[w_4]$).

Case (i)
If $w_1 \in D$, then as $(D - w_2) \cup \{w_4\}$ is a γ_t-set of H, it follows that $u \in D$. As both $w_1, u \in D, N(u) \cap D \cap V(G) = \phi$. [Otherwise $D - w_1$ itself a total dominating set for H]. For any $v \in PN_t(u, D), v \in N(u) \cap V(G)$. Hence $D' = (D - (w_1, w_2, w_3)) \cup \{v\}$ is a TVE γ_t-set for G. Then D' is TVE and $u \in D'$, $v \in N(v) \cap V(G)$ such that $PN_t(v, D) = \{u\}$.

Case(ii)
Let $w_1 \notin D$. As $(D - \{w_2, w_3\}) \cup \{w_4\}$ is a γ_t-set of H and as $w_1 \notin D$, it follows that $u \in D$. As $w_1 \notin D$ and $u \in D$, $N(u) \cap D \cap V(G) \neq \phi$. As D is TVE γ_t-set of H, there exist $x \in D$ such that $D^* = \{D - x\} \cup \{w_1\}$ is a γ_t-set of H. Clearly $x \notin \{w_2, w_3\}$. We claim that $x \neq u$. If $x = u$, then $D - \{u, w_2, w_3\}$ is a total dominating set for G, which is a contradiction. Therefore $x \neq u$. If $x \notin N[u]$, then $D - \{w_3, w_2, x\}$ is a total dominating set for G, a contradiction. There fore $x \in N[u]$. Clearly $PN_t(D', x) = \{u\}$.

Conversely, let G be a total very excellent such that there is a TVE γ_t-set $D$of G such that (i) $u \in D$ (ii) there exist $v \in N(u) \cap D$ such that $PN_t(v, D) = \{u\}$. Then $D \cup \{w_2, w_3\}$ is a TVE γ_t-set of H. □

Theorem 3.5 *Let G be a total very excellent graph and $u \in V(G)$ such that $\gamma_t(G - N[u])$ exists and $\gamma_t(G - N[u]) \geq \gamma_t(G) - 1$ and $\gamma_t(G) = \gamma_t(G - u)$. Then the graph H is obtained from G by attaching a path P_3 at u, is total very excellent.*

Proof H is obtained by joining the vertex u of G and the vertex w_1 of the path $w_1w_2w_3$. Let D be any γ_t-set of G. Then $D \cup \{w_2, w_1\}$ is a total dominating set for H. Therefore $\gamma_t \leq \gamma_t(G)+2$. Let D' be any γ_t-set of H. Then $|D \cap \{w_2, w_1, w_0\}| = 2$. Let $D^* = D' \cap V(G)$. If $u \notin D^*$, then D^* is a total dominating set for $G - u$, and $|D^*| \geq \gamma_t(G - u) = \gamma_t(G)$. If $u \in D^*$ and u is not isolated in D^*, then D^* is the a total dominating set for G. If $u \in D^*$ but it is an isolated vertex in D^*, then $D^* - u$ is a total dominating set for $(G - N[u])$, and $|D^* - u| = |D^*| - 1 \geq \gamma_t[G - N(u)] \geq \gamma_t(G) - 1$. Thus, $\gamma_t(H) = \gamma_t(G) + 2$. If D is TVE γ_t-set of G, then $D \cup \{w_2, w_1\}$ is a γ_t-set of H. In fact it is a TVE γ_t-set of H. □

Theorem 3.6 *Let G be a TVE graph and u be a vertex of G such that $\gamma_t(G - u)$ exists and $\gamma_t(G - u) \geq \gamma_t(G)$. Attach a path $uw_1w_2w_3$ at u. Let the resulting graph be H. If there is a γ_t-set D for H such that either $u \notin D$ or u is not an isolated vertex in $\langle D \cap V(G)\rangle$, then H is also a TVE.*

Proof From the first part of the proof of Theorem 3.5, $\gamma_t(H) = \gamma_t(G) + 2$. If D^* is a TVE γ_t-set of G, then $D^* \cup \{w_1, w_2\}$ is a TVE γ_t-set of G. □

4 Characterization of TVE Trees

In this section, we characterize the TVE trees. First we introduce four types of operations.

Type I Operation:
Let G be a TVE graph and v be a vertex of G adjacent to a pendant vertex w. Attach $k(k \geq 1)$ more pendant vertices at v. Then the resulting graph H is also TVE graph. We say that H is obtained from G by using the operation of Type I.

Type II Operation:
Let G be a TVE graph and D be a TVE γ_t-set of G. Let $u \in D$ and there exist $x \in D$ such that $PN(x, D) = \{u\}$. Let H be the graph obtained from G by attaching a path $uw_1w_2w_3w_4$ at u. Then by Theorem 3.4, H is also TVE. We say that H is obtained from G by using the operation of Type II.

Type III Operation:
Let G be a TVE graph and u be a vertex of G. Let $\gamma_t(G - u) \geq \gamma_t(G)$. Attach a path $uw_3w_2w_1$ at u. Let the resulting graph be H. If there is a γ_t-set D of H such that either $u \notin D$ or u is not isolated in $\langle D \cap V(x) \rangle$, then H is said to be obtained from G with the operation of the type III. (By Theorem 3.6, H is aalso TVE).

Type IV Operation
Let G be a TVE graph and D be a TVE γ_t-set of G. Let $x \in D$ such that $PN_t(x, D) = \{u\}$. Attach a new path $w_1w_2w_3w_4u_3u_2u_1$ by joining w_4 and u. The resulting graph H is also a TVE graph. We say that H is obtained from G by using the operation of Type IV.

The following theorem characterize TVE trees.

Theorem 4.1 *A tree T with $n \geq 2$ vertices is TVE tree if and only if it can be obtained from P_2 by applying finite sequence of operations I, II, III, and IV.*

Proof We prove the result by induction of the order of T. Clearly the result is true if order is of $T \leq 5$.

Assume that the result is true for all TVE trees of order $m < n$ for some $n \geq 5$. Let T be a TVE tree with n vertices. If there is a vertice $u \in V(T)$ which is adjacent to two pendant vertices, say u_1 and u_2. Then the tree $T - u_1$ is a TVE tree of order $n - 1$ and hence by our assumption the result is true for $T - u_1$ and hence for T.

Assume that every vertex of T is adjacent to at most one pendant vertex x. $\cdots\cdots\cdots (*)$

Let $u_1, u_2, \ldots, u_k$ be the longest path in T. Then by $(*)$, $k \geq 4$ and $\deg u_1 = \deg u_k = 1$; while $\deg u_2 = \deg u_{k-1} = 2$. By lemma 3.1, the vertex u_3 is not adjacent to any pendant vertex and by lemma 3.3, there is no path $u_3w_2w_1$ where $w_2 \neq u_2, u_3$. In other words $\deg(u_3) = 2$ in T.

Let D be a γ_t–set of T. Without loss of generality, we can assume that D contains no pendant vertex.

The vertex u_4 is not adjacent to a pendant vertex $w \neq u_5$. [For if a pendant vertex $w \neq u_5$ is adjacent to u_4, there exist an element $x \in D$ such that $(D - x) \cup \{w\}$ is a γ_t-set. Clearly $x \neq u_2, u_3, u_4$. Now $(D - x)$ is also a total dominating set of T, which is a contradiction].

Case(i) Let $\deg(u_4) = 2$. We claim that $u_5 \in D$. As D is a TVE γ_t-set, $(D-x) \cup \{u_1\}$ is a γ_t-set of T for some $x \in D$. As $PN_t(D, x) \subseteq N[u_1]$, it follows that $x = u_3$. Let $D^* = (D - x) \cup \{u_1\}$. As D^* is a γ_t-set of T, $\deg(u_4) = 2$ and $u_3 \notin D^*$, we have $u_5 \in D^*$(the vertex u_4 may or may not be in D^*). Thus $u_5 \in D$.

Subcase(i) If $u_4 \notin D$, let $D' = D - \{u_2, u_3\}$. Then D' is a γ_t-set for the tree $T_5 = T - \{u_1, u_2, u_3, u_4\}$. In fact D' is a TVE γ_t-set for T_5. As D is a TVE γ_t-set for T, and as $u_4 \notin D$, there is one $y \in D$ such that $(D - y) \cup \{u_4\}$ is a γ_t-set for T. Clearly $y \notin \{u_2, u_3\}$, and hence $y \in D'$. Now $PN_t(y, D') = \{u_5\}$; $y \neq u_5$ and yu_5 is an edge in T_5. Thus T_5 is a TVE tree of order $n - 4$, $u_5 \in D'$, a TVE γ_t-set for T_5; and $PN_t(y, D') = \{u_5\}$ for some $y \in D'$. T can be viewed as it is obtained from T_5 by attaching a path P_4 at u_5. In other words T is obtained from T_5 by using the operation of type II. By the induction hypothesis the result is true for T_5 and for T.

Subcase (ii) Both $u_4, u_5 \in D$. Note that $PN_t(D, u_5) \neq \phi$, select a vertex $w \in PN_t(u_5, D)$. As $N(u_4) \cap D = \{u_3, u_5\}$, $w \neq u_4$. Also $w \notin D$, (otherwise $D - u_4$ is a γ_t-set for T). Let $D' = (D - \{u_2, u_3, u_4\}) \cup \{w\}$. Then D' is a TVE γ_t—set for T_5, also $PN_t(D', w) = \{u_5\}$. As T can be viewed to be obtained from T_5 using the operation of type II, the result is true for T also.

Case(ii) Let $\deg(u_4) \geq 3$. Note that u_4 is not adjacent to any pendant vertex of T.

Subcase (i) Let $u_4 w_2 w_1$ be a path in T with $\deg w_1 = 1$ and $w_2 \neq u_5$. Then $w_2, u_4 \in D$ and let $D' = D - \{u_2, u_3\}$. Then D' is a TVE γ_t-set for T_4 (where $T_4 = T - \{u_1, u_2, u_3\}$). Note that $\gamma_t(T_4 - u_4) \not< \gamma_t(T_4)$. Hence T can be viewed as obtained by using the operation of Type III. As T_4 is a TVE tree of order less than n by the induction hypothesis, the result is true for T_4 and hence for T.

Subcase(ii) Let $u_4 w_3 w_2 w_1$ be a path of length three in T, where $w_3 \neq u_3, u_5$ and no free P_2 is attached at u_4. By Lemmas 3.1 and 3.3, $\deg(w_3) = 2$.

(a) If $N(u_4) \cap (D - \{u_3, w_3\}) \neq \phi$, Then $D - \{u_2, u_3\}$ is a TVE γ_t-set for T_4, where $T_4 = T - \{u_1, u_2, u_3\}$ and $\gamma_t(T - u_4) \geq \gamma_t(T_4) = \gamma_t(T) - 2$. In this case T is obtained from T_4 by using the operation of type III.

(b) If $N[u_4] \cap (D - \{u_3, u_4\}) = \{u_4\}$ then $\deg(u_4) = 3$; $u_5 \notin D$ and $PN_t(D, u_4) = \{u_5\}$. It follows that $N[u_5] \cap D = \{u_4\}$ and hence $u_6 \notin D$. But u_6 is dominated by $D - \{u_2, u_3, u_4, w_2, w_3\}$. Let $D' = (D - \{u_2, u_3, u_4, w_2, w_3\}) \cup \{u_6\}$. Then D' is a TVE γ_t-set for T_5 (The tree component of $T - u_4u_5$ that contains u_5) and $PN_t(D', u_6) = \{u_5\}$. The result is true for T_5 and T is obtained from T_5 using the operation of type IV.

If $N[u_4] \cap (D - \{u_3, w_3\}) = \phi$, then $d(u_4) = 3$ in T and $u_4, u_5 \neq D$. There exist $x \in D - \{u_2, u_3, w_2, w_3\}$ such that $(D - x) \cup \{u_4\}$ is a γ_t-set for T. As $PN_t(x, D) \subseteq N[u_4]$, we have $PN_t(x, D) = \{u_5\}$. Note that $x \neq u_5$. Let T_5 be the tree component of $T - u_4u_5$ that contains the vertex u_5. As D is a TVE γ_t-set of T and $u_4, u_5 \notin D$, it follows that $\gamma_t(T_5) = \gamma_t(T) - 4$ and $D - \{u_2, u_3, w_2, w_3\}$ is a TVE γ_t-set of T. In this case also T is obtained from T_5 by using the operation of type IV. □

Illustration: We illustrate the theorem by the following example given in the figure.

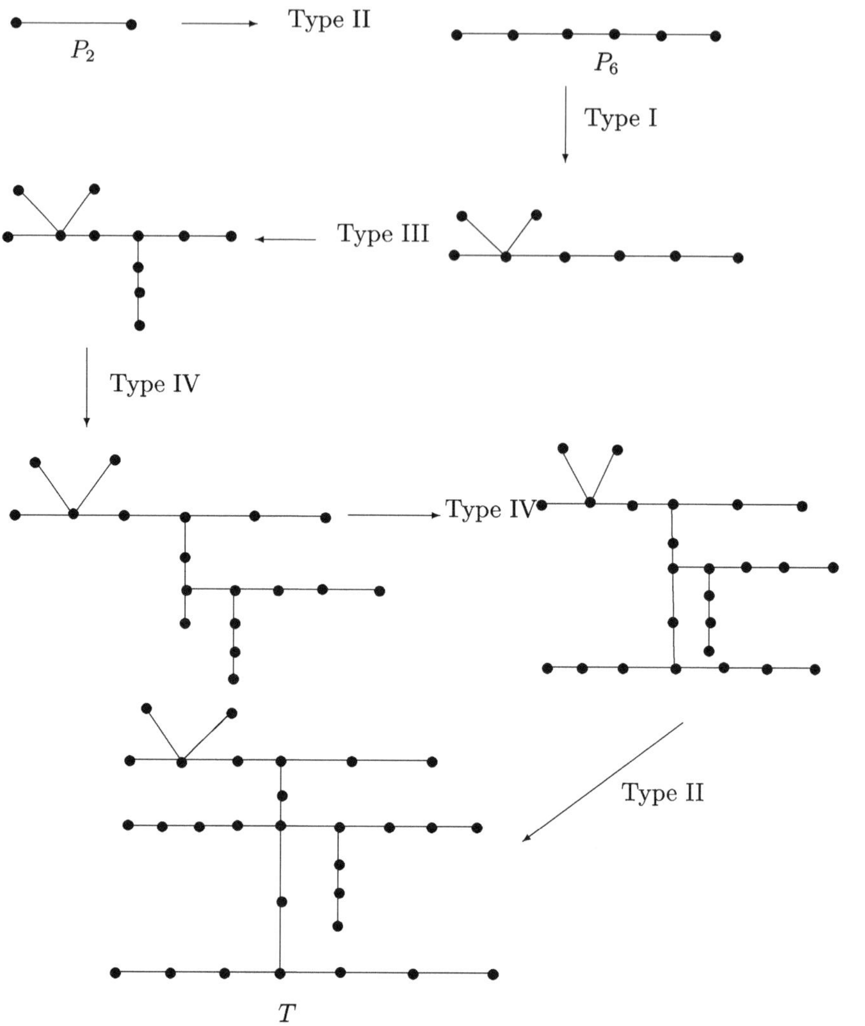

An issustration to obtain a TVE tree T from P_2 using the theorem 4.1.

Theorem 4.2 *If T is a TVE tree, then $\gamma_t(T)$ is even.*

Proof Note that $\gamma_t(P_2) = 2$. Let T be a TVE tree of order $n \geq 2$. Then by the Theorem 4.1, T is obtained from P_2 by applying finite sequence of operations I, II, III or IV. So we get a sequence $P_2 = T_1, T_2, \ldots, T_k = T$ of TVE trees such that each T_{i+1} is obtained from T_i by using one of the four operations. Note that

(1) $\gamma_t(T_i) = \gamma_t(T_{i+1})$ if T_{i+1} is obtained from T_i by using the operation of type I.
(2) $\gamma_t(T_{i+1}) = \gamma_t(T_i) + 2$ if T_{i+1} is obtained from T_i by using the operation of type either II or III.
(3) $\gamma_t(T_{i+1}) = \gamma_t(T_i) + 4$ if T_{i+1} is obtained from T_i by using the operation of type IV.

Thus $\gamma_t(T_i)$ is even iff $\gamma_t(T_{i+1})$ is even.
As $\gamma_t(T_1) = \gamma_t(P_2) = 2$, it follows that $\gamma_t(T)$ is even. □

Remark 4.3 For every integer $m \geq 1$, there exists a TVE tree T_m with $\gamma_t(T_m) = 2m$. If $m = 1$, take $T_1 = P_2$. If $m > 1$, consider the star $K_{1,m}$. Now subdivide exactly one edge of $K_{1,m}$ once and subdivide all other edges of $K_{1,m}$ twice. Let T_m be the resulting tree. Clearly $\gamma_t(T_m) = 2m$ and T_m is a TVE tree. (For $m = 4$, T_4 is shown in the following figure).

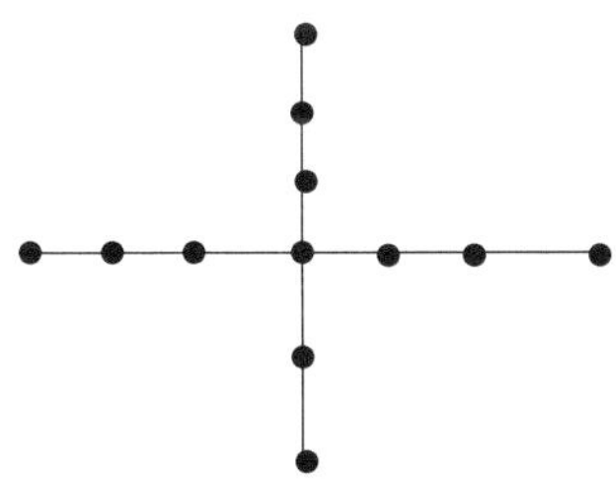

Tree T_4 which is a TVE tree and $\gamma_t(T_4) = 8$.

Conjecture: If G is a TVE-graph, then $\gamma_t(G)$ is even.

Acknowledgments This work was supported by the Department of Science and Technology, Government of India through Project SR/S4/MS:357/06 to the first and second authors. The authors thank the referees for their valuable comments which helped to improve the paper.

References

1. Balakrishnan, R., Ranganathan, K., A text book of graph theory, Springer (2000)
2. Fricke, G.H., Haynes, T.W., Hedetniemi, S.T., Hedetniemi, S.M., Laskar, R.C.: Excellent trees. Bull. Int. Combin. Appl. **34**, 27–38 (2002)
3. Haynes, T.W., Hedetniemi, S.T., Slater, P. J.: Fundamentals of domination in graphs. Marcel Dekker Inc, (1998)
4. Henning, M.A., Haynes, T.W.: Total domination excellent trees. Discrete Appl. Math. **263**, 93–104 (2003)
5. Sridharan, N., Yamuna, M.: Every γ-excellent, γ-flexible graph is γ_{bi}-excellent. J. Discrete Math. Sci. Cryptogr. **7**(1),103–110 (2004)
6. Sridharan, N., Yamuna, M.: A note on excellent graphs. Ars Combinatoria **78**, 267–276 (2006)
7. Sridharan, N., Yamuna, M.: Very excellent graphs and rigid very excellent graphs. AKCE J. Graphs. Combin. **4**, 211–221 (2007)
8. Yamuna, M.: Excellent—Just excellent—Very excellent graphs. Ph.D thesis, Alagappa University (2003)

Quadratic Residue Cayley Graphs on Composite Modulus

Angsuman Das

Abstract In this paper, we initiate the study of quadratic residue Cayley graphs Γ_N modulo $N = pq$, where p, q are distinct primes of the form $4k + 1$. It is shown that Γ_N is a regular, symmetric, Eulerian, and Hamiltonian graph. Also, the vertex connectivity, edge connectivity, diameter, and girth of Γ_N are studied and their relationship with the forms of p and q are discussed. Moreover, we specify the forms of primes for which Γ_N is triangulated or triangle-free and provide some bounds for the order of the automorphism group of Γ_N, $Aut(\Gamma_N)$ and domination number of Γ_N.

Keywords Cayley graph · Quadratic residue · Pythagorean prime

1 Introduction

The Cayley graph was first considered for finite groups by Arthur Cayley in 1878. Since then, a lot of research has been done on various families of Cayley graphs, e.g., unitary Cayley graphs, Paley graphs, Dihedral Cayley graphs, quadratic residue Cayley graphs, etc. In this paper, we will focus on quadratic residue Cayley graphs, i.e., where the generating set is the set of all quadratic residues in the group. Many works exist in the literature on Cayley graphs on quadratic residues on prime and prime power modulus. In fact, the family of Cayley graphs also contain another important subfamily of Paley graphs, where the generating set is the set of all quadratic residues in the finite field $\mathbb{F}_q, q = p^n$ with a prime p of the form $4k + 1$. In [1], the authors studied quadratic residue modulo 2^n Cayley graphs. However, as far as our

The author's research is supported in part by National Board of Higher Mathematics, Department of Atomic Energy, Government of India (No 2/48(10)/2013/NBHM(R.P.)/R&D II/695).

A. Das
Department of Mathematics, St. Xavier's College,
Kolkata, 30, Park Street, Kolkata 700016, India
e-mail: angsumandas054@gmail.com

R.N. Mohapatra et al. (eds.), *Mathematics and Computing*,
Springer Proceedings in Mathematics & Statistics 139,
DOI 10.1007/978-81-322-2452-5_19

knowledge, quadratic residue Cayley graphs on modulus of the form pq, where p and q are distinct primes remained unexplored till date.

In this paper, we study the quadratic residue Cayley graphs Γ_N modulo $N = pq$, where p, q are distinct Pythagorean primes, i.e., primes of the form $4k+1$. It is shown that Γ_N is a regular, Eulerian, Hamiltonian, and arc-transitive graph. Also, the vertex connectivity, edge connectivity, diameter, and girth of Γ_N are studied. Moreover, the conditions under which Γ_N is triangulated and triangle-free are discussed. We also provide some bounds for the order of $Aut(\Gamma_N)$ and domination number of Γ_N.

2 Preliminaries

In this section, for convenience of the reader and also for later use, we recall some definitions and notations concerning integers modulo N and quadratic residues in elementary number theory. For undefined terms and concepts in graph theory the reader is referred to [2] and [5]. Throughout this paper, graphs are undirected, simple, and without loops.

An odd prime p is called a Pythagorean prime if $p \equiv 1(\text{mod } 4)$. Throughout this paper, even if it is not mentioned, a prime p always means a Pythagorean prime and $N = pq$ means the product of two distinct Pythagorean primes. By $\mathbb{Z}_N, \mathbb{Z}_N^*, \mathcal{QR}_N, \mathcal{QNR}_N, \mathcal{J}_N^{+1}, \mathcal{J}_N^{-1}$, we mean the set of all integers modulo N, the set of all units in integers modulo N, the set of all quadratic residues and nonquadratic residues, which are also units in integers modulo N, the set of all units in integers modulo N with Jacobi symbol $+1$ and -1 respectively. For the sake of convenience, $a \equiv b(\text{mod } n)$ is sometimes written as $a = b$, in places where the modulus is clear from the context. We can conclude the following lemma from the results which can be found in any elementary number theory book, e.g., [4].

Lemma 1 *If $N = pq$, then the following are true:*

- $\mathcal{J}_N^{+1}$ *is a subgroup of* $\mathbb{Z}_N^*$ *and* $\mathcal{QR}_N$ *is a subgroup of* $\mathcal{J}_N^{+1}$.
- $|\mathbb{Z}_N^*| = \phi(N) = (p-1)(q-1)$, $|\mathcal{J}_N^{+1}| = |\mathcal{J}_N^{-1}| = \frac{(p-1)(q-1)}{2}$ *and* $|\mathcal{QR}_N| = \dfrac{(p-1)(q-1)}{4}$, *where ϕ denotes the Euler's Phi function.*
- $x \in \mathcal{QR}_N \iff x \in \mathcal{QR}_p \cap \mathcal{QR}_q$.
- $x \in \mathcal{J}_N^{+1} \setminus \mathcal{QR}_N \iff x \in \mathcal{QNR}_p \cap \mathcal{QNR}_q$.
- $x \in \mathcal{J}_N^{-1} \iff x \in \mathcal{QNR}_p \cap \mathcal{QR}_q$ *or* $x \in \mathcal{QR}_p \cap \mathcal{QNR}_q$. □

Lemma 2 *If p, q are two distinct primes of the form $p \equiv q \equiv 1(\text{mod } 4)$, then -1 is a quadratic residue in $\mathbb{Z}_N$.*

Proof To show that -1 is a quadratic residue in $\mathbb{Z}_N$, we need to show that $x^2 \equiv -1(\text{mod } N)$ has a solution. But,

$$x^2 \equiv -1 (\text{mod } N) \Leftrightarrow x^2 \equiv -1 (\text{mod } p) \text{ and } x^2 \equiv -1 (\text{mod } q)$$

Now, as p and q are Pythagorean primes, -1 is a square in both $\mathbb{Z}_p$ and $\mathbb{Z}_q$. Thus, $x^2 \equiv -1 (\text{mod } N)$ have a solution in $\mathbb{Z}_N$. □

3 Quadratic Residue Cayley Graph Modulo N

We now define the quadratic residue Cayley graphs Γ_N modulo $N = pq$ and study some of their basic properties.

Definition 1 *Quadratic Residue Cayley Graph modulo N* For $N = pq$, quadratic residue modulo N Cayley graphs Γ_N is given by $\Gamma_N = (V, E)$, where $V = \mathbb{Z}_N$ and $(a, b) \in E \Leftrightarrow a - b \in \mathcal{QR}_N$.

Remark 1 Γ_N is a Cayley Graph (G, S) where $G = (\mathbb{Z}_N, +)$ and $S = \mathcal{QR}_N$. Observe that as $-1 \in \mathcal{QR}_N$ and $\mathcal{QR}_N$ is a group with respect to modular multiplication, $\mathcal{QR}_N$ is also closed with respect to additive inverse, i.e., $S = -S$ and $0 \notin S$.

Theorem 1 *Γ_N is Hamiltonian and hence connected.*

Proof Since, $1 \in \mathcal{QR}_N$, the vertex set $\{0, 1, 2, \ldots, N - 1\}$, taken in order, can be thought of as a Hamiltonian path. Hence, the theorem. □

Theorem 2 *Γ_N is regular with valency $\phi(N)/4$ and hence Eulerian.*

Proof Let $x \in \mathbb{Z}_N$. By $N(x)$, we mean the set of vertices in Γ_N which are adjacent to x, i.e., $N(x) = \{z \in \mathbb{Z}_N : x - z \in \mathcal{QR}_N\}$. If possible, let $\exists z_1, z_2 \in N(x)$ with $z_1 \neq z_2$ such that $x - z_1 = x - z_2$. But, $x - z_1 = x - z_2 = s(\text{say}) \in \mathcal{QR}_N \Rightarrow z_1 = x - s = z_2$, a contradiction. Thus, $\forall s \in \mathcal{QR}_N, \exists$ a unique $z \in \mathbb{Z}_N$ such that $x - z = s$. Thus, degree or valency of $x = |N(x)| = |\mathcal{QR}_N| = \phi(N)/4$. Now, let $p = 4k + 1, q = 4l + 1$. Since, degree of each vertex $= \frac{\phi(N)}{4} = \frac{(p-1)(q-1)}{4} = \frac{4k \cdot 4l}{4} = 4kl$ is even, Γ_N is Eulerian. □

Note However, Γ_N is not strongly regular (See Remark 3).

Remark 2 Γ_N is not self-complementary: A necessary condition for a self-complementary graph G with n vertices is that number of edges in G equals $\frac{n(n-1)}{4}$. But, the number of edges in Γ_N with N vertices is $\frac{N \cdot \phi(N)}{8} < \frac{N(N-1)}{4}$. However, the next theorem shows that Γ_N has a homomorphic image of itself as a subgraph of its complement graph.

Theorem 3 *Γ_N has a homomorphic image of itself as a subgraph of its complement graph Γ_N^c.*

Proof Let $n \in \mathbb{Z}_N^* \setminus \mathcal{QR}_N$. We define a function $\psi : \Gamma_N \to \Gamma_N^c$ given by $\psi(x) = nx$. For injectivity, $\psi(x_1) = \psi(x_2) \Rightarrow nx_1 = nx_2 \Rightarrow x_1 = x_2$, as n is a unit in $\mathbb{Z}_N$. For homomorphism, x, y adjacent in $\Gamma_N \Rightarrow x - y \in \mathcal{QR}_N \Rightarrow n(x-y) \notin \mathcal{QR}_N \Rightarrow nx$ and ny are not adjacent in Γ_N, i.e., $\psi(x)$ and $\psi(y)$ are adjacent in Γ_N^c. □

4 Symmetricity of Γ_N

In this section, we study the action of $Aut(\Gamma_N)$ and its consequences.

Theorem 4 *Γ_N is vertex-transitive.*

Proof As Γ_N is a Cayley graph, it is vertex transitive. (by Theorem 3.1.2 in [2]) However, we show the existence of such automorphisms explicitly, which will be helpful later.

Choose $a \in \mathcal{QR}_N$ and $b \in \mathbb{Z}_N$ and define a function $\varphi : \mathbb{Z}_N \to \mathbb{Z}_N$ given by $\varphi(x) = ax + b, \forall x \in \mathbb{Z}_N$. We show that φ is an automorphism. φ is injective, for

$$\varphi(x_1) = \varphi(x_2) \Rightarrow ax_1 + b = ax_2 + b \Rightarrow a(x_1 - x_2) = 0 \Rightarrow x_1 = x_2 \text{ as } a \in \mathbb{Z}_N^*$$

For surjectivity, $\forall y \in \mathbb{Z}_N, \exists x = a^{-1}y - a^{-1}b \in \mathbb{Z}_N$ such that $\varphi(x) = a(a^{-1}y - a^{-1}b) + b = y$. Moreover, φ is a graph homomorphism, as x and y are adjacent in Γ_N $\Leftrightarrow x - y \in \mathcal{QR}_N \Leftrightarrow a(x-y) + b - b \in \mathcal{QR}_N \Leftrightarrow (ax+b) - (ay+b) \in \mathcal{QR}_N \Leftrightarrow \varphi(x) - \varphi(y) \in \mathcal{QR}_N \Leftrightarrow \varphi(x)$ and $\varphi(y)$ are adjacent in Γ_N. Thus, $\varphi \in Aut(\Gamma_N)$.

Now, let $u, v \in \mathbb{Z}_N$ be two vertices of Γ_N. We take $a = 1 \in \mathcal{QR}_N$ and $b = v - u \in \mathbb{Z}_N$. Then the map $\varphi : \mathbb{Z}_N \to \mathbb{Z}_N$ given by $\varphi(x) = ax + b$ is an automorphism on Γ_N such that $\varphi(u) = v$. Thus, $Aut(\Gamma_N)$ acts transitively on $\mathbb{Z}_N$, i.e., $V(\Gamma_N)$. □

Theorem 5 *Γ_N is arc-transitive and hence edge transitive.*

Proof Let $\{u_1, v_1\}, \{u_2, v_2\}$ be two edges (considered as having a direction) in Γ_N. Therefore, $u_1 - v_1, u_2 - v_2 \in \mathcal{QR}_N$. We take $a = (u_2 - v_2)(u_1 - v_1)^{-1} \in \mathcal{QR}_N$ and $b = u_2 - au_1 \in \mathbb{Z}_N$ and construct the automorphism $\varphi(x) = ax + b$ as in Theorem 4. Since $\varphi(u_1) = u_2$ and $\varphi(v_1) = v_2$, Γ_N is arc transitive, and hence edge transitive. □

Corollary 1 $|Aut(\Gamma_N)| \geq \frac{N\phi(N)}{4}$.

Proof In Theorem 4, it was shown that $\varphi : \mathbb{Z}_N \to \mathbb{Z}_N$ given by $\varphi(x) = ax + b, \forall x \in \mathbb{Z}_N$ is an automorphism for $a \in_R \mathcal{QR}_N$ and $b \in_R \mathbb{Z}_N$. Thus, $|Aut(\Gamma_N)| \geq \frac{N\phi(N)}{4}$. □

Corollary 2 *Edge connectivity of Γ_N is $\phi(N)/4$.*

Proof Since Γ_N is connected and vertex-transitive, by Lemma 3.3.3 in [2], its edge connectivity is equal to its valency. □

Lemma 3 [2] *The vertex connectivity of a connected edge transitive graph is equal to its minimum valency.* □

Corollary 3 *Vertex connectivity of Γ_N is $\phi(N)/4$.*

Proof Since, Γ_N is a connected edge-transitive graph with valency $\phi(N)/4$, by Lemma 3, Γ_N has vertex connectivity $\phi(N)/4$. □

5 Diameter, Girth, and Triangles of Γ_N

In this section, we find the diameter and girth of Γ_N. It is noted that Γ_N is of dual nature when it comes to diameter and girth. To be more specific, it depends on whether 5 is a factor of N or not. If 5 is one of the two factors of N, we call it Γ_N of Type-I and else call it Γ_N of Type-II. First, we prove two lemmas which will be used later.

Lemma 4 *Let p be a prime of the form $4k + 1$ and $c \in \mathbb{Z}_p$. Then, the number of ways in which c can be expressed as difference of two quadratic residues in $\mathbb{Z}_p^*$ are*

$$(1)\ \tfrac{p-1}{2} \text{ if } c \equiv 0(\text{mod } p).\ (2)\ \tfrac{p-5}{4} \text{ if } c \in \mathcal{QR}_p.\ (3)\ \tfrac{p-1}{4} \text{ if } c \in \mathcal{QNR}_p.$$

Proof

1. If $c \equiv 0(\text{mod } p)$, then for all $r \in \mathcal{QR}_p$, c can be expressed as $r - r$. Thus, the number in this case, is equal to the number of elements in $\mathcal{QR}_p$, i.e., $\frac{p-1}{2}$.
2. For this case, assume that $c \not\equiv 0(\text{mod } p)$, i.e., $c \in \mathbb{Z}_p^*$. Let $c = a^2 - b^2 = (a+b)(a-b)$, where $a, b \in \mathbb{Z}_p^*$. Now, for all $p - 1$ values of $d \in \mathbb{Z}_p^*$, letting $a+b = d$; $a-b = \frac{c}{d}$, we get all possible solutions of the equation $c = a^2 - b^2$. From this, we get $a = \frac{1}{2}\left(d + \frac{c}{d}\right)$ and $b = \frac{1}{2}\left(d - \frac{c}{d}\right)$. However, we need to ensure that $a, b \in \mathbb{Z}_p^*$, i.e., $d \pm \frac{c}{d} \not\equiv 0(\text{mod } p)$, i.e., $d^2 \not\equiv \pm c(\text{mod } p)$.
   Now, if $c \in \mathcal{QR}_p$, then $-c \in \mathcal{QR}_p$. (as -1 is a quadratic residue in $\mathbb{Z}_p^*$). In this case, there exist two square roots of c and two other square roots of $-c$. Thus, we lose 4 possible values of d. Thus, the number of solutions is reduced to $p - 5$. Moreover, it is observed that the 4 solutions of $(a + b, a - b)$, namely $(d, \frac{c}{d})$, $(-d, \frac{c}{-d})$, $(\frac{c}{d}, d)$, $(\frac{c}{-d}, -d)$ lead to the same solution $a^2 = \frac{1}{4}\left(d + \frac{c}{d}\right)^2$; $b^2 = \frac{1}{4}\left(d - \frac{c}{d}\right)^2$. (As p is odd, $d \neq -d$). Thus, the number of distinct solutions is reduced to $\frac{p-5}{4}$.
3. The proof for $c \in \mathcal{QNR}_p$ follows exactly using the same arguments except the fact that in this case, we do not lose those four solutions as $c \not\equiv \pm d^2$. Thus, the number of ways c can be expressed as difference of quadratic residues is $\frac{p-1}{4}$. □

Lemma 5 *Let $N = pq$, where p, q are Pythagorean primes. Then*

1. *If $c \in \mathcal{QR}_N$, then the number of ways in which c can be expressed as difference of two quadratic residues, i.e., $c = x^2 - y^2$, $x, y \in \mathbb{Z}_N^*$ is $\frac{(p-5)(q-5)}{16}$.*

2. *If* $c \in \mathcal{J}_N^{+1} \setminus \mathcal{QR}_N$, *then the number of ways in which c can be expressed as difference of two quadratic residues is* $\frac{(p-1)(q-1)}{16}$.
3. *If* $c \in \mathcal{J}_N^{-1}$, *then the number of ways in which c can be expressed as difference of two quadratic residues is either* $\frac{(p-1)(q-5)}{16}$ [*if* $c \in \mathcal{QR}_q$, *but* $c \notin \mathcal{QR}_p$] *or* $\frac{(p-5)(q-1)}{16}$ [*if* $c \in \mathcal{QR}_p$, *but* $c \notin \mathcal{QR}_q$].
4. *If* $c(\neq 0) \in \mathbb{Z}_N \setminus \mathbb{Z}_N^*$, *i.e., c is a nonzero, nonunit in* $\mathbb{Z}_N$, *then*

   a. *If* $c \equiv 0 (mod\ q)$ *and* $c \in \mathcal{QR}_p$, *then the number of ways in which c can be expressed as difference of two quadratic residues is* $\frac{(p-5)(q-1)}{8}$.
   b. *If* $c \equiv 0 (mod\ q)$ *and* $c \in \mathcal{QNR}_p$, *then the number of ways in which c can be expressed as difference of two quadratic residues is* $\frac{(p-1)(q-1)}{8}$.
   c. *If* $c \equiv 0 (mod\ p)$ *and* $c \in \mathcal{QR}_q$, *then the number of ways in which c can be expressed as difference of two quadratic residues is* $\frac{(q-5)(p-1)}{8}$.
   d. *If* $c \equiv 0 (mod\ p)$ *and* $c \in \mathcal{QNR}_q$, *then the number of ways in which c can be expressed as difference of two quadratic residues is* $\frac{(q-1)(p-1)}{8}$.

Proof

1. If $c \in \mathcal{QR}_N$, then $c \in \mathcal{QR}_p$ and $c \in \mathcal{QR}_q$. Thus, the result follows from the Chinese Remainder Theorem and second part of Lemma 4.
2. If $c \in \mathcal{J}_N^{+1} \setminus \mathcal{QR}_N$, then $c \in \mathcal{QNR}_p$ and $c \in \mathcal{QNR}_q$. Thus, the result from the Chinese Remainder Theorem and third part of Lemma 4.
3. If $c \in \mathcal{J}_N^{-1}$, then either of two cases may arise, namely $c \in \mathcal{QR}_q;\ c \in \mathcal{QNR}_p$ or $c \in \mathcal{QR}_p;\ c \in \mathcal{QNR}_q$.
   If $c \in \mathcal{QR}_q;\ c \in \mathcal{QNR}_p$, then by applying the second part of Lemma 4 for q and the third part of Lemma 4 and Chinese Remainder Theorem, we get the count as $\frac{(p-1)(q-5)}{16}$. Similarly, the case $c \in \mathcal{QR}_p;\ c \in \mathcal{QNR}_q$ follows.
4. As $c \in \mathbb{Z}_N \setminus \mathbb{Z}_N^*$, either $p \mid c$ or $q \mid c$ [not both, as that would imply $c \equiv 0 (\text{mod } N)$].
   If $q \mid c$ and $p \nmid c$, two cases arise, namely (a) $c \equiv 0 (\text{mod } q)$ and $c \in \mathcal{QR}_p$, and (b) $c \equiv 0 (\text{mod } q)$ and $c \in \mathcal{QNR}_p$. In both the cases, the lemma follows from the Chinese Remainder Theorem and Lemma 4.
   Similarly, if $q \nmid c$ and $p \mid c$, two cases arise, namely (c) $c \equiv 0 (\text{mod } p)$ and $c \in \mathcal{QR}_q$ and (d) $c \equiv 0 (\text{mod } p)$ and $c \in \mathcal{QNR}_q$. Again, these cases follow similarly. □

5.1 Quadratic Residue Cayley Graph of Type-I

Lemma 6 *If* $N = 5q$, *then* $x, y \in \mathcal{QR}_N \Rightarrow x - y \notin \mathcal{QR}_N$.

Proof Since $x, y \in \mathcal{QR}_N, \exists a, b \in \mathbb{Z}_N^*$ such that $x \equiv a^2 (\text{mod } N)$ and $y \equiv b^2 (\text{mod } N)$. If possible, let $x - y \in \mathcal{QR}_N$. Then, $\exists c \in \mathbb{Z}_N^*$ such that $x - y \equiv c^2 (\text{mod } N)$. Therefore, $a^2 - b^2 \equiv c^2 (\text{mod } N) \Rightarrow a^2 \equiv b^2 + c^2 (\text{mod } N) \Rightarrow$

$a^2 \equiv b^2 + c^2 \pmod 5$. Now, as $a, b, c \in \mathbb{Z}_N^*$, a, b, c are relatively prime to 5. But $a^2 \equiv b^2 + c^2 \pmod 5$ has no solution in $\mathbb{Z}_5^*$, which is a contradiction. □

Theorem 6 *If* $N = 5q$, *then* Γ_N *is triangle-free.*

Proof If possible, let $x, y, z \in \mathbb{Z}_N$ be vertices of a triangle in Γ_N. Then, $x - y, z - y, x - z \in \mathcal{QR}_N$. However, $x - z \equiv (x - y) - (z - y) \pmod N$, a contradiction to Lemma 6. Thus, Γ_N is triangle-free. □

Lemma 7 [2] *If* G *is an abelian group and* S *is an inverse-closed subset of* $G \setminus \{e\}$ *with* $|S| \geq 3$, *then the Cayley graph* (G, S) *has girth at most* 4. □

Corollary 4 *If* $N = 5q$, *then* $girth(\Gamma_N) = 4$.

Proof Since Γ_N is triangle-free, $girth(\Gamma_N) \geq 4$. However, as Γ_N is a Cayley graph with $G = \mathbb{Z}_N$ and generating set $S = \mathcal{QR}_N$ such that $|S| = q - 1 \geq 3$, by Lemma 7, $girth(\Gamma_N)$ is at most 4. Thus, $girth(\Gamma_N) = 4$. □

Now, with the help of the following two lemmas, we prove that if $N = 5q$, where q is a Pythagorean prime, then $diam(\Gamma_N) = 3$.

Lemma 8 *If* $N = 5q$, *where* q *is a Pythagorean prime, then the number of vertices at distance* 2 *from the vertex* $0 \in \Gamma_N$ *is* $3q - 1$.

Proof Let x be a vertex at distance 2 from 0. Clearly, $x \neq 0$. Since, $d(0, x) \neq 1$, it follows that $x \notin \mathcal{QR}_N$. Also, as $d(0, x) = 2$, $\exists u \in \Gamma_N$ such that $0, u$ are adjacent and u, x are adjacent, i.e., $u, u - x \in \mathcal{QR}_N$, i.e., $x = u - (u - x)$ can be expressed as difference of two quadratic residues modulo N. Thus, the number of vertices x at distance 2 from the vertex 0 is equal to the number of $x \notin \mathcal{QR}_N$ which can be expressed as difference of two quadratic residues. Now, we finish the proof by appealing to Cases 2,3, and 4 of Lemma 5 with $p = 5$.

Case 2: The number of such $x \in \mathcal{J}_N^{+1} \setminus \mathcal{QR}_N$, i.e., $|\mathcal{J}_N^{+1} \setminus \mathcal{QR}_N|$ is $\frac{(p-1)(q-1)}{4} = q - 1$.

Case 3: In $\mathcal{J}_N^{-1}$, only those x's, for which $x \in \mathcal{QR}_q$ but $x \notin \mathcal{QR}_5$, can be expressed as difference of two quadratic residues. Note that the other type of x's cannot be expressed as difference of quadratic residues as $p = 5$. Thus, the number of $x \in \mathcal{J}_N^{-1}$ which can be expressed as difference of two quadratic residues is $|\{x \in \mathcal{J}_N^{-1} : x \in \mathcal{QR}_q \;\&\; x \notin \mathcal{QR}_5\}| = \left(\frac{q-1}{2}\right) 2 = q - 1$.

Case 4: If x is a nonzero, nonunit element in $\mathbb{Z}_N$, out of the four cases in Lemma 5, the last three cases are applicable. Note that in the first case x cannot be expressed as difference of quadratic residues as $p = 5$. Thus, the number of x which can be expressed as difference of two squares in this category is

$$\begin{aligned} &= |\{x : x \equiv 0 \pmod q) \;\&\; x \in \mathcal{QNR}_5\}| \;+\; |\{x : x \equiv 0 \pmod 5) \;\&\; x \in \mathcal{QR}_q\}| \\ &\quad + |\{x : x \equiv 0 \pmod 5) \;\&\; x \in \mathcal{QNR}_q\}| \\ &= \frac{5-1}{2} + \frac{q-1}{2} + \frac{q-1}{2} = q + 1 \end{aligned}$$

Combining all these cases, we get the total number of vertices at a distance 2 from the vertex 0 as $(q-1)+(q-1)+(q+1)=3q-1$. □

Lemma 9 *If $N=5q$, where q is a Pythagorean prime, then the number of vertices at distance 3 from the vertex $0 \in \Gamma_N$ is $q+1$.*

Proof From the proof of Lemma 8, it is evident that x's which are not at a distance 1 or 2 from the vertex 0 fall under either of the two categories: (i) $x \in \mathcal{J}_N^{-1}$, with $x \in \mathcal{QR}_5$, but $x \notin \mathcal{QR}_q$ or (ii) x is a nonzero, nonunit in $\mathbb{Z}_N$ such that $x \equiv 0 (\text{mod } q)$ and $x \in \mathcal{QR}_5$. Observe that in both the cases, $x \in \mathcal{QR}_5$.

We now construct a path of length 3 from 0 to x. Consider the vertex 1 and x. Now, $x-1 \notin \mathcal{QR}_5$, otherwise, we get two consecutive integers $x, x-1 \in \mathcal{QR}_5$, which is a contradiction. Thus, by Lemma 5, $d(x,1)=d(x-1,0)=2$ or 1. Also, $d(1,x) \neq 1$ as that would give a path $0, 1, x$ of length 2 from 0 to x, a contradiction. Hence, $d(1,x)=2$. Let the shortest path from 1 to x be $1, u, x$. Then, $0, 1, u, x$ is a path from 0 to x and hence, $d(0,x) \leq 3$. On the other hand, $d(0,x) \neq 1, 2$. Thus, $d(0,x)=3$.

Now, the number of such x's at a distance 3 from 0 is

$$|\{x \in \mathcal{J}_N^{-1} : x \in \mathcal{QR}_5; x \notin \mathcal{QR}_q\}| + |\{x \in \mathbb{Z}_N : x \equiv 0 (\text{mod } q); x \in \mathcal{QR}_5\}|$$

$$= 2\left(\frac{q-1}{2}\right) + \frac{5-1}{2} = (q-1)+2 = q+1.$$

□

Theorem 7 *If $N=5q$, where q is a Pythagorean prime, then $diam(\Gamma_N)=3$.*

Proof Since Γ_N is regular with degree $\phi(N)/4=q-1$, number of vertices adjacent to 0, i.e., at distance 1 from 0 is $q-1$. By Lemma 8, Lemma 9 and counting the point 0 itself, we get the number of all points at distance 0, 1, 2, 3 from the vertex 0 as $1+(q-1)+(3q-1)+(q+1)=5q=N$. Thus, it exhausts all the vertices in Γ_N, i.e., all the points, apart from 0 itself, are at either distance 1, 2 or 3 from 0. Since, Γ_N is symmetric, the maximum distance between any two vertex is 3, i.e., $diam(\Gamma_N)=3$. □

5.2 Quadratic Residue Cayley Graph of Type-II

Theorem 8 *If $N=pq$ where $5 \nmid N$, then Γ_N is triangulated and $girth(\Gamma_N)=3$.*

Proof Let $x \in \mathbb{Z}_N$ be any vertex in Γ_N. Consider $x, x+3^2, x+5^2 \in \mathbb{Z}_N$. These three vertices form a triangle as 9, 16, 25 are relatively prime to N and belong to $\mathcal{QR}_N$. Thus, every vertex $x \in \Gamma_N$ is a vertex of a triangle in Γ_N. Hence, Γ_N is triangulated. Now, existence of triangle in Γ_N ensures its girth to be 3. □

Lemma 10 *Let $N = pq$ where $5 \nmid N$. If $0, x \in \mathbb{Z}_N$ be nonadjacent vertices in Γ_N, then $\exists u \in \mathbb{Z}_N$ such that 0 and u are adjacent and u and x are adjacent.*

Proof Since $0, x \in \mathbb{Z}_N$ be nonadjacent vertices in Γ_N, x is not a quadratic residue in $\mathbb{Z}_N$. Also, $N = pq$ with $5 \nmid N$ implies $p, q > 5$. Therefore, by Lemma 5, x can always be expressed as difference of two quadratic residues, say $u, v \in \mathcal{QR}_N$ such that $x = u - v$. Since, $u \in \mathcal{QR}_N$, 0 and u are adjacent in Γ_N. Also, $u - x = v$ is a quadratic residue, i.e., u and x are adjacent in Γ_N. □

Theorem 9 *If $N = pq$ where $5 \nmid N$, then $diam(\Gamma_N) = 2$.*

Proof Let $x, y \in \mathbb{Z}_N$. If $x - y \in \mathcal{QR}_N$, then $d(x, y) = 1$. If $x - y$ is not a quadratic residue, then 0 and $x - y$ are non-adjacent vertices in Γ_N. Therefore, by Lemma 10, $\exists u \in \mathbb{Z}_N$ such that 0 is adjacent to u and u is adjacent to $x - y$. So using a translation of y, we get y is adjacent to $u + y$ and $u + y$ is adjacent to x in Γ_N. Thus, $d(x, y) = 2$ and hence $diam(\Gamma_N) = 2$. □

Now, we turn toward a special property of Γ_N of Type-II. Earlier, we have mentioned that Γ_N, both Type-I and II, are not strongly regular. However, in Γ_N of Type-II, if x, y are two adjacent vertices, then there are a fixed number of vertices (depending only on N and not on x, y) in Γ_N which are adjacent to both x and y.

Theorem 10 *Let $N = pq$, where $p, q > 5$ are primes with $p = 4k+1, q = 4l+1$. If x, y are two adjacent vertices in Γ_N, then there are exactly $(k - 1)(l - 1)$ vertices in Γ_N which are adjacent to both x and y.*

Proof Since x, y are two adjacent vertices in Γ_N, $x - y \in \mathcal{QR}_N$. By Lemma 5, the number of ways in which $x - y$ can be expressed as difference of two quadratic residues is $\frac{(p-5)(q-5)}{16} = \frac{(4k-4)(4l-4)}{16} = (k - 1)(l - 1)$. Let $x - y = u - v$ where $u, v \in \mathcal{QR}_N$. Therefore, 0, u are adjacent (as $u \in \mathcal{QR}_N$) and $u, x - y$ are adjacent (as $u - (x - y) = v \in \mathcal{QR}_N$) in Γ_N. Thus, by using a translation by y and symmetricity of Γ_N, $y, u + y$ are adjacent and $u + y, x$ are adjacent. Hence, there are exactly $(k - 1)(l - 1)$ vertices in Γ_N which are adjacent to both x and y. □

Remark 3 By Theorem 2 and 10, it follows that Γ_N of Type-II is regular and any two neighbours in Γ_N have equal number of common neighbours. However, any two nonadjacent vertices may not have equal number of common neighbors. Thus, Γ_N is not strongly regular.

In Theorem 8, it was shown that Γ_N of Type-II is triangulated. Now, by using Theorem 10, we count the number of triangles in Γ_N of Type-II.

Theorem 11 *If $N = pq$ with $p = 4k+1, q = 4l+1$ being primes >5, then number of triangles in Γ_N is $\frac{2}{3}Nk(k - 1)l(l - 1)$.*

Proof Let x be a vertex in Γ_N. The number of vertices adjacent to x is $\phi(N)/4$. Let y be one of those vertices adjacent to x. Now, by Theorem 10, there are $(k - 1)(l - 1)$ vertices z_i's in Γ_N which are adjacent to both x and y, thereby forming a triangle.

Thus, the count of triangles with x as a vertex, comes to $\frac{\phi(N)}{4}(k-1)(l-1)$. However, this number is twice the actual number of triangles with x as a vertex, since we could have also started with choosing z_i instead of y and get y as the common neighbor of x and z_i. Thus, the actual number of triangles with x as a vertex is $\frac{\phi(N)}{8}(k-1)(l-1)$. Now, varying x over the vertex set of Γ_N, the count becomes $\frac{\phi(N)}{8}N(k-1)(l-1)$. Again, this count is to be divided by 3, as if x, y, z are vertexes of a triangle, then the triangle is counted thrice once with respect to each vertex. Thus, the actual number of triangles in Γ_N is $= \frac{\phi(N)}{24}N(k-1)(l-1)$

$$= \frac{(p-1)(q-1)}{24}N(k-1)(l-1) = \frac{4k.4l}{24}N(k-1)(l-1) = \frac{2}{3}Nk(k-1)l(l-1).$$

□

Remark 4 Note that one of $k-1, k, k+1$ is divisible by 3. But as $p = 4k+1 = 3k + (k+1)$, $k+1$ is not divisible by 3, thus $k(k-1)$ is divisible by 3. As a result, the number of triangles is a positive integer.

6 Domination Number of Γ_N

In this section, we use some existing theorems in the literature, to find a bound on the domination number γ of Γ_N. First, we state some results in graph domination which we will use, without proof. (See [3])

Theorem 12 [3] *Let G be a graph with n vertices. Then the following are true:*

1. *If G has a degree sequence $d_1, d_2, \ldots, d_n$ with $d_i \geq d_{i+1}$, then*

$$\gamma(G) \geq \min\{k : k + (d_1 + d_2 + \ldots + d_k) \geq n\}.$$

2. *If G has no isolated vertex and has minimum degree $\delta(G)$, then*

$$\gamma(G) \leq \frac{n}{\delta(G)+1} \sum_{j=1}^{\delta(G)+1} \frac{1}{j}.$$

Theorem 13 *If $N = pq$, then $\gamma(\Gamma_N) \geq 5$. Specifically, if $N = 5q$, then*

$$5 \leq \gamma(\Gamma_N) \leq 5\sum_{j=1}^{q} \frac{1}{j}.$$

Proof For the first part, we assume that $p = 4l + 1$. Since, Γ_N is regular with degree $\frac{\phi(N)}{4} = \frac{(p-1)(q-1)}{4} = l(q-1)$, we have $\gamma(\Gamma_N) \geq \min\{k : k + kl(q-1) \geq (4l+1)q\} = 5$.

For the second part, i.e., $N = 5q$, we put $l = 1$. Also, as Γ_N has no isolated vertex,

$$\gamma(\Gamma_N) \leq \frac{5q}{(q-1)+1}\sum_{j=1}^{q}\frac{1}{j} = 5\sum_{j=1}^{q}\frac{1}{j}.$$

□

Remark 5 A similar upper bound could have been given for the general case, however, the expression being messy, may not provide meaningful insight.

7 Conclusion and Future Work

In this paper, we introduced a special class of quadratic residue Cayley graphs and proved some basic features of this family. However, a lot of questions are still unresolved. The chromatic number and domination number of this family of graphs can be interesting topics for further research.

Acknowledgments The author is thankful to Avishek Adhikari of Department of Pure Mathematics, University of Calcutta, India for some fruitful suggestions and careful proofreading of the manuscript.

References

1. Giudici, R.E., Olivieri, A.A.: Quadratic modulo 2^n Cayley graphs. Discrete Math. **215**, 73–79 (2000) (Elsevier)
2. Godsil, C., Royle, G.: Algebraic Graph Theory. Graduate Texts in Mathematics. Springer, Heidelberg (2001)
3. Haynes, T.W., Hedetniemi, S.T., Slater, P.J.: Fundamentals of Domination in Graphs. Marcel Dekker Inc., New York (1998)
4. Rosen, K.H.: Elementary Number Theory and its Applications. Addison-Wesley, Reading (1984)
5. West, D.B., (2001) Introduction to Graph Theory. Prentice Hall, New York (2001)

A Dynamic Programming Algorithm for Solving Bi-Objective Fuzzy Knapsack Problem

V.P. Singh and D. Chakraborty

Abstract This paper considers bi-objective knapsack problem with fuzzy weights, says bi-objective fuzzy knapsack problem (BOFKP). Here we introduce an index which gives the possibility of choosing the item (weights and knapsack availability are fuzzy in nature) for knapsack with crisp capacity such that both the objective value are optimized. A methodology using dynamic programming technique has been introduced in this paper with an algorithm which gives the optimal solution for single objective fuzzy knapsack problem (FKP) with some possibility. Using this methodology an algorithm is given to find the Pareto frontier in case of bi-objective fuzzy knapsack problem. Compromise ratio method for decision-making under fuzzy environment has been used to find the compromise solution. The possibility index gives an idea to choose the solution according to decision-maker's choice. An illustrative example is given to demonstrate the methodology.

Keywords Bi-objective fuzzy knapsack problem · Triangular fuzzy number · Dynamic programming · Possibility index · Compromise ratio method

1 Introduction

The bi-objective fuzzy knapsack problem is an extension of fuzzy knapsack problem [7]. Fuzzy knapsack problem is a knapsack problem where the weight of the items are fuzzy in nature. Knapsack problem is one of the most relevant mathematical programming problem with numerous applications in different areas. The knapsack problem [8] is a problem where a decision-maker is searching for a combination of different items for filling the knapsack. The objective is to optimize the total

V.P. Singh (✉) · D. Chakraborty
Department of Mathematics, Indian Institute of Technology Kharagpur, Kharagpur 721302, India
e-mail: vishnupratapsingh56@gmail.com

D. Chakraborty
e-mail: debjani@maths.iitkgp.ernet.in

R.N. Mohapatra et al. (eds.), *Mathematics and Computing*,
Springer Proceedings in Mathematics & Statistics 139,
DOI 10.1007/978-81-322-2452-5_20

utility value of all chosen items by the decision-maker subject to the capacity of knapsack. In bi-objective fuzzy knapsack problem the objective is to optimize both the objectives subject to the capacity of knapsack. The knapsack may correspond to a ship, truck, or a resource. Knapsack problem has a simple structure which permits to study combinatorial optimization problems.

In most real world situations, decisions are often taken by the decision-maker in the presence of conflicting objectives. Some researchers used fuzzy theory to solve this type of problem. Zadeh [14] proposed fuzzy set theory, using this theory [10] described multiple choice knapsack problem with fuzzy coefficients. Kasperski and Kulej [6] solve the 0-1 knapsack problem with fuzzy data. Lin and Yao [7] described FKP by taking each weight $w_i, i = 1, 2...n$ as imprecise value. They consider $\tilde{w_i} = (w_i - \Delta_{i1}, w_i, w + \Delta_{i2})$ be the fuzzy number, thus the decision-maker should determine an acceptable range of values for each $\tilde{w}_i$, which is the interval $[w_i - \Delta_{i1}, w_i + \Delta_{i2}], 0 \leq \Delta_{i1} < w_i$, and $0 \leq \Delta_{i2}$. Then the decision-maker chooses a value from the interval $[w_i - \Delta_{i1}, w_i + \Delta_{i2}]$ as an estimate of each weight. Estimate is exactly $\tilde{w}_i$ if the acceptable grade is 1, otherwise, the acceptable grade will get smaller when the estimate approaches either $w_i - \Delta_{i1}$ or $w_i + \Delta_{i2}$. To calculate an estimate of the fuzzy weight defuzzification of the fuzzy number $\tilde{w}_i$ from the interval $[w_i - \Delta_{i1}, w_i + \Delta_{i2}]$ has been used.

The main idea behind this paper is to solve the bi-objective knapsack problem in fuzzy environment. Since in real world situations, the decision-makers often face the problem of uncertainty in selecting the amount of data which is used in packaging the knapsack. To overcome this difficulty we have taken fuzzy data to solve this type of problem without defuzzification. There are varieties of applications available for bi-objective fuzzy knapsack problem such as various packing problem, cargo loading, cutting stock, or economic planning. For example, the problem of making investment decisions in which the size of an investment is based on the amount of money required, the knapsack capacity is the amount of available money to invest, the investment profit, and rate of investment profit are the expected return. The notable features of our approach are as follows:

- We develop a new possibility index for calculating the possibility of putting fuzzy weight into a knapsack of crisp capacity as well as fuzzy capacity. Possibility index gives an opportunity to the decision-makers to select the fuzzy weight according to their choice.
- We introduced a dynamic programming algorithm to solve fuzzy knapsack problem which gives the optimal solution with some possibility index. The selection of possibility index may vary according to the choice of decision-makers.
- For solving bi-objective fuzzy knapsack problem, the Pareto optimal frontier is generated using the optimal values of each objective. Then the compromise ratio method for decision-maker under fuzzy environment has been used for selecting the best compromise solution.

In this paper, the weight as triangular fuzzy number has been used and solve it without defuzzification. Defuzzification of fuzzy number gives a real value corresponding to that fuzzy number with some loss of information. Defuzzification of fuzzy number, converts the fuzzy knapsack problem into crisp knapsack problem. Since the weights are fuzzy in nature we can fill the weights with some possibility, having any value between [0, 1]. Sengupta and Pal [11] introduced an acceptability index to order two intervals in terms of value. Similarly, we introduced a possibility index for calculating the possibility [3] of putting fuzzy weight within a knapsack. Proposed possibility index provides the measure whether the knapsack can hold fuzzy weight. There are three types of decision-makers [9] who want to get the solution. If the possibility index is 1 weight can be filled completely in the knapsack and if it is zero it cannot fill. If the possibility index lies between [0,1] weight can be filled with this much possibility. Possibility index may be near 1 and also may be closer to zero. It depends on the decision-maker how he chooses the weight. There are Pessimistic decision-maker, Optimistic decision-maker, and Moderate decision-maker. An optimistic decision-maker can take the worst case for optimizing the solution, i.e., he tolerates the less possibility index for expected higher utility value, on the other hand pessimistic decision-maker always chooses the highest possibility index. A moderate decision-maker can choose the middle value of the possibility.

An algorithm has been introduced based on dynamic programming [1, 5, 12, 13] to solve fuzzy knapsack problem, which gives the optimal solution with some possibility index. The possibility index gives the possibility of choosing fuzzy weights out of available weight (knapsack capacity). The possibility index is introduced in Sect. 3. For solving bi-objective fuzzy knapsack problem, first we optimize each objective function using proposed dynamic programming algorithm. We select number of copies for both the objective functions separately and the Pareto optimal frontier is generated by using these numbers of copies. Then the compromise ratio method [4] for decision-maker under fuzzy environment has been used for selecting the best compromise solution.

The rest of the paper is organized as follows, Sect. 2 outlines the preliminaries and definition of the fuzzy compromise ratio method for multi-attribute decision-making. In Sect. 3, we develop possibility index for selecting the fuzzy weight into the knapsack of crisp (or fuzzy) capacity. In Sect. 4, bi-objective fuzzy knapsack problem by multistage decision process has been defined and a methodology along with algorithms has been given to calculate the Pareto frontier. In Sect. 5, a numerical example has been demonstrated by our proposed algorithm. Finally, we state our conclusion in Sect. 6.

2 Preliminaries: Concepts and Definitions

Zadeh [14] in 1965, introduced the concept of a fuzzy set. A fuzzy set $\widetilde{A}$ in X is characterized by a membership function $\mu_{\widetilde{A}}(x)$ which associates with each points in X a real number in the interval [0,1], with the value of $\mu_{\widetilde{A}}(x)$ at x representing the "grade of membership" of x in $\widetilde{A}$. where X is a space of points (objects), with a generic element of X denoted by x.

2.1 LR-type Fuzzy Number

Definition 1 A fuzzy number $\widetilde{M}$ is of $LR - type$ if there exist reference functions L(for left), R(for right), and scalars $\alpha > 0, \beta > 0$ with

$$\mu_{\widetilde{M}}(x) = \begin{cases} L(\frac{m-x}{\alpha}), & \text{for } x \leq m \\ R(\frac{x-m}{\beta}), & \text{for } x \geq m \end{cases} \tag{1}$$

m, called the mean value of $\widetilde{M}$, is a real number and α and β are called the left and right spreads, respectively. Symbolically $\widetilde{M}$ is denoted by $(m, \alpha, \beta)_{LR}$.

For reference function L, different function can be chosen. Dubois and Prade in 1988 mention, for instance, $L(x) = \max(0, 1 - x)^p$, $L(x) = \max(0, 1 - x^p)$, with $p > 0$, $L(x) = e^{-x}$ or $L(x) = e^{-x^2}$. If m is not a real number but an interval $[\underline{m}, \overline{m}]$ then the fuzzy set $\widetilde{M}$ is not a fuzzy number but a fuzzy interval.

For LR fuzzy number the computations necessary for the arithmetic operations are considerably simplified: the exact formulas that can be given for $\oplus$ and $\ominus$. Let $\widetilde{A} = (a, \alpha, \beta)_{LR}$, $\widetilde{B} = (b, \gamma, \delta)_{LR}$ be two fuzzy number of LR-type. Then,

1. $(a, \alpha, \beta)_{LR} \oplus (b, \gamma, \delta)_{LR} = (a + b, \alpha + \gamma, \beta + \delta)_{LR}$
2. $-(a, \alpha, \beta)_{LR} = (-a, \beta, \alpha)_{LR}$
3. $(a, \alpha, \beta)_{LR} \ominus (b, \gamma, \delta)_{LR} = (a - b, \alpha + \delta, \beta + \gamma)_{LR}$

2.2 Triangular Fuzzy Number

Definition 2 It is a fuzzy number represented with three points as follows:

$$\widetilde{A} = (a_1, a_2, a_3)$$

this representation is interpreted as membership functions:

$$\mu_{\widetilde{A}}(x) = \begin{cases} 0, & x < a_1 \\ \frac{x-a_1}{a_2-a_1}, & a_1 \leq x \leq a_2 \\ \frac{a_3-x}{a_3-a_2}, & a_2 \leq x \leq a_3 \\ 0, & x > a_3 \end{cases} \tag{2}$$

2.3 Fuzzy Compromise Ratio Method for MADM

There are n possible alternatives $s_1, s_2, \ldots, s_n$ from which the decision-maker has to choose on the basis of m attributes $c_1, c_2, \ldots, c_m$. Here it has been assumed that

m attributes have equal weights. The ratings of the alternatives $s_1, s_2, \ldots, s_n$ on attribute $c_1, c_2, \ldots, c_m$, as given by the decision-maker be $\widetilde{f}_{ij} = (m_{ij}; \alpha_{ij}, \beta_{ij})$. So the fuzzy decision matrix is given by

$$\widetilde{Y} = (\widetilde{f}_{ij})_{m\times n} = \begin{pmatrix} \widetilde{f}_{11} & \widetilde{f}_{12} & .. & \widetilde{f}_{1n} \\ \widetilde{f}_{21} & \widetilde{f}_{22} & .. & \widetilde{f}_{2n} \\ . & . & .. & . \\ . & . & .. & . \\ \widetilde{f}_{m1} & \widetilde{f}_{m2} & .. & \widetilde{f}_{mn} \end{pmatrix}$$

Since the m attributes may be measured in different ways, the decision matrix Y needs to be normalized. The linear scale transformation has been used here to transform the various attribute scale into a comparable scale. After normalization we get

$$\widetilde{r}_{ij} = \Big(\frac{m_{ij}}{d_{ij}^{max}}; \frac{\alpha_{ij}}{d_{ij}^{max}}, \frac{\beta_{ij}}{d_{ij}^{max}}\Big) \text{ for } c_i \in C^1$$

and

$$\widetilde{r}_{ij} = \begin{cases} \Big(\frac{a_i^{min}}{m_{ij}}; \frac{a_i^{min}.\beta_{ij}}{m_{ij}.(m_{ij}+\beta_{ij})}, \frac{a_i^{min}.\alpha_{ij}}{m_{ij}.(m_{ij}-\alpha_{ij})}\Big), & \text{for } a_i^{min} \neq 0, c_i \in C^2 \\ \Big(1 - \frac{m_{ij}}{d_{ij}^{max}}; \frac{\alpha_{ij}}{d_{ij}^{max}}, \frac{\beta_{ij}}{d_{ij}^{max}}\Big), & \text{for } a_i^{min} = 0, c_i \in C^2 \end{cases}$$

Where

$$d_i^{max} = \max_{1<j<n} \{m_{ij} + \beta_{ij} \mid \widetilde{f}_{ij} = (m_{ij}; \alpha_{ij}, \beta_{ij})\} \text{ and}$$

$$a_i^{min} = \min_{1<j<n} \{m_{ij} - \alpha_{ij} \mid \widetilde{f}_{ij} = (m_{ij}; \alpha_{ij}, \beta_{ij})\}$$

The normalization method mentioned above is to preserve the property that the range of a normalized triangular fuzzy number belongs to the closed interval [0, 1]. Then the fuzzy decision matrix can be transformed into normalized fuzzy decision matrix:

$$\widetilde{R} = (\widetilde{r}_{ij})_{m\times n} = \begin{pmatrix} \widetilde{r}_{11} & \widetilde{r}_{12} & .. & \widetilde{r}_{1n} \\ \widetilde{r}_{21} & \widetilde{r}_{22} & .. & \widetilde{r}_{2n} \\ . & . & .. & . \\ . & . & .. & . \\ \widetilde{r}_{m1} & \widetilde{r}_{m2} & .. & \widetilde{r}_{mn} \end{pmatrix}$$

Then, the fuzzy positive ideal solution s^+ and the fuzzy negative ideal solution s^- have been defined, whose weighted normalized vectors are $\widetilde{a}^+ = (\widetilde{a}_1^+, \widetilde{a}_2^+ \ldots \widetilde{a}_m^+)$ and $\widetilde{a}^- = (\widetilde{a}_1^-, \widetilde{a}_2^- \ldots \widetilde{a}_m^-)$, respectively. Difference between each alternative

s_j $(j = 1, 2 \ldots n)$ and the positive ideal solution and the fuzzy negative ideal solution have been measured by using fuzzy distance between two fuzzy numbers. It is given by

$$\widetilde{D}(s_j, s^+) = \sum_{i=1}^{m} \widetilde{d}(\widetilde{r}_{ij}, \widetilde{a}_i^+)$$
$$\widetilde{D}(s_j, s^-) = \sum_{i=1}^{m} \widetilde{d}(\widetilde{r}_{ij}, \widetilde{a}_i^-)$$

Now the smaller $\widetilde{D}(s_j, s^+)$, the better s_j.

3 The Possibility Index

Let us consider two fuzzy numbers $\widetilde{A} = (a_1, a_2, a_3)$ and $\widetilde{B} = (b_1, b_2, b_3)$ whose membership functions can be calculated by Eq. 2. Now if we take a knapsack of capacity $\widetilde{B}$ and we want to fill the weight $\widetilde{A}$ into the knapsack of capacity $\widetilde{B}$ then we have three possibilities for filling the weight in the knapsack which are classified as follows.

1. $\widetilde{A}$ can be completely filled into the knapsack of capacity $\widetilde{B}$, i.e., possibility is one.
2. $\widetilde{A}$ cannot be filled into the knapsack of capacity $\widetilde{B}$, i.e., possibility is zero.
3. $\widetilde{A}$ can be filled with some possibility into the knapsack of capacity $\widetilde{B}$, i.e., possibility lies between zero and one.

If $\widetilde{A} = (a_1, a_2, a_3)$ and $\widetilde{B} = (b_1, b_2, b_3)$ are two fuzzy numbers then the possibility index for filling fuzzy weights in given capacity is denoted by $PI(\widetilde{A} \blacktriangle \widetilde{B})$, i.e., the possibility of filling $\widetilde{A}$ in $\widetilde{B}$ and given by

$$PI(\widetilde{A} \blacktriangle \widetilde{B}) = \begin{cases} 1 - y_1 \frac{(a_3 - b_3)}{(a_3 - a_1)}, & \text{if } b_3 < a_3 \text{ and } a_2 \leq b_2 \\ y_2 \frac{(b_3 - a_1)}{(a_3 - a_1)}, & \text{if } b_3 < a_3 \text{ and } a_2 > b_2 \\ 1, & \text{if } b_3 \geq a_3 \\ 0, & \text{if } b_3 \leq a_1 \end{cases} \tag{3}$$

where $y_1 = \{\mu_{\widetilde{D}}(x) | \mu_{\widetilde{A}}(x) = \mu_{\widetilde{B}}(x) \ for \ x \geq b_2\}$, $y_2 = \{\max \mu_{\widetilde{D}}(x) | \mu_{\widetilde{D}}(x) = \min(\mu_{\widetilde{A}}(x), \mu_{\widetilde{B}}(x)\}$ and $\mu_{\widetilde{D}}(x)$ represents the membership value of fuzzy set $\widetilde{D} = \widetilde{A} \cap \widetilde{B}$. Figure 1 shows some sets which defines above condition for calculating the possibility index.

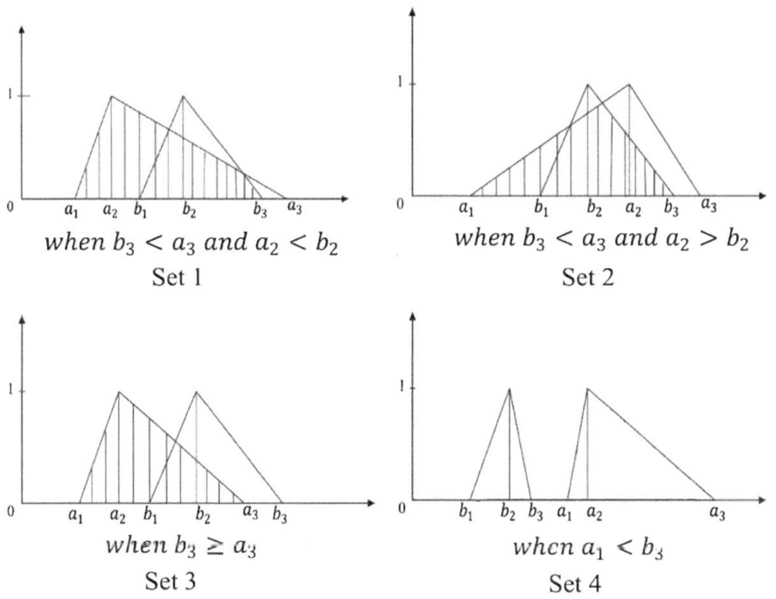

Fig. 1 Sets of fuzzy numbers

When $b_3 < a_3$ and $a_2 < b_2$, the possibility index can be calculated as:

$$PI(\widetilde{A}\blacktriangle\widetilde{B}) = \frac{\text{Area occupied by } \widetilde{A} \text{ in the knapsack of capacity } \widetilde{B}}{\text{Total area of } \widetilde{A}} = 1 - y_1\frac{(a_3 - b_3)}{(a_3 - a_1)}$$

where $y_1 = \{\mu_{\widetilde{D}}(x) | \mu_{\widetilde{A}}(x) = \mu_{\widetilde{B}}(x)\ for\ x \geq b_2\}$ and $\mu_{\widetilde{D}}(x)$ represents the membership value of fuzzy set $\widetilde{D} = \widetilde{A} \cap \widetilde{B}$.

When $b_3 < a_3$ and $a_2 > b_2$ the possibility index can be calculated as:

$$PI(\widetilde{A}\blacktriangle\widetilde{B}) = \frac{\text{Area occupied by } \widetilde{A} \text{ in the knapsack of capacity } \widetilde{B}}{\text{Total area of } \widetilde{A}} = y_2\frac{(b_3 - a_1)}{(a_3 - a_1)}$$

where $y_2 = \{\max \mu_{\widetilde{D}}(x) | \mu_{\widetilde{D}}(x) = \min(\mu_{\widetilde{A}}(x), \mu_{\widetilde{B}}(x)\}$ and $\mu_{\widetilde{D}}(x)$ represents the membership value of fuzzy set $\widetilde{D} = \widetilde{A} \cap \widetilde{B}$.

When $b_3 \geq a_3$ the possibility index can be calculated as:

$$PI(\widetilde{A}\blacktriangle\widetilde{B}) = \frac{\text{Area occupied by } \widetilde{A} \text{ in the knapsack of capacity } \widetilde{B}}{\text{Total area of } \widetilde{A}} = 1$$

When $a_1 \geq b_3$ the possibility index can be calculated as:

$$PI(\widetilde{A}\blacktriangle\widetilde{B}) = \frac{\text{Area occupied by } \widetilde{A} \text{ in the knapsack of capacity } \widetilde{B}}{\text{Total area of } \widetilde{A}} = 0$$

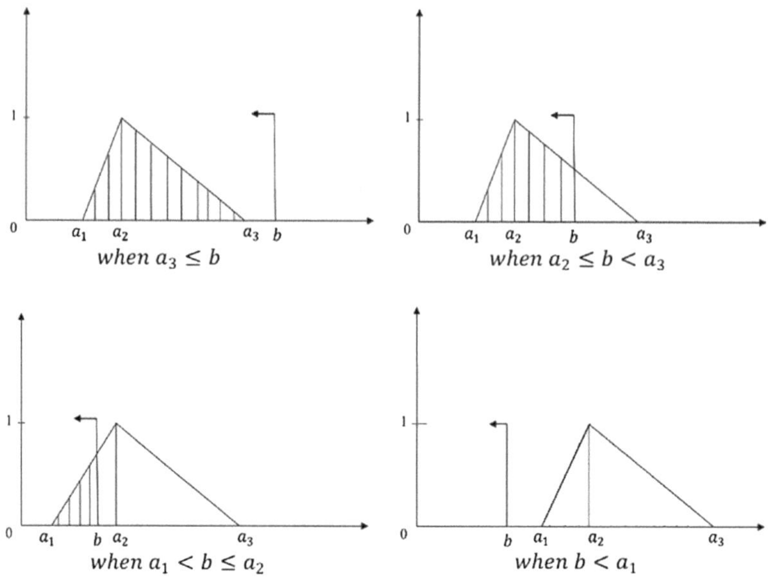

Fig. 2 PI of $\widetilde{A} = (a_1, a_2, a_3)$ for different value of b(crisp)

Since the knapsack capacity is crisp value. Let $\widetilde{B} = (b, b, b)$, i.e., a crisp value b is knapsack capacity which is shown in Fig. 2. Now our possibility index can be given as:

$$PI(\widetilde{A} \blacktriangle b) = \begin{cases} 1, & \text{if } a_3 \leq b \\ 1 - \mu_{\widetilde{A}}(b)\frac{(a_3-b)}{(a_3-a_1)}, & \text{if } a_2 \leq b < a_3 \\ \mu_{\widetilde{A}}(b)\frac{(b-a_1)}{(a_3-a_1)}, & \text{if } a_1 < b < a_2 \\ 0, & \text{if } b \leq a_1 \end{cases} \tag{4}$$

Please note that the first line of text that follows a heading is not indented, whereas the first lines of all subsequent paragraphs are.

4 Bi-Objective Fuzzy Knapsack Problem by Multistage Decision Process

A bi-objective fuzzy knapsack problem may be viewed as n stage decision process where the fuzzy stage transformation equation unite all the stages (Fig. 3). In a dynamic programming structure of bi-objective fuzzy knapsack problem, the stage transformation equation transforms input state variable and decision variable to an

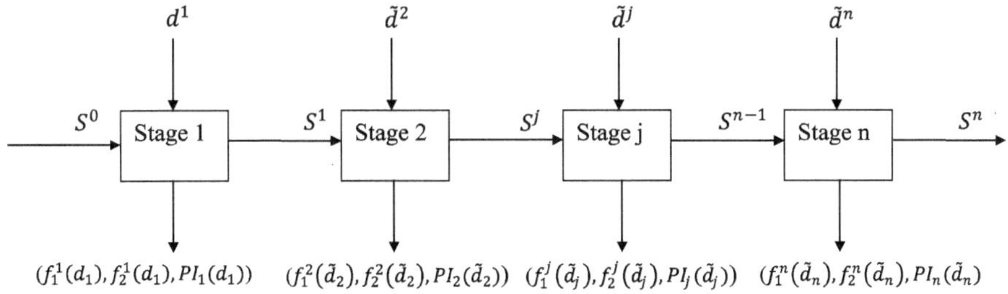

Fig. 3 Multistage decision process in fuzzy environment

output state which works as an input state variable for its next stage and this process continues upto nth stage. If S^0 is the input state variable for first stage and d^1 is the decision variable then stage 1 will consume some part of input and decision variable and it will give an immediate return in the form of utility value for both the objective function f_1^1 and f_2^2 correspond to their possibility index PI_1 at first stage. At first stage decision variable has a crisp value, while from 2nd stage onwards it becomes fuzzy. If we have S^j as input state fuzzy variable at jth stage which is output from the $(j-1)$ stage and $\widetilde{d}^j$ for $(j > 1)$ is decision variable then the immediate return at stage j is given by (f_j^1, f_j^2, PI_j). Similarly at nth stage we get an optimal return f_n^1 and f_n^2 for first and second objective, respectively, with possibility PI_n. This optimal value that corresponds to each objective function will be of optimal utility value for first and second objectives, respectively, and by moving in backward direction with respect to the corresponding decision variable, we calculate the nondominated set of items. Here possibility index plays an important role at each stage. Since the decision-makers have values of possibility index at each stage so that they can select the optimal value according to their tolerance limits. The selection of possibility index will change the solution according to DM's choices.

4.1 Methodology

In the classical bi-objective knapsack problem all the weights and item profits are assumed to be crisp in nature. Mathematically, it is defined as,

$$\max f_k(x) = \sum_{i=1}^{n} u_i^k x_i \text{ for } k = 1, 2$$

$$s.t \sum_{i=1}^{n} w_i x_i < W, i = 1, \ldots, n.$$

If the item are n in number then u_i^1 and u_i^2 ($i = 1, 2, \ldots n$)represents the utility values of each item for first and second objective, respectively, and w_i represents the crisp weight of each item with knapsack capacity W. In the formulation of bi-objective fuzzy knapsack problem, the used weights are fuzzy in nature. In practice, we see many knapsack problems that involve items whose weights or price value are imprecise. Here we consider the problem in which weights of the items are triangular fuzzy number $\widetilde{w}_i = (w_{1i}, w_{2i}, w_{3i})$, knapsack capacity W and utility values are crisp. After fuzzifying the crisp value W we obtain $\widetilde{W} = (W_1, W_2, W_3)$. Now the bi-objective fuzzy knapsack problem as a linear programming model is described by

$$\max f_k(x) = \sum_{i=1}^{n} u_i^k x_i \text{ for } k = 1, 2$$

$$s.t \sum_{i=1}^{n} \widetilde{w}_i x_i \preceq W, i = 1, \ldots, n.$$

Now a dynamic programming technique of decision-making in fuzzy environment [2] is given to solve FKP using the possibility index introduced in the previous section. The solution obtained by this method depends upon the DM who chooses the profit with respect to possibility index in each stage. Since in the dynamic programming we divide an n-stage problem into n single stage problem and then used backward recursive approach to get the solution. Following steps are given to solve fuzzy knapsack problem.

Step 1: First we formulate the problem by defining the symbol given below-

x_i: Number of copies of an item i selected for knapsack.

y_k^i: Upper bound of an item i for kth objective.

d_i: State variable (Available weight in each stage i).

u_i^k: Value of an item i for kth objective (k= 1, 2) selected for knapsack.

$F_k^i(x_i)$: Value in stage i given x_i number of copies for kth objective (k= 1, 2).

$f_k^{i,N_t}(d_i)$: Maximum possible utility value selected at stage i to n for kth objective (k= 1, 2) according to the decision-makers' tolerance limit N_t for $t = 1, 2, 3$.

$PI_i(x_i)$: Possibility index in stage i for weights selected in the knapsack.

Step 2: We start from nth item and calculate the optimal value for each objective function and possibility index of weight for this item. In the next stage we take nth and $(n-1)$th item and again calculate the optimal value and possibility index of weights. Continuing in this manner at ith stage we have i number of items, for calculating the optimal value and possibility index which is selected by the DM in each stage we require optimal value and possibility index from previous stage. So we defined stage transformation equations for profit and the possibility index as

$$F_k^i(x_i, d_i) = x_i * u_i + f_k^{i+1,N_t}(d_i) \text{ for k} = 1, 2 \tag{5}$$

$$PI_i(x_i, d_i) = \frac{1}{2}(PI((x_i * \widetilde{w}_i)\blacktriangle d_i) + PI(\widetilde{ow}_i \blacktriangle \widetilde{rw}_i)) \tag{6}$$

$$f_k^{i,N_t}(d_i) = \max\{F_k^i(x_i, d_i) | \max\{PI_i(x_i, d_i)\} \in N_t \text{ for } t = 1, 2, 3\} \tag{7}$$

Here $d' = \lceil \widetilde{rw}_{i3} \rceil$, $\widetilde{rw}_i = d_i - (x_i * \widetilde{w}_i)$ is the remaining fuzzy weight at stage i and $\widetilde{ow}_i = \sum_{j=i+1}^{n} x_j^* \widetilde{w}_j$ is the optimal weight at stage i due to its all previous stages where x_i^* is the optimal value at stage i. $PI_i(\widetilde{ow}_i \blacktriangle \widetilde{rw}_i)$ represent the possibility index of two fuzzy number which is calculated by Eq. 3. Initial values are given by the equations.

$$f_k^{n+1,N_t}(d_i) = 0 \tag{8}$$

$$PI_{n+1}(\widetilde{ow}_n \blacktriangle \widetilde{rw}_n) = PI((x_i * \widetilde{w}_i)\blacktriangle d_i) \tag{9}$$

Step 3: Once calculating the value of $f_k^{i,N_t}(d_i)$, $PI_i(x_i)$ at first stage it depends upon the decision-maker to choose the optimal value in the next stage. Let N_1, N_2, N_3 are the tolerance limit for the optimistic, moderate, and pessimistic decision-makers, respectively. Then the selection of $PI_i(x_i)$ for the next stage has been showed in Fig. 4 depending on the selected decision-maker and the corresponding $f_k^{i,N_t}(d_i)$ value will be optimal value for profit.

Step 4: Now we have utility values and possibility index for all the stages. Moving backward by considering the optimal value (chosen by DM) corresponding to remaining weight from first stage to nth stage will give the solution and the selected number of copies of item for that decision-maker and the selected number of copies. Similarly for other decision-makers they can select there tolerance limit.

Step 5: According to the decision-maker's tolerance limit select the upper bound of each item for each objective. Here we select only those values of x_i for which the possibility index $PI_i \in N^t$ at $d_i = W$. Now stage transformation equations for uniting all the stages in case of mixed approach

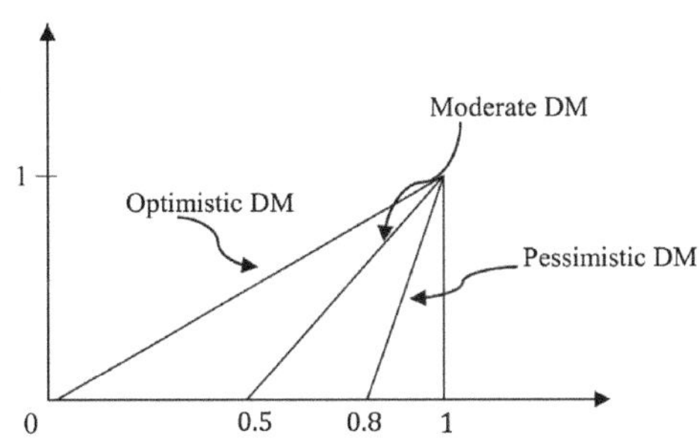

Fig. 4 Selection of possibility index by DM

$$F_k^i(x_i, d_i) = x_i * u_i^k + f_k^{i+1,N_t}(d')$$
$$f_k^{i,N_t}(d_i) = \max\{F_k^i(x_i, d_i) \mid \max\{PI_i(x_i, d_i)\} \in N_t\}$$

Step 6: Now, we select number of copies for both the objectives by moving in backward direction with maximum objectives values provided the possibility index for $d_i = W$ at last stage ($i = 10$) should not be less than from possibility index at $d_{i-1} = W$.

4.2 Algorithms for Fuzzy Knapsack Problem

Algorithm 1 Calculating functional value for each objective function at each state of a stage with possibility index

Require: $\widetilde{w}_i = (w_{1i}, w_{2i}, w_{3i}), W, u_i^k, n, k$
Ensure: $f_k^i(d_i), PI_i'(d_i), y_k^i$
$\widetilde{w}_i, i = 1, 2, \ldots n$
for $i := n \to 1$ **do**
  $l := 0$
  while $w_{2i} * l \le W$ **do**
    $l := l + 1$
  end while
  for $x_i := 0 \to l$ **do**
    for $d_i := 0 \to W$ **do**
      $x_{n+1}^*(d_i) := 0$
      if $(d_i < w_{1i})$ **then**
        $f_k^n(d_i) := 0$
        $PI_n(d_i) := 0$
      end if
      if $(w_{3i} < d_i < x_i * w_{3i})$**or**$(d_i \ge x_i * w_{3i})$ **then**
        $f_k^{n+1,N_t}(d_i) := 0$ for all decision-makers, *i.e.*, $t = 1, 2, 3$
        $\widetilde{rw}_i := d_i - (x_i * \widetilde{w}_i)$
        $\widetilde{ow}_i := \sum_{j=i+1}^{n} x_j^* \widetilde{w}_j$
        $PI(\widetilde{ow}_n \blacktriangle \widetilde{rw}_n) := PI((x_i * \widetilde{w}_i) \blacktriangle d_i)$
        $F_k^i(x_i, d_i) := x_i * u_i^k + f_k^{i+1,N_t}(d_i)$
        $PI_i(x_i, d_i) := \frac{1}{2}(PI((x_i * \widetilde{w}_i) \blacktriangle d_i) + PI(\widetilde{ow}_i \blacktriangle \widetilde{rw}_i))$
        if $(PI_i(x_i, d_i) \in N_t)$ **then**
          $y_k^i = x_i$
        end if
        $f_k^{i,N_t}(d_i) := \max\{F_k^i(x_i, d_i) \mid \max\{PI_i(x_i, d_i)\} \in N_t$ for $t = 1, 2, 3\}$
        $x_i^*(d_i) := x_i$ for which $\max\{F_k^i(x_i, d_i) \mid \max\{PI_i(x_i, d_i)\} \in N_t\}$ is selected.
        {Here N_1, N_2, N_3 are the tolerance limits for optimistic, moderate, pessimistic decision-maker respectively.}
      end if
    end for
  end for
end for

Algorithm 2 Generating Pareto frontier

Require: $\widetilde{w}_i = (w_{1i}, w_{2i}, w_{3i}), W, u_i^k, n, k$
Ensure: $X = (x_1, x_2, \ldots x_n)$
$\widetilde{w}_i, i = 1, 2, \ldots n$
for $i := n \rightarrow 1$ **do**
  for $x_i := 0 \rightarrow y_k^i$ **do**
    for $d_i := 0 \rightarrow W$ **do**
      Repeat Steps 9 to 20 of Algorithm 1.
      $PI_i^*(x_i, d_i) := \max\{PI_i(x_i, d_i)\} \in N_t \mid \max\{F_k^i(x_i, d_i) \text{ for } t = 1, 2, 3\}$
    end for
  end for
end for
for $x_i := 0 \rightarrow y_k^i$ **do**
  if $(PI_1(x_i, W)) \geq PI_2^*(x_i, W)$ **then**
    $X := X \cup \{x_i\} \cup \{x_{i-1}^*\} \cup \ldots \cup \{x_1^*\}$
  end if
end for

5 Numerical Example

Let us consider that there is a truck with 10 tons loading capacity. The decision-makers have two types of items A and B with fuzzy weights. Here f_1 represents the profits on the items of type A and B, respectively, f_2 represent the amount used

Table 1 Data for knapsack problem

Weight	$\widetilde{2} = (1.5, 2, 3)$	$\widetilde{2} = (1.5, 2, 3)$
f_1	6	10
f_2	20	3
Type	A	B

Table 2 Solution for first objective at stage 1

d_2	$x_2 = 0$	$x_2 = 1$	$x_2 = 2$	$x_2 = 3$	$x_2 = 4$	$x_2 = 5$	max	x_2^*
1	–	–	–	–	–	–	–	–
2	–	(10, 0.34)	–	–	–	–	(10, 0.34)	1
3	–	(10, 1)	–	–	–	–	(10, 1)	1
4	–	(10, 1)	(20, 0.34)	–	–	–	(10, 1)	1
5	–	(10, 1)	(20, 0.83)	(30, 0.03)	–	–	(20, 0.83)	2
6	–	(10, 1)	(20, 1)	(30, 0.34)	–	–	(20, 1)	2
7	–	(10, 1)	(20, 1)	(30, 0.7)	(40, 0.08)	–	(30, 0.7)	3
8	–	(10, 1)	(20, 1)	(30, 0.92)	(40, 0.34)	(50, 0.01)	(30, 0.92)	3
9	–	(10, 1)	(20, 1)	(30, 1)	(40, 0.62)	(50, 0.12)	(40, 0.62)	4
10	–	(10, 1)	(20, 1)	(30, 1)	(40, 0.83)	(50, 0.34)	(40, 0.83)	4

Table 3 Solution for first objective at stage 2

d_1	$x_1 = 0$	$x_1 = 1$	$x_1 = 2$	$x_1 = 3$	$x_1 = 4$	$x_1 = 5$	max	x_1^*	x_2^*
1	–	–	–	–	–	–	–		
2	(10, 0.34)	(6, 0.34)	–	–	–	–	(10, 0.34)	0	1
3	(10, 1)	(6, 1)	–	–	–	–	(10, 1)	0	1
4	(10, 1)	(16, 0.84)	(12, 0.34)	–	–	–	(16, 0.84)	1	1
5	(20, 0.83)	(16, 1)	(22, 0.55)	(18, 0.03)	–	–	(22, 0.55)	2	1
6	(20, 1)	(26, 0.75)	(22, 1)	(18, 0.34)	–	–	(26, 0.75)	1	2
7	(30, 0.7)	(26, 0.97)	(22, 1)	(28, 0.66)	(24, 0.08)	–	(30, 0.7)	0	3
8	(30, 0.92)	(36, 0.72)	(32, 0.83)	(28, 0.96)	(34, 0.2)	(30, 0.01)	(36, 0.72)	1	3
9	(40, 0.62)	(36, 0.9)	(32, 1)	(38, 0.65)	(34, 0.81)	(30, 0.12)	(40, 0.62)	0	4
10	(40, 0.83)	(46, 0.78)	(42, 0.77)	(38, 0.96)	(34, 0.91)	(40, 0.11)	(46, 0.78)	1	4

Table 4 Solution for second objective at stage 1

d_2	$x_2 = 0$	$x_2 = 1$	$x_2 = 2$	$x_2 = 3$	$x_2 = 4$	$x_2 = 5$	max	x_2^*
1	–	–	–	–	–	–	–	–
2	–	(3, 0.34)	–	–	–	–	(3, 0.34)	1
3	–	(3, 1)	–	–	–	–	(3, 1)	1
4	–	(3, 1)	(6, 0.34)	–	–	–	(3, 1)	1
5	–	(3, 1)	(6, 0.83)	(9, 0.03)	–	–	(6, 0.83)	2
6	–	(3, 1)	(6, 1)	(9, 0.34)	–	–	(6, 1)	2
7	–	(3, 1)	(6, 1)	(9, 0.7)	(12, 0.08)	–	(9, 0.7)	3
8	–	(3, 1)	(6, 1)	(9, 0.92)	(12, 0.34)	(15, 0.01)	(9, 0.92)	3
9	–	(3, 1)	(6, 1)	(9, 1)	(12, 0.62)	(15, 0.12)	(12, 0.62)	4
10	–	(3, 1)	(6, 1)	(9, 1)	(12, 0.83)	(15, 0.34)	(12, 0.83)	4

in manufacturing the items of the type A and B, respectively. Our objective is to optimize both the objective functions subject to knapsack capacity ($W = 10$) and find number of copies per item which gives the compromise solution for both the objectives (Table 1).

If a moderate decision-maker is trying to find the solution then the selected possibility index should be greater than 0.5. Here the solution is given for moderate decision-maker. First we solve it by taking each objective separately using our proposed dynamic programming algorithm.

From Tables 2, 3, 4, 5, 6, and 7 it is clear that we select $x_2 = 0, 1, 2, 3, 4$ and $x_1 = 0, 1, 2, 3, 4$ since the possibility index at $X_1 = X_2 = 5$ is less than 0.5.

The following Pareto frontier can be generated for the above problem

- (40,12)$\longmapsto$ (0,4)
- (38,66)$\longmapsto$ (3,2)
- (34,83)$\longmapsto$ (4,1)

Now fuzzy decision matrix is given by

$$\widetilde{Y} = (\widetilde{f}_{ij})_{2\times 3} = \begin{pmatrix} (0;0;0) & (6;1.5;3) & (8;2;4) \\ (8;2;4) & (4;1;2) & (2;0.5;1) \end{pmatrix}$$

After calculating normalized fuzzy decision matrix fuzzy positive ideal solution and fuzzy negative ideal solution are given by $\widetilde{a}^+ = \{(0.67; 0.16; 0.34), (0.67; 0.16; 0.34)\}$ and $\widetilde{a}^- = \{(0; 0; 0), (0.16; 0.04; 0.08)\}$, respectively.

So using compromise ratio method we get the solution which is 3 copies of type A and 2 copies of type B with the possibility index 0.96. Where decision at each stage for selecting the possibility depends upon the tolerance limit of moderate decision-maker.

Table 5 Solution for second objective at stage 2

d_1	$x_1 = 0$	$x_1 = 1$	$x_1 = 2$	$x_1 = 3$	$x_1 = 4$	$x_1 = 5$	max	x_1^*	x_2^*
1	–	–	–	–	–	–	–		
2	(3, 0.34)	(20, 0.34)	–	–	–	–	(20, 0.34)	1	0
3	(3, 1)	(20, 1)	–	–	–	–	(20, 1)	1	0
4	(3, 1)	(23, 0.84)	(40, 0.34)	–	–	–	(23, 0.84)	1	1
5	(6, 0.83)	(23, 1)	(43, 0.55)	(60, 0.03)	–	–	(43, 0.55)	2	1
6	(6, 1)	(29, 0.75)	(43, 1)	(60, 0.34)	–	–	(43, 1)	2	1
7	(9, 0.7)	(29, 0.97)	(43, 1)	(63, 0.66)	(80, 0.08)	–	(63, 0.66)	3	1
8	(9, 0.92)	(38, 0.72)	(46, 0.83)	(63, 0.96)	(83, 0.2)	(100, 0.01)	(63, 0.96)	3	1
9	(12, 0.62)	(38, 0.9)	(46, 1)	(66, 0.65)	(83, 0.81)	(100, 0.12)	(83, 0.81)	4	1
10	(12, 0.83)	(50, 0.78)	(49, 0.77)	(66, 0.96)	(83, 0.91)	(103, 0.11)	(83, 0.91)	4	1

Table 6 Solution for both objectives at stage 1

d_2	$x_2 = 0$	$x_2 = 1$	$x_2 = 2$	$x_2 = 3$	$x_2 = 4$	$max f_1$
1	–	–	–	–	–	–
2	–	(10, 3, 0.34)	–	–	–	(10, 3, 0.34)
3	–	(10, 3, 1)	–	–	–	(10, 3, 1)
4	–	(10, 3, 1)	(20, 6, 0.34)	–	–	(10, 3, 1)
5	–	(10, 3, 1)	(20, 6, 0.83)	(30, 9, 0.03)	–	(20, 6, 0.83)
6	–	(10, 3, 1)	(20, 6, 1)	(30, 9, 0.34)	–	(20, 6, 1)
7	–	(10, 3, 1)	(20, 6, 1)	(30, 9, 0.7)	(40, 12, 0.08)	(30, 9, 0.7)
8	–	(10, 3, 1)	(20, 6, 1)	(30, 9, 0.92)	(40, 12, 0.34)	(30, 9, 0.92)
9	–	(10, 3, 1)	(20, 6, 1)	(30, 9, 1)	(40, 12, 0.62)	(40, 12, 0.62)
10	–	(10, 3, 1)	(20, 6, 1)	(30, 9, 1)	(40, 12, 0.83)	(40, 12, 0.83)

Table 7 Solution for both objectives at stage 2

d_1	$x_1 = 0$	$x_1 = 3$	$x_1 = 4$	$max f_1$	$max f_2$
10	(40, 12, 0.83)	(38, 66, 0.96)	(34, 83, 0.91)	(40, 12, 0.83)	(34, 83, 0.91)

6 Conclusion

From the crisp knapsack problem when it is extended into fuzzy knapsack problem and solving it without defuzzification the resulting value of optimal profit varies with the selection of possibility index by DM's choices at each stage. The possibility index of selecting fuzzy weights in available fuzzy weights gives the opportunity to DM to select the optimal value. The result of single objective is used while solving for both the objective together and find the Pareto frontier. After calculating the Pareto frontier we use compromise ratio method for fuzzy to find the best solution.

Acknowledgments Authors are grateful to the anonymous reviewers for their constructive comments and valuable suggestions.

References

1. Bellman, R.: Dynamic programming and lagrange multipliers. Proc. Nat. Acad. Sci. U.S.A. **42**(10), 767 (1956)
2. Bellman, R.E., Zadeh, L.A.: Decision-making in a fuzzy environment. Manag. Sci. **17**(4), B-141 (1970)
3. Dubois, D., Prade, H.: Possibility Theory. Springer (1988)
4. Guha, D., Chakraborty, D.: Compromise ratio method for decision making under fuzzy environment using fuzzy distance measure, a a. **1**, 2 (2008)
5. Horowitz, E., Sahni, S., Rajasekaran, S.: Computer algorithms C++: C++ and pseudocode versions. Macmillan (1997)

6. Kasperski, A., Kulej, M.: The 0–1 knapsack problem with fuzzy data. Fuzzy Optim. Decis. Making **6**(2), 163–172 (2007)
7. Lin, F.-T., Yao, J.-S.: Using fuzzy numbers in knapsack problems. Eur. J. Oper. Res. **135**(1), 158–176 (2001)
8. Martello, S., Toth, P.: Knapsack Problems. Wiley, New York (1990)
9. North, D.W.: A tutorial introduction to decision theory. IEEE Trans. Syst. Sci. Cybern. **4**(3), 200–210 (1968)
10. Okada, S., Gen, M.: Fuzzy multiple choice knapsack problem. Fuzzy Sets Syst. **67**(1), 71–80 (1994)
11. Sengupta, A., Pal, T.K.: On comparing interval numbers. Eur. J. Oper. Res. **127**(1), 28–43 (2000)
12. Toth, P.: Dynamic programming algorithms for the zero-one knapsack problem. Computing **25**(1), 29–45 (1980)
13. Yoshida, Y.: Dynamical aspects in Fuzzy decision making, Vol. 73, Springer (2001)
14. Zadeh, L.A.: Fuzzy sets. Inf. Control **8**(3), 338–353 (1965)

A Fuzzy Random Periodic Review Inventory Model Involving Controllable Back-Order Rate and Variable Lead-Time

Sushil Kumar Bhuiya and Debjani Chakraborty

Abstract In this paper, a fuzzy random periodic review inventory model with controllable back-order rate and variable lead-time has been considered where the annual demand is treated as a fuzzy random variable. The shortage is partially backlogged and the back-order rate is dependent on the back-order discount and the length of the lead-time. The lead-time crashing cost is being introduced as a negative exponential function of the lead-time. We develop a methodology to find the optimal review period, optimal target level, and optimal lead-time. A numerical example is provided to illustrate the model.

Keywords Inventory · Fuzzy random variable · Possibilistic mean value

1 Introduction

In any real-life inventory system, the occurrence of shortage is a natural phenomenon. In such a situation, some customers are willing to wait for back orders and some customers become impatient and turn to other firms. Thus, the inventory model, which considers both back orders and lost sales cases, is more realistic than the ones based on the individual cases. Montogomery et al. [14] developed continuous and periodic review inventory models with a mixture of back orders and lost sales. Abad [1, 11, 17], and many other researchers also studied on the problem of mixture of back orders and lost sales. In fuzzy environment, Ouyang and Yao [16] developed a mixture inventory model with fuzzy demand and random lead-time demand. Chang et al. [5] analyzed a mixture inventory model with fuzzy random variable lead-time demand. Vijayan and Kumaran [20] developed the mixed model for both the continuous and

S.K. Bhuiya (✉) · D. Chakraborty
Department of Mathematics, Indian Institute of Technology Kharagpur, Kharagpur 721302, India
e-mail: skbhuiya@maths.iitkgp.ernet.in

D. Chakraborty
e-mail: debjani@maths.iitkgp.ernet.in

R.N. Mohapatra et al. (eds.), *Mathematics and Computing*,
Springer Proceedings in Mathematics & Statistics 139,
DOI 10.1007/978-81-322-2452-5_21

periodic review inventory model with fuzzy cost. Dey [6] presented a fuzzy random continuous review inventory model under mixture of back orders and lost sales with imprecise budgetary constraint and with constant lead-time. Throughout the above model, the back rate is considered constant. However, in realistic situations, the back-order rate is dependent on the back-order discount and the length of the lead-time. The supplier can always provide a price discount on the stock-out item in order to increase more back orders. Thus, the rate of back is proportional to the discount offered by the production house. Pan and Hsiao [18] derived a continuous review inventory model with back-order discount and variable lead-time. The back-order rate is also dependent on the lead-time. Since bigger lead-time might lead to longer shortage period, many customers may not like to wait for back orders. Ouyang and Chuang [15] have considered the controllable back-order rate, which is dependent on the length of the lead-time as a function of the amount of shortage. In this study, we consider the rate of back order as a mixture of back-order discount proportion and the length of the lead-time through an exponential function of the amount of shortage.

In most inventory models, lead-time is considered as a predetermined constant, which is not a control variable. However, after implementation of Just In Time [13], lead-time has received extensive attention in every manufacturing company. By reducing the lead-time, we can minimize the safety stock and can improve the level of service to a customer. Liao and Shyu [13] first derived a probabilistic model with variable lead-time where order quantity was preassumed. Ben-Daya and Raouf [2] extended the [13] model by considering the order quantity as a decision variable. The model [2] extended by [17] by including stock-out cost. Many studies (for instance [4, 5, 18]) presented the problem of the lead-time reduction. Wu et al. [21] developed a computational procedure for optimal inventory policy with negative exponential crashing cost and variable lead-time demand. Recently, Dey and Chakraborty [9] proposed a fuzzy random periodic inventory system. They introduced the lead-time crashing cost as a negative exponential function of the lead-time.

In this paper, we propose a model to analyze fuzzy random periodic review inventory system with mixture of back orders and lost sales. We consider the lead-time as a decision variable. The crashing cost is introduced as a negative exponential function of the lead-time. The back-order rate is an amalgamation of back-order discount proportion and the length of the lead-time through the reciprocal of the exponential function of the amount of the shortage. The purpose of this paper is to develop a periodic review inventory model involving controllable back-order rate and variable lead-time under fuzzy random variable demand. A methodology is established such that the total cost is minimized in the fuzzy sense. A solution procedure is demonstrated to find the optimal policy. A numerical example is also presented to illustrate the proposed methodology.

The paper is organized as follows: In Sect. 2, we mention the basic concepts from fuzzy sets, which we need for our purpose. In Sect. 3, we build the model that represents our problem and find the solution. We then give an algorithm to compute optimal solution. In Sect. 4, we consider a numerical example to which we apply our algorithm and interpret the solution. Finally, we derive our conclusion and list relevant references.

2 Preliminary Concepts

We begin with a few basic concepts that we will need.

2.1 Triangular Fuzzy Numbers

A normalized triangular fuzzy number $\tilde{A} = (\underline{a}, a, \overline{a})$ is a fuzzy subset of the real line $\mathbb{R}$, whose membership function $\mu_{\tilde{A}}(x)$ satisfies the following conditions [22]:
(i) $\mu_{\tilde{A}}(x)$ is a continuous function from $\mathbb{R}$ to the closed interval [0, 1],
(ii) $\mu_{\tilde{A}}(x) = L(x) = \frac{x-\underline{a}}{a-\underline{a}}$ is strictly increasing function on $[\underline{a}, a]$,
(iii) $\mu_{\tilde{A}}(x) = 1$ for $x = a$,
(iv) $\mu_{\tilde{A}}(x) = R(x) = \frac{\overline{a}-x}{\overline{a}-a}$ is strictly decreasing function on $[a, \overline{a}]$,
(v) $\mu_{\tilde{A}}(x) = 0$ elsewhere,
where $\underline{a}, a, \overline{a}$, are real numbers. $L(x)$ and $R(x)$ are the left and right shape continuous functions, respectively. A triangular fuzzy number $\tilde{A} = (\underline{a}, a, \overline{a})$ can be represented by its α-cuts as $A = [A_\alpha^-, A_\alpha^+]$, where $\alpha \in [0, 1]$, $A_\alpha^- = \underline{a} + \alpha(a - \underline{a})$, and $A_\alpha^+ = \overline{a} - \alpha(\overline{a} - a)$.

Without any loss of generality, all fuzzy quantities are assumed to be triangular fuzzy numbers throughout this paper.

2.2 Possibilistic Mean Value of a Fuzzy Number

Let $\tilde{A}$ be a fuzzy number, then the interval-valued possibilistic mean is defined as $M(\tilde{A}) = \left[M_*(\tilde{A}), M^*(\tilde{A})\right]$, where $M_*(\tilde{A})$, and $M^*(\tilde{A})$ are the lower and upper possibilistic mean values of $\tilde{A}$ [3] and are, respectively, defined by

$$M_*(\tilde{A}) = \frac{\int_0^1 \alpha A_\alpha^- \mathrm{d}\alpha}{\int_0^1 \alpha \mathrm{d}\alpha} \text{ and } M^*(\tilde{A}) = \frac{\int_0^1 \alpha A_\alpha^+ \mathrm{d}\alpha}{\int_0^1 \alpha \mathrm{d}\alpha}.$$

The possibilistic mean value of $\tilde{A}$ is then defined by

$$\overline{M}(\tilde{A}) = \frac{M_*(\tilde{A}) + M^*(\tilde{A})}{2}.$$

In other words, the possibilistic mean value of $\tilde{A}$ can be written as

$$\overline{M}(\tilde{A}) = \int_0^1 \alpha(A_\alpha^- + A_\alpha^+)\mathrm{d}\alpha.$$

Let $\tilde{A}$ and $\tilde{B}$ be two fuzzy numbers where $A_\alpha = [A^-_\alpha, A^+_\alpha]$, and $B_\alpha = [B^-_\alpha, B^+_\alpha]$, $\alpha \in [0, 1]$, then, for ranking fuzzy numbers, we have $\tilde{A} \leq \tilde{B} \iff \overline{M}(\tilde{A}) \leq \overline{M}(\tilde{B})$.

2.3 Fuzzy Random Variable and Its Expectation

Kwakernaak [12] first introduced the basic concept of fuzzy random variables and later [19] further developed the concept. We mention below what we need for our work in subsequent sections.

Let us consider $F_c(\mathbb{R})$, the set of all fuzzy numbers. Now, let $(\Omega, \mathscr{B}, \mathscr{P})$ be a probability space, then a mapping $\chi : \Omega \to F_c(\mathbb{R})$ is said to be a fuzzy random variable (or FRV for short) if for all $\alpha \in [0, 1]$, the two real-valued mapping $inf\, \chi_\alpha : \Omega \to \mathbb{R}$ and $sup\, \chi_\alpha : \Omega \to \mathbb{R}$ (define so that for all $\omega \in \Omega$ we have that $\chi_\alpha(\omega) = [inf(\chi(\omega))_\alpha, sup\,(\chi(\omega))_\alpha]$) are real-valued random variables. If $\tilde{X}$ is a fuzzy random variable then the fuzzy expectation of $\tilde{X}$ is a unique fuzzy number. It is defined by $E(\tilde{X}) = \int \tilde{X} \mathrm{d}P = \left\{ \left(\int X^-_\alpha \mathrm{d}P, \int X^+_\alpha \mathrm{d}P \right) : 0 \leq \alpha \leq 1 \right\}$,
where the α-cut of fuzzy random variable is $[X]_\alpha = [X^-_\alpha, X^+_\alpha]$ for all $\alpha \in [0, 1]$. The α-cut representation of fuzzy expectation is given by $\upsilon_\alpha = [E(\tilde{X})]_\alpha = E[X]_\alpha = [E(X^-_\alpha), E(X^+_\alpha)], \quad \alpha \in [0, 1]$.

3 Methodology

In this section, we build a mathematical model, which we will need to represent our model and then find its solution, which will be interpreted in the context. The analysis interpretations is similar to those of [7–9].

3.1 Model and Assumptions

In real world, the most widely used operating doctrine for periodic review system is the order up to target level R doctrine. This system requires the inventory level to be reviewed periodically and a sufficient order is placed to bring the inventory position up to R.

The following notations have been used:

T	time between review
R	target inventory level
J	cost of making a review
C_0	fixed ordering cost per order
A	$= J + C_0$

h	holding cost per unit per year
π	stock-out cost per unit stock-out
π_0	marginal profit per unit
π_x	back-order price discount offered by supplier
β	fraction of demand back-ordered during the stock-out period, $(0 \leq \beta \leq 1)$
L	lead-time(in years)
$R(L)$	lead-time crashing cost
$\tilde{d}(\omega)$	annual demand ($\omega \in \Omega$ where $(\Omega, \mathscr{B}, \mathscr{P})$ is a probability space)
$\tilde{d}_L(\omega)$	lead-time demand ($\omega \in \Omega$)
$\tilde{d}_{L+T}(\omega)$	lead-time plus one period demand ($\omega \in \Omega$)
x^+	$\max\{0, x\}$

There is no time-dependent back-order cost. The back orders are incurred in very small quantities so that when an order arrives, it is usually sufficient to meet any outstanding back orders.

In periodic review inventory system, the safety stock or buffer stock is defined as the difference between target level R and the lead-time plus one period demand. In order to maintain the nonnegative safety level, we assumed $R \geq \overline{M}(\tilde{d}_{L+T})$ where $\overline{M}(\tilde{d}_{L+T})$ denotes the expected lead-time plus one period demand in possibilistic sense and defined by

$$\overline{M}(\tilde{d}_{L+T}) = \int_0^1 \alpha \left[d^-_{L+T,\alpha} + d^+_{L+T,\alpha} \right] \mathrm{d}\alpha. \tag{1}$$

In order to incorporate fuzziness and randomness simultaneously [7], the annual demand is treated as a discrete fuzzy random variable $\tilde{d}(\omega)$ ($\omega \in \Omega$ where $(\Omega, \mathscr{B}, \mathscr{P}$ is a probability space). Let us suppose that the annual customer demand $\tilde{d}(\omega)$ is of the form $\{(\tilde{d}_1, p_1), (\tilde{d}_2, p_2), (\tilde{d}_3, p_3), \cdots\cdots, (\tilde{d}_n, p_n)\}$, where each of $\tilde{d}_i$'s are triangular fuzzy numbers of the form $(\underline{d}_i, d_i, \overline{d}_i)$ with corresponding probabilities p_i's, $i = 1, 2, 3, \cdots\cdots, n$. Further, let $\mu_{\tilde{d}_i}(x)$ denote the membership function corresponding to each $\tilde{d}_i$, $i = 1, 2, 3, \cdots\cdots, n$, and defined by

$$\mu_{\tilde{d}_i}(x) = \begin{cases} L_i(x), & \underline{d}_i \leq x \leq d_i \\ R_i(x), & d_i \leq x \leq \overline{d}_i \end{cases}$$

with $[\underline{d}_i, \overline{d}_i]$ as the support of each $\tilde{d}_i$. Where d_i is the modal of fuzzy number $\tilde{d}_i$, L_i, and R_i are the left and right reference functions, respectively. The lead-time demand and the lead-time plus one period demand are considered to be connected to the annual demand through the length of the lead-time and period in the following form:

$\tilde{d}_L(\omega) = \tilde{d}(\omega) \times L$

$\tilde{d}_{L+T}(\omega) = \tilde{d}(\omega) \times (L + T)$

Annual demand $\tilde{d}(\omega)$ is a fuzzy random variable of the form $\tilde{d}_i = (\underline{d}_i, d_i, \overline{d}_i)$

with probability p_i, for all $i = 1, 2, 3 \cdots\cdots, n$, thus, the lead-time demand and the lead-time plus one period demand are also fuzzy random variable of the form $\tilde{d}_{L,i} = (\underline{d}_{L,i}, d_{L,i}, \overline{d}_{L,i})$, and $\tilde{d}_{L+T,i} = (\underline{d}_{L+T,i}, d_{L+T,i}, \overline{d}_{L+T,i})$, respectively, with probability p_i, for all $i = 1, 2, 3 \cdots\cdots, n$. The expectation of fuzzy random variable is a unique fuzzy number. The triangular form of the expected lead-time demand and lead-time plus one period demand are given by $E(\tilde{d}_L(\omega)) = \tilde{d}_L = (\underline{d}_L, d_L, \overline{d}_L)$, and $E(\tilde{d}_{L+T}(\omega)) = \tilde{d}_{L+T} = (\underline{d}_{L+T}, d_{L+T}, \overline{d}_{L+T})$, respectively. The α-cut representation of the expected lead-time demand and lead-time plus one period demand are defined as follows:

$d^-_{L,\alpha}(\omega) = d^-_{L,\alpha}(\omega) \times L$ and $d^+_{L,\alpha}(\omega) = d^+_{L,\alpha}(\omega) \times L$

$$\Rightarrow \begin{cases} E\left(d^-_{L,\alpha}(\omega)\right) = \sum\limits_{i=1}^{n} d^-_{i,\alpha} p_i \times L \\ E\left(d^+_{L,\alpha}(\omega)\right) = \sum\limits_{i=1}^{n} d^+_{i,\alpha} p_i \times L \end{cases}$$

and

$d^-_{L+T,\alpha}(\omega) = d^-_{L+T,\alpha}(\omega) \times (L+T)$ and $d^+_{L+T,\alpha}(\omega) = d^+_{L+T,\alpha}(\omega) \times (L+T)$

$$\Rightarrow \begin{cases} E\left(d^-_{L+T,\alpha}(\omega)\right) = \sum\limits_{i=1}^{n} d^-_{i,\alpha} p_i \times (L+T) \\ E\left(d^+_{L+T,\alpha}(\omega)\right) = \sum\limits_{i=1}^{n} d^+_{i,\alpha} p_i \times (L+T) \end{cases}$$

The total cost function in fuzzy sense is given by

$$\begin{aligned} &\tilde{\mathscr{C}}(R, T, L) \\ &= \left[h\left\{R - \tilde{d}(\omega)L - \frac{\tilde{d}(\omega)}{2}T\right\} + \left\{h(1-\beta) + \frac{\{\pi + \pi_0(1-\beta)\}}{T}\right\}\overline{M}(\tilde{d}_{L+T} - R)^+\right] \\ &+ \left[\frac{A + R(L)}{T}\right] \end{aligned} \tag{2}$$

where $\overline{M}(\tilde{d}_{L+T} - R)^+$ denoted the possibilistic value of the expected shortage at each cycle and calculated by

$$\overline{M}\left(\tilde{d}_{L+T} - R\right)^+ = \int\limits_0^1 \alpha\left[\left((\tilde{d}_{L+T} - R)^+\right)^-_\alpha + \left((\tilde{d}_{L+T} - R)^+\right)^+_\alpha\right]\mathrm{d}\alpha.$$

The lead-time is a decision variable in our proposed model. We consider a negative exponential crashing cost function as in [9]. The total crashing cost is of the following form

Crashing cost $R(L) = \lambda e^{-\mu L}$.

Some Known values of lead-time crashing cost for values of the lead-time L are used to calculate the parameter λ and μ.

Now as explained earlier, the back-order rate is proportional to back-order discount offered by supplier and it is also dependent on the lead-time through the amount of shortage. We define the back-order rate as a combination of back-order discount proportion and the length of the lead-time through the reciprocal of the exponential function of the amount of shortage. We define the back-order rate as

Back-order rate $\beta = k_1\frac{\pi_x}{\pi_0} + k_2 e^{-\overline{M}(\tilde{d}_{L+T}-R)^+}$ where $0 \leq \pi_x \leq \pi_0$ and $0 \leq k_1 + k_2 \leq 1$. Thus the total cost function (2) is rewritten as

$$\tilde{\mathscr{C}}(R, T, L)$$
$$= \left[h\left\{R - \tilde{d}(\omega)L - \frac{\tilde{d}(\omega)}{2}T\right\} + \left\{h\left(1 - k_1\frac{\pi_x}{\pi_0} - k_2 e^{-\overline{M}(\tilde{d}_{L+T}-R)^+}\right)\right\}\overline{M}(\tilde{d}_{L+T} - R)^+\right]$$
$$+ \left[\left\{\frac{\{\pi + \pi_0(1 - k_1\frac{\pi_x}{\pi_0} - k_2 e^{-\overline{M}(\tilde{d}_{L+T}-R)^+})\}}{T}\right\}\overline{M}(\tilde{d}_{L+T} - R)^+ + \frac{A + \lambda e^{-\mu L}}{T}\right] \tag{3}$$

To find the optimal solution, we need to determine

(i) the exact expression of $\overline{M}(\tilde{d}_{L+T} - R)^+$;
(ii) the fuzzy expected value of the total cost function and, then defuzzified the expected value;
(iii) derive the optimal values of the target level R^*, review of period T^* and the lead-time L^* such that minimize the expected total annual cost.

3.2 Determination of Expected Shortage $\overline{M}(\tilde{d}_{L+T} - R)^+$

The expected shortage in possibilistic sense is determined [9, 10] as follows:

Situation 1. For R lying between $\underline{d}_{L+T}$ and d_{L+T}, we have the α-level set of the lead-time plus one period demand as

$$(\tilde{d}_{L+T})_\alpha = \begin{cases} [R, d^+_{L+T,\alpha}], & \alpha \leq L(R) \\ [d_{L+T,\alpha}, d^+_{L+T,\alpha}], & \alpha > L(R) \end{cases}$$

which implies

$$\left((\tilde{d}_{L+T} - R)^+\right)_\alpha = \begin{cases} [0, d^+_{L+T,\alpha} - R], & \alpha \leq \mathrm{L(R)} \\ [d^-_{L+T,\alpha} - R, d^+_{L+T,\alpha} - R], & \alpha > \mathrm{L(R)} \end{cases} \tag{4}$$

Therefore, the possibilistic mean is calculated as follows:

$$\overline{M}\left(\tilde{d}_{L+T} - R\right)^{+} = \int_{0}^{1} \alpha\left[\left((\tilde{d}_{L+T} - R)^{+}\right)_{\alpha}^{-} + \left((\tilde{d}_{L+T} - R)^{+}\right)_{\alpha}^{+}\right]\mathrm{d}\alpha$$

$$= \int_{0}^{L(R)} \alpha(d_{L+T,\alpha}^{+} - R)\mathrm{d}\alpha + \int_{L(R)}^{1} \alpha\{(d_{L+T,\alpha}^{-} - R) + (d_{L+T,\alpha}^{+} - R)\}\mathrm{d}\alpha$$

$$= \int_{0}^{1} \alpha d_{L+T,\alpha}^{+}\mathrm{d}\alpha + \int_{L(R)}^{1} \alpha d_{L+T,\alpha}^{-}\mathrm{d}\alpha - R(1 - 0.5L^{2}(R)) \tag{5}$$

Situation 2. For R lying between d_{L+T} and $\overline{d}_{L+T}$, the α-level set of the lead-time plus one period demand is given by

$$(\tilde{d}_{L+T})_{\alpha} = \begin{cases} [R, d_{L+T,\alpha}^{+}], & \alpha \leq \mathrm{R(R)} \\ \phi, & \alpha > \mathrm{R(R)} \end{cases}$$

which implies

$$\left((\tilde{d}_{L+T} - R)^{+}\right)_{\alpha} = \begin{cases} [0, d_{L+T,\alpha}^{+} - R], & \alpha \leq \mathrm{R(R)} \\ \phi, & \alpha > \mathrm{R(R)} \end{cases} \tag{6}$$

Therefore, the possibilistic mean is calculated as follows:

$$\overline{M}\left(\tilde{d}_{L+T} - R\right)^{+} = \int_{0}^{1} \alpha\left[\left((\tilde{d}_{L+T} - R)^{+}\right)_{\alpha}^{-} + \left((\tilde{d}_{L+T} - R)^{+}\right)_{\alpha}^{+}\right]\mathrm{d}\alpha$$

$$= \int_{0}^{R(R)} \alpha(d_{L+T,\alpha}^{+} - R)\mathrm{d}\alpha$$

$$= \int_{0}^{R(R)} \alpha d_{L+T,\alpha}^{+}\mathrm{d}\alpha - 0.5RR^{2}(R) \tag{7}$$

3.3 Possibilistic Mean Value of the Fuzzy Expected Total Cost Function

The total cost function (3) is given by

$$\begin{aligned}
&\tilde{\mathscr{C}}(R, T, L)\\
&= \left[h\left\{R - \tilde{d}(\omega)L - \frac{\tilde{d}(\omega)}{2}T\right\} + \left\{h\left(1 - k_1\frac{\pi_x}{\pi_0} - k_2 e^{-\overline{M}(\tilde{d}_{L+T}-R)^+}\right)\right\}\overline{M}(\tilde{d}_{L+T} - R)^+\right]\\
&+ \left[\left\{\frac{\{\pi + \pi_0(1 - k_1\frac{\pi_x}{\pi_0} - k_2 e^{-\overline{M}(\tilde{d}_{L+T}-R)^+})\}}{T}\right\}\overline{M}(\tilde{d}_{L+T} - R)^+ + \frac{A + \lambda e^{-\mu L}}{T}\right] \quad (8)
\end{aligned}$$

where the possibilistic mean value of the expected shortage at each cycle, $\overline{M}(\tilde{d}_{L+T} - R)^+$ is given by either (5) or (7) depending on the position of $R \in [\underline{d}_{L+T}, \overline{d}_{L+T}]$. As mentioned before, the total cost function is a fuzzy random variable. The expectation of the fuzzy random variable is a unique fuzzy number. For computational purpose, the expectation of the total annual cost is defuzzified using its possibilistic mean value.

The α-level set of the total cost function is given by

$$\begin{aligned}
&\mathscr{C}_\alpha^-\\
&= h\left[R - d_\alpha^+ L - \frac{d_\alpha^+}{2}T\right] + \left[h\left(1 - k_1\frac{\pi_x}{\pi_0} - k_2 e^{-\overline{M}(\tilde{d}_{L+T}-R)^+}\right)\right]\overline{M}(\tilde{d}_{L+T} - R)^+\\
&+ \left[\left\{\frac{\{\pi + \pi_0(1 - k_1\frac{\pi_x}{\pi_0} - k_2 e^{-\overline{M}(\tilde{d}_{L+T}-R)^+})\}}{T}\right\}\overline{M}(\tilde{d}_{L+T} - R)^+ + \frac{A + \lambda e^{-\mu L}}{T}\right]
\end{aligned}$$

$$\begin{aligned}
&\mathscr{C}_\alpha^+\\
&= h\left[R - d_\alpha^- L - \frac{d_\alpha^-}{2}T\right] + \left[h\left(1 - k_1\frac{\pi_x}{\pi_0} - k_2 e^{-\overline{M}(\tilde{d}_{L+T}-R)^+}\right)\right]\overline{M}(\tilde{d}_{L+T} - R)^+\\
&+ \left[\left\{\frac{\{\pi + \pi_0(1 - k_1\frac{\pi_x}{\pi_0} - k_2 e^{-\overline{M}(\tilde{d}_{L+T}-R)^+})\}}{T}\right\}\overline{M}(\tilde{d}_{L+T} - R)^+ + \frac{A + \lambda e^{-\mu L}}{T}\right]
\end{aligned}$$

The α-level set of expected value of total cost function is the given by

$$\begin{aligned}
&E(\mathscr{C}_\alpha^-)\\
&= \sum_{i=1}^{n}\left\{h\left[R - d_\alpha^+ L - \frac{d_\alpha^+}{2}T\right] + \left[h\left(1 - k_1\frac{\pi_x}{\pi_0} - k_2 e^{-\overline{M}(\tilde{d}_{L+T}-R)^+}\right)\right]\overline{M}(\tilde{d}_{L+T} - R)^+\right.\\
&+ \left.\left[\frac{\{\pi + \pi_0(1 - k_1\frac{\pi_x}{\pi_0} - k_2 e^{-\overline{M}(\tilde{d}_{L+T}-R)^+})\}}{T}\right]\overline{M}(\tilde{d}_{L+T} - R)^+ + \frac{A + \lambda e^{-\mu L}}{T}\right\}p_i
\end{aligned}$$

$$E(\mathscr{C}_\alpha^+)$$
$$= \sum_{i=1}^{n} \left\{ h\left[R - d_\alpha^- L - \frac{d_\alpha^-}{2} T \right] + \left[h\left(1 - k_1 \frac{\pi_x}{\pi_0} - k_2 e^{-\overline{M}(\tilde{d}_{L+T} - R)^+} \right) \right] \overline{M}(\tilde{d}_{L+T} - R)^+ \right.$$
$$\left. + \left[\frac{\{\pi + \pi_0 (1 - k_1 \frac{\pi_x}{\pi_0} - k_2 e^{-\overline{M}(\tilde{d}_{L+T} - R)^+})\}}{T} \right] \overline{M}(\tilde{d}_{L+T} - R)^+ + \frac{A + \lambda e^{-\mu L}}{T} \right\} p_i$$

Therefore, the possibilistic mean value of the fuzzy expected total cost is calculated by

$$\overline{M}(R, L, T) = \int_0^1 \alpha \left(E(\mathscr{C}_\alpha^-) + E(\mathscr{C}_\alpha^+) \right) d\alpha$$
$$= \left[\frac{A + \lambda e^{-\mu L}}{T} + Rh \right] + \left[\left(1 - k_1 \frac{\pi_x}{\pi_0} - k_2 e^{-\overline{M}(\tilde{d}_{L+T} - R)^+} \right) (h + \frac{\pi_0}{T}) + \frac{\pi}{T} \right] \overline{M}(\tilde{d}_{L+T} - R)^+$$
$$- h\left(L + \frac{T}{2} \right) \left\{ \frac{1}{6} \sum_{i=1}^{n} (\underline{d}_i + \overline{d}_i) p_i + \frac{2}{3} \sum_{i=1}^{n} d_i p_i \right\} \tag{9}$$

3.4 Optimal Solution

The optimal solution minimizes the total expected cost per year in the fuzzy sense. To find the optimal solution we follow the following steps:

Step 1: Input the values of $A, C_0, h, J, \pi, \pi_0, \pi_x, k_1, k_2, \lambda$ and μ;
Step 2: Calculate the possibilistic mean value of the fuzzy expected shortage using either (5) or (7) with the condition $0 \leq L(R) \leq 1$ or $0 \leq R(R) \leq 1$, respectively;
Step 3: Determine the safety stock criteria, i.e., $R - \overline{M}(d_{L+T}) \geq 0$;
Step 4: Find the possibilistic mean value of the fuzzy expected total cost from (9);
Step 5: Solve the following minimization problem;

$$\begin{aligned} &\text{Min } \overline{M}(R, T, L) \\ &\text{Subject to} \\ &R - \overline{M}(d_{L+T}) \geq 0 \\ &R - \overline{d}_{L+T} \leq 0 \\ &R, T, L \geq 0 \end{aligned}$$

Step 6: Stop.

We next mention an example to which we will apply the algorithm mentioned here as an illustration.

4 Numerical Example

A small watchband company sells a certain type of watch. The company uses an '*order up to R*' policy. The cost of placing an order and review is Rs. 100. The inventory carrying cost is Rs. 40. The penalty cost for stock-out is Rs. 135 and the marginal profit per item is Rs 300. At the stock-out period, the back-order price discount is Rs. 100. It is estimated that the lead-time crashing cost is $\lambda e^{-\mu L}$(L is in year) where $\lambda = 150$ and $\mu = 70$. The annual demand information is given in the following table

Demand	Probability
(560, 580, 605)	.18
(540, 570, 595)	.17
(400, 415, 435)	.24
(450, 475, 500)	.16
(435, 460, 490)	.25

Here, we have $A = 100$, $h = 40$, $\pi = 135$, $\pi_0 = 300$, and $\pi_x = 100$.
The expected lead-time plus one period demand is $(469.35, 491.9, 516.95) \times (L+T)$.
The possibilistic mean value of the expected annual cost is

$$\overline{M}(R, T, L) = \left[\left(1 - \frac{k_1}{3} - k_2 e^{-\overline{M}(\tilde{d}_{L+T} - R)^+}\right)\left(40 + \frac{300}{T}\right) + \frac{135}{T}\right]\overline{M}(\tilde{d}_{L+T} - R)^+$$
$$+ \left[\frac{100 + 150e^{-70L}}{T} + 40R\right] - \left(L + \frac{T}{2}\right)19692.67$$

The safety or buffer stock is calculated by

$$SS = R - \left\{\tfrac{1}{6}\sum_{i=1}^{n}(\underline{d}_i + \overline{d}_i)p_i + \tfrac{2}{3}\sum_{i=1}^{n} d_i p_i\right\}(L + T) = R - 492.3167(L + T).$$

The final form of the optimization problem reduces to

$$\begin{aligned}
&\text{Min } \overline{M}(R, T, L)\\
&\text{Subject to}\\
&R \geq 492.3167(L + T)\\
&R - 491.9(L + T) \geq 0\\
&R - 516.95(L + T) \leq 0\\
&R, T, L \geq 0
\end{aligned}$$

The changes in the values for the parameters k_1 and k_2 can appear due to uncertainties and dynamic market conditions. In a decision-making process, the sensitivity analysis can be useful for implications of these changes in the values of parameters.

Table 1 Optimal solution of the example when the values of k_2 change

k_1	k_2	β	R	L(in yr)	T(in yr)	Total cost
0	0	0	85.20904	0.06849983	0.09744154	2146.329
0	0.1	0.0998180	85.22673	0.06854816	0.09746703	2145.792
0	0.2	0.1995910	85.24659	0.06860238	0.09749552	2145.191
0	0.3	0.2993042	85.26911	0.06866375	0.09752765	2144.513
0	0.4	0.3989359	85.29491	0.06873396	0.09756423	2143.740
0	0.5	0.4984539	85.32483	0.06881528	0.09760635	2142.849
0	0.6	0.5978084	85.36003	0.06891086	0.09765550	2141.804
0	0.7	0.6969198	85.40215	0.06902519	0.09771371	2140.560
0	0.8	0.7956542	85.45360	0.06916484	0.09778398	2139.045
0	0.9	0.8937754	85.51789	0.06933979	0.09787006	2137.147

Table 2 Optimal solution of the example when the values of k_1 change

k_1	k_2	β	R	L(in yr)	T(in yr)	Total cost
0.1	0	0.03333333	85.21474	0.06851541	0.09744978	2146.156
0.2	0	0.06666670	85.22066	0.06853156	0.09745832	2145.977
0.3	0	0.10000000	85.22682	0.06854834	0.09746718	2145.791
0.4	0	0.13333330	85.23321	0.06856578	0.09747638	2145.598
0.5	0	0.16666670	85.23988	0.06858393	0.09748594	2145.397
0.6	0	0.20000000	85.24682	0.06860283	0.09749589	2145.188
0.7	0	0.23333330	85.25407	0.06862254	0.09750626	2144.971
0.8	0	0.27777770	85.26164	0.06864312	0.09751707	2144.744
0.9	0	0.30000000	85.26956	0.06866463	0.09752837	2144.508

Table 3 Optimal solution of the example when the values of k_1 and k_2 change

k_1	k_2	β	R	L(in yr)	T(in yr)	Total cost
0.1	0.1	0.1331443	85.23312	0.06856559	0.09747622	2145.599
0.2	0.2	0.2662222	85.26137	0.06864258	0.09751664	2144.747
0.3	0.3	0.3992013	85.29510	0.06873435	0.09756454	2143.730
0.4	0.4	0.5320265	85.33629	0.06884615	0.09762249	2142.515

A sensitivity analysis of the above problem is carried out. The optimal solutions of the above problem for different values of k_1 and k_2 are presented in the Tables 1, 2 and 3, respectively.

Tables 1, 2 and 3 show that as the back-order parameter increases, the back-order rate increases and with the increases of back-order rates, the total cost decreases. It is also observed that when the back-order rate increases, target inventory level, lead-time, and period of review are increasing. Based on the optimal solutions for different values of k_1 and k_2, management judged the review period, target inventory level, and lead-time. If k_1 is zero then there will be no back-order discount, the optimal expected minimum cost is 2137.147, which attains at $T^* = 0.09787006$, $R^* = 85.51789$, $L^* = 0.06933979$, and $\beta^* = 0.8937754$. For $k_2 = 0$, the rate of back order is independent of the lead- time. The optimal values of the review period, target inventory level, lead-time, and rate of back order are $T^* = 0.09752837$, $R^* = 85.26956$, $L^* = 0.06866463$, and $\beta^* = 0.30000000$, respectively, and corresponding cost is 2144.508. If both the k_1 and k_2 are nonzero, the optimal expected cost is 2142.515, which attains at $T^* = 0.09762249$, $R^* = 85.33629$, $L^* = 0.06884615$, and $\beta^* = 0.5320265$.

5 Conclusion

In this paper, a fuzzy random periodic review inventory model with mixture of back order and lost sales is considered where the annual demand is assumed as a fuzzy random variable. The review period, target inventory level, lead-time, and back-order rate are considered as decision variables. The back-order rate has been introduced by the mixture of back-order discount proportion and the length of the lead-time through the reciprocal of the exponential function of the amount of shortage. A methodology has been developed such that the total cost is minimized in the fuzzy sense. We present a solution procedure to find the optimal policy. Finally, a numerical example is solved by our proposed methodology. The sensitivity of the solution for changes in the values of parameters k_1 and k_2 has been discussed.

In future research on this model, it would be interesting to deal with imprecise probabilities and treat the back-order discount proportion as a decision variable. On the other hand, a possible extension of this model can be derived by considering the service level constraint.

Acknowledgments The authors are most grateful to the Editors and referees for their helpful and constructive comments for improvement of this paper.

References

1. Abad, P.L.: Optimal lot size for a perishable good under conditions of finite production and partial backordering and lost sale. Comput. Ind. Eng. **38**(4), 457–465 (2000)
2. Ben-Daya, M., Raouf, A.: Inventory models involving lead time as a decision variable. J. Oper. Res. Soc. 579–582 (1994)
3. Carlsson, C., Fullér, R.: On possibilistic mean value and variance of fuzzy numbers. Fuzzy Sets Syst. **122**(2), 315–326 (2001)
4. Chang, H.-C., Yao, J.-S., Ouyang, L.-Y.: Fuzzy mixture inventory model with variable lead-time based on probabilistic fuzzy set and triangular fuzzy number. Math. Comput. Model. **39**(2), 287–304 (2004)

5. Chang, H.-C., Yao, J.-S., Ouyang, L.-Y.: Fuzzy mixture inventory model involving fuzzy random variable lead time demand and fuzzy total demand. Eur. J. Oper. Res. **169**(1), 65–80 (2006)
6. Dey, O.: Amalgamation of fuzziness and randomness in developing mathematical formalism to some inventory problems, PhD thesis, Indian Institute of Techonology Kharagpur (2010)
7. Dey, O., Chakraborty, D.: Fuzzy periodic review system with fuzzy random variable demand. Eur. J. Oper. Res. **198**(1), 113–120 (2009)
8. Dey, O., Chakraborty, D.: A fuzzy random continuous review inventory system. Int. J. Prod. Econ. **132**(1), 101–106 (2011)
9. Dey, O., Chakraborty, D.: A fuzzy random periodic review system with variable lead-time and negative exponential crashing cost. Appl. Math. Model. **36**(12), 6312–6322 (2012)
10. Dutta, P., Chakraborty, D., Roy, A.: Continuous review inventory model in mixed fuzzy and stochastic environment. Appl. Math. Comput. **188**(1), 970–980 (2007)
11. Kim, D.H., Park, K.S.: (q, r) inventory model with a mixture of lost sales and time-weighted backorders. J. Oper. Res. Soc. 231–238 (1985)
12. Kwakernaak, H.: Fuzzy random variablesi. definitions and theorems. Inform. Sci. **15**(1), 1–29 (1978)
13. Liao, C.-J., Shyu, C.-H.: An analytical determination of lead time with normal demand. Int. J. Oper. Prod. Manage. **11**(9), 72–78 (1991)
14. Montogomery, D., Bazara, M., Keswani, A.K.: Inventory model with a mixture of back-orders and lost sales. Naval Res. Logistic Q. **20**(2), 225–263 (1973)
15. Ouyang, L.-Y., Chuang, B.-R.: Mixture inventory model involving variable lead time and controllable backorder rate. Comput. Ind. Eng. **40**(4), 339–348 (2001)
16. Ouyang, L.Y., Yao, J.S.: A minimax distribution free procedure for mixed inventory model involving variable lead-time with fuzzy demand. Comput. Oper. Res. **29**, 471–487 (2002)
17. Ouyang, L.-Y., Yeh, N.-C., Wu, K.-S.: Mixture inventory model with backorders and lost sales for variable lead time. J. Oper. Res. Soc. 829–832 (1996)
18. Pan, J.C.-H., Hsiao, Y.-C.: Integrated inventory models with controllable lead time and backorder discount considerations. Int. J. Prod. Econ. **93**, 387–397 (2005)
19. Puri, M.L., Ralescu, D.A.: Fuzzy random variables. J. Math. Anal. Appl. **114**(2), 409–422 (1986)
20. Vijayan, T., Kumaran, M.: Inventory models with a mixture of backorders and lost sales under fuzzy cost. Eur. J. Oper. Res. **189**(1), 105–119 (2008)
21. Wu, J.-W., Lee, W.-C., Tsai, H.-Y.: Computational algorithmic procedure of optimal inventory policy involving a negative exponential crashing cost and variable lead time demand. Appl. Math. Comput. **184**(2), 798–808 (2007)
22. Zimmermann, H.J.: Fuzzy Sets Theory and its Applications, Kluwer Academic Publishers (1991)

Supplier Selection Using Fuzzy Risk Analysis

Kartik Patra and Shyamal Kumar Mondal

Abstract In this paper, three different multi-item supplier selection model have been developed. The selection has been made optimizing the profit and risk which are considered as objective functions for all models. The optimization has been done under some constraints. It is considered that each supplier has an limited capacity to supply any item. The purchasing cost of each item from different supplier as well as associative risk is known. Also total space and budget are constant of a retailer. All the parameters have been considered as crisp in Model-I. The demand has been considered as fuzzy Model II. Necessity and possibility measures have been introduced in this paper to defuzzyfy the fuzzy constraints. The risk values have been considered as fuzzy in Model III in addition to the fuzzy demand. To defuzzyfy the fuzzy objective two different methods, credibility measure and α-cut method have been introduced. Multi-Objective Genetic Algorithm (*MOGA*) has been used to illustrate all the models numerically.

Keywords Risk · Supplier selection · Possibility · Necessity · *MOGA*

1 Introduction

Supplier selection is one of the most widely researched areas in supply chain management. One of the most significant business decisions faced by a retailer in a supply chain is the selection of appropriate suppliers while trying to satisfy multi-criteria based on price, quality, demand and delivery. Hence supplier selection is a multi-

K. Patra (✉)
Department of Mathematics, Sikkim Manipal Institute of Technology,
Sikkim Manipal University, East Sikkim 737136, India
e-mail: kpatrakp@gmail.com

S.K. Mondal
Department of Applied Mathematics with Oceanology and Computer Programming,
Vidyasagar University, Midnapore 721102, India
e-mail: shyamal_260180@yahoo.com

R.N. Mohapatra et al. (eds.), *Mathematics and Computing*,
Springer Proceedings in Mathematics & Statistics 139,
DOI 10.1007/978-81-322-2452-5_22

criteria decision making problem. The objective of supplier selection is to identify suppliers with the highest potential for meeting a retailers needs consistently.

In this paper three different multi-objective and multi-items supplier selection models have been developed in crisp and fuzzy environments. All parameters have been considered as crisp in first model. In real world problems, the demand of a commodity is not always certain. Generally it is vague in nature. So demand of the items has been considered as fuzzy in the second model. As a result the constraints becomes fuzzy. As a fuzzy constraint represents a fuzzy event, it should be satisfied in the some predefined possibility and necessity (cf. [5, 6, 13, 14]). Analogous to chance constrained programming with stochastic parameters, in fuzzy environment, it is assumed that some constraints will hold with a least possibility, η_1. Again some constraints may be satisfied with some predefined necessity, η_2. These possibility and necessity constraints may be imposed as per the demand of the situation. Also the risk in any system are not always certain, so the risk and demand of the items are considered as fuzzy in the third model. The total available space and budget are constant for a retailer. Each items purchased from different suppliers have different risk depending on their purchasing cost, time of delivery etc. Now a retailer always wants to maximize their total profit and minimize their risk in the business. So in this paper the profit function is maximized and risk is minimized for all the models. Also to convert the fuzzy objective to crisp objective two different methods such as α-cut method and credibility measure method have been used.

To get the optimality of the proposed model, Multi-Objective Genetic Algorithm (*MOGA*) has been introduced. Genetic Algorithm manipulates a family of solutions in the search of an optimal solution. So a retailer can take any optimal value from the set of solutions to buy a item from a supplier as per his/her need.

2 Literature Review

Different supplier selection models have been established by different researchers in different times in crisp or fuzzy environments. Lin [12] introduced an integrated model for supplier selection under a fuzzy situation. Arikan [1] presented a fuzzy solution approach for multi-objective supplier selection. Ruiz-Torres et al. [20] described a supplier selection model with contingency planning for supplier failures. Shirkouhi et al. [21] presented a supplier selection and order allocation problem using a two-phase fuzzy multi-objective linear programming. Kilic [11] presented an integrated approach for supplier selection in multi-item/multi-supplier environment. Rezaei and Davoodi [18] presented a multi item inventory model with imperfect quality.

The fuzzy set theory is one of the best tools to handle impreciseness and vagueness. It was first introduced by Zadeh [24]. Goguen [8, 9] showed the intention of the authors to generalize the classical notion of a set. Zadeh [25] also introduced the concept of linguistic variable and its application to approximate reasoning. Dubois and Prade [4] presented theory and application on fuzzy set theory.

Just like in most real-world decision making problems, uncertainty is another important property of supplier selection problems. So risk is an important factors in any business. Different risk analysis problems have been introduced by different researchers. Chen et al. [2] introduced fuzzy risk analysis based on ranking generalized fuzzy numbers with different left and right heights. Patra and Mondal [17] presented a new ranking method of generalized trapezoidal fuzzy numbers and applied it to evaluate the risk in diabetes problems.

Genetic algorithm approach was first proposed by Holland [10]. Because of its generality, it has been successfully applied to many optimization problems, for its several advantages over conventional optimization methods. There are several approaches using genetic algorithms to deal with the multi-objective optimization problems. These algorithms can be classified into two types-(i) Non-Elitist *MOGA* and (ii) Elitist *MOGA*. Among Non-Elitist *MOGA* Fonseca and Fleming's *MOGA* [7], Srinivas and Deb's NSGA [22] enjoyed more attention. Among Elitist *MOGA*s one can refers Rudolph's Elitist Multi-objective evolutionary algorithm (Rudolph [19]), Deb et al.'s [3] Elitist Non-dominated Shorting Multi-objective Genetic Algorithm. These algorithms normally select solution from parent population for cross-over and mutation randomly. After these operations parent and child population are combined together and among them better solutions are selected for next iteration.

3 Preliminaries

Triangular Fuzzy Number (*TFN*): A *TFN* $\widetilde{A}$ is specified by the triplet (a_1, a_2, a_3) and is defined by its continuous membership function $\mu_{\widetilde{A}}(x) : F \to [0, 1]$ as follows:

$$\mu_{\widetilde{A}}(x) = \begin{cases} \dfrac{x - a_1}{a_2 - a_1} & \text{if } a_1 \leq x \leq a_2 \\ \dfrac{a_3 - x}{a_3 - a_2} & \text{if } a_2 \leq x \leq a_3 \\ 0 & \text{otherwise} \end{cases} \tag{1}$$

α-cut of Fuzzy Number:
The α-cut / α - level set of a fuzzy number $\widetilde{A}$ is a crisp set which is defined as $\widetilde{A}_\alpha = \{x \in R : \mu_{\widetilde{A}}(x) \geq \alpha\}$ where $\alpha \in [0, 1]$.

3.1 Possibility, Necessity and Credibility

Any fuzzy subset $\tilde{a}$ of R (where R represents a set of real numbers) with membership function $\mu_{\tilde{a}}(x) : R \to [0, 1]$ is called a fuzzy number. Let $\tilde{a}$ and $\tilde{b}$ be two fuzzy numbers with membership functions $\mu_{\tilde{a}}(x)$ and $\mu_{\tilde{b}}(x)$, respectively. According to Dubois and Prade [5, 6], Zadeh [26], Liu and Iwamura [13, 14],Wang and Shu [23],

Liu and Iwamura [14]

$$Pos\left(\tilde{a} * \tilde{b}\right) = \left\{sup(min(\mu_{\tilde{a}}(x), \mu_{\tilde{b}}(y))),\ \ x, y \in R \text{ and } x * y\right\} \tag{2}$$

$$Nes\left(\tilde{a} * \tilde{b}\right) = \left\{inf(max(1 - \mu_{\tilde{a}}(x), \mu_{\tilde{b}}(y))),\ \ x, y \in R \text{ and } x * y\right\} \tag{3}$$

where the abbreviation "Pos" and "Nes" represent possibility and necessity respectively. Also, $*$ is any of the relations $>, <, =, \leq, \geq$.

On the other hand necessity measure of an event $\tilde{a} * \tilde{b}$ is a dual of possibility measure. The grade of necessity of an event is the grade of impossibility of the opposite event and is defined as

$Nes(\tilde{a} * \tilde{b}) = 1 - Pos(\overline{\tilde{a} * \tilde{b}})$.

Also necessity measures satisfy the condition

$Min(Nes(\tilde{a} * \tilde{b}), Nes(\overline{\tilde{a} * \tilde{b}})) = 0$

If $\tilde{a}, \tilde{b} \in R$ and $\tilde{c} = f(\tilde{a}, \tilde{b})$ where $f : R \times R \rightarrow R$ be a binary operation then membership function $\mu_{\tilde{c}}$ of $\tilde{c}$ is defined as

$\mu_{\tilde{c}}(z) = sup\{(min(\mu_{\tilde{a}}(x), \mu_{\tilde{b}}(y))),\ \ x, y \in R\ \ and\ \ z = f(x, y) \forall z \in R\}$

Let $\tilde{a} = (a_1, a_2, a_3)$ and $\tilde{b} = (b_1, b_2, b_3)$ be two triangular fuzzy numbers. Then for these fuzzy numbers, following (Wang and Shu [23], Liu and Iwamura [14]) Lemmas 1–2 can be derived.

Lemma 1 *When b is a crisp number,* $Pos(\tilde{a} \leq b) < \eta$ *iff* $\delta = \frac{b-a_1}{a_2-a_1} < \eta$

Proof Let $Pos(\tilde{a} \leq b) < \eta$

From Fig. 1 it is clear that $Pos(\tilde{a} \leq b) = \frac{b-a_1}{a_2-a_1}$

Therefore, $Pos(\tilde{a} \leq b) < \eta$ iff $\delta = \frac{b-a_1}{a_2-a_1} < \eta$

Lemma 2 *When b is a crisp number,* $Pos(\tilde{a} \geq b) > \eta$ *iff* $\delta = \frac{a_3-b}{a_3-a_2} > \eta$

Proof Let $Pos(\tilde{a} \geq b) > \eta$

From Fig. 1 it is clear that $Pos(\tilde{a} \geq b) = \frac{a_3-b}{a_3-a_2}$

Therefore, $Pos(\tilde{a} \geq b) > \eta$ iff $\delta = \frac{a_3-b}{a_3-a_2} > \eta$

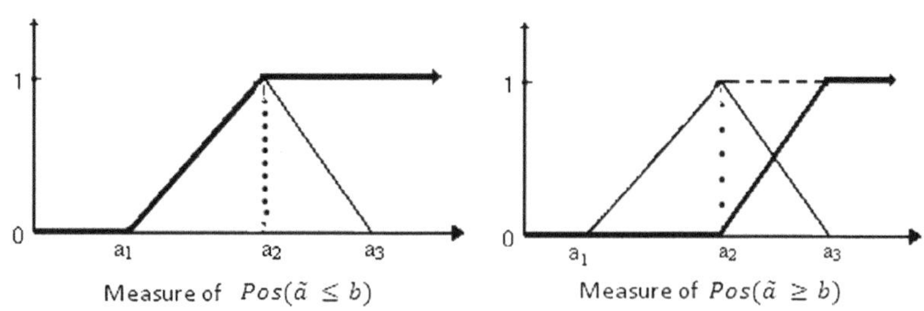

Fig. 1 Measure of $Pos(\tilde{a} \leq b)$ and $Pos(\tilde{a} \geq b)$

Based on possibility measure and necessity measure the the third set function Cr, called credibility measure, was analyzed by Liu and Liu [15], Maity et al. [16]. They defined the credibility measure in the following form

$$Cr(A) = [\rho Pos(A) + (1 - \rho)Nec(A)] \tag{4}$$

where A be a fuzzy subset of R and $0 < \rho < 1$.

Using this credibility the expected value of any fuzzy number $\tilde{A} = (a_1, a_2, a_3)$ can be calculated as

$$E(\tilde{A}) = \frac{1}{2}[(1 - \rho)a_1 + a_2 + \rho a_3] \tag{5}$$

4 Multi-objective Programming Problems Under Possibility and Necessity Constraints

A general multi-objective mathematical programming problem with fuzzy parameters should have the following form:

$$\begin{aligned} & Max\, f_1(u, \xi) \\ & Min\, f_2(u, \xi) \\ & s.t.\ g_j(u, \xi) \leq b, \quad j = 1, 2, ..., n. \end{aligned} \tag{6}$$

where u is the decision vector, ξ is a vector of fuzzy parameter, $f_1(u, \xi)$ and $f_2(u, \xi)$ are the objective functions, $g_j(u, \xi)$ are constraint functions, $j = 1, 2, ..., n$. To convert the fuzzy objectives and constraints to their crisp equivalents, Liu and Iwamura [14] proposed a method to convert the above problem into an equivalent fuzzy programming problem under possibility constraints. Similarly we can convert the above problem to following fuzzy programming problem under possibility/necessity constraints

$$\begin{aligned} & Max\, f_1(u, \xi) \\ & Min\, f_2(u, \xi) \\ & s.t.\ Nes\{\xi | g_j(u, \xi) \leq b\} > \eta_{1j}\ \ and/or\ \ Pos\{\xi | g_j(u, \xi) \leq b\} > \eta_{2j} \end{aligned} \tag{7}$$

where η_{1j} and $\eta_{2j}, j = 1, 2, ..., n$ are predetermined confidence level for fuzzy constraints. $Nes\{.\}$ denotes the necessity of the event in $\{.\}$. So a point ξ is feasible if and only if necessity of the set $\{\xi | g_j(u, \xi) \leq b\}$ is at least η_{1j}. $Pos\{.\}$ denotes the possibility of the event in $\{.\}$. So a point ξ is feasible if and only if possibility of the set $\{\xi | g_j(u, \xi) \leq b\}$ is also at least $\eta_{2j}, j = 1, 2, ..., n$.

5 Mathematical Formulation of a Supplier Selection Model in Crisp and Fuzzy Environment

5.1 Notations

To develop the proposed model following notations have been used.

- S_i: Selling price of ith item .
- p_{ij}: The purchase cost of ith item from jth supplier.
- T_j: The transaction cost for jth supplier.
- D_i: The demand of ith item.
- r_{ij}: The risk value for ith item supplies by jth supplier.
- C_{ij}: The capacity of ith item which can be supplied by jth supplier.
- ω_i: A storage space needed by product i
- W: Available total storage space.
- X_{ij}: Number of ith items supplied from jth supplier.
- TP: Total profit in the business.
- R: Total risk in the business.
- B: Available total budget of a retailer.

5.2 Assumptions

The proposed model have been formulated under the following assumptions.

- Shortages and backordering are not allowed.
- Each supplier has an limited capacity for each item.
- Available total storage space for a retailer is limited.
- Total budget of a retailer is limited.
- For each item a risk value has been considered for a supplier due to various factors such as (*i*) shipment in delay, (*ii*) purchasing cost, (*iii*) economic dealing.
- A supplier dependent transaction cost has been considered.
- Each item needs a storage space.

5.3 Proposed Supplier Selection Model in Crisp Environment: Model I

In this paper a supplier selection model has been considered in which there are m different approved suppliers and each supplier may supply n different products with limited capacity. There exists a risk value r_{ij} for jth supplier who supplies ith product to a retailer whose demand (D_i) is known over a finite planing horizon. The retailer have a selling price S_i for ith item. Here storage space and budget constraints have been considered for the retailer. The purchasing cost of each item varies from supplier to supplier. Also there exist different transaction cost for different suppliers. Now the retailer want to procure each required amount of item from a supplier such that the total profit of the retailer is maximum as well as total risk value is minimum.

To formulate the above problem it is suppose that the retailer procures ith item of amount X_{ij} from jth supplier. Therefore the total procurement cost (TC) for n items is given by

$$TC = \sum_{i=1}^{n}\sum_{j=1}^{m} X_{ij}p_{ij} + \sum_{j=1}^{m} T_j Y_j \tag{8}$$

where Y_j ($j = 1, 2, \ldots, m$) be calculated as follows:

$$Y_j = \begin{cases} 1, & \text{for } X_{ij} > 0 \\ 0, & \text{for } X_{ij} = 0 \end{cases}$$

The retailer sells these n items to the customers. After selling all items he/she collects the total revenue (TR) which is given by

$$TR = \sum_{i=1}^{n}\sum_{j=1}^{m} X_{ij}S_i \tag{9}$$

Therefore from this business the retailer earns the total profit (TP) that is given by

$$TP = \sum_{i=1}^{n}\sum_{j=1}^{m} X_{ij}S_i - \sum_{i=1}^{n}\sum_{j=1}^{m} X_{ij}p_{ij} - \sum_{j=1}^{m} T_j Y_j \tag{10}$$

Simultaneously the retailer wants to minimize the total risk to collect all these items from the suppliers. Now the total risk R is given by

$$R = \sum_{i=1}^{n}\sum_{j=1}^{m} X_{ij}r_{ij} \Big/ \sum_{i=1}^{n}\sum_{j=1}^{m} X_{ij} \tag{11}$$

Therefore the above problem can be described in the following form in crisp environment:

$$Max\ TP = \sum_{i=1}^{n}\sum_{j=1}^{m} X_{ij}S_i - \sum_{i=1}^{n}\sum_{j=1}^{m} X_{ij}p_{ij} - \sum_{j=1}^{m} T_j Y_j$$

and

$$Min\ R = \sum_{i=1}^{n}\sum_{j=1}^{m} X_{ij}r_{ij} \Big/ \sum_{i=1}^{n}\sum_{j=1}^{m} X_{ij}$$

subject to

$$\begin{cases} \sum_{j=1}^{m} X_{ij} - D_i \geq 0 & i = 1, 2, ..., n \\ \sum_{i=1}^{n} \omega_i (\sum_{j=1}^{m} X_{ij} - D_i) \leq W & \\ 0 \leq X_{ij} \leq C_{ij} & i = 1, 2, .., n \ \& \ j = 1, 2, ..., m \\ Y_j = \begin{cases} 1, & \text{for } X_{ij} > 0 \\ 0, & \text{for } X_{ij} = 0 \end{cases} & j = 1, 2, ..., m \\ \sum_{i=1}^{n} \sum_{j=1}^{m} X_{ij} p_{ij} \leq B & \end{cases}$$

This is a multi objective decision making problem where a retailer wants to maximize the profit (*TP*) and minimize the risk (*R*). To solve the above problem Multi-Objective Genetic Algorithm (*MOGA*) has been applied. *GA* manipulates a family of solution in the search of an optimal solution. This is an advantage of *GA* which is better than another methods. So different retailer may have choose different optimal value as per their strategy in the business.

5.4 Proposed Supplier Selection Model in Fuzzy Environment: Model II

In this model retailer's demand for each item has been considered fuzzy which is triangular. All other constraints are same as in Model I. Therefore under this fuzzy environment the model can be depicted as follows:

$$Max\ TP = \sum_{i=1}^{n} \sum_{j=1}^{m} X_{ij} S_i - \sum_{i=1}^{n} \sum_{j=1}^{m} X_{ij} p_{ij} - \sum_{j=1}^{m} T_j Y_j$$

and

$$Min\ R = \sum_{i=1}^{n} \sum_{j=1}^{m} X_{ij} r_{ij} / \sum_{i=1}^{n} \sum_{j=1}^{m} X_{ij}$$

subject to

$$\begin{cases} \sum_{j=1}^{m} X_{ij} - \tilde{D}_i \geq 0 & i = 1, 2, ..., n \\ \sum_{i=1}^{n} \omega_i (\sum_{j=1}^{m} X_{ij} - \tilde{D}_i) \leq W & \\ 0 \leq X_{ij} \leq C_{ij} & i = 1, 2, ..., n \ \& \ j = 1, 2, ..., m \\ Y_j = \begin{cases} 1, & \text{for } X_{ij} > 0 \\ 0, & \text{for } X_{ij} = 0 \end{cases} & j = 1, 2, ..., m \\ \sum_{i=1}^{n} \sum_{j=1}^{m} X_{ij} p_{ij} \leq B & \end{cases}$$

where $\sim$ indicates the fuzzyness of the parameter.

Here the two fuzzy constraints actually stand for fuzzy relation. There are several representations of fuzzy relation. Here these relations are interpreted in the form of possibility theory in which fuzzy numbers are interpreted by a degree of uncertainty. It is considered that there are n items that are supplied by m suppliers. According to Liu and Iwamura [13], first two constraints reduce to following respective necessary and possibility constraints. There may be two different combinations of the fuzzy constraints depending on the different scenarios such as

Scenario 1:

$$Nes\{\tilde{D}_1 > \sum_{j=1}^{m} x_{1j}\} < \eta_{11}$$
$$Nes\{\tilde{D}_2 > \sum_{j=1}^{m} x_{2j}\} < \eta_{12}$$
$$\ldots$$
$$Nes\{\tilde{D}_n > \sum_{j=1}^{m} x_{nj}\} < \eta_{1n}$$
$$Nes\{(\sum_{i=1}^{n}\sum_{j=1}^{m} \omega_i x_{ij} - W) < \sum_{i=1}^{n} \omega_i \tilde{D}_i\} > \eta_{1,n+1}$$

Scenario 2:

$$Pos\{\tilde{D}_1 \geq \sum_{j=1}^{m} x_{1j}\} < \eta_{11}$$
$$Pos\{\tilde{D}_2 \geq \sum_{j=1}^{m} x_{2j}\} < \eta_{12}$$
$$\ldots$$
$$Pos\{\tilde{D}_n \geq \sum_{j=1}^{m} x_{nj}\} < \eta_{1n}$$
$$Pos\{(\sum_{i=1}^{n}\sum_{j=1}^{m} \omega_i x_{ij} - W) \leq \sum_{i=1}^{n} \omega_i \tilde{D}_i\} > \eta_{2,n+1}$$

Equivalent Crisp Representation of Model II

Let $\tilde{D}_i = (D_{i1}, D_{i2}, D_{i3})$ be a triangular fuzzy number represented in Fig. 1. So $\sum_{i=1}^{m} \omega_i \tilde{D}_i = (D'_1, D'_2, D'_3)$ is also a triangular fuzzy number by it's properties. Therefore on the basis of Lemma 1–2 the fuzzy Model II reduces to following multi objective crisp model:

$$Max\ TP = \sum_{i=1}^{n}\sum_{j=1}^{m} X_{ij}S_i - \sum_{i=1}^{n}\sum_{j=1}^{m} X_{ij}p_{ij} - \sum_{j=1}^{m} T_jY_j$$

and

$$Min\ R = \sum_{i=1}^{n}\sum_{j=1}^{m} X_{ij}r_{ij} / \sum_{i=1}^{n}\sum_{j=1}^{m} X_{ij}$$

subject to constraint for all scenarios

$$\begin{cases} 0 \leq X_{ij} \leq C_{ij} & i = 1, 2, \ldots, n \ \&\ j = 1, 2, \ldots, m \\ Y_j = \begin{cases} 1, & \text{for } X_{ij} > 0 \\ 0, & \text{for } X_{ij} = 0 \end{cases} & j = 1, 2, \ldots, m \\ \sum_{i=1}^{n}\sum_{j=1}^{m} X_{ij}p_{ij} \leq B \end{cases}$$

and

Scenario 1:

$$\frac{(\sum_{j=1}^{m} x_{1j} - D_{11})}{(D_{12} - D_{11})} > (1 - \eta_{11})$$
$$\frac{(\sum_{j=1}^{m} x_{2j} - D_{21})}{(D_{22} - D_{21})} > (1 - \eta_{12})$$
$$\ldots$$
$$\frac{(\sum_{j=1}^{m} x_{nj} - D_{n1})}{(D_{n2} - D_{n1})} > (1 - \eta_{1n})$$
$$\frac{(\sum_{i=1}^{n} \sum_{j=1}^{m} \omega_i x_{ij} - W) - D'_1}{(D'_2 - D'_1)} < (1 - \eta_{1,n+1})$$

Scenario 2:

$$\frac{(D_{13} - \sum_{j=1}^{m} x_{1j})}{(D_{13} - D_{12})} < \eta_{11}$$
$$\frac{(D_{23} - \sum_{j=1}^{m} x_{2j})}{(D_{23} - D_{22})} < \eta_{12}$$
$$\ldots$$
$$\frac{(D_{n3} - \sum_{j=1}^{m} x_{nj})}{(D_{23} - D_{22})} < \eta_{1n}$$
$$\frac{D'_3 - (\sum_{i=1}^{n} \sum_{j=1}^{m} \omega_i x_{ij} - W)}{(D'_3 - D'_2)} > \eta_{2,n+1}$$

5.5 *Proposed Supplier Selection Model with Fuzzy Risk: Model III*

Here the risk value of each item supplied from each supplier has been considered as fuzzy which are taken as triangular fuzzy number i.e., $\tilde{r}_{ij} = (r_{ij1}, r_{ij2}, r_{ij3})$ and demand is also taken as fuzzy as in Model II. Therefore the proposed model can be described as follows

$$Max\ TP = \sum_{i=1}^{n} \sum_{j=1}^{m} X_{ij} S_i - \sum_{i=1}^{n} \sum_{j=1}^{m} X_{ij} p_{ij} - \sum_{j=1}^{m} T_j Y_j$$

and

$$Min\ \tilde{R} = \sum_{i=1}^{n} \sum_{j=1}^{m} X_{ij} \tilde{r}_{ij} / \sum_{i=1}^{n} \sum_{j=1}^{m} X_{ij}$$

subject to

$$\begin{cases} \sum_{j=1}^{m} X_{ij} - \tilde{D}_i \geq 0 & i = 1, 2, \ldots, n \\ \sum_{i=1}^{n} \omega_i (\sum_{j=1}^{m} X_{ij} - \tilde{D}_i) \leq W & \\ 0 \leq X_{ij} \leq C_{ij} & i = 1, 2, \ldots, n \ \&\ j = 1, 2, \ldots, m \\ Y_j = \begin{cases} 1, & \text{for } X_{ij} > 0 \\ 0, & \text{for } X_{ij} = 0 \end{cases} & j = 1, 2, \ldots, m \\ \sum_{i=1}^{n} \sum_{j=1}^{m} X_{ij} p_{ij} \leq B & \end{cases}$$

Lemma 3: Since all the risk values ($\tilde{r}_{ij}$) are triangular fuzzy number so the total risk $\tilde{R}$ is also a triangular fuzzy number such that

$$\begin{aligned}\tilde{R} &= \sum_{i=1}^{n}\sum_{j=1}^{m} X_{ij}\tilde{r}_{ij} / \sum_{i=1}^{n}\sum_{j=1}^{m} X_{ij} \\ &= (\sum_{i=1}^{n}\sum_{j=1}^{m} X_{ij}r_{ij1} / \sum_{i=1}^{n}\sum_{j=1}^{m} X_{ij}, \sum_{i=1}^{n}\sum_{j=1}^{m} X_{ij}r_{ij2} / \sum_{i=1}^{n}\sum_{j=1}^{m} X_{ij}, \sum_{i=1}^{n}\sum_{j=1}^{m} X_{ij}r_{ij3} / \sum_{i=1}^{n}\sum_{j=1}^{m} X_{ij}) \\ &= (R_1, R_2, R_3)\end{aligned}$$

where $R_k = \sum_{i=1}^{n}\sum_{j=1}^{m} X_{ij}r_{ijk} / \sum_{i=1}^{n}\sum_{j=1}^{m} X_{ij}, k = 1, 2, 3.$

Since one of the objective functions is fuzzy in nature, hence to solve the model the fuzzy objective function converted into the crisp objective functions. Here two methods for defuzzyfications of the objective function have been given as follows:

α-cut Method

Now the fuzzy objective function is converted to a crisp objective function using the α-cut of the objective function. Let $(\tilde{R})_\alpha = [R^L_\alpha, R^R_\alpha]$. Now our aim is to minimize both the R^L_α and R^R_α and maximize TP with the given constraints. Here the fuzzy constraints are converted to crisp constraints using necessity measure as given in the previous model. So the problem becomes

$$Max\ TP = \sum_{i=1}^{n}\sum_{j=1}^{m} X_{ij}S_i - \sum_{i=1}^{n}\sum_{j=1}^{m} X_{ij}p_{ij} - \sum_{j=1}^{m} T_jY_j$$

$$Min\ R^L_\alpha = R_1 + \alpha(R_2 - R_1)$$

and

$$Min\ R^R_\alpha = R_3 - \alpha(R_3 - R_2)$$

subject to

$$\begin{cases} \dfrac{(\sum_{j=1}^{m} x_{ij} - D_{11})}{(D_{i2} - D_{i1})} > (1 - \eta_{1i}) & i = 1, 2, ..., n \\ \dfrac{(\sum_{i=1}^{n}\sum_{j=1}^{m} \omega_i x_{ij} - W) - D'_1}{(D'_2 - D'_1)} < (1 - \eta_{1,n+1}) & \\ \sum_{i=1}^{n} \omega_i(\sum_{j=1}^{m} X_{ij} - \tilde{D}_i) \le W & \\ 0 \le X_{ij} \le C_{ij} & i = 1, 2, ..., n\ \&\ j = 1, 2, ..., m \\ Y_j = \begin{cases} 1, & \text{for } X_{ij} > 0 \\ 0, & \text{for } X_{ij} = 0 \end{cases} & j = 1, 2, ..., m \\ \sum_{i=1}^{n}\sum_{j=1}^{m} X_{ij}p_{ij} \le B & \end{cases}$$

Credibility Measure Method

On the basis of credibility measure in by Liu and Iwamura [13, 14], Maity et al. [16] the expected value of fuzzy risk objective $\tilde{R}$ is given by

$$E(\tilde{R}) = \frac{1}{2}((1-\rho)R_1 + R_2 + \rho R_3) \text{ where } 0 < \rho < 1$$

6 Procedure of MOGA

Step-1: Generate initial population P_1 of size N.
Step-2: $i \leftarrow 1$ [i represent the number of current generation.]
Step-3: Select solution from P_i for crossover.
Step-4: Made crossover on selected solution to get child set C_1.
Step-5: Select solution from P_i for mutation.
Step-6: Made mutation on selected solution to get solution set C_2.
Step-7: Set $P_i' = P_i \bigcup C_1 \bigcup C_2$
Step-8: Partition P_i' into subsets $F_1, F_2, \cdots, F_k$, such that each subset contains non-dominated solutions of P_i' and every solutions of F_i dominates every solu.s of F_{i+1} for $i = 1, 2, \cdots, k-1$.
Step-9: Select largest possible integer l, so that no of solu.s in the set $F_1 \bigcup F_2 \bigcup \cdots \bigcup F_l \leq N$.
Step-10: Set $P_{i+1} = F_1 \bigcup F_2 \bigcup \cdots \bigcup F_l$.
Step-11: Sort F_{l+1} in decreasing order by crowding distance.
Step-12: Set M = number of solutions in P_{i+1}.
Step-13: Select first $N - M$ solutions from set F_{l+1}.
Step-14: Insert these solution in solution set P_{i+1}.
Step-15: Set $i \leftarrow i + 1$.
Step-16: If termination condition does not hold, goto step-3.
Step-17: Output P_i.
Step-18: End.

7 Numerical Illustration

To illustrate the Model I, Model II and Model III it is considered that there are two suppliers who supply two different items. Now a retailer has to decide what amount of items will be taken from which supplier such as the profit and risk will be optimized. To study the feasibility of both models following input values of the parameters have been taken.

Table 1 Solution set of Model I

X_{11}	X_{12}	X_{21}	X_{22}	R	$TP(\$)$
4.31	177.69	92.06	8.72	0.1097	3485.41
14.34	171.04	89.75	10.27	0.1117	3572.36
16.43	169.03	89.75	10.26	0.1120	3578.36
40.86	141.0	72.74	28.08	0.1201	3585.75
61.31	122.5	67.83	34.74	0.1251	3606.63

Input data for Model I:

$m = 2, n = 2, S_1 = 50\$, S_2 = 40\$, p_{11} = 25\$, p_{12} = 27\$, p_{21} = 30\$, p_{22} = 32\$, r_{11} = 0.15, r_{12} = 0.1, r_{21} = 0.12, r_{22} = 0.18, S_1 = 50\$, S_2 = 40\$, T_1 = 1000\$, T_2 = 700\$, w_1 = 0.2, w_2 = 0.18, C_{11} = 150, C_{12} = 200, C_{21} = 100, C_{22} = 80, W = 200, B = 8000\$, D_1 = 130, D_2 = 100.$

Output data for Model I:

Optimizing the Model I by multi objective Genetic Algorithm with respect to the above input values, the results has been shown in Table 1 as follows:

From Table 1 it is observed that the minimum risk is 0.1097 and to get the minimum risk, the profit value is 3485.41\$. To obtain these risk and profit values, the retailer buys the item 1 and 2 of amount 4.31 and 92.06 respectively from supplier 1 and 177.69 and 8.72 from supplier 2. It also noticed that the maximum profit is 3606.63 and at that time the risk is 0.1251. So, if a retailer wants to get higher profit then he/she has to take higher risk and if he/she wants lesser risk then he/she will have less profit.

Input data for Model II:

Here the demand of the items are considered as $\tilde{D}_1 = (110, 130, 150), \tilde{D}_2 = (90, 100, 110), \eta_{11} = 0.95, \eta_{12} = 0.9, \eta_{13} = 0.7, \eta_{21} = 0.15, \eta_{22} = 0.2, \eta_{23} = 0.8$. All other input values are same as in Model I. Since $n = 2$, so the number of fuzzy constrained is 3 which are discussed numerically as follows.

Scenario 1:

$$\frac{(\sum_{j=1}^{2} x_{1j} - D_{11})}{(D_{12} - D_{11})} > (1 - \eta_{11})$$
$$\frac{(\sum_{j=1}^{2} x_{2j} - D_{21})}{(D_{22} - D_{21})} > (1 - \eta_{12})$$
$$\frac{(\sum_{i=1}^{2}\sum_{j=1}^{2} \omega_i x_{ij} - W) - D_1'}{(D_2' - D_1')} < (1 - \eta_{13}) \tag{12}$$

Output data for Scenario 1:

The optimal results of Model II in scenario 1 are obtained by *MOGA* in Table 2 as follows: Scenario 2:

Table 2 Solution set of Model II in scenario 1

X_{11}	X_{12}	X_{21}	X_{22}	R	$TP(\$)$
39.35	73.97	48.50	46.87	0.1320	1845.29
39.35	73.97	48.22	48.04	0.1323	1851.79
39.35	73.97	48.22	47.46	0.1321	1847.18
39.35	73.97	48.22	47.61	0.1322	1848.39
39.35	73.97	49.85	42.07	0.1308	1820.36

Table 3 Solution set of Model II in scenario 2

X_{11}	X_{12}	X_{21}	X_{22}	R	$TP(\$)$
121.72	25.29	60.03	48.06	0.1436	2909.71
121.72	25.29	60.03	48.51	0.1437	2913.37
121.72	25.29	60.03	48.94	0.1438	2916.80
121.72	25.29	59.32	50.29	0.1440	2920.57
120.95	26.07	60.02	47.97	0.1435	2907.55

$$\frac{(D_{13} - \sum_{j=1}^{2} x_{1j})}{(D_{13} - D_{12})} < \eta_{11}$$
$$\frac{(D_{23} - \sum_{j=1}^{2} x_{2j})}{(D_{23} - D_{22})} < \eta_{12}$$
$$\frac{D'_3 - (\sum_{i=1}^{2} \sum_{j=1}^{2} \omega_i x_{ij} - W)}{(D'_3 - D'_2)} > \eta_{23} \tag{13}$$

Output data for Scenario 2:

The optimal results of Model II in scenario 2 are obtained by *MOGA* in Table 3 as follows: From the consideration of necessity and/or possibility constraints in above two scenarios it is observed that the total demand for a planning horizon for scenarios 1 and 2 belongs to $[D_{i1}, D_{i2}]$ and $[D_{i2}, D_{i3}]$ respectively. From these it may be concluded that necessity and possibility constraints demand the lower and upper range of the values of demand. Hence if a decision maker desires to impose the demand constraints in possibility sense, he/she should be expected to happen the imprecise demand at higher level (i.e., $[D_{i2}, D_{i3}]$). On the other hand, for necessary constraint, he/she will expect the demand at lower level. This feature is reflected from the results of scenarios 1 and 2 as the scenario 1 involving three necessary constraints furnishes lowest profit.

Input data for Model III:

Here the risk values are considered as $\tilde{r}_{11}=(0.1, 0.15, 0.2)$, $\tilde{r}_{12}=(0.07, 0.1, 0.13)$, $\tilde{r}_{21} = (0.08, 0.12, 0.16)$, $\tilde{r}_{22} = (0.14, 0.18, 0.21)$. All other parameters are remain same as in the previous model.

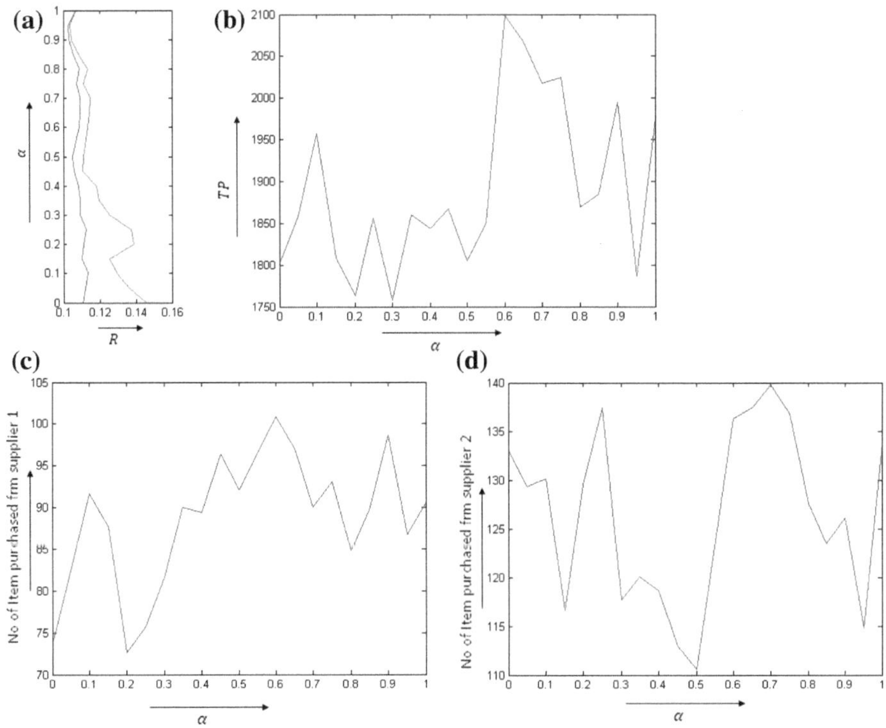

Fig. 2 Risk, Profit, no. of item purchased from different suppliers with different α-cuts

Output data for Model III by α-cut method:

Taking different α-cut of the fuzzy risk the obtained optimized results are given in the following Fig. 2

From these Fig. 2a–d, the amount of risk and profit as well as the amount of quantities collected from supplier 1 and supplier 2 can be obtained very easily for any α-cut. For example, when $\alpha = 0.5$ at that time the minimum risk value lies in the interval [0.1049 0.1110] along with the maximum profit 1805.02$, also the amount of item purchased from supplier 1 and supplier 2 are 92.02 and 110.58 respectively.

Output data for Model III by credibility measure method:

Now the optimal result has been obtained using the credibility measure of the fuzzy risk objective function taking the value of $\rho = 0.5$. Here the maximum profit is 1708.95$ and the minimum risk is 0.1051 when the amount of items purchased from supplier 1 is 39.29 and 45.42 respectively and the amount of items purchased from supplier 2 is 74.04 and 33.65 respectively.

Comparison of result obtained by α-cut and credibility measure methods:

From the above two results it is seen that the risk calculated by the credibility measure is 0.1051 and it lies in the interval [0.1049 0.1110] which is found by α-cut method. But it is seen that the maximum profit in the α-cut method is more than the credibility measure method. So the α-cut method gives the better solution than the credibility measure method.

8 Conclusion

Different multi item supplier selection by a retailer has been considered. The selection has been done maximizing the profit and minimizing the risk. All parameters associated with the suppliers and retailer have been considered as crisp in the first model. In the second model the demand of the items for a retailer has been considered as fuzzy and in the third model risk of taking an item from a supplier as well as demand of the items for a retailer have been considered also as fuzzy. The necessity and possibility constraint are used to convert fuzzy constraints to crisp constraints and the fuzzy objective function has been converted to crisp objective using α-cut and credibility measure methods. The objective functions of all models have been optimized simultaneously using Multi-Objective Genetic Algorithm. Finally all the models are illustrated using numerical example.

References

1. Arikan, F.: A fuzzy solution approach for multi objective supplier election. Expert Syst. Appl. **40**, 947–952 (2013)
2. Chen, S.M., Munif, A., Chen, G.S., Liu, H.C.,Kuo, B.C.: Fuzzy risk analysis based on ranking generalized fuzzy numbers with different left heights and right heights. Expert Syst. Appl. **39**, 6320–6334 (2012)
3. Deb, K., Pratap, A., Agarawal, S., Meyarivan, T.: A fast and elitist multiobjective genetic algorithm: NSGA-II. IEEE Trans. Evol. Comput. 182–197 (2002)
4. Dubois, D., Prade, H.: Fuzzy sets and systems. Theory and applications. Academic Press. Inc., New York (1980)
5. Dubois, D., Prade, H.: Ranking fuzzy numbers in the setting of possibility theory. Inf. Sci. **30**,183–224 (1983)
6. Dubois, D., Prade, H.: Possibility theory. Academic Press, New York (1988)
7. Fonseca, C.M. Fleming, P.J.: Genetic Algorithms for Multiobjective Optimization, Formulation, Discussion and Generalization. In: Forrest, S. (ed.) Proceedings of the Fifth International Conference on Genetic Algorithm, San Mateo, CA: Morgan Kauffman, 416–423 (1993)
8. Goguen, J.A.: L-Fuzzy sets. JMAA **18**, 145–174 (1967)
9. Goguen, J.A.: The logic of inexact concepts. Synthese **19**, 325–373 (1969)
10. Holland, H.J.: Adaptation in Natural and Artificial Systems, University of Michigan (1975)
11. Kilic, H.S.: An integrated approach for supplier selection in multi-item/ multi-supplier environment. Appl. Math. Model. **37**, 7752–7763 (2013)
12. Lin, R.H.: An integrated model for supplier selection under a fuzzy situation. Int. J. Prod. Econ. **138**, 55–61 (2012)
13. Liu, B., Iwamura, K.B.: Chance constraint programming with fuzzy papameters. Fuzzy Sets Syst. **94**, 227–237 (1998)
14. Liu, B., Iwamura, K.B.: A note on chance constrainted programming with fuzzy coefficients. Fuzzy Sets Syst. **100**, 229–233 (1998)

15. Liu, B., Liu, Y.K.: Expected value of fuzzy variable and fuzzy expected value model. IEEE Trans. Fuzzy Syst. **10**(4), 445–450 (2002)
16. Maity, A.K., Maity, K., Maity, M.: A production recycling inventory system with imprecise holding costs. Appl. Math. Model. **32**, 2241–2253 (2008)
17. Patra, K., Mondal, S.K.: Risk analysis in diabetes prediction based on a new approach of ranking of generalized trapezoidal fuzzy numbers. Int. J. Cybern. Syst. **43**(8), 623–650 (2012)
18. Rezaei, J., Davoodi, M.: A deterministic multi item inventory model with supplier selection and imperfect quality. Appl. Math. Model. **32**, 2106–2116 (2008)
19. Rudolph, G.: Evolutionary Search under Partially Ordered Fitness Sets. In: Proceedings of the International Symposium on Information Science Innovations in Engineering of Natural and Artificial Intelligent Systems (ISI), 818–822 (2001)
20. Ruiz-Torres, A.J., Mahmoodi, F., Zeng, A.Z.: Supplier selection model with contingency planning for supplier failures. Comput. Ind. Eng. **66**, 374–382 (2013)
21. Shirkouhi, S.N., Shakouri, H., Javadi, B., Keramati, A.: Supplier selection and order allocation problem using a two-phase fuzzy multi-objective linear programming. Appl. Math. Model. **37**, 9308–9323 (2013)
22. Srinivas, N., Deb, K.: Multi-objective optimization using nondominated sorting in genetic algorithms. J. Evol. Comput. **2**(3), 221–248 (1994)
23. Wang, J., Shu, Y.F.: Fuzzy decision modelling for supply chain management. Fuzzy Sets Syst. **150**, 107–127 (2005)
24. Zadeh, L.A.: Fuzzy Sets. Inf. Control. **8**, 338–356 (1965)
25. Zadeh, L.A.: The concept of linguistic variable and its application to approximate reasoning. I, II, III. Inf. Sci. **8**,199–249 (1975), **9**, 43–58 (1976)
26. Zadeh, L.A.: Fuzzy sets as a basis of possibility. Fuzzy Sets Syst. **1**, 3–24 (1978)

The Control for Prey–Predator System with Time Delay and Refuge

Shashi Kant and Vivek Kumar

Abstract In this paper, a prey–predator model with control and refuge is proposed and analyzed. Linear functional response is used. Time delay is the gestation period. Linear stability analysis is performed. The stability result is proved by assuming a suitable Lyapunov function. The main contribution of this paper is to propose a new model and derive an expression for the control of model. The control has many applications in sustainable development process. This research is not a case study, hence real data is not available for numerical simulation. However, the system is simulated by using an artificial set of parameters to validate our theoretical formulation.

Keywords Prey–predator system · Time delay · The control · Refuge

1 Introduction

Study of prey–predator systems is a current research area in ecology. Modeling of such systems is very challenging and crucial because it involves number of parameters such as environment, etc. However, a good literature is available. Basic variables in any prey–predator system are prey and predator.

To include the impact of environment on the prey–predator system, researchers introduced the concept of time delay. By this prey–predator ecosystem became more scientific and complicated. Few examples include:

S. Kant (✉) · V. Kumar
Department of Applied Mathematics, Delhi Technological University,
Delhi 110042, India
e-mail: onlineskmishra@gmail.com

V. Kumar
e-mail: vivekkumar.ag@gmail.com

R.N. Mohapatra et al. (eds.), *Mathematics and Computing*,
Springer Proceedings in Mathematics & Statistics 139,
DOI 10.1007/978-81-322-2452-5_23

1. Prey–predator system with continuous time delay [1]

$$\begin{cases} \dot{x_1}(t) &= x_1(t)[r_1 - a_{11}x_1(t-\tau) - \frac{a_{12}x_2(t)}{1+m_1x_1(t)}], \\ \dot{x_2}(t) &= x_2(t)[-r_2 + \frac{a_{21}x_1(t-\tau)}{1+m_1x_1(t-\tau)} - \frac{a_{23}x_3(t)}{1+m_2x_2(t)}], \\ \dot{x_3}(t) &= x_3(t)[-r_3 + \frac{a_{32}x_2(t-\tau)}{1+m_2x_2(t-\tau)}]. \end{cases} \tag{1.1}$$

2. Ratio-dependent predator–prey system with time delay [2]

$$\begin{cases} \frac{dx}{dt} &= x(a - bx) - \frac{cxy}{my+x}, \\ \frac{dy}{dt} &= y(-d + \frac{f(x(t-\tau))}{my(t-\tau)+x(t-\tau)}). \end{cases} \tag{1.2}$$

3. Viral Infection model with delayed non-lytic immune response [3]

$$\begin{cases} \dot{x} &= s - dx + kx(1 - \frac{x}{x_{max}}) - \frac{\beta xy}{1+qz}, \\ \dot{y} &= \frac{\beta xy}{1+qz} - ay, \\ \dot{z} &= ce^{-a\tau}y(t-\tau) - bz. \end{cases} \tag{1.3}$$

4. A prey–predator Model with continuous time delay [4]

$$\begin{cases} \dot{x} &= x(t)[\varepsilon_1 - \alpha_1 x(t) - \gamma_1 y(t)], \\ \dot{y} &= y(t)[-\varepsilon_2 - \alpha_2 x(t) + \gamma_2 \int_{-\infty}^{t} F(t-\tau)x(\tau)d\tau]. \end{cases} \tag{1.4}$$

5. A Stage- structured prey–predator model with time delay [5]

$$\begin{cases} \dot{x} &= \alpha(t)y(t) - rx(t) - \Omega(t)x(t) - \eta(t)x^2(t), \\ \dot{y} &= \Omega(t)x(t) - Q(t)y(t) - \rho(t)y^2(t) - a(t)y(t)z(t), \\ \dot{z} &= z(t)[-r_1(t) + \lambda(t)a(t)y(t) - c(t)]z(t) - \beta(t)\int_{-\tau}^{0} k(s)z(t+s)ds]. \end{cases} \tag{1.5}$$

It is also important to mention here that delays are used for different purposes, for example, gestation period, maturation period, etc. The concept of prey refuge is adopted from the study of Sahabuddin Sarwardi et al. [6]. The concept of control is adopted from the study of Li Yi-min and Zhu Yan [7]. Motivated by the study of G.-P Hu and X.-L Li [8] linear functional response is considered.

2 The Model

By the above discussion, in this study we proposed the following delayed prey–predator system:

$$\begin{cases} \frac{dx}{dt} = & x(t)[r_1 - a_{11}x(t)] - p_1(1-m)y(t)x(t), \\ \frac{dy}{dt} = & qp_1(1-m)y(t-\tau)x(t-\tau) - r_2y(t), \end{cases} \tag{2.1}$$

where $x(t)$ and $y(t)$ denote prey and predator population densities at time t respectively. r_1 is the growth rate of prey population, r_2 is the death rate of predator population. p_1 is predation coefficient, q is conversion coefficient, m is prey refuge, τ is time delay the gestation period of predator and a_{11} is a positive constant. By refuge, we mean that prey has a defense by means of habitat structure, etc. Therefore, only $(1-m)x$ prey population is available for predation.

Remark 1 In the study of Li Yi-min and Zhu Yan [7], the time delay τ is the time taken by predator from infant stage to the ripe stage.

The main concern in ecology is to find the stability of coexistence. Therefore, we will also pay stress on the coexisting equilibrium of the model (2.1), which admit one positive equilibrium point $E^*(x^*, y^*)$. The components of this positive equilibrium point must be

$$\begin{cases} x^* = \frac{r_2}{qp_1(1-m)}, \\ y^* = \frac{r_1qp_1(1-m)-a_{11}r_2}{qp_1^2(1-m)^2}. \end{cases}$$

First, let us consider the system without delay. Let

$$\begin{cases} X = x(t)[r_1 - a_{11}x(t)] - p_1(1-m)y(t)x(t), \\ Y = qp_1(1-m)y(t)x(t) - r_2y(t). \end{cases}$$

Jacobian of (2.1) at $E^*(x^*, y^*)$ takes the form

$$J = \begin{pmatrix} \frac{\partial X}{\partial x} & \frac{\partial X}{\partial y} \\ \frac{\partial Y}{\partial x} & \frac{\partial Y}{\partial y} \end{pmatrix}_{(x^*,y^*)} = \begin{pmatrix} c_{11} & c_{12} \\ c_{21} & c_{22} \end{pmatrix}, \tag{2.2}$$

where

$$\begin{cases} c_{11} & = r_1 - 2a_{11}x^* - p_1(1-m)y^*, \\ c_{12} & = -p_1(1-m)x^*, \\ c_{21} & = qp_1(1-m)y^*, \\ c_{22} & = -r_2 + qp_1(1-m)x^*. \end{cases}$$

The characteristics equation of jacobian matrix (2.2) is given by

$$\lambda^2 - (c_{11} + c_{22})\lambda + (c_{11}c_{22} - c_{12}c_{21}) = 0. \tag{2.3}$$

Second, we consider the system with delay viz. system (2.1). Let $M(t) = x(t) - x^*$, $N(t) = y(t) - y^*$ be the perturbed variables. The system (2.1) can be expressed in matrix form after the process of linearization as:

$$\frac{d}{dt}\begin{pmatrix} M(t) \\ N(t) \end{pmatrix} = A_1 \begin{pmatrix} M(t) \\ N(t) \end{pmatrix} + A_2 \begin{pmatrix} M(t-\tau) \\ N(t-\tau) \end{pmatrix}, \tag{2.4}$$

where A_1 and A_2 are derived from the model system. Indeed, the jacobian of the system (2.1) at positive equilibrium point takes the form;

$$J^* = \begin{pmatrix} r_1 - 2a_{11}x^* - p_1(1-m)y^* & -p_1(1-m)x^* \\ 0 & -r_2 \end{pmatrix} + \begin{pmatrix} 0 & 0 \\ qp_1(1-m)y^* & qp_1(1-m)x^* \end{pmatrix} e^{-\lambda\tau}$$

or

$$J^* = \begin{pmatrix} r_1 - 2a_{11}x^* - p_1(1-m)y^* & -p_1(1-m)x^* \\ qp_1(1-m)y^*e^{-\lambda\tau} & qp_1(1-m)x^*e^{-\lambda\tau} - r_2 \end{pmatrix}. \tag{2.5}$$

Hence, in the general form the characteristics equation of (2.5) may be written as

$$P(\lambda) + Q(\lambda)e^{-\lambda\tau} = 0, \tag{2.6}$$

where

$$\begin{cases} P(\lambda) = \lambda^2 + m_1\lambda + m_2, \\ Q(\lambda) = n_1\lambda + n_2, \\ m_1 = -(r_1 - 2a_{11}x^* - p_1(1-m)y^* - r_2), \\ m_2 = -r_2(r_1 - 2a_{11}x^* - p_1(1-m)y^*), \\ n_1 = -qp_1(1-m)x^*, \\ n_2 = \left((r_1 - 2a_{11}x^* - p_1(1-m)y^*)(qp_1(1-m)x^*) + qp_1^2(1-m)^2x^*\right). \end{cases}$$

Hence, by putting $\lambda = i\omega$ in (2.6) and separating the real and imaginary parts and following standard process, it is easy to find the value of τ_0, a crucial point for Hopf bifurcation. The point τ_0 is called Hopf bifurcation point. The detail may be seen in a standard book on delay differential equations (DDE). We skip the detail of that process. The value of τ_0 is calculated as;

$$\tau_0 = \frac{1}{\omega}\arccos\frac{\omega^2(n_2 - m_1n_1) - n_2m_2}{n_2^2 + n_1^2\omega^2}. \tag{2.7}$$

Hence, by the above discussion and using the Routh–Hurwitz criteria and Hopf bifurcation theorem, we can state the following lemma:

Lemma 2.1 *If* $c_{11} + c_{22} < 0$ *and* $c_{11}c_{22} - c_{12}c_{21} > 0$, *then the equilibrium* $E^*(x^*, y^*)$ *of system* (2.1) *is asymptotically stable for* $\tau \in [0, \tau_0)$ *and unstable for* $\tau > \tau_0$. *Hopf bifurcation occurs when* $\tau = \tau_0$, *where* $c_{11} = r_1 - 2a_{11}x^* - p_1(1 - m)y^*$, $c_{12} = -p_1(1 - m)x^*$, $c_{21} = qp_1(1 - m)y^*$, $c_{22} = -r_2 + qp_1(1 - m)x^*$.

3 The Control in the Responding System with Ideal Time Delay

The responding system with ideal time delay takes the form

$$\begin{cases} \frac{dx_1}{dt} = x_1(t)[r_1 - a_{11}x_1(t)] - p_1(1 - m)y_1(t)x_1(t), \\ \frac{dy_1}{dt} = qp_1(1 - m)y_1(t - \tau)x_1(t - \tau) - r_2y_1(t). \end{cases} \tag{3.1}$$

If we add the control function u to the equation governing the predator population, we have

$$\begin{cases} \frac{dx_1}{dt} = x_1(t)[r_1 - a_{11}x_1(t)] - p_1(1 - m)y_1(t)x_1(t), \\ \frac{dy_1}{dt} = qp_1(1 - m)y_1(t - \tau)x_1(t - \tau) - r_2y_1(t) + u. \end{cases} \tag{3.2}$$

The control function u has many applications in sustainable development. For example, if we are interested to maintain the population of a particular area to a standard level and prey population is much higher than predator population. In this condition, predator cannot predate their prey sufficiently. People can rise the mortality of prey to maintain the equilibrium level.

If we denote error between the systems (3.2) and (2.1) as

$$\begin{cases} e_1 = x_1 - x, \\ e_2 = y_1 - y. \end{cases} \tag{3.3}$$

The error equations takes the form

$$\begin{cases} \frac{de_1}{dt} = r_1e_1 - a_{11}e_1(x_1(t) + x(t)) - p_1(1 - m)(e_2x_1(t) + y(t)e_1), \\ \frac{de_2}{dt} = -r_2e_2 + qp_1(1 - m)[y_1(t - \tau)x_1(t - \tau) - y(t - \tau)x(t - \tau)] + u. \end{cases} \tag{3.4}$$

Now, we are going to state and prove the main theorem of this paper:

Theorem 3.1 *If the conditions in Lemma* (2.2) *are satisfied, a controller exists in* (3.2) *of the form*
$u = qp_1(1-m)[y(t-\tau)x(t-\tau) - y_1(t-\tau)x_1(t-\tau)] + p_1(1-m)e_1x_1(t)$
and system (3.4) *is stable when* $e_1 = 0$ *and* $e_2 = 0$.

Proof Let us consider a Lyapunov function of the form:

$$V = \frac{1}{2}(e_1^2 + e_2^2) > 0. \tag{3.5}$$

Therefore, the derivative of V w.r.t. time t is

$\dot{V} = e_1\dot{e_1} + e_2\dot{e_2}$,
$\dot{V} = e_1^2[r_1 - a_{11}(x_1(t)+x(t)) - p_1(1-m)y(t)] - p_1(1-m)e_1e_2x_1(t) + e_2^2[-r_2] + e_2[qp_1(1-m)\{y_1(t-\tau)x_1(t-\tau) - y(t-\tau)x(t-\tau)\}] + ue_2$
$= e_1^2[r_1 - a_{11}(x_1(t)+x(t)) - p_1(1-m)y(t)] + e_2^2[-r_2] + e_2[qp_1(1-m)\{y_1(t-\tau)x_1(t-\tau) - y(t-\tau)x(t-\tau)\} - p_1(1-m)e_1x_1(t) + u]$.
Hence,

$$\begin{cases} r_1 - a_{11}(x_1(t)+x(t)) - p_1(1-m)y(t) & < 0, \\ -r_2 & < 0, \\ (qp_1(1-m)\{y_1(t-\tau)x_1(t-\tau) - y(t-\tau)x(t-\tau)\} - p_1(1-m)e_1x_1(t) + u) & = 0. \end{cases} \tag{3.6}$$

$\therefore u = qp_1(1-m)[y(t-\tau)x(t-\tau) - y_1(t-\tau)x_1(t-\tau)] + p_1(1-m)e_1x_1(t)$.

Therefore, by Eq. (3.6) we have

$\dot{V} = e_1^2[r_1 - a_{11}(x_1(t)+x(t)) - p_1(1-m)y(t)] - p_1(1-m)e_1e_2x_1(t) + e_2^2[-r_2] < 0.$

The proof is completed.

4 Numerical Example

For numerical simulation, we take the following set of parameters:

$a_{11} = \frac{1}{4}, r_1 = 1, p_1 = \frac{1}{2}, (1-m) = \frac{1}{3}, qp_1 = \frac{1}{4}, r_2 = \frac{1}{4}.$

The model (2.1) takes the form;

$$\begin{cases} \frac{dx}{dt} & = x(t)[1 - \frac{1}{4}x(t)] - \frac{1}{6}y(t)x(t), \\ \frac{dy}{dt} & = \frac{1}{12}y(t-\tau)x(t-\tau) - 1/4y(t), \end{cases} \tag{4.1}$$

and the initial values are taken as $x(0) = 0.6$, $y(0) = 2$. The positive equilibrium point for (4.1) is calculated as $E^*(3, \frac{3}{2})$. The value of τ_0 is calculated as 0.43. The responding system adding the controller takes the form;

$$\begin{cases} \frac{dx_1}{dt} &= x_1(t)[1 - \frac{1}{4}x_1(t)] - \frac{1}{6}y_1(t)x_1(t), \\ \frac{dy_1}{dt} &= \frac{1}{12}y_1(t-\tau)x_1(t-\tau) - 1/4y_1(t) + u, \end{cases} \tag{4.2}$$

By theorem (3.1), the controller for (4.1) is given by

$$u = \frac{1}{12}[y(t - 0.43)x(t - 0.43) - y_1(t - 0.43)x_1(t - 0.43)] + \frac{1}{6}e_1x_1 \tag{4.3}$$

The system behavior of (4.1) is represented in Figs. 1, 2, 3, 4.

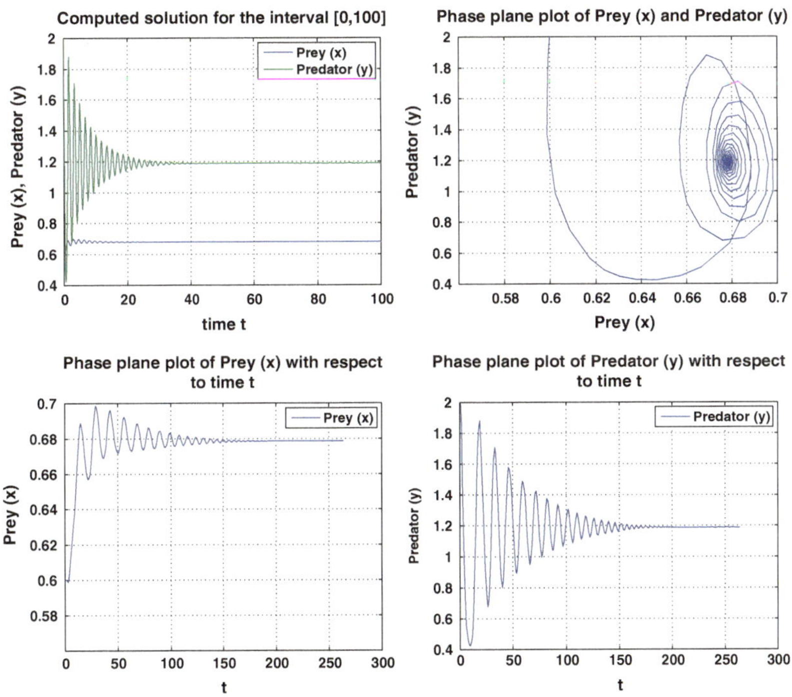

Fig. 1 Solution of (4.1) for $\tau = 0.40 < \tau_0$

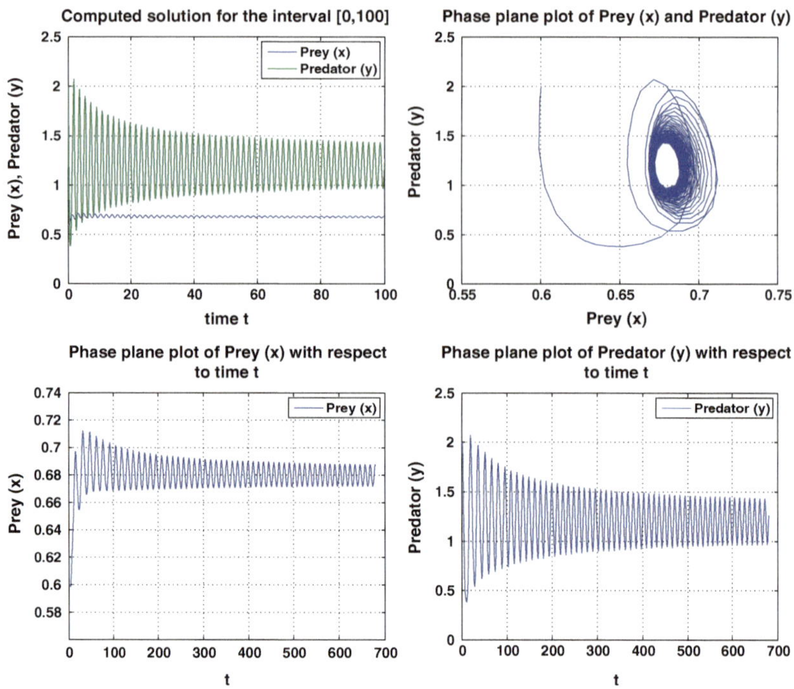

Fig. 2 Solution of (4.1) for $\tau_0 = 0.43$

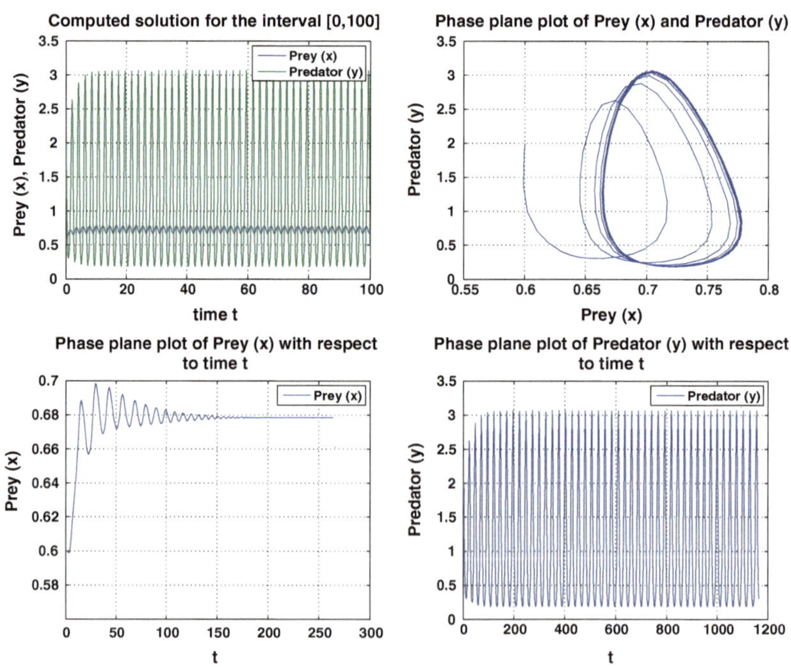

Fig. 3 Solution of (4.1) for $\tau = 0.50 > \tau_0$

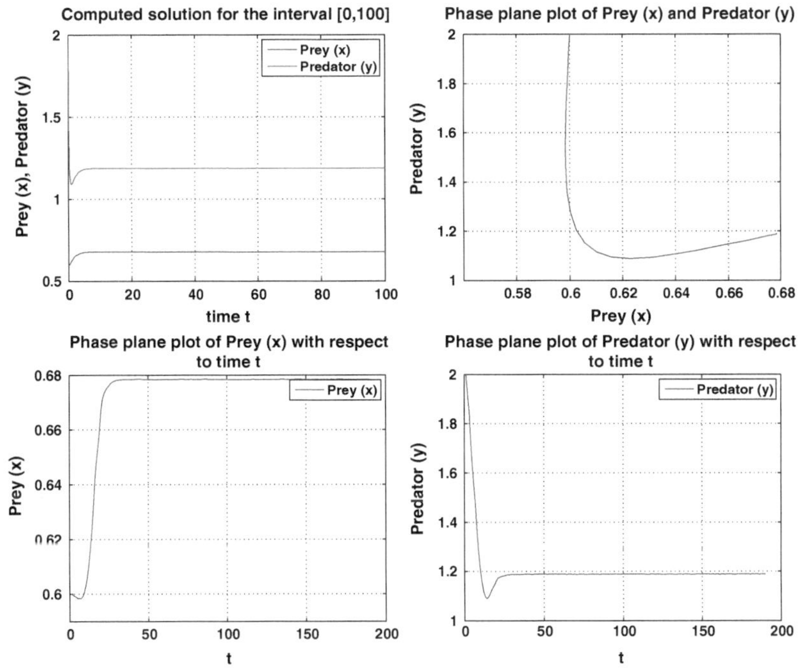

Fig. 4 The control of the system (4.1)

5 Discussion

If we ignore the prey refuge the system (2.1) takes the form

$$\begin{cases} \frac{dx}{dt} &= x(t)[r_1 - a_{11}x(t)] - p_1 y(t)x(t), \\ \frac{dy}{dt} &= qp_1 y(t-\tau)x(t-\tau) - r_2 y(t), \end{cases} \tag{5.1}$$

the similar analysis may be done. In this study we derive an expression for the control in Theorem (3.1) which is the main contribution of this paper. The control has many applications in the the development process with conserve. This is not a case study, hence real data is not available for the purpose of numerical simulation. However, system (2.1) has been simulated by considering an artificial set of parameters.

References

1. Changjin, X., Maoxin, L.: Bifurcation behaviors in a delayed three-species food chain model with Holling Type-II functional response. Appl. Anal. **92**(12), 2468–2486 (2013) http://dx.doi.org/10.1080/00036811.2012.742187
2. Maiti, A., Jana, M.M., Samanta, G.P.: Determinitic and stochatic analysis of a ratio-dependent predator-prey system with delay. Nonlinear Anal.: Model. Control **12**(3), 383–398 (2007)
3. Song, X., Wang, S. Zhou, X.: Stablilty and Hopf bifurcation for a viral infection model with delayed non-lytic immune response. J. Appl. Math. Comput. **33**, 251–265 (2010) http://dx.doi.org/10.1007/s12190-009-0285-y
4. Wang, J.L.: A kind of prey-predator ecological system with continous time delay. J. Biomath. **13**(4), 472–478 (1998)
5. Wang, S.W. et al.: A nonautonomous prey-predator system with stage structure and time delay. J. Math. Technol. **17**(06), 1–5 (2001)
6. Sarwardi, S., Mandal, P.K., Ray, S.: Dynamical behaviour of a two-predator model with prey refuge. J. Biol. Phys. **39**(4), 701–722 (2013) http://dx.doi.org/10.1007/s10867-013-9327-7
7. Li, Y.M., Zhu, Y.: The control and the reconfigurable control for prey-prdator ecosystem with time delay. Appl. Math. Model. **33**, 148–160 (2009) http://dx.doi.org//10.1016/j.amc.2014.01.025
8. Guang -Ping, H., Xiao-Ling, L.: Stability and Hopf bifurcation for a delayed predator-prey model with disease in the prey. Chaos Soliton Fract. **45**, 229–237 (2012) http://dx.doi.org//10.1016/j.chaos.2011.11.011

Evaluation of Solving Time for Multivariate Quadratic Equation System Using XL Algorithm Over Small Finite Fields on GPU

Satoshi Tanaka, Chen-Mou Cheng and Kouichi Sakurai

Abstract The security of multivariate public-key cryptography is largely determined by the complexity of solving multivariate quadratic equations over finite fields, a.k.a. *the MQ problem*. XL (eXtended Linearization) is an efficient algorithm for solving the MQ problem, so its running time is an important indicator for the complexity of solving the MQ problem. In this work, we implement XL on graphics processing unit (GPU) and evaluate its solving time for the MQ problem over several small finite fields, namely, GF(2), GF(3), GF(5), and GF(7). Our implementations can solve MQ instances of 74 equations in 37 unknowns over GF(2) in 36,972 s, 48 equations in 24 unknowns over GF(3) in 933 s, 42 equations in 21 unknowns over GF(5) in 347 s, as well as 42 equations in 21 unknowns over GF(7) in 387 s. Moreover, we can also solve the MQ instance of 48 equations in 24 unknowns over GF(7) in 34,882 s, whose complexity is about $O(2^{67})$ with exhaustive search.

Keywords Multivariate public-key cryptography · XL · GPGPU

1 Introduction

The problem of finding roots of nonlinear multivariate polynomial equations over finite fields lies at the core of the security for multivariate public-key cryptography (MPKC). Many MPKCs, e.g., Unbalanced Oil and Vinegar (UOV) [8], Hidden Field

S. Tanaka (✉) · K. Sakurai
Institute of Systems, Information Technologies and Nanotechnologies, Fukuoka, Japan
e-mail: tanasato@itslab.inf.kyushu-u.ac.jp

K. Sakurai
e-mail: sakurai@csce.kyushu-u.ac.jp

S. Tanaka · C.-M. Cheng · K. Sakurai
Kyushu University, Fukuoka, Japan

C.-M. Cheng
National Taiwan University, Taipei, Taiwan
e-mail: ccheng@imi.kyushu-u.ac.jp

R.N. Mohapatra et al. (eds.), *Mathematics and Computing*,
Springer Proceedings in Mathematics & Statistics 139,
DOI 10.1007/978-81-322-2452-5_24

Equations (HFE) [12], and the QUAD stream cipher [5], base their security on the quadratic case of such problems, which we will refer to as the MQ problem. Therefore, estimating the complexity of solving the MQ problem is of crucial importance for determining the security of these MPKCs.

To this date, there are two kinds of efficient algorithms for solving the MQ problem. One is the Gröbner basis method, and the other, the eXtended Linearization (XL) algorithm. Both algorithms generate new equations from the original systems. Although XL is shown to be a redundant variant of a Gröbner basis algorithm F_4 [3], it does have the advantage of having a smaller memory footprint in practice [15].

The bottleneck computation in XL is the solving of linearized systems. For sparse systems generated by XL, the Wiedemann algorithm can be used to efficiently solve an $N \times N$ nonsingular system with row sparsity k in $O(kN^2)$ complexity in terms of multiplications and additions. Here N is determined by the *degree of regularity* for the MQ problem, which we will explain in more detail later in this paper.

There are several implementations of the XL-Wiedemann algorithm. Yang et al. estimated the solving time for MQ instances in 6–15 unknowns based on a C++ implementation [15]. Moreover, they showed that the expected time for solving an MQ instance of 40 equations in 20 unknowns over GF(256) is about 2^{45} CPU cycles. Cheng et al. implemented the XL-Wiedemann algorithm on a cluster of 8 PCs of NUMA architecture [6]. As a result, they solved MQ instances of 36 equations in 36 unknowns over GF(2) in 46,944 s, 64 equations in 32 unknowns over GF(16) in 244,338 s, as well as 58 equations in 29 unknowns over GF(31) in 12,713 s.

Also, Mohamed et al. discussed how to solve systems derived from the HFE Challenge 2 [10]. They use the MXL_3 algorithm [9], which is essentially the XL algorithm with the "mutant" strategy. They solved the HFE challenge 2 system with 128 equations and 16 hidden equations in 144 unknowns, with appropriate guessing of variables. Their implementation of MXL_3 solved such a system in 365,801 s with guessing 52 variables on a PC with 4 quad-core AMD Opteron 8356 Processors and 128 GB memory. Moreover, according to their estimation, it would require approximately 100,000 GB of memory in order to break the full version of HFE Challenge 2 using MXL_3.

So far, we have not seen any implementation of the XL-Wiedemann algorithm on GPU, which is a candidate for further speed-up because several steps of the XL-Wiedemann algorithm can be parallelized. Therefore, we consider accelerating XL-Wiedemann on GPU. However, GPU implementation poses a set of very different limitations from its CPU counterpart. Hence, in this paper we shall detail these challenges and how we have dealt with them.

Our contributions include the following. We present several GPU implementations of the XL-Wiedemann algorithm, in which multiplication of a sparse matrix with a dense vector is parallelized on GPU. Moreover, we benchmark an implementation based on the cuSPARSE library using floating-point arithmetic. Finally, we show the experimental results of solving MQ instances over GF(2), GF(3), GF(5), and GF(7). Our implementation can solve MQ instances of 74 equations in 37 unknowns over

GF(2) in 36,972 s, 48 equations in 24 unknowns over GF(3) in 933 s, as well as 42 equations in 21 unknowns over GF(5) in 347 s. The largest instance we have solved is 48 equations in 24 unknowns over GF(7) in 34,883 s, whose complexity is around $O(2^{67})$ if we use a brute-force kind of approach.

The cuSPARSE library only supports floating-point arithmetic, not integer arithmetic, let alone finite field arithmetics. Therefore, we need to use cuSPARSE functions to implement finite field arithmetics via additional operations such as the modular operations.

2 The MQ Problem and the XL-Wiedemann Algorithm

The security of MPKC is largely based on the complexity of solving a system of multivariate nonlinear equations over finite fields. The MQ problem is a quadratic case of this problem. Generic MQ is known to be NP-complete [4].

Let $q = p^k$, where p is a prime, and $\boldsymbol{x} = \{x_1, \ldots, x_n\}$ ($\forall i, x_i \in \mathrm{GF}(q)$). Generally, multivariate quadratic polynomial equations in n unknowns over GF(q) can be described as follows:

$$f(\boldsymbol{x}) = \sum_{1 \leq i \leq j \leq n} \alpha_{i,j} x_i x_j + \sum_{1 \leq i \leq n} \beta_i x_i + \gamma = 0, \tag{1}$$

where $\forall i, j, \alpha_{i,j}, \beta_i, \gamma \in \mathrm{GF}(q)$. The MQ problem consists solving quadratic polynomial equations given by $\boldsymbol{y} = \{f_1(\boldsymbol{x}), \ldots, f_m(\boldsymbol{x})\}$

The original XL algorithm was proposed by Courtois et al. in [7]. The idea of XL is based on a linearization technique, in which new unknowns representing nonlinear terms, e.g., $y_{1,2} = x_1 x_2$, are generated and treated as an independent variable. If the number of equations is greater than the number of variables in the resulted linearized system, then we can solve it by, e.g., Gaussian elimination. If not, we can generate new equations from the original ones by raising to a higher degree. For the sake of completeness, the XL algorithm is described in Algorithm 1. Simply put, the *degree of regularity* D is the minimal degree at which the number of linearly independent equations exceeds the number of unknowns in the linearized system.

The XL algorithm generates sparse equations in Step 1 of Algorithm 1. The number of nonzero terms of an equation is only $\binom{n+2}{2}$ out of all possible $\binom{n+D}{D}$ terms, since the generated equations are just a product of the original equations and some monomials. However, the Gaussian elimination is not suited for solving such sparse linear systems, as it cannot take advantage of the sparsity. The XL-Wiedemann algorithm [11] addresses this problem of the original XL by replacing the Gaussian elimination with the Wiedemann algorithm [14], which is more efficient for solving systems of sparse linear equations.

Algorithm 1 The XL algorithm [7]

Require: m quadratic polynomial equations $F = \{f_1, \ldots, f_m\}$, m-th vector $\boldsymbol{y} = F(\boldsymbol{x})$, and the degree of regularity D.
Ensure: The n-th unknown vector $\boldsymbol{x} = \{x_1, \ldots, x_n\}$.
1: Multiply: Generate products between all polynomial equations and all unknowns of the form $\prod_{j=1}^{D-2} x_{i_j}$.
2: Linearize: Treat each monomial in x_i of degree $\leq D$ as a new, independent unknown and perform an elimination algorithm on the linearized equations obtained in Step 1 to derive a univariate equation.
3: Solve: Solve the univariate equations obtained in Step 1 over GF(q).
4: Back-substitute: Find the values of the other unknowns by back-substitution into the linearized system.

The Wiedemann algorithm [14] is a solving method for a system of linear sparse equations over finite fields. Let A be an $N \times N$ nonsingular matrix over GF(q). The Wiedemann algorithm finds a nonzero vector $\mathbf{x}$, where $\mathbf{y} = \mathbf{Ax}$. The Wiedemann algorithm is described in Algorithm 2.

Algorithm 2 The Wiedemann algorithm [14]

Require: $N \times N$ nonsingular matrix A and vector $\boldsymbol{b}$, where $A\boldsymbol{x} = \boldsymbol{b}$.
Ensure: The unknown solution vector $\boldsymbol{x}$.
1: Set $\boldsymbol{b}_0 = \boldsymbol{b}$, $k = 0$, $\boldsymbol{y}_0 = 0$, and $d_0 = 0$.
2: Compute the matrix sequence $s_i = \boldsymbol{u}_{k+1} A^i \boldsymbol{b}_k$ for $0 \leq i \leq 2(N - d)$, with a random vector u_{k+1}.
3: Set $f(\lambda)$ to the minimum polynomial of the sequence of s_i using the Berlekamp–Massey algorithm.
4: Set $\boldsymbol{y}_{k+1} = \boldsymbol{y}_k + f^-(A)\boldsymbol{b}_k$, where $f^-(\lambda) := \frac{f(\lambda)-f(0)}{\lambda}$, $\boldsymbol{b}_{k+1} = \boldsymbol{b}_0 + A\boldsymbol{y}_{k+1}$, and $d_{k+1} = d_k + \deg f(\lambda)$.
5: If $\boldsymbol{b}_{k+1} = 0$, then the solution is $\boldsymbol{x} = \boldsymbol{y}_k$
6: Set $k = k + 1$ and go to Step 2.

3 CUDA and Its Linear Algebra Libraries

Provided by NVIDIA, CUDA is a development environment for GPU based on C language. Proprietary tools for using GPU have existed before CUDA; such tools often need to tweak OpenGL and/or DirectX and disguise computation as graphics rendering commands. Therefore, these tools are not easy to use, whereas CUDA is efficient because it can use GPU's computational cores directly.

In CUDA, hosts correspond to PC, and devices correspond to GPU. CUDA works by making the host control, the device via kernels. Because only one kernel can be executed at a time, we need to parallelize processing inside a kernel. A kernel handles some blocks in parallel. A block also handles some threads in parallel. Therefore, a kernel can handle many threads simultaneously.

NVIDIA provides several libraries for linear algebra. For example, the cuBLAS library provides functions of the Basic Linear Algebra Subprograms (BLAS) library. BLAS is classified into three levels of functionalities. Level 1 functions provide operations on vectors, level 2 operations on vectors and matrices, while level 3 allows matrix–matrix operations. The cuSPARSE library is actually the sparse version of the cuBLAS library. Therefore, cuSPARSE also provides these three levels of functions.

We assume that D is the degree of regularity for the XL algorithm. Then, XL constructs an $\binom{n+D}{D} \times \binom{n+D}{D}$ linearized matrix from the MQ instances of m equations in n unknowns over GF(q). However, quadratic polynomial equations in n unknowns have only $\binom{n+2}{2}$ terms. Therefore, we can reduce computations of matrix-vector product as well as the memory footprint if we store the matrix in sparse form.

Let N be the degree of row and column in a matrix, and num_{NZ} be the number of nonzero elements in the matrix. Sparse matrix forms have value, row-index, and column-index data of nonzero elements in a matrix. There are some sparse matrix formats such as the following [1]

- The COO (coordinate) format is the most basic one. It simply holds value, row-index, and column-index data of nonzero elements in the matrix. Therefore, it requires $3\mathrm{num}_{NZ}$ for the memory space.
- The CSR (compressed storage row) assumes that the data vector is ordered by the row-index. It differs only row-index from the COO formats, in which it holds the head number of nonzero terms in each row-vector of the matrix instead of row-index data. Then, it requires $2\mathrm{num}_{NZ} + N$ memory.
- The ELL (Ellpack-Itpack) format uses two dense $N \times \mathrm{max}_{NZ}$ matrices, where max_{NZ} is the maximal number of nonzero terms in a row-vector. One matrix shows the value of nonzero matrix, and the other shows the column-index.

Figure 1 shows examples of each of the three formats.

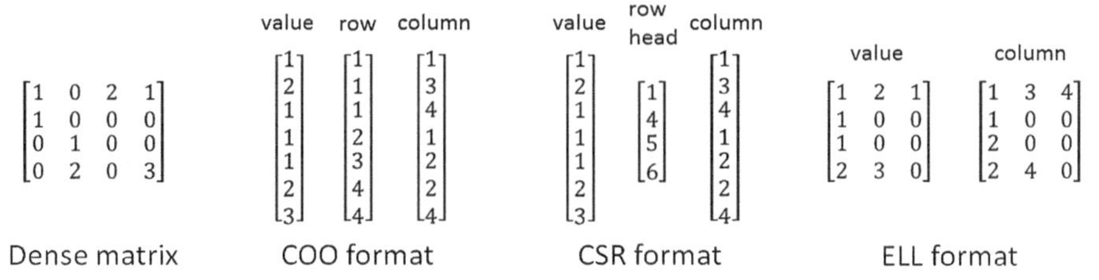

Fig. 1 Sparse matrix formats

4 Implementing XL-Wiedemann on GPU

4.1 Degrees of Regularity Over Small Fields

The bottleneck of the XL-Wiedemann algorithm is the linear algebra part that solves an $N \times N$ matrix system. Here N is determined by the degree of regularity D as $N = \binom{N+D}{D}$. The degree of regularity is the minimal degree where the number of linearly independent equations exceeds the number of linearized unknowns. We can figure out the number of linearized unknowns N for the degree d as $N = \binom{N+d}{d}$ easily. Rønjon and Raddum gave an upper bound for the number of linearly independent equations I, which can be decided using the following formula [13]:

$$I = \sum_{i=0}^{\frac{D_m}{D_e}} (-1)^i \binom{m+i}{i+1} \sum_{j=0}^{D_m - i \cdot D_e} \binom{n}{j}. \tag{2}$$

Here, D_m is the maximal degree of the monomials, and D_e is the degree of the original equations. For the MQ problem, $D_m = D - 2$ and $D_e = 2$. Therefore, we can find the minimal degree D, where $I \geq N (= \binom{N+D}{D})$ by Formula (2). Figure 2 shows degrees of regularity for MQ instances of $2n$ equations in n unknowns over GF(2), GF(3), GF(5), and other prime fields for $n \leq 64$. The cases of GF(5) and other larger prime fields are actually quite similar. Only GF(2) and GF(3) differ from the other cases because we need to take into consideration field equations $\alpha^q = \alpha$.

From the definition of the degree of regularity, it is obvious that $I \geq N$. However, for the Wiedemann algorithm to work, we need to reduce to N from I. The simplest way is to randomly remove certain equations, which is our strategy in our implementation.

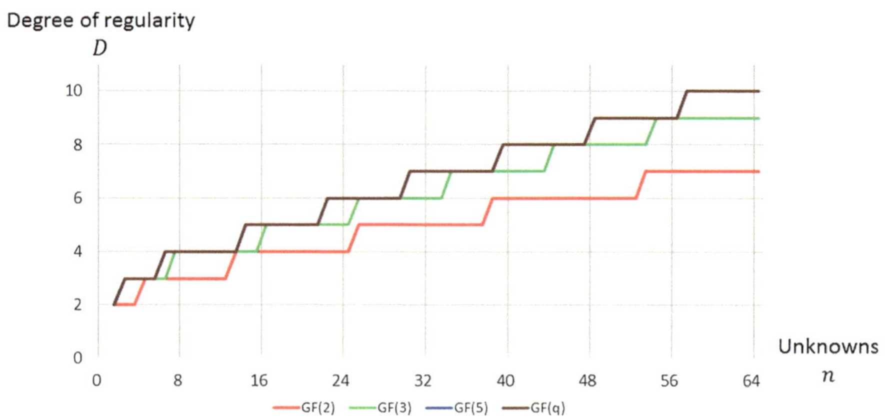

Fig. 2 The degrees of regularity for $m = 2n$ cases for $n \leq 64$

4.2 The Wiedemann Algorithm

The Wiedemann algorithm has three separate steps. The first step is to generate the sequence $\{(\boldsymbol{u}, A^i\boldsymbol{b})\}_{i=0}^{2N}$ for an $N \times N$ matrix A and a vector $\boldsymbol{b}$, where $A\boldsymbol{x} = \boldsymbol{b}$, as well as a random vector $\boldsymbol{u}$. The second step is to find the minimal polynomial of the generated sequence $f(\lambda)$ using the Berlekamp-Massey algorithm. The final step is to compute $f^{-}(A)\boldsymbol{b}$, where $f^{-}(\lambda) = \frac{f(\lambda)-f(0)}{\lambda}$. In this work, we only implement the first step and the final step on GPU. This is because the Berlekamp–Massey algorithm is sequential in nature, and hence might not benefit from parallelization. For example, it has many conditional branches, which are not suitable for GPU implementation. Therefore, we implement the second step on CPU.

4.3 Generating Sequence $\{(U, A^i b)\}_{i=0}^{2N}$

This step requires multiplying the sparse matrix A and the dense vector $A^{i-1}\boldsymbol{b}$, as well as taking dot product $(\boldsymbol{u}, A^i\boldsymbol{b})$. However, we can choose the random vector $\boldsymbol{u}$ as $\boldsymbol{u} = \{1, 0, \ldots, 0\}$. Therefore, taking dot product amounts to looking up the first coordinate in the vector $A^i\boldsymbol{b}$. Hence, we should consider only multiplication of the sparse matrix A and the dense vector $A^{i-1}\boldsymbol{b}$.

Multiplying the sparse matrix A and the dense vector $A^{i-1}\boldsymbol{b}$ takes two steps. The first one is multiplying nonzero elements in the matrix with the elements in the vector. The other is summing the results of the partial multiplications for each row.

We choose the ELL format for representing sparse matrices. One advantage is that every column width is the same in a matrix, and the multiplication result also has such width. In CUDA kernels, the column width corresponds to the number of threads, while the row height corresponds to the number of blocks. To achieve maximal efficiency, each block should have the same number of threads. Therefore, the ELL format is best suited for GPU implementation.

In summing the partial multiplication results, we use the parallel reduction technique [2]. This technique reduces sequential algorithms to the parallel version. Basically, it handles minimal computations of algorithms in parallel with several processors. Then, their results are inputs of next steps. The parallel reduction method iterates to generate the final result. For summations, additions reduces the number of terms into half in each step. Therefore, such a technique allows computing summation of n items in $O(\log n)$ steps. Figure 3 shows the image and an example of parallel reduction techniques for summations. It computes a summation of 8 terms over GF(7) in 3 steps with 4 processors.

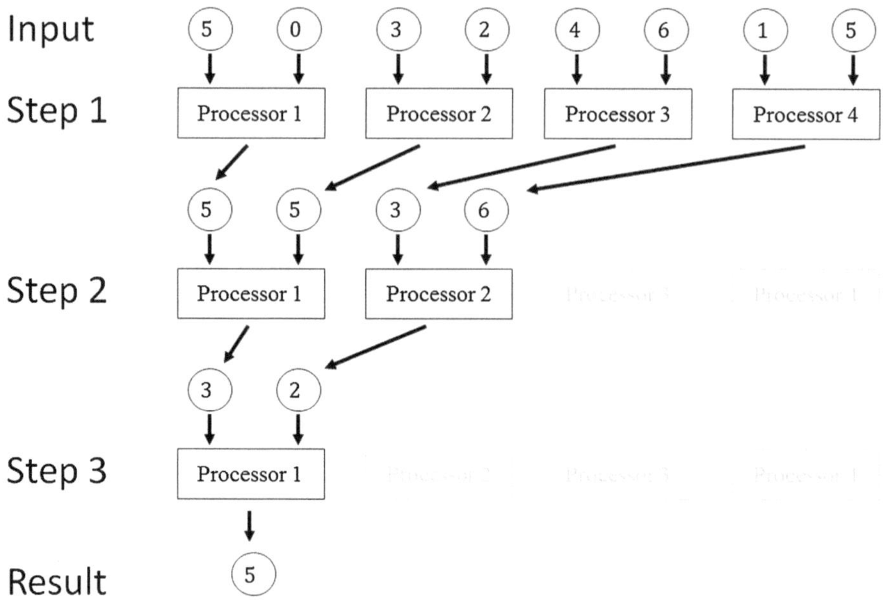

Fig. 3 Parallel reduction of summing 8 terms over GF(7)

4.4 Computing $f^{-}(A)\boldsymbol{b}$

Since $f^{-}(A)\boldsymbol{b} = \sum_{i=1}^{d} c_i A^{i-1}\boldsymbol{b}$, where d is the degree of $f(\lambda)$, this step amounts to summing $c_i A^{i-1}\boldsymbol{b}$, using the same partial sums from the previous step. Hence, there are two strategies for computing $A^i\boldsymbol{b}$. The first one is to store the result of $A^i\boldsymbol{b}$ on GPU. This strategy can avoid recomputing $A^i\boldsymbol{b}$. However, it needs about $O(N^2)$ memory for storing $A^i\boldsymbol{b}$, where $0 \le i \le N$ (since $d \le N$). Therefore, this strategy can only work for smaller matrices.

The other strategy is to recompute $A^i\boldsymbol{b}$ on the fly. Although it repeats the computation of d products of $A^i\boldsymbol{b}$, it only requires $O(A^{i-1}\boldsymbol{b})$ memory to hold the last vector of $A^i\boldsymbol{b}$. Therefore, this strategy is more suitable for large matrices.

4.5 cuSPARSE

The cuSPARSE library [1] provides functions that multiply a sparse matrix with a dense vector. Therefore, we consider using cuSPARSE as an alternative implementation for computing A and $A^{i-1}\boldsymbol{b}$. There are two important issues with implementations. First, the interface is fixed and opaque. The cuSPARSE library only provides this function for CSR format: $\boldsymbol{y} \leftarrow \alpha A\boldsymbol{x} + \beta \boldsymbol{y}$, where A is a matrix, $\boldsymbol{x}$, $\boldsymbol{y}$ are vectors, and α, β are scalars. Therefore, we set $\beta = 0$ for the first step. Moreover, we are stuck with CSR format for representing sparse matrices when we use cuSPARSE library.

The second issue is the restriction of the unknown type. The cuSPARSE library only supports floating-point arithmetic, not integer arithmetic, let alone finite field arithmetics. Therefore, we need to use cuSPARSE functions to implement finite field arithmetics via additional operations such as the modular operations.

5 Experimental Results

We implement the XL-Wiedemann algorithm on GPU using two strategies, integer version and cuSPARSE (floating-point) version. We experiment with solving the largest cases for $D = 4, 5$ over GF(2), GF(3), GF(5), and GF(7) by both implementation strategies and summarize these instances in Table 1.

Table 2 shows the overall experimental results, and Table 3 shows the profiling results of the Wiedemann algorithm. Despite the overhead brought by the two issues mentioned previously, the cuSPARSE version seems to outperform integer version for larger cases. In our experiments, the Berlekamp–Massey algorithm can occupy a significant portion of the total running time and hence may be worth further optimization. We can also use high-quality, state-of-the-art implementations from commercial computer algebra systems like MAGMA.

Finally, we solve the largest case of $D = 6$ over GF(7), which has a system of 24 unknowns and 48 polynomials. We choose the version of using the cuSPARSE library as a solver of the MQ instance, because of the result of $D = 5$ cases. Table 4 shows the construction and experimental result of solving the MQ instance.

Table 1 MQ instances in our experiments

Field GF(q)	GF(2)		GF(3)		GF(5)		GF(7)	
Degree of regularity D	4	5	4	5	4	5	4	5
Unknowns n	24	37	15	24	13	21	13	21
Equations m	48	74	30	48	26	42	26	42
Matrix								
Linearized terms	12,950	510,415	3,635	110,954	2,379	65,758	2,379	65,779
Nonzero terms	301	704	136	325	105	253	105	253

Table 2 Running time of XL-Wiedemann on GPU

	Field GF(q)	GF(2)		GF(3)		GF(5)		GF(7)	
	Degree of regularity D	4	5	4	5	4	5	4	5
	Unknowns n	24	37	15	24	13	21	13	21
	Equations m	48	74	30	48	26	42	26	42
Integer	Solving time (s)	14.7358	83,782.11	0.5847	2,089.30	0.4415	601.124	0.4856	670.963
	Extension (s)	0.1248	130.98	0.0116	7.29	0.0059	3.347	0.0053	2.913
	Wiedemann (s)	14.6101	83,651.08	0.5729	2,082.01	0.4355	597.777	0.4802	668.049
cuSPARSE	Solving time (s)	8.8982	36,971.85	0.8684	932.95	0.4852	346.571	0.5063	387.121
	Extension (s)	0.0885	128.28	0.0098	8.00	0.0050	3.366	0.0050	3.354
	Wiedemann (s)	8.8077	36,843.49	0.8583	924.95	0.4800	343.204	0.5012	383.764

Table 3 Profiling results for the Wiedemann algorithm

	Field GF(q)	GF(2)		GF(3)		GF(5)		GF(7)	
	Degree of regularity D	4	5	4	5	4	5	4	5
	Unknowns n	24	37	15	24	13	21	13	21
	Equations m	48	74	30	48	26	42	26	42
Integer	Running time (s)								
	Wiedemann	14.6101	83,651.08	0.5729	2,082.01	0.4355	597.777	0.4802	668.049
	Generating Sequence	9.5806	49,719.75	0.3030	1,104.82	0.2131	302.236	0.2304	336.292
	Berlekamp–Massey	4.9253	9,035.16	0.2379	439.1057	0.19	148.328	0.2195	167.483
	Computing $f^{-}(A)\boldsymbol{b}$	0.0937	24,895.43	0.0305	537.99	0.0273	147.188	0.0295	164.249
	Memory usage (MB)								
	Matrix	29.74	2741.49	5.66	412.67	2.86	190.39	2.86	190.46
	Stream	1279.47	0	100.81	0	43.22	0	43.22	0
cuSPARSE	Running time (s)								
	Wiedemann	8.8077	36,843.49	0.8583	924.94	0.4800	343.204	0.5012	387.764
	Generating sequence	3.8079	22,215.69	0.4284	325.75	0.2418	108.0073	0.2393	114.814
	Berlekamp–Massey	4.8855	9,059.83	0.4284	325.75	0.1999	183.685	0.2223	214.049
	Computing $f^{-}(A)\boldsymbol{b}$	0.1045	5,567.20	0.0403	160.77	0.0372	51.473	0.0386	54.863
	Memory usage (MB)								
	Matrix	44.66	4114.18	5.67	413.10	2.87	190.64	2.87	190.71
	Stream	1279.47	0	100.81	0	43.22	0	43.22	0

Table 4 Solving the MQ instance of 48 equations in 24 unknowns over GF(7)

Constructions	Unknowns n	24
	Equations m	48
Matrix	Linearized terms	593,774
	Nonzero terms	325
Memory (MB)	Matrix	2,208.44
Running time	XL-Wiedemann Linearization Wiedemann	34,881.637 580.406 34,301.231
Wiedemann algorithm	Generating sequence Berlekamp–Massey Compute $f^{-}(A)\boldsymbol{b}$ (s)	11,046.464 17,698.748 5,555.593

6 Conclusion

We provide GPU implementations of the XL-Wiedemann algorithm using both integer and floating-point arithmetic via the cuSPARSE library. Our implementation can solve MQ instances of 74 equations in 37 unknowns over GF(2) in 36,972 s, 48 equations in 24 unknowns over GF(3) in 933 s, as well as 42 equations in 21 unknowns over GF(5) in 347 s by using the cuSPARSE library. Finally, we can solve the largest case of $D = 7$ over GF(7), the MQ instance of 48 equations in 24 unknowns. By using the cuSPARSE library, it takes 34,882 s. Our next goal is to estimate the expected solving time for larger degree cases.

Acknowledgments This work is partly supported by "Study on Secure Cryptosystem using Multivariate polynomial," No. 0159-0091, Strategic Information and Communications R&D Promotion Programme (SCOPE), the Ministry of Internal Affairs and Communications, Japan and Grant-in-Aid for Young Scientists (B), Grant number 24740078.

References

1. cuSPARSE::cuda toolkit documentation. http://docs.nvidia.com/cuda/cusparse, Accessed Aug 2014
2. Optimizing parallel reduction in cuda. http://developer.download.nvidia.com/compute/cuda/1.1-Beta/x86_website/projects/reduction/doc/reduction.pdf, Accessed Aug 2014
3. Ars, G,, Faugere, J.-C., Imai, H., Kawazoe, M., Sugita, M.: Comparison between XL and Gröbner basis algorithms. In; Advances in cryptology-ASIACRYPT, pp. 338–353. Springer, Berlin (2004)
4. Bard, G.V.: Algebraic cryptanalysis, Springer, Berlin (2009)
5. Berbain, C., Gilbert, H., Patarin, J.: Quad: a practical stream cipher with provable security. In: Advances in cryptology-EUROCRYPT, pp. 109–128. Springer, Berlin (2006)
6. Cheng, C.-M., Chou, T., Niederhagen, R., Yang, B.-Y.: Solving quadratic equations with XL on parallel architectures. In: Cryptographic hardware and embedded systems-CHES, pp. 356–373. Springer, Berlin (2012)

7. Courtois, N., Klimov, A., Patarin, J., Shamir, A.: Efficient algorithms for solving overdefined systems of multivariate polynomial equations. In: Advances in cryptology-EUROCRYPT, pp. 392–407. Springer, Berlin (2000)
8. Kipnis, A., Patarin, J., Goubin, L.: Unbalanced oil and vinegar signature schemes. In: Advances in cryptology-EUROCRYPT 99, pp. 206–222. Springer, Berlin (1999)
9. Mohamed, M.S.E,, Cabarcas, D., Ding, J., Buchmann, J,, Bulygin, S.: Mxl3: An efficient algorithm for computing Gröbner bases of zero-dimensional ideals. In: Information, security and cryptology-ICISC, pp. 87–100. Springer, Berlin (2010)
10. Mohamed, M.S.E., Ding, J., Buchmann, J.: Towards algebraic cryptanalysis of hfe challenge 2. In: The 5th international conference on information security and assurance, CCIS, vol. 200, pp. 123–131. Springer, Berlin (2011)
11. Mohamed, W.S.A., Ding, J., Kleinjung, T., Bulygin, S., Buchmann, J.: Pwxl: a parallel Wiedemann-XL algorithm for solving polynomial equations over $gf(2)$. SCC **89–100**, 2010 (2010)
12. Patarin, J.: Hidden fields equations (HFE) and isomorphisms of polynomials (IP): two new families of asymmetric algorithms. In: Advances in cryptology-Eurocrypt 96, pp. 33–48. Springer, Berlin (1996)
13. Rønjom, S., Raddum, H.: On the number of linearly independent equations generated by XL. In: Sequences and their applications-SETA, pp. 239–251. Springer, Berlin (2008)
14. Wiedemann, D.: Solving sparse linear equations over finite fields. IEEE Trans. Inf. Theory **32**(1), 54–62 (1986)
15. Yang, B.-Y., Chen, O.C.-H., Bernstein, D.J., Chen, J.-M.: Analysis of quad. In: Fast software encryption, pp. 290–308. Springer, Berlin (2007)

Hierarchical Visual Secret Sharing Scheme Using Steganography

Biswapati Jana, Amita Samanta and Debasis Giri

Abstract The rapid growth of computer networks and technology constructs a favorable environment that can tolerate the multiusers in a hierarchy based. In any organization the personnel are frequently organized in the form of a hierarchy and there is the requirement that information is distributed over the hierarchy on a "need-to-know" basis. In this paper, we propose a new hierarchical visual secret sharing scheme, where steganographic technique has been used to maintain hierarchy and detect fake share using weight matrix-based embedding method. In this approach, we have used a key matrix (K) and a weight matrix (W) to hide critical information (M) into the share on each level of our proposed scheme. The basic ideas are: (i) to use an EXclusive-OR operator to protect the key matrix (K) and (ii) to use a weight matrix (W) to increase the data hiding rate while maintaining high quality of the share image in each level in hierarchy of the scheme. The share generator or Trusted Authority (TA) generates weight matrix W_i for share S_i and each level modifies weight matrix W_{i+1} using the formula $W_{i+1} = (W_i \times 5 \bmod 8) + 1$ where $i = 0, 1, 2, \ldots, n$ to keep track of level and maintain the hierarchical structure in proposed scheme. The experimental results are demonstrated and tested using Peak Signal-to-Noise Ratio (PSNR) value and relative entropy. It shows that our scheme is superior in terms of PSNR compared to existing schemes.

Keywords Visual cryptography · Visual secret sharing · Extended visual secret sharing · Hierarchical visual secret sharing · Steganography · Weight matrix

---

B. Jana (✉) · A. Samanta
Department of Computer Science, Vidyasagar University, West Bengal 721102, India
e-mail: biswapati.jana@mail.vidyasagar.ac.in

A. Samanta
e-mail: amitasamanta1@gmail.com

D. Giri
Department of Computer Science and Engineering, Haldia Institute of Technology, Haldia 721657, India
e-mail: debasis_giri@hotmail.com

R.N. Mohapatra et al. (eds.), *Mathematics and Computing*,
Springer Proceedings in Mathematics & Statistics 139,
DOI 10.1007/978-81-322-2452-5_25

1 Introduction

Visual Cryptography (VC) is one kind of Visual Secret Sharing (VSS), introduced by Naor and Shamir [1] in 1994. In VC, a secret image is encrypted into several shares which is completely unrecognizable. While the shares are separate, the secret image is completely incoherent. Each share holds different pieces of image and the secret image comes out only by stacking a sufficient number of shares together [2]. VC eliminates complex mathematical computation to recover the secret. The encrypted message can be decrypted directly by the Human Visual System (HVS). In (k, n) basic model of VC any k shares will decode the secret image out of n shares which reduces security level. Any subset of k or more qualified shares can decrypt the secret image but no information can be obtained by stacking lesser number of qualified shares or by stacking disqualified shares. Chen et al. [3] pointed out that cheating is possible in (k, n) VC when $k < n$. There are two types of cheaters in VC. One is a malicious participant (MP) who is also a legitimate participant, namely MP $\in P$ (Qualified participant), uses his original share to create a Fake Share (FS) to cheat the other qualified participant and the other is a malicious outsider (MO), where MO $\notin P$, will create FS by using some random images as input to decode the original image. The MO will try to create FS of different sizes because the size of the original share may vary. Cheating may also happen in Extended Visual Cryptographic Schemes (EVCS) by MP.

In this paper, we proposed Hierarchical Visual Secret Sharing (HVSS) scheme to protect secret in any hierarchical organizational environment then we propose a Steganographic approach to prevent cheating by detecting FS in HVSS and then revealed secret image from original share by simple stacking the valid shares. We have used a secret key K and a weight matrix W to hide critical information M into the share on each level of hierarchical secret sharing scheme. Here we consider the share generator as the Trusted Authority (TA). When any share holder wants to know whether a share is original or not, he sends a request to the trusted authority (TA) to check both the shares. Then TA extracts the embedding message from both shares. If TA retrieves the original embedded message from any share, the share will be acceptable, otherwise it will be rejected.

The rest of the paper is organized as follows. Section 2 reviews some required primitives including related work. Overview of Visual Secret Sharing (VSS) is discussed in Sect. 3. Data hiding techniques are discussed in Sect. 4. Our proposed method is discussed in Sect. 5. Performance evaluation and security analysis of the proposed scheme are presented in Sect. 6. Finally, some conclusions are given in Sect. 7.

2 Related Work

The main principle of VC was first introduced by Naor and Shamir [1]. In 1996, Ateniese et al. [4] proposed an elegant VC scheme for general access structures based on the cumulative array method. In 1997, Naor and Pinkas [5] showed some methods of authentication and identification for VC. Their scenario focuses on authentication and identification between two participants. In 1999, Yang and Laih [6] presented two cheating prevention VC schemes to break the misleading secrets forged by dishonest participants. The first method generates an additional verification share to check the validity of each share, where the verification share should be hold by the trusted authority (TA) to verify the validity of each share. The second method transforms a conventional VC scheme to another cheating prevention VC scheme with greater pixel expansion in each generated shares. The stacking of any two shares reveals the verification image which can be inspected by user to check the validity of the shares. In 2002, Hu and Tzeng [7] proposed a new definition for VC, in which the secret image can be either darker or lighter than the background. In 2006, Horng et al. [8] proposed a cheating method against some VC schemes. In their cheating method, the cheater needs to know the exact distribution of black and white subpixels of the shares of honest participants. They demonstrated a process of collusive cheating by $n+1$ participants to the other user in $(2, n)$ VC schemes, and presented two simple possible solutions to address the problem. The first method generates a dedicated verification share to each participant which can be applied to investigate the genuine of the shares gathered from other participants. The second one uses a $(2, n+l)$ VC scheme instead of $(2, n)$ scheme in a 2-out-of-n coding instance, that frustrates the malicious user in predicting the structure of the transparencies possessed by other participants. In 2007, Hu and Tzeng [9] presented three robust methods to improve the weaknesses of previously cheating prevention VC schemes, two for conventional VC and another for extended VC. However, like the previous cheating prevention VC schemes in [8, 9], additional verification share or greater pixel expansion is required to endow the ability about resisting cheating against malicious participants. In 2011, Chen et al. [3, 10] proposed another cheating prevention method where the method can divide the cheating prevention schemes into two classes. One is based on share authentication where another share (transparency) is used to authenticate other shares (transparencies) and the other is based on blind authentication where some property of the image is used to authenticate the reconstructed secret image. In 2007, Tassa [11, 12] proposed a new secret sharing scheme based on Birkhoff interpolation to deal with hierarchical threshold access structures. However, unlike Shamir's secret sharing scheme, Tassa's scheme is not able to use all potentials of underlying polynomial to share multiple secrets. Using Tassa's scheme to share more than t_0 secrets makes it possible for some nonauthorized subset of participants to recover some of the secrets. In 2012, Guo et al. [13] proposed a Hierarchical Threshold Secret Image Sharing (HTSIS) based on Steganography and Birkhoff interpolation. In this technique nonauthorized participants are able to recover the secret image. To overcome this weakness, Nasrollah et al. [14] proposed a secret image sharing scheme with a hierarchical threshold access structure using cellular automata and Birkhoff interpolation. In their technique participants are able to detect tampering of the recovered secret image. A data hiding scheme for binary images has been proposed by Tseng et al. [15]. In 2013, Fan et al. [16] proposed an improved efficient data hiding scheme for gray scale images using weight matrix. Here, we propose Hierarchical Visual Secret Sharing scheme that provides cheating prevention in hierarchical structure using a weight matrix and a key matrix.

3 Visual Secret Sharing

In Visual Secret Sharing (VSS), the shares are presented into transparencies. After taking the Secret Image (SI), the transparencies are generated; each white and black pixel of SI is handled separately. The structure appears as a collection of m black and white subpixels in each of the n transparencies. So one pixel of the SI corresponds to $n \times m$ subpixels, denoted by an $n \times m$ Boolean matrix, called as base matrix (B), such that $B_{ij} = 1$ if and only if the j-th subpixel of the i-th share is black and $B_{ij} = 0$ if and only if the j-th subpixel of the i-th share is white. The gray level of the stack of k shared blocks is determined by the Hamming Weight $H(V)$ of the "or" ed m-vector V of the corresponding k rows in B. This gray level is interpreted by the visual system of the users as black if $H(V) \geq d$ and as white if $H(V) \leq d - \alpha \times m$ for some fixed threshold d and relative difference α. According to Naor and Shamir [1], a solution to the (k, n)-VSS consists of two collections C_0 (for white) and C_1 (for black) of $n \times m$ base matrices. The solution is considered valid if the following conditions hold:

1. A block of a stacking result represents the color is white by the HVS when the "or" V of any k of the n rows satisfies that $H(V)$ is less than or equal to $d - \alpha \times m$ for any matrix B_0 in C_0.
2. A block of a stacking result represents the color is black by the HVS when the "or" V of any k of the n rows satisfies that $H(V)$ is more than or equal to d for any matrix B_1 in C_1.
3. For any subset $\{i_1, i_2, \ldots i_q\}$ of $\{1, 2, \ldots, n\}$ with $q < k$, the two collections D_0, D_1 of $q \times m$ matrices obtained by restricting each $n \times m$ matrix in C_0, C_1 to rows i_1, i_2, ..., i_q are indistinguishable in the sense that they contain the same matrices with the same frequencies.

3.1 Cheating in VSS

There are three types of cheating in VSS which are described below:

1. **Cheating a VSS by an MP (CA-1)**:
   As the cheater is an MP, it is possible to use genuine share as a template to construct a set of fake share which are indistinguishable from its genuine share.

The stacking of these FS and S_i (from which FS generated) reveals the fake image of perfect blackness. As the cheaters set a cheating image and the goal is to generate fake transparency and make victim to accept the cheating image, CA-1 is a meaningful cheating.

2. **Cheating a VSS by an MO (CA-2)**:
Malicious Outsider (MO) does not hold any genuine transparency; the MO only knows the transparency construction technique. As MO is the outsider, he does not know the right transparency size for the fake transparency. For this, Hu and Tzeng [7] give one solution, that is, to try all possible transparency sizes. At first, MO chooses a fake image and encode the fake image into two fake transparencies FT_1 and FT_2 with the optimal (2, 2)-VSS. Then enough pairs of fake transparencies $FT_{1,i}$ and $FT_{2,i}$ with various sizes and subpixel distributions are generated, where $1 \leq i \leq r$ for some value of r. Now the stacking of two fake transparency $FT_{1,c}$, $FT_{2,c}$, and T_v (Victim's transparency) shows the fake image for some c, where $1 \leq c \leq r$.
3. **Cheating an EVSS by an MP**:
The VSS in which the shares are meaningful or identifiable to every participant, is called Extended VSS (EVSS). The qualified participant creates the FS from the genuine share by interchanging the black pixels by the white pixels which leads to less contrast of the reconstructed image. The less contrast in reconstructed image will be hard to see the image.The fake image in the stacking of the fake shares has enough contrast against the background since the fake image is recovered in perfect blackness. There are two phases when cheating process happens against a VSS. The phases are: (1) *Fake share construction phase*: In this phase the fake shares are generated. (2) *Image reconstruction phase*: The genuine share and fake share are stacked together and then the fake image appears.

4 Data Hiding Using Weight Matrix

A data hiding scheme for binary images using a key matrix (K) and a weight matrix (W) has been proposed by Tseng et al. [13]. Fan et al. [14] proposed an improved efficient data hiding scheme for grayscale images. In the following, we describe Tseng's data hiding scheme which exploits HVSS. The sender extracts the LSB plane S from the grayscale image SI as the embedding domain and then S is partitioned into nonoverlapping blocks S_i of size $m \times n$. For simplicity, we assume that the size of S is a multiple of $m \times n$. A key matrix K of the same size as S_i is created with a key shared by sender and receiver. Next, an $m \times n$ integer weight matrix W which will be shared by sender and receiver is generated. The criterion of choosing W is that each entry of matrix is randomly assigned a value from the aggregate $\{1, 2, \ldots, 2^r - 1\}$ and each element of aggregate $\{1, 2, \ldots, 2^r - 1\}$ appears at least once in W, where r denotes the number of secret bits that will be embedded into each block S_i. As $2^r - 1 \leq m \times n$ and there are many choices for W, it can first pick $2^r - 1$ elements from W and assign $\{1, 2, \ldots, 2^r - 1\}$ to them. The remaining $m \times n - (2^r - 1)$

elements can be assigned randomly. Next, it will embed two data bits, say b_1b_2 into S_i. Calculate SUM$((S_i \oplus K) \otimes W)$. Where $\oplus$ denotes bitwise EXclusive-OR and $\otimes$ denotes entry-wise multiplication operator. The function of SUM() represents the modular summation of all the entries of matrix $(S_i \oplus K) \otimes W$. Then, calculate weight difference: $d = (b_1, b_2, \ldots, b_r) - \text{SUM}((Si \oplus K) \otimes W) \bmod 2^r$. If d is equal to zero modulo 2^r then S_i is intact; otherwise, modify S_i to S_i' to satisfy the following invariant:

$$\text{SUM}(S_i' \otimes W) = b_1b_2 \bmod 2^r \tag{1}$$

With this invariant, the receiver can derive b_1b_2 by computing SUM$(S_i' \otimes W)$ mod 2^r. Here, we have used this method for share authentication by embedding critical information M in high embedding rate. There exists high-risk security vulnerability in special case, because an attacker will be able to estimate the form of random key matrix and weight matrix by using brute-force attack. In order to overcome the drawbacks of the above scheme, an improved embedding strategy is developed in this paper by changing the weight matrix sequence using the formula $W_{i+1} = (W_i \times 5 \bmod 8) + 1$, where $i = 0, 1, 2, \ldots, n$ number of shares in each level.

5 Proposed Scheme

5.1 Hierarchical Visual Secret Sharing (HVSS)

In this section, we first describe our proposed hierarchical visual secret sharing scheme using steganography and then describe how cheating cannot be mounted in this scheme. Encrypt the secret for different levels. In a hierarchical structure, a user in a security class has access to information items of security classes of lower levels, but not of upper levels. Hierarchical structures are used in many applications including military, government, schools and colleges, private corporations, computer network systems, etc. The block diagram of proposed work is shown in Fig. 1. Here a secret image (SI) can be distributed into n number of secret shares S_i, each of these are unique subset of original secret. Each shareholder of level-1 keeps a copy of his share. Now for each share of level-1, a unique secret image have to be choosen and modified level-1 shares would be generated by OR-ing an original level-1 share with its corresponding unique secret. At the time of stacking the level-2 shares, the corresponding unique secret reveals which was chosen by the parent level-1 share. The participant can cheat other. If the cheater is a Malicious Participant (MP), he uses his genuine share as a template to construct a set of fake shares which are indistinguishable from its genuine share. The stacking of these fake shares and the original share reveals the fake image.The Algorithm-1 is for share generation, Algorithm -2 is for level-1 share modification, and Algorithm-3 for generating fake shares are shown below:

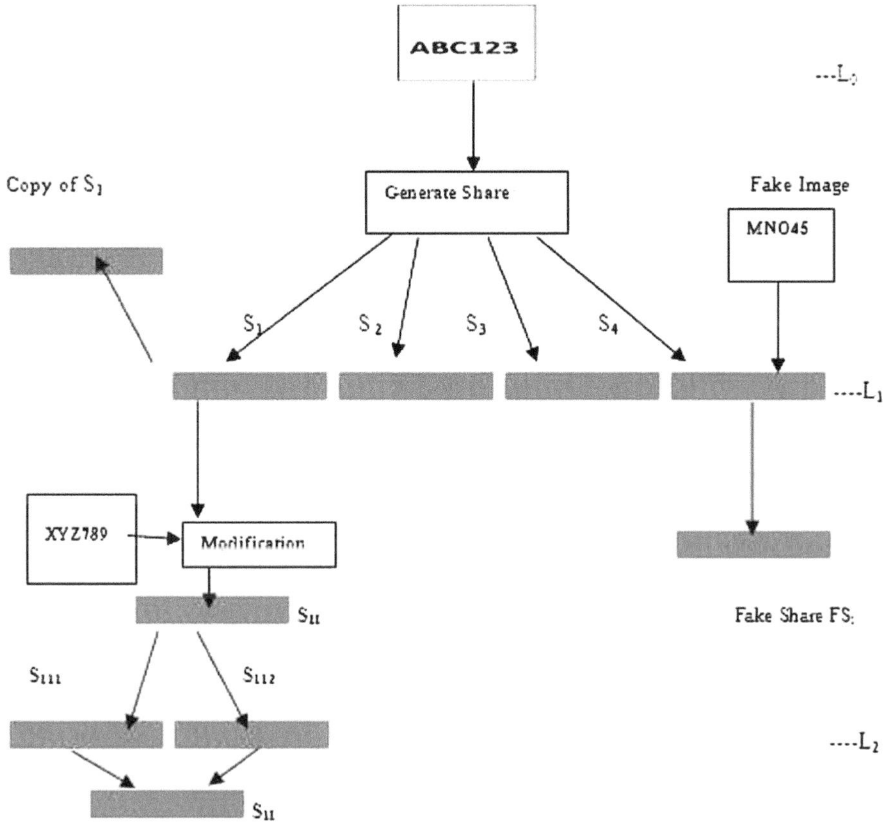

Fig. 1 Block diagram of Hierarchical Visual Secret Sharing (HVSS)

Algorithm-1 (Algorithm for share generation in HVSS):

Step 1: Take a secret image (SI[i][j]) and convert it into binary form.
Step 2: Construct a general matrix GM using formula $J_i^0 = 0^{i-1}10^{k-i}$ for $1 \leq i < k$ and $J_k^0 = 1^{k-1}0$, where k is the number of secret share and $i = 0, 1, 2, \ldots, k$.
Step 3: Choose two matrices C_0 (for white pixels) and C_1 (for black pixels) obtained by permuting the columns of the corresponding matrix GM in all possible ways.
Step 4: Take each pixel from binary image SI[i][j] and generate sub pixel of $k \times k$ matrices.
Step 5: For $i = 1$ to k

```
        For x = 1 to length(SI[i][j])
          For y = 1 to length(SI[i][j])
            If (SI[x][y] = 0),
              then select row of matrix C0 (Starting from row 1)
```

Else
select row of matrix C_1 (Starting from row 1)
EndIf
Share $S(i) =$ Split pixel according to matrix value of C_0 or C_1
$i = i + 1$;
End For
End For
End For

Step 6: End

Algorithm-2 (Algorithm for Level-1 share modification):

Step 1: Input original share S_i, and choose another secret image SI_i.
Step 2: Assume that each pixel of S_i has m black and n white subpixels.
Step 3: For each white pixel of the secret image SI_i, copy the corresponding subpixels of the pixel in S_i to modified share S_i'.
Step 4: For each black pixel of the secret image SI_i, randomly assign m black and n white subpixels to modified share S_i' such that the pixel in the stacking of modified share S_i' with original share S_i is perfect black.
Step 5: Generate modified share S_i'.
Step 6: End

Algorithm-3 (Algorithm for generating fake share):

Step 1: Input original share S_i, and a fake image FI, which has the same size of secret image SI.
Step 2: Assume that each pixel of S_i has x black and y white subpixels.
Step 3: For each white pixel of the fake image FI, copy the corresponding subpixels of the pixel in S_i to fake share FS.
Step 4: For each black pixel of the fake image FI, randomly assign x black and y white subpixels to fake share FS such that the pixel in the stacking of fake share FS with original share S_i is perfect black.
Step 5: Generate fake share FS.
Step 6: End

5.2 Cheating prevention in hierarchical visual secret sharing (HVSS)

A Steganographic approach is proposed to prevent cheating by detecting fake share in HVSS scheme and then revealed secret image from original share. The attacks are to reveal fake images which cheat honest participants. In this approach, we have used a key matrix K and a weight matrix W to hide critical information M into the share on each level of HVSS. The basic ideas are: (i) to use a different binary operator Exclusive-OR to protect the key matrix K from being compromised, and

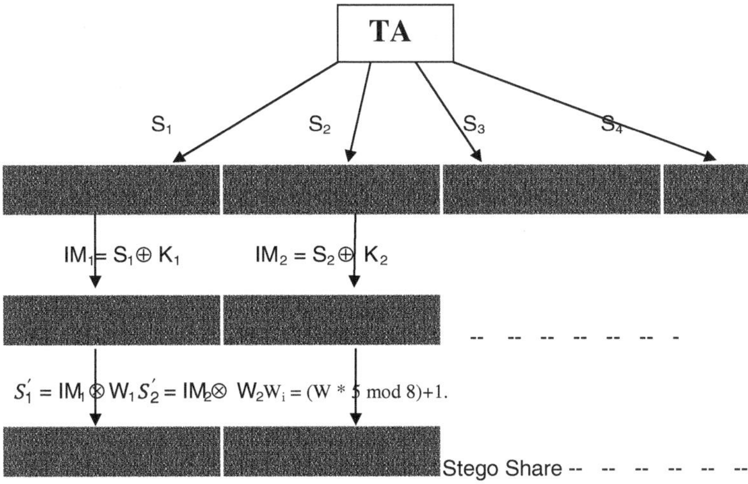

Fig. 2 Block Diagram of generating stego share in HVSS using weight matrix

(ii) to use a weight matrix W to increase the data hiding rate while maintaining high quality of the host image. The block diagram of the scheme is depicted in Fig. 2. The share generator or Trusted Authority (TA) generates different key matrix to keep track of the level of each share which is predefined and generates different weight matrix W_i for each share using the formula $W_{i+1} = (W_i \times 5 \bmod 8) + 1$ and embedded the critical information (M) into each share with different weight matrix. After generating shares (S_i), TA generates an Intermediate matrix (IM) by EXclusive-ORing between the share and key matrix (K), $IM_i = S_i \oplus K_i$ and keep a copy of the Intermediate matrix (IM). The modified share (S_i') is generated by changing 1 bit modification from S_i (using Algorithm-4). TA keeps a table of Intermediate matrix (IM), its corresponding key matrix (K) and weight matrix (W). When any share holder wants to know whether a share is original or not, he sends a request to the TA to check both the shares. Then TA Exclusive-OR the modified share with all predefined key matrix and check with all intermediate matrix in one bit tolerant process. In this way TA check the modified share to decide the corresponding weight matrix. Now using the corresponding weight matrix, the embedding message from the share is extracted. If TA retrieves the original embedded message from any share, the share will be accepted; otherwise, it will be rejected (Fig. 3).

Algorithm-4 (Embedding Process):

Step 1: Consider original share S_i, a critical message M, a 3×3 key matrix K, and a 3×3 Weight matrix W.

Step 2: Divide each character of message into two (4 bits each) parts and store into message msg.

Step 3: Copy S into T.

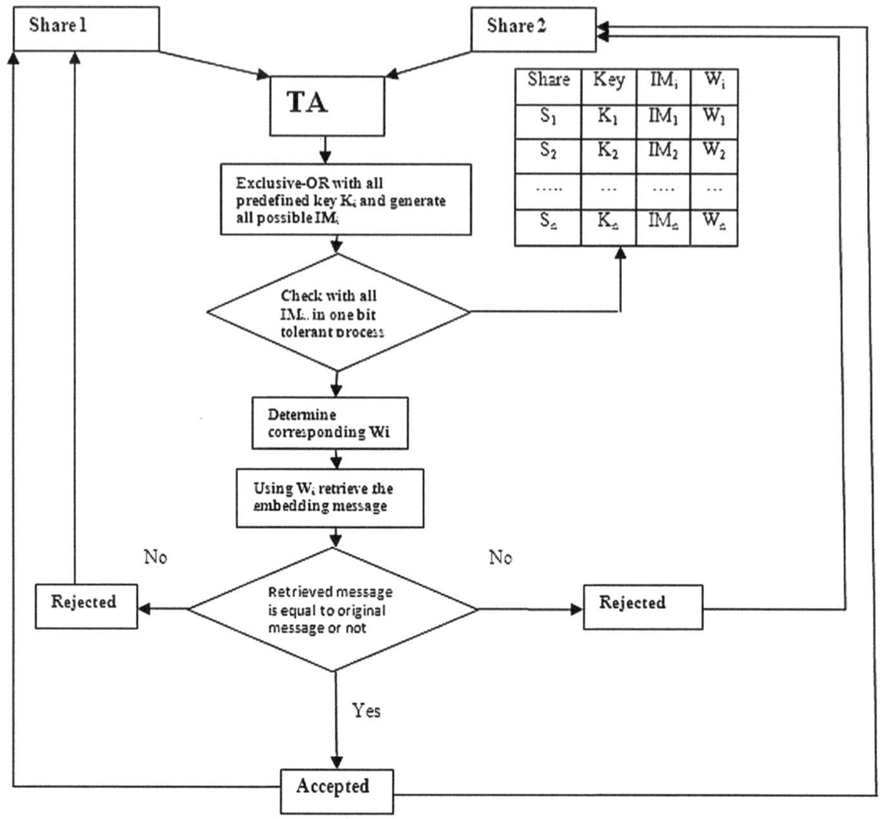

Fig. 3 Block diagram of cheating prevention in HVSS using weight matrix

Step 4: To compute Intermediate matrix (IM), consider a 3×3 matrix of original share (S) that is, S_i and EXclusive-OR each pixel value with corresponding position value of 3×3 key matrix. Hence IM_i is computed as $IM_i = S_i \oplus K_i$.

Step 5: Keep a copy of IM_i.

Step 6: Multiply each pixel value of IM_i with corresponding position value of 3×3 weight Matrix W.

Step 7: Calculate modulo value of $SUM(IMi \otimes W_i)(\text{mod}2^r)$, where $r = 4$.

Step 8: Compute the weight difference: $d = (b_1, b_2, \ldots, b_r) - SUM(IM_i \bigotimes W_i)$.

Step 9: If $d = 0$, keep F_i intact, otherwise complement dth position of W_i in S_i.

Step 10: Choose next 3×3 matrix of S and repeat until entire message is embedded.

Step 11: End

Algorithm-5 (Extracting Process):

Step 1: Consider Stego share T, all predefined 3×3 key matrix K and all copy of Intermediate matrix (IM_i).
Step 2: Consider 3×3 matrix of T and EXclusive-OR with all predefined key matrix (K_i) and check with all intermediate matrix (IM_i) in one bit tolerant process and determine weight matrix (W_i).
Step 3: Multiply each pixel value of T with corresponding position value of W_i.
Step 4: Calculate Sum of each product.
Step 5: Extract message by Sum (mod2^r).
Step 6: Concatenate two messages into one and then convert ASCII value of each character into string.
Step 7: End

Let us consider the following original share S (Ref. Table 1).

TA generates several key matrices K_i ($i = 1, 2, \ldots, n$) to secure the system. Use K_1 for share 1, K_2 for share 2, ... K_n for share n. Key Matrices are predefined, like (Figs. 4 and 5)

$$K_1 = \begin{bmatrix} 1\,0\,0 \\ 1\,0\,1 \\ 1\,1\,0 \end{bmatrix}, \quad K_2 = \begin{bmatrix} 1\,0\,1 \\ 0\,0\,1 \\ 0\,1\,0 \end{bmatrix}, \quad K_3 = \begin{bmatrix} 1\,1\,0 \\ 1\,0\,0 \\ 0\,1\,0 \end{bmatrix} \quad \text{and so on.}$$

Weight matrices are generated by the formula $W_{i+1} = (W_i \times 5 \bmod 8) + 1$, $i = 0, 1, 2, \ldots, n$. We use W_1 for share 1, W_2 for share 2, ..., W_n for share n, where

$$W_1 = \begin{bmatrix} 1\,2\,3 \\ 4\,5\,6 \\ 7\,8\,2 \end{bmatrix}, \quad W_2 = \begin{bmatrix} 6\,3\,8 \\ 5\,2\,7 \\ 4\,1\,3 \end{bmatrix}, \quad W_3 = \begin{bmatrix} 7\,8\,1 \\ 2\,3\,4 \\ 5\,6\,8 \end{bmatrix} \quad \text{and so on.}$$

Message Embedding:

Suppose, we want to embed a string of two characters, say 'is', where ASCII values of 'i' and 's' are 105 and 115, respectively.

Now, binary representations of 105 and 115 by 8-bit word are 01101001 and 01110011. In 01101001, first 4-bit represents 6 in decimal and next 4-bit represents 9. Similarly, in 01110011, first 4-bit represents 7 in decimal and next 4-bit represents

Table 1 Binarization of host image F

1	0	1	1	0	0
1	1	0	1	1	1
0	0	1	0	0	0
0	1	1	1	0	1
1	0	0	0	1	1
0	1	0	1	1	0

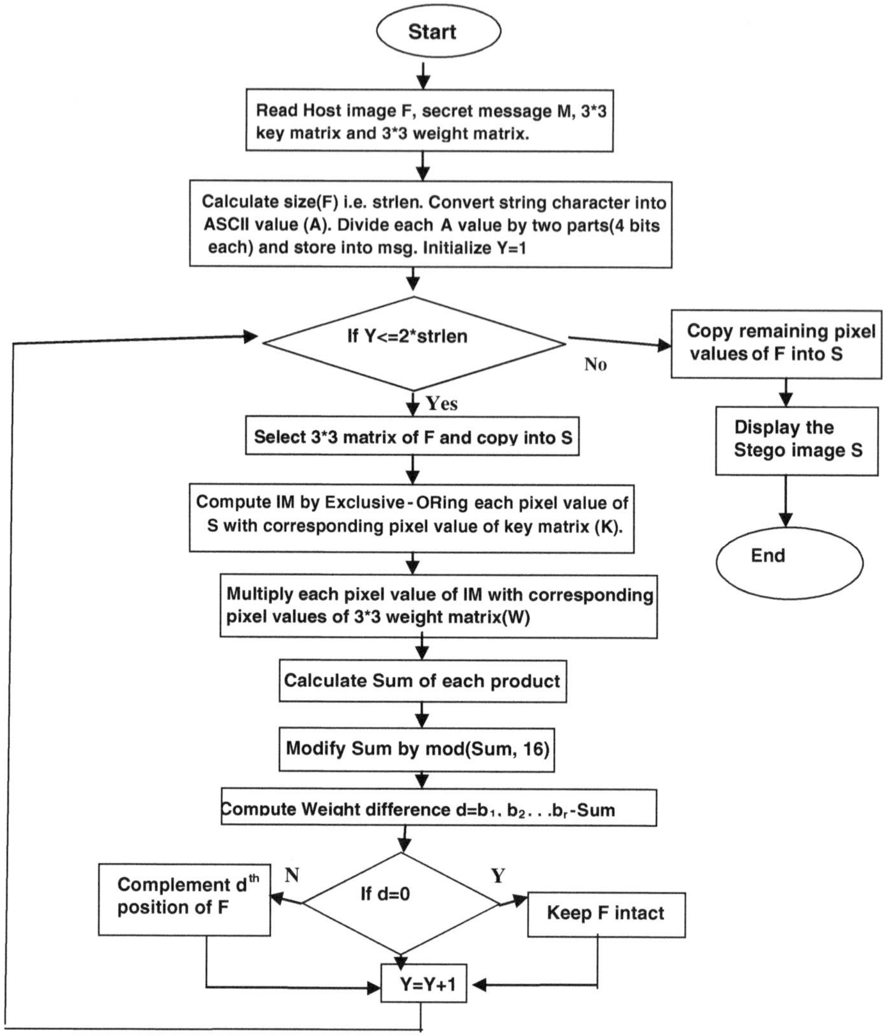

Fig. 4 Flow chart of embedding process

3. So message, $msg = 6, 9, 7, 3$. For simplicity, we consider the embed string, say 'i'. To generate Intermediate matrix IM_1, we choose first 3×3 matrix of S_1 and EXclusive-ORing by key matrix K_1. That is,

$$\mathrm{IM}_1 = F_1 \oplus K_1 = \begin{bmatrix} 1 & 0 & 1 \\ 1 & 1 & 0 \\ 0 & 0 & 1 \end{bmatrix} \oplus \begin{bmatrix} 1 & 0 & 0 \\ 1 & 0 & 1 \\ 1 & 1 & 0 \end{bmatrix} = \begin{bmatrix} 0 & 0 & 1 \\ 0 & 1 & 1 \\ 1 & 1 & 1 \end{bmatrix}$$

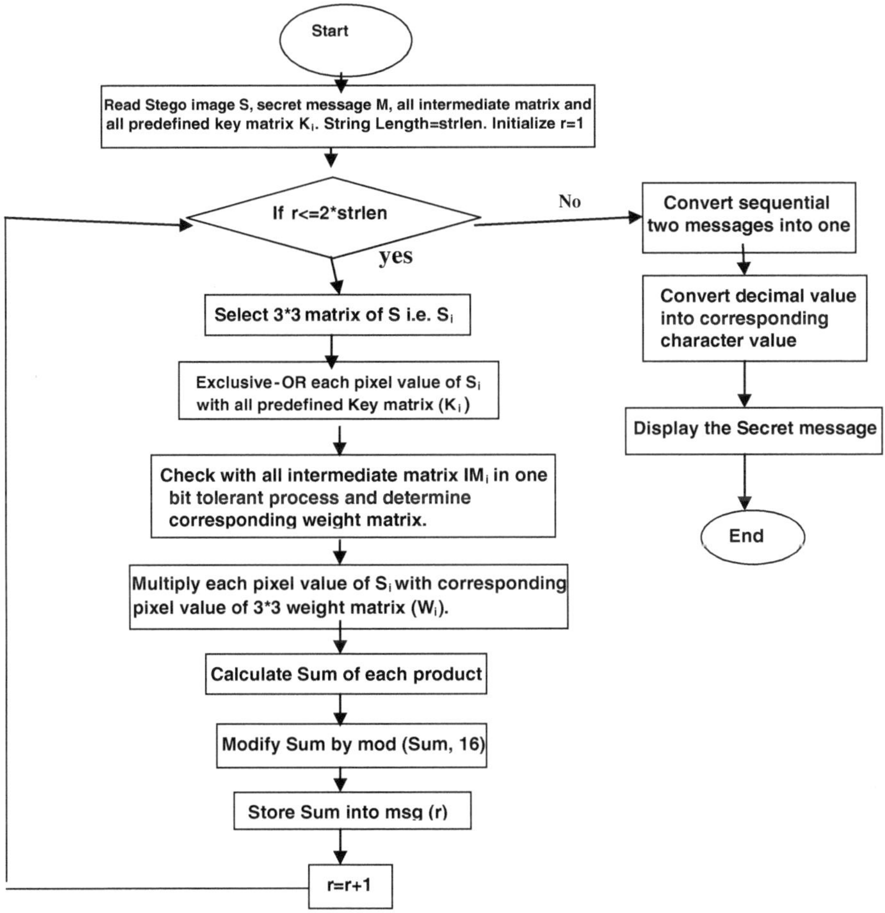

Fig. 5 Flow chart of extracting process

Now, $(\mathrm{IM}_1 \otimes W_1)$ should be calculated as

$$\begin{bmatrix} 0 & 0 & 1 \\ 0 & 1 & 1 \\ 1 & 1 & 1 \end{bmatrix} \otimes \begin{bmatrix} 1 & 2 & 3 \\ 4 & 5 & 6 \\ 7 & 8 & 2 \end{bmatrix} = \begin{bmatrix} 0 & 0 & 3 \\ 0 & 5 & 6 \\ 7 & 8 & 2 \end{bmatrix}$$

And so as $\mathrm{SUM}(\mathrm{IM}_1 \otimes W_1) = 31$.
Now, d can be calculated as

$$\begin{aligned} d &= (b_1, b_2, \ldots, b_r) - \mathrm{SUM}(IM_i \otimes W_i) \bmod 2^r, \text{ where } r = 4 \\ &= 6 - 31 \bmod 16 \\ &= 7 \end{aligned}$$

So we modify seventh position of S_1. Modified S_1, that is, S_1' will be

$$\begin{bmatrix} 1 & 0 & 1 \\ 1 & 1 & 0 \\ 1 & 0 & 1 \end{bmatrix}$$

Similarly, we can choose second 3×3 matrix of S_i and EXclusive-ORing by key matrix K_1. Then we can get IM_2 as

$$\mathrm{IM}_2 = S_2 \oplus K_1 = \begin{bmatrix} 1 & 0 & 0 \\ 1 & 1 & 1 \\ 0 & 0 & 0 \end{bmatrix} \oplus \begin{bmatrix} 1 & 0 & 0 \\ 1 & 0 & 1 \\ 1 & 1 & 0 \end{bmatrix} = \begin{bmatrix} 0 & 0 & 0 \\ 0 & 1 & 0 \\ 1 & 1 & 0 \end{bmatrix}$$

Now $(\mathrm{IM}_2 \otimes W_1)$ should be calculated as

$$\begin{bmatrix} 0 & 0 & 0 \\ 0 & 1 & 0 \\ 1 & 1 & 0 \end{bmatrix} \otimes \begin{bmatrix} 1 & 2 & 3 \\ 4 & 5 & 6 \\ 7 & 8 & 2 \end{bmatrix} = \begin{bmatrix} 0 & 0 & 0 \\ 0 & 5 & 0 \\ 7 & 8 & 0 \end{bmatrix}.$$

Then we calculate $\mathrm{SUM}(\mathrm{IM}_2 \otimes W_1) = 20$.

$$\begin{aligned} d &= (b_1, b_2, \ldots, b_r) - \mathrm{SUM}(\mathrm{IM}i \otimes W_i) \bmod 2^r \\ &= 9 - 20 \bmod 16 \\ &= 5 \end{aligned}$$

So, we modify fifth position of S_2. Modified S_2, that is, S_2' will be

$$\begin{bmatrix} 1 & 0 & 0 \\ 1 & 0 & 1 \\ 0 & 0 & 0 \end{bmatrix}$$

Message Extraction:

Receiver have to know the corresponding weight matrix W to retrieve the secret message, but receiver has no information about the weight matrix, which is used for that particular share. So, one has to Ex-OR all predefined key matrix (K_i) with modified share S_i' and compare with all intermediate matrix (IM_i) in one bit tolerant. As soon as one gets proper intermediate matrix, then the corresponding weight matrix will be determined.

Now, we calculate Ex-OR between predefined key matrix (K_i) and modified share F_i' as follows:

$$S_1' \oplus K_1 = \begin{bmatrix} 1 & 0 & 1 \\ 1 & 1 & 0 \\ 1 & 0 & 1 \end{bmatrix} \oplus \begin{bmatrix} 1 & 0 & 0 \\ 1 & 0 & 1 \\ 1 & 1 & 0 \end{bmatrix} = \begin{bmatrix} 0 & 0 & 1 \\ 0 & 1 & 1 \\ 0 & 1 & 1 \end{bmatrix}$$

$$S_1' \oplus K_2 = \begin{bmatrix} 1\,0\,1 \\ 1\,1\,0 \\ 1\,0\,1 \end{bmatrix} \oplus \begin{bmatrix} 1\,0\,1 \\ 0\,0\,1 \\ 0\,1\,0 \end{bmatrix} = \begin{bmatrix} 0\,0\,1 \\ 1\,1\,1 \\ 1\,1\,1 \end{bmatrix}$$

$$S_1' \oplus K_3 = \begin{bmatrix} 1\,0\,1 \\ 1\,1\,0 \\ 1\,0\,1 \end{bmatrix} \oplus \begin{bmatrix} 1\,1\,0 \\ 1\,0\,0 \\ 0\,1\,0 \end{bmatrix} = \begin{bmatrix} 0\,1\,1 \\ 0\,1\,0 \\ 1\,1\,1 \end{bmatrix}$$

Now, one can check with all intermediate matrix (IM_i) in one bit tolerant process and determine the weight matrix W_1.

$$\text{SUM}(S_1' \otimes W_1) \bmod 2^r = \text{SUM} \begin{bmatrix} 1\,0\,1 \\ 1\,1\,0 \\ 1\,0\,1 \end{bmatrix} \otimes \begin{bmatrix} 1\,2\,3 \\ 4\,5\,6 \\ 7\,8\,2 \end{bmatrix} \bmod 16 = \begin{bmatrix} 1\,0\,3 \\ 4\,5\,0 \\ 7\,0\,2 \end{bmatrix}$$

$$= 22 \bmod 16 = 6$$

To extract next message we consider next block S_2'. Ex-OR all predefined key matrix (K_i) with the modified share S_i' as follows:

$$S_2' \oplus K_1 = \begin{bmatrix} 1\,0\,0 \\ 1\,0\,1 \\ 0\,0\,0 \end{bmatrix} \oplus \begin{bmatrix} 1\,0\,0 \\ 1\,0\,1 \\ 1\,1\,0 \end{bmatrix} = \begin{bmatrix} 0\,0\,0 \\ 0\,0\,0 \\ 1\,1\,0 \end{bmatrix}$$

$$S_2' \oplus K_2 = \begin{bmatrix} 1\,0\,0 \\ 1\,0\,1 \\ 0\,0\,0 \end{bmatrix} \oplus \begin{bmatrix} 1\,0\,1 \\ 0\,0\,1 \\ 0\,1\,0 \end{bmatrix} = \begin{bmatrix} 0\,0\,1 \\ 1\,0\,0 \\ 0\,1\,0 \end{bmatrix}$$

$$S_2' \oplus K_3 = \begin{bmatrix} 1\,0\,0 \\ 1\,0\,1 \\ 0\,0\,0 \end{bmatrix} \oplus \begin{bmatrix} 1\,1\,0 \\ 1\,0\,0 \\ 0\,1\,0 \end{bmatrix} = \begin{bmatrix} 0\,1\,0 \\ 0\,0\,1 \\ 0\,1\,0 \end{bmatrix}$$

Now, we check with all intermediate matrix (IM_i) in one bit tolerant process and determine the weight matrix W_1.

$$\text{SUM}(S_2' \otimes W_1) \bmod 2^r = \text{SUM} \begin{bmatrix} 1\,0\,0 \\ 1\,0\,1 \\ 0\,0\,0 \end{bmatrix} \otimes \begin{bmatrix} 1\,2\,3 \\ 4\,5\,6 \\ 7\,8\,2 \end{bmatrix} \bmod 16 = \begin{bmatrix} 1\,0\,0 \\ 4\,0\,6 \\ 0\,0\,0 \end{bmatrix}$$

$$= 11 \bmod 16 = 9$$

Now, one concatenates binary values of two messages 6(=0110 in binary) and 9 (=1001 in binary), which will be 01101001 in binary and equivalent in decimal will be 105. Hence, receiver can convert 105 into corresponding character '*i*'.

Lemma 5.1 *Maximum possible weight matrix of r bit will be* $(2^r+1)!-2^r$.

Proof Consider r bit message need to embed, the possible combinations of r bit is 2^r. As per requirement weight matrix W will be $(1, 2, \ldots, 2^r\text{-}1)$. Number of values within W is 2^r+1. The all possible combinations of W will be $(2^r+1)!-2^r$. □

Lemma 5.2 *Maximum possible child of any node of HVSS will be* $(2^r+1)!-2^r$.

Proof As per requirement of HVSS, the possible number of weight matrix is $(2^r+1)!-2^r$. Every child will carry information using different weight matrix of same parent. So, maximum number of possible children will be $(2^r+1)!-2^r$. □

Lemma 5.3 *Maximum possible size of share for each level will be* $(2^r+1)!-2^r$.

Proof As per lemma 2, maximum number of children will be $(2^r+1)!-2^r$. To generate share one can use Base matrix B_0. Here for $(2^r+1)!-2^r$ number of child share one may have to use is $(2^r+1)!-2^r \times (2^r+1)!-2^r$ Base matrix. So possible maximum size of share will increase $(2^r+1)!-2^r$. □

Lemma 5.4 *Maximum possible key matrix will be* $2^{r \times r}$.

Proof Key matrix is used to enhance security of HVSS. For $r+1$ bit message embedding, we use key matrix (K) of $r \times r$ size. The key matrix have only 1 or 0. So, maximum possible key matrix will be $2^{r \times r}$.

Lemma 5.5 *Number of Intermediate Matrix to be stored at Trusted Authority is equal to the length of message bit.*

Lemma 5.6 *Predefined Key matrix* (K) *will be chosen depending upon the variation of bit which may be different* $H \geq 2$ *(*H *= Hamming Distance).*

The main aim of HVSS is secret message sharing through generating and distributing share which is perfect secure technique. In this scheme, key matrix K and weight matrix W are necessary to protect share from intruders. This K is predefined and W is also predefined for each share. This K and W and intermediate matrix IM are stored as a table in Trusted Authority. None can get easily. So we conclude that our scheme is secure. If one can use size invariant visual cryptography then the main problem will be the increasing share size at a big rate in each level, can be removed.

6 Experimental Results and Discussion

We consider an original image shown in Fig. 6 and the corresponding four shares or transparencies are shown in Fig. 7 using Algorithm-1. The transparencies are usually shared by four participants so that each participant is expected to keep one transparency. The secret image can be retrieved if k shares out of k shares are stacked together as it is a (k, k)-scheme, which are shown in Fig. 7 for $k = 4$. However, the

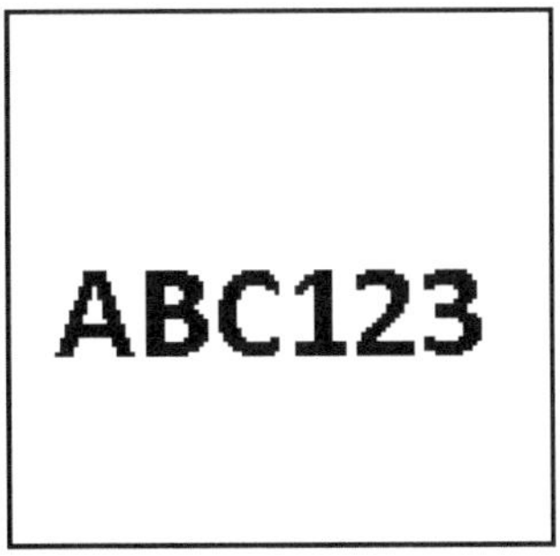

Fig. 6 Original image

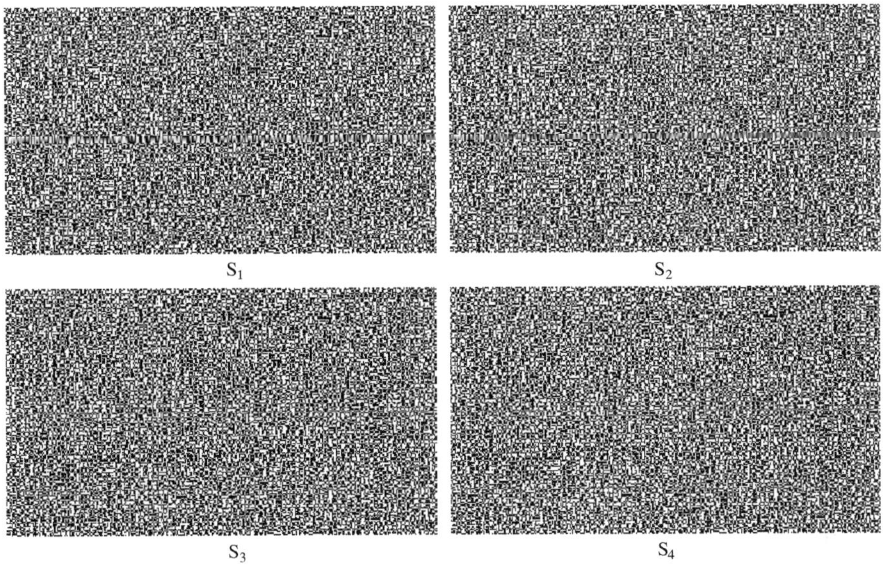

Fig. 7 Generation of share S_1, S_2, S_3, S_4

secret image is totally invisible if fewer than k transparencies are stacked is shown in Figs. 8 and 9. By stacking all the shares, one can get the original image in perfect black shown in Fig. 10. It is implemented in NetBeans IDE 7.3.1 (Figs. 11, 12, 13, 14, 15, 16).

Consider share from level-1. Suppose S_1 and another secret image I_1. Now generate the modified share S_{11} by OR-ing S1 with SI_1. S_{11} is distributed into two shares S_{111}, S_{112}. By overlapping the two shares, we can get back S_{11} as shown below.

Cheating in Hierarchical Visual Secret Sharing (HVSS):
A malicious participant (MP) may cheat by creating a FS by taking another fake image (FI) shown in Fig. 17 and giving it to other participant when asked for the

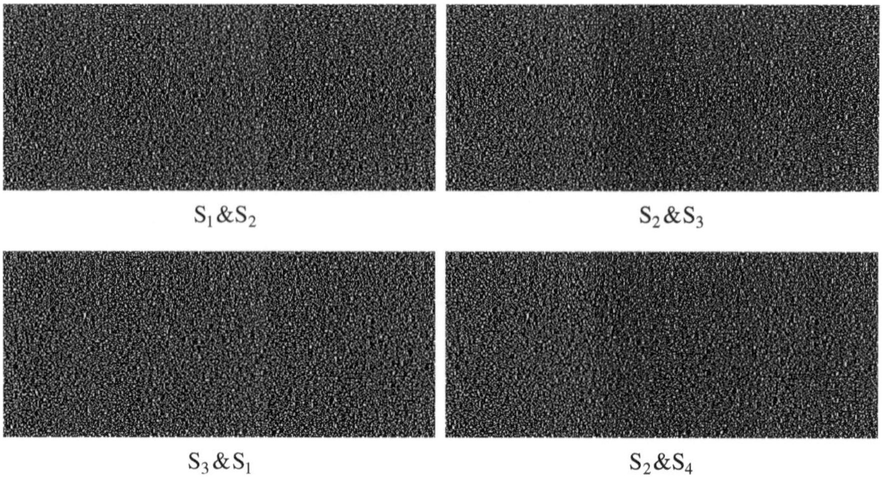

Fig. 8 Stacking of two share S_1 & S_2, S_2 & S_3, S_3 & S_1 and S_2 & S_4, no visual information can be retrieved

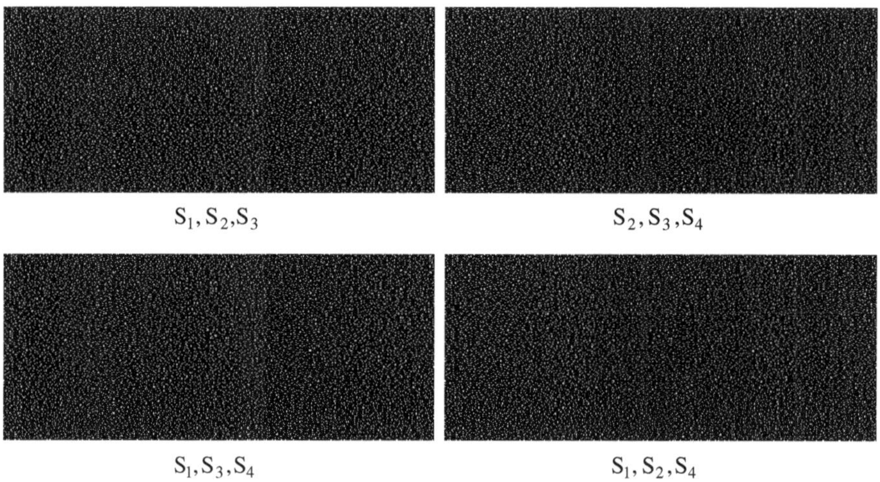

Fig. 9 Stacking of any three shares. Visual information cannot be retrieved because it is a (k, k)-scheme

share. The FS is created with the help of the original share S_i using Algorithm-2 shown in Fig. 18. It would be hard to detect it with a normal look that it is a FS_1 and not the original one.

Overlapped result of the FS_1 with the share S_1 is shown in Fig. 19 which only shows fake image. Also overlapping the FS_1 with all other shares including original share S1 are shown in Fig. 20 which only shows fake image. In Fig. 21, we present

Fig. 10 Result of stacking all shares

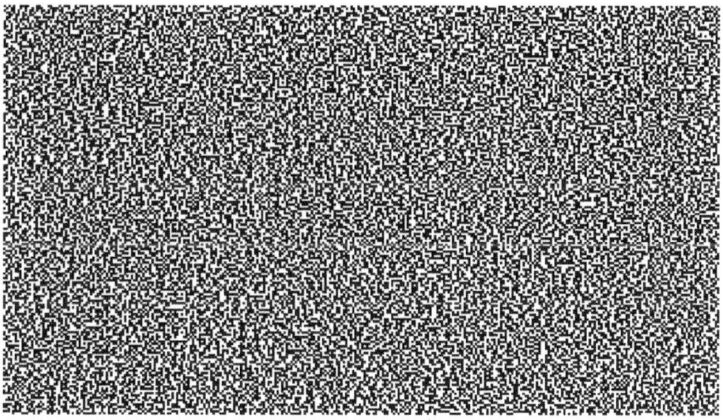

Fig. 11 S_1

Fig. 12 I_1

Fig. 13 Modified share S_i'

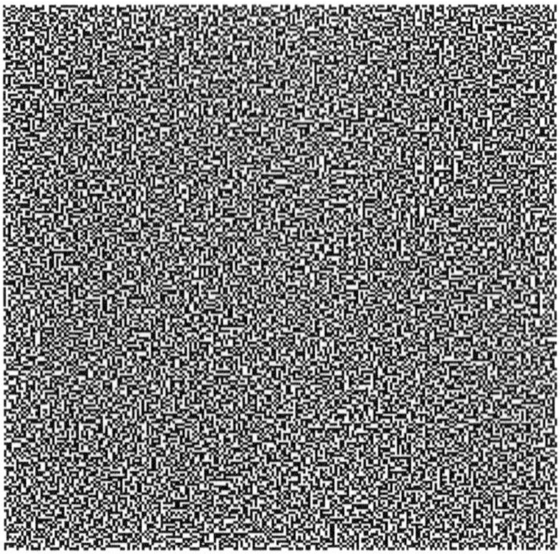

Fig. 14 S'_{11}

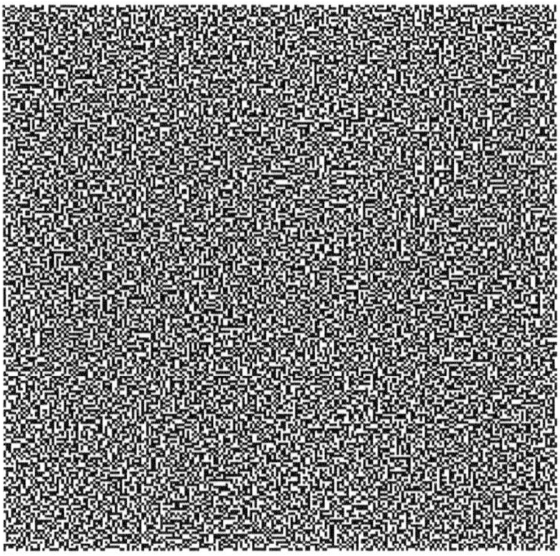

Fig. 15 S'_{12}

the result of stacking of fake share with any one share excluding S_1. When stack FS_1 with all the shares excluding the S_1, one can get both the images in an overlapped manner which will create confusion, called Partial Cheating. This is known as partial cheating as it creates a kind of confusion between the participants about the original image.

It is noted that the critical message M is embedded into the share image in sender side, and then in the receiver side, the critical message M which was embedded can be extracted. If two messages are equal then the image is original, it can be accepted; otherwise, it will be rejected. In this way, cheating can be prevented in our HVSS scheme using data hiding techniques using matrix embedding method.

Fig. 16 Result of overlapping $S'_{11} S'_{12}$

Fig. 17 Fake image for generation of FS

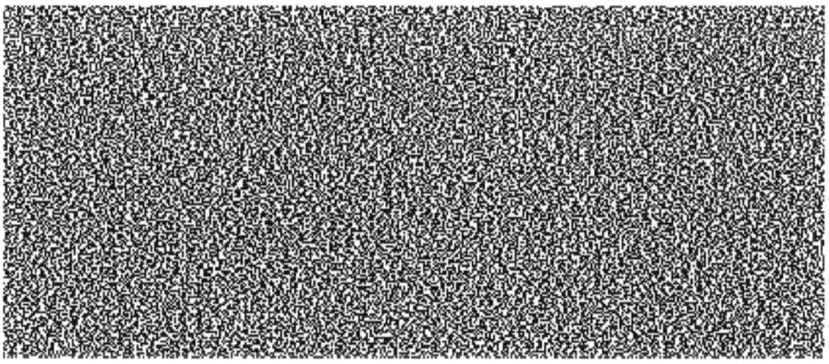

Fig. 18 Generation of Fake Share using Algorithm-2. (by Share S_1)

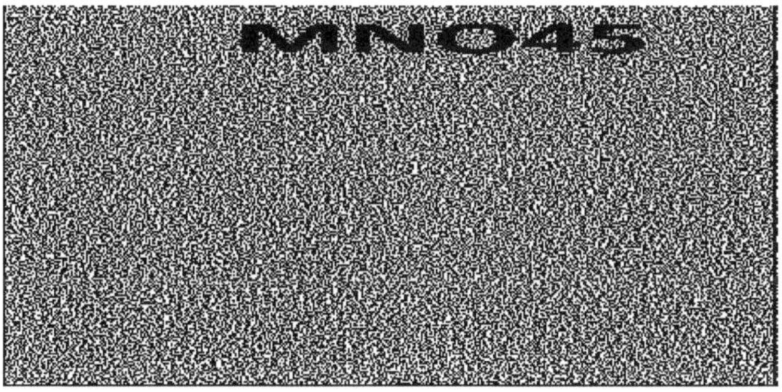

Fig. 19 Stacking of Fake Share and S_1, generate fake image

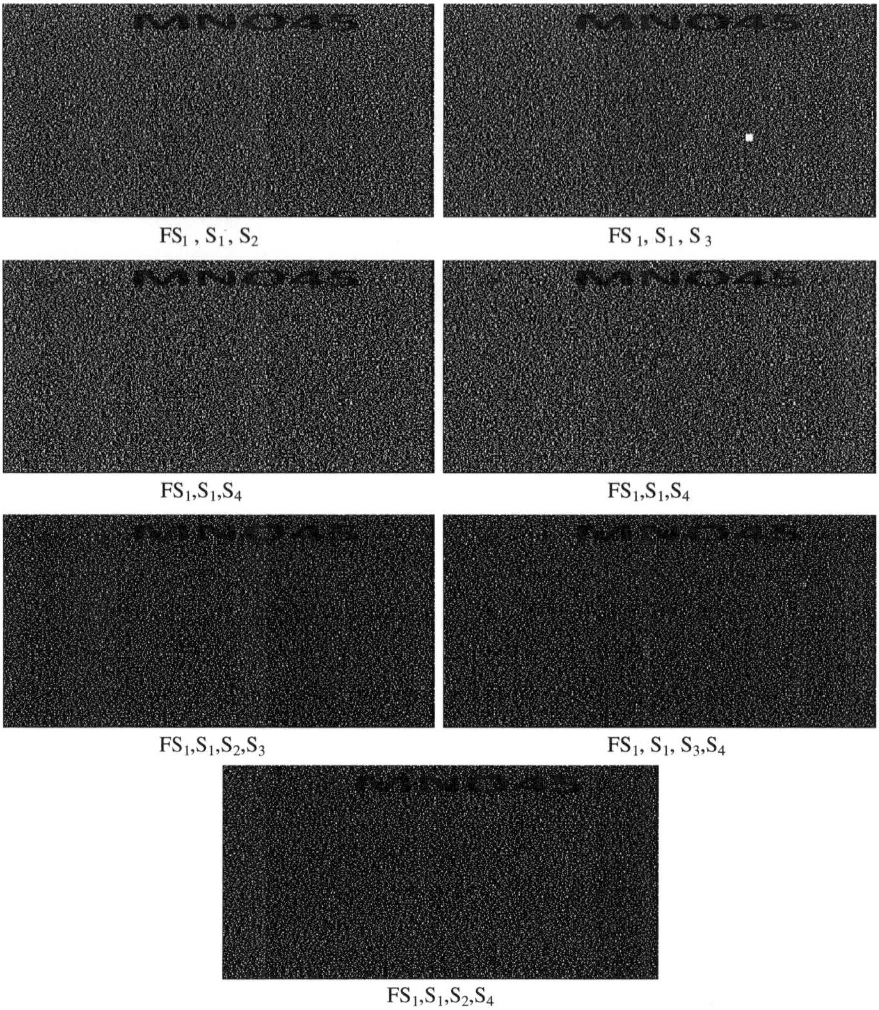

Fig. 20 Overlapping the Fake Share with all other shares Including original share (S_1) which shown fake image

Distortion is measured by means of two parameters namely, Mean Square Error (MSE) and Peak Signal-to-Noise Ratio (PSNR). The MSE is calculated using (2),

$$\mathrm{MSE} = \frac{1}{MN}\sum_{i=1}^{M}\sum_{j=1}^{N}[X(i, j) - Y(i, j)] \tag{2}$$

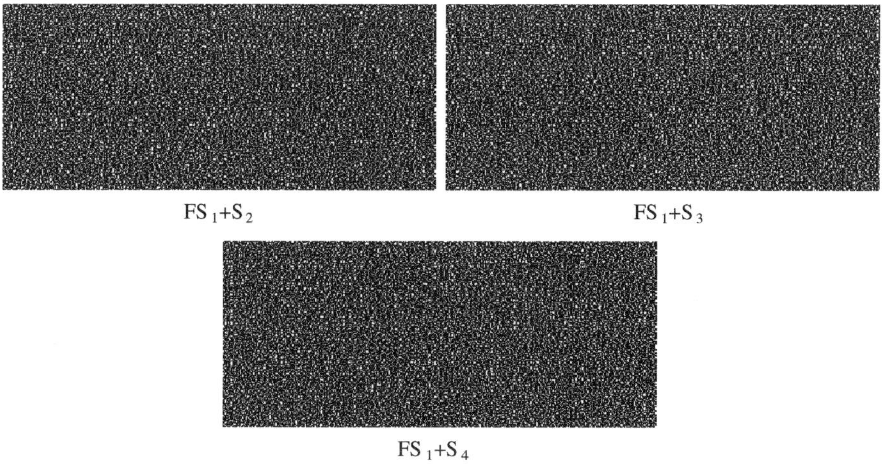

Fig. 21 Overlapping the fake share with all other shares excluding original share (S_1) which shown overlapped image for partial cheating. (Here we using only two share, so no information can be retrieved)

where M and N denote the total number of pixels in the horizontal and the vertical dimensions of the image. $X(i, j)$ represents the pixels in the original image and $Y(i, j)$ represents the pixels of the Stego image (Fig. 22).

Fig. 22 Overlapping the Fake Share with all other shares excluding original share (S_1) which shown overlapped image for partial cheating

The PSNR is calculated using the Eq. 3,

$$\mathrm{PSNR} = 10\log_{10}\left(\frac{I_max^2}{\mathrm{MSE}}\right) \mathrm{d}B, \quad (3)$$

where $I_{\max}$ is the intensity value of each pixel. The analysis in terms of PSNR of original share and Stego share has given promising result. The PSNR of Stego shares with respect to number of bits in secret message are shown in Table 2 and corresponding graph shown in Fig. 23. It is found that when the number of bits in secret message is increasing, the PSNR of Stego share is decreasing. The PSNR varies from 88.7722 to 69.1343 when the number of bits in secret message varies from 1 to 100 for share size of 93 × 239 pixel and the PSNR varies from 94.1617 to 74.4769 when the number of bits in secret message varies from 1 to 100 for share size of 135 × 198 pixel.

To test the security in our proposed method, we have calculated relative entropy (the differences) between the probability distributions of the original share and the Stego share has been calculated by (4). Let p_m and q_n be probability measures for original share, Mo and Stego share, Ns, respectively. The relative entropy distance $D(Ns||Mo)$ (also known as Kullback–Leibler distance) is defined as follows:

$$D(Ns||Mo) = \sum q_n(x) log \frac{q_n(x)}{p_m(x)}. \quad (4)$$

When relative entropy between two probability distribution functions is zero then the system is perfectly secure. $D(Ns||Mo)$ is a nonnegative continuous function and

Table 2 PSNR versus embedded message

Share size	Message length	PSNR
S_1 (93 × 239)	1	88.7722
	5	81.7825
	10	78.9950
	20	76.2195
	50	72.2886
	80	71.9850
	100	69.1343
S_2 (135 × 198)	1	94.1617
	5	88.1411
	10	84.8675
	20	81.6090
	50	77.6296
	80	75.7091
	100	74.4769

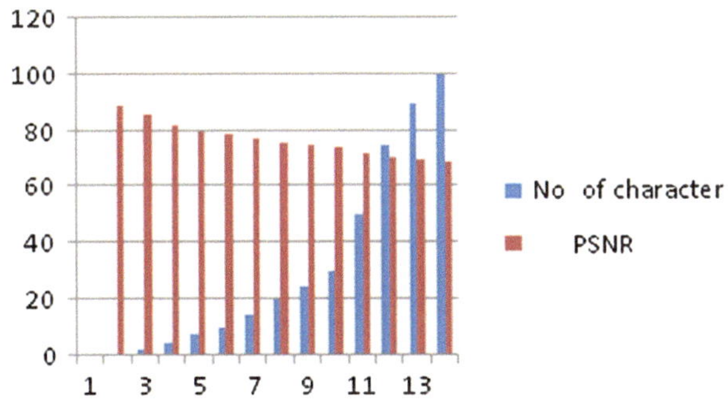

Fig. 23 Comparison graph of number of character versus PSNR

equals to zero if and only if p_m and q_n coincide. Thus $D(Ns||Mo)$ can be naturally viewed as a distance between the measures p_m and q_n.

Relative entropy of the probability distribution of the original share and the Stego share varies depending upon number of character of secret message. In our experiment, it is shown that when the number of characters in the secret message is increasing, the relative entropy in Stego share is also increasing which is shown in Table 3 and the corresponding graph is shown in Fig. 24.

Tables 4 and 5 show the comparison of our proposed scheme for binary share with Guo et al.'s scheme [13] and Pakniat et al.'s scheme [14] in term of average PSNR values.

From Table 4, it is concluded that our proposed scheme has higher PSNR values than previous schemes. Here we have used share as binary image, as a cover image, but Pakniat et al.'s scheme uses grayscale image. The comparison of Pakniat et al.'s scheme with our proposed scheme with respect to level-1, level-2, and level-3 shares has been shown in Table 5 and it is clear that our scheme is superior in terms of PSNR compared to existing techniques.

Table 3 Relative entropy between the probability distribution of the original share and the Stego share

	No. of character	Entropy of original share	Entropy of stego share	Entropy difference
S_1 (93 × 239)	1	7.4719	7.4719	0.0000
	2	7.4719	7.4722	0.0003
	5	7.4719	7.4724	0.0005
	10	7.4719	7.4728	0.0009
	15	7.4719	7.4730	0.0011
	25	7.4719	7.4735	0.0016
	50	7.4719	7.4746	0.0027
	80	7.4719	7.4746	0.0027
	100	7.4719	7.4747	0.0028

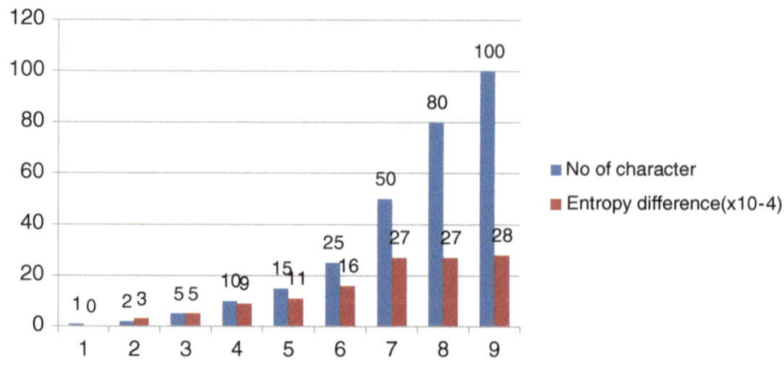

Fig. 24 Comparison graph between number of character and Entropy difference (in 10^{-4})

7 Conclusion

In this paper, we have proposed a new protocol for hierarchical visual secret sharing using Steganography. We have shown that cheating cannot be mounted in our scheme. We have shown that in our scheme, PSNR varies from 88.7722 to 69.1343 when the number of characters varies from 1 to 100 of share size is 93×239 pixel. Relative entropy of the probability distribution of the original share and the Stego share varies depending upon number of character of secret message. In our experiment, it is shown that when the number of characters in the secret message is increasing then the relative entropy in Stego share is also increasing. No verification share is required to prevent the cheating in our scheme. The relative entropy of the probability distribution of the original share and Stego share is minimum which implies that our system is assumed to be more secure. In comparison with Pakniat et al.'s scheme with our proposed scheme in terms of PSNR, our scheme gives 78.95. The size of the share will increase proportionally with the level number in HVSS. So, one can develop size invariant HVSS. Further, we have compared our scheme with existing hierarchical secret sharing schemes and it showed that our scheme is better than the existing schemes.

Table 4 Comparisons of average PSNR between the proposed scheme for level-1 share and Guo et al.'s scheme and Pakniat et al.'s scheme for grayscale image

Scheme	Average PSNR
Guo et al.'s scheme (grayscale image)	38.19
Pakniat et al.'s scheme (grayscale image)	51.23
Proposed scheme (binary share)	78.953

Table 5 Comparisons of average PSNR in level-1, level-2, and level-3 between the proposed scheme for binary share and Pakniat et al.'s scheme for grayscale image

Scheme	PSNR in level-1	PSNR in level-2	PSNR in level-3
Pakniat et al.'s scheme (grayscale image)	51.23	51.23	51.23
Proposed scheme(binary share)	78.95	81.94	88.96

References

1. Naor, M., Shamir, A.: Visual cryptography. In: Proceedings of Advances in Cryptology, vol. 950, LNCS, pp. 1–12 (1994)
2. Shamir, A.: How to share a secret. Commun. ACM **22**, 612–613 (1979)
3. Chen, Y.C., Tsai, D.S., Horng, G.: Comment on 'cheating prevention in visual cryptography'. IEEE Trans. Image Process. **21**(7), 3319–3323 (2012)
4. Ateniese, G., Blundo, C., De Santis, A., Stinson, D.R.: Visual cryptography for general access structures. Inf. Comput. **129**(2), 86–106 (1996)
5. Naor, M., Pinkas, B.: Visual authentication and identification. In: Proceedings of Advances in Cryptology, vol. 1294, LNCS, pp. 322–336 (1997)
6. Yang, C.N., Laih, C.S.: Some new types of visual secret sharing schemes. Proc. Nat. Comput. Symp. **3**, 260–268 (1999)
7. Tzeng, W.-G., Hu, C.-M.: A new approach for visual cryptography. Des. Codes Crypt. **27**(3), 207–227 (2002)
8. Horng, G.B., Chen, T.H., Tsai, D.S.: Cheating in visual cryptography. Des. Codes Crypt. **38**, 219–236 (2006)
9. Hu, C.M., Tzeng, W.G.: Cheating prevention in visual cryptography. IEEE Trans. Image Process. **16**(1), 36–45 (2007)
10. Chen, Y.C., Horng, G., Tsai, D.S.: Cheating prevention in visual cryptography. In: Visual Cryptography and Secret Image Sharing. CRC Press/Taylor & Francis, Boca Raton (2011)
11. Tassa, T.: Hierarchical threshold secret sharing. In: The First Theory of Cryptography Conference, TCC, LNCS, vol. 2951, pp. 473–490. MIT, Cambridge (2004)
12. Tassa, T., Dyn, N.: Multipartite secret sharing by bivariate interpolation. In: The 33rd International Colloquium on Automata, Languages and Programming, ICALP, Part II, LNCS, vol. 4052, pp. 288–299. Venice (2006)
13. Guo, C., Chang, C.C., Qin, C.: A hierarchical threshold secret image sharing. Pattern Recognit. Lett. **33**(1), 83–91 (2012)
14. Pakniat, N., Noroozi, M., Eslami, Z.: Secret image sharing scheme with hierarchical threshold access structure. J. Vis. Commun. Image Represent. **25**(5), 1093–1101 (2014)
15. Yu-Chee, Tseng, Yu-Yuan, Chen, Hsiang-Kuang, Pan: A secure data hiding scheme for binary images. IEEE Trans. Commun. **50**(8), 1227–1231 (2002)
16. Fan, Li, Gao, Tiegang, Cao, Yanjun: Improving the embedding efficiency of weight matrix-based steganography for grayscale images. Comput. Electr. Eng. **39**, 873–881 (2013)

Covering Arrays of Strength Four and Software Testing

Yasmeen Akhtar, Soumen Maity and Reshma C. Chandrasekharan

Abstract A covering array $t - \mathrm{CA}(n, k, g)$, of size n, strength t, degree k, and order g, is a $k \times n$ array on g symbols such that every $t \times n$ subarray contains every $t \times 1$ column on g symbols at least once. Covering arrays have been studied for their applications on software testing, hardware testing, drug screening, and in areas where interactions of multiple parameters are to be tested. We define the *coverage measure* $\mu_t(A)$ of an array A by the ratio between the number of distinct t-tuples contained in the column vectors of A and the total number of t-tuples given by $\binom{k}{t}g^t$. Given fixed values of t, k, g, and n, our objective is to construct an array A of size atmost n having largest possible coverage measure. This problem is called *covering arrays with budget constrains*. In this article, we present an algebraic construction method for strength four covering arrays with budget constraints.

Keywords Covering arrays · Combinatorics · Group action · Software testing

1 Introduction

This article focuses on constructing new strength four covering arrays with very good coverage measure. A covering array $t - \mathrm{CA}(n, k, g)$, of size n, strength t, degree k, and order g, is a $k \times n$ array on g symbols such that every $t \times n$ subarray contains every $t \times 1$ column on g symbols at least once. The covering array number $t - \mathrm{CAN}(k, g)$ is the smallest n for which a $t - \mathrm{CA}(n, k, g)$ exists. For example, a $4 - CA(22, 5, 2)$ is shown below [4]:

Y. Akhtar (✉) · S. Maity · R.C. Chandrasekharan
Indian Institute of Science Education and Research, Pune 411008, India
e-mail: yasmeensa@students.iiserpune.ac.in

S. Maity
e-mail: soumen@iiserpune.ac.in

R.C. Chandrasekharan
e-mail: reshmac@students.iiserpune.ac.in

R.N. Mohapatra et al. (eds.), *Mathematics and Computing*,
Springer Proceedings in Mathematics & Statistics 139,
DOI 10.1007/978-81-322-2452-5_26

$$\begin{pmatrix} 0&1&1&0&1&0&1&0&0&0&1&1&1&1&0&0&0&0&1&1&0&1 \\ 0&1&0&1&0&1&0&1&0&1&0&1&0&1&0&1&0&0&1&0&0&1 \\ 0&0&1&1&0&0&1&1&0&0&1&1&0&0&1&1&0&1&1&0&1&1 \\ 0&0&0&0&1&1&1&1&0&0&0&0&1&1&1&1&1&0&1&0&1&1 \\ 0&0&0&0&0&0&0&0&1&1&1&1&1&1&1&1&1&1&1&1&0&0 \end{pmatrix}$$

There is a vast array of literature [1, 5] on covering arrays, and the problem of determining small covering arrays has been studied under many guises over the past 30 years. In [5], Hartman and Raskin discussed several generalizations of the problem of creating small covering arrays motivated by their applications in the realm of software testing. When testing a software system with k parameters, each of which must be tested with g values, the total number of possible test cases is g^k. For instance, if there are 20 parameters and three values for each parameter, then the number of input combinations or test cases of this system is $3^{20} = 3486784401$. A fundamental problem with software testing is that testing under all combinations of inputs is not feasible, even with a simple product [7, 11]. This means that the number of defects in a software product can be vast, and defects that occur infrequently are difficult to find in testing. Software developers cannot test everything, but they can use combinatorial test design to identify the minimum number of tests needed to get the coverage they want. Combinatorial test design enables users to get greater test coverage with fewer tests. A study conducted by NIST in 2002 reports that software bugs cost the U.S. economy \$59.5 billion annually. More than a third of this cost could be avoided if better software testing was performed. The goal of most combinatorial testing research is to create test suites that find a large percentage of errors of a system while having a small number of tests required. Covering arrays prove useful in locating a large percentage of errors in software systems [2, 13]. The test cases are the columns of a covering array $t - \mathrm{CA}(n, k, g)$. This is one of the five natural generalizations in [5].

Covering arrays with budget constraints: A practical limitation in the realm of testing is budget. In most software development environments, time, computing, and human resources needed to perform the testing of a component is strictly limited. To model this situation, we consider the problem of creating the best possible test suite (covering the maximum number of t-tuples) within a fixed number of test cases (fixed number of columns of the array). The coverage measure $\mu_t(A)$ of a testing array A is defined by the ratio between the number of distinct t-tuples contained in the column vectors of A and the total number of t-tuples given by $\binom{k}{t} g^t$. Our objective is to construct a testing array A of size at most n having largest possible coverage measure, given fixed values of t, k, g, and n. This problem is called *covering arrays with budget constrains*.

Pairwise or two-way interaction testing and three-way interaction testing are known for its effectiveness in different types of software testing [2, 8, 9]. However, software failures may be caused by interactions of more than two parameters. A recent NIST study indicates that failures can be triggered by interactions up to 6 parameters [11]. Here we consider the problem of four-way interaction testing of the parameters. In this article, we present an algebraic construction method for strength four covering arrays with budget constraints.

1.1 Fractional Linear Group

We summarize the results from group theory that we use. Let F be a Galois field GF(s) where $s = p^m$ and p is prime. We adjoin to F the symbol ∞: it may be helpful to think of the resulting set

$$X = F \cup \{\infty\}$$

as the projective line consisting of $s + 1$ points. Define

$$L(s) = \{\alpha : X \mapsto X \mid x\alpha = \frac{ax + b}{cx + d}, \text{ where } a, b, c, d \in F \text{ and } ad - bc \neq 0\}$$

Here it is understood that the symbol ∞ is subject to such formal arithmetic rules as $x + \infty = \infty$, $\frac{\infty}{\infty} = 1$, etc. It is easy to verify that $L(s)$ is a group with respect to functional composition: indeed $L(s)$ is isomorphic with projective general linear group $PGL_2(F)$. $L(s)$ is called fractional linear group and $|L(s)| = |PGL_2(F)| = \frac{(s^2-1)(s^2-s)}{(s-1)} = (s+1)s(s-1)$. It is known that the group $L(s)$ is sharply 3-transitive on $F \cup \{\infty\}$ with degree $s + 1$. For the undefined terms and more details we refer the reader to Robinson [12]; Chap. 7.

2 The Construction

Given fixed values of t, k, g, and n we are to construct an array of size maximum n having largest possible coverage measure. Here we only consider $t = 4$. The elements of $X = GF(g - 1) \cup \{\infty\}$ are the symbols of covering array. We choose g so that $g - 1$ is a prime or prime power. Group construction involves selecting a group G and finding a vector $v \in X^k$. Set M to be the $k \times k$ circulant matrix generated from starter vector v. Here we take $G = L(g - 1)$. For each $a \in G$, let M_a be the matrix formed by the action of a on the elements of M. Let C be the $k \times g$ matrix that has a constant column with each entry equal to x, for each $x \in X$. A vector $v \in X^k$ is said to be a *starter vector* for a $4 - CA(n, k, g)$ if any $4 \times k$ subarray of the circulant matrix M has at least one representative from each non-constant orbit of $L(g - 1)$ acting on 4-tuples from X. If v is a starter vector and $k|L(g - 1)| + g \leq n$, then the array formed by concatenating C, M_a, $a \in G$ is a covering array with coverage measure one. To see this, consider any four rows x_1, x_2, x_3, and x_4. The patterns with all equal entries occur on rows x_1, x_2, x_3, x_4 since they occur in C. All other patterns appear in rows x_1, x_2, x_3, x_4 since every $4 \times k$ submatrix of M contains at least one representative from each of the orbits 2–5 given below and G is sharply 3-transitive on $X = GF(g - 1) \cup \{\infty\}$.

If starter vector is not found, we look for a vector that produces an array with maximum possible coverage measure. Such vector is called *vector with good coverage*. This group construction follows the technique used in [10]. Since $g - 1$ is prime power, the group $G = L(g-1)$ is sharply 3-transitive on the projective line $X = GF(g-1) \cup \{\infty\}$. Under this group action, there are precise $g+11$ orbits of 4-tuples. These $g+11$ orbits are determined by the pattern of entries in their 4-tuples:

1. One orbit of patterns with four equal entries or constant orbit:
   $\{[a,a,a,a] : a \in X\}$
2. Four orbits of patterns with three equal entries and one different:
   $\{[a,a,a,b] : a,b \in X, a \neq b\}$, $\{[a,a,b,a] : a,b \in X, a \neq b\}$,
   $\{[a,b,a,a] : a,b \in X, a \neq b\}$, $\{[b,a,a,a] : a,b \in X, a \neq b\}$
3. Three orbits of patterns with two copies of two different entries:
   $\{[a,a,b,b] : a,b \in X, a \neq b\}$, $\{[a,b,a,b] : a,b \in X, a \neq b\}$,
   $\{[a,b,b,a] : a,b \in X, a \neq b\}$
4. Six orbits of patterns with two equal entries and two different entries:
   $\{[a,a,b,c] : a,b,c \in X, a \neq b \neq c\}$, $\{[b,a,a,c] : a,b,c \in X, a \neq b \neq c\}$,
   $\{[a,b,a,c] : a,b,c \in X, a \neq b \neq c\}$, $\{[b,a,c,a] : a,b,c \in X, a \neq b \neq c\}$,
   $\{[a,b,c,a] : a,b,c \in X, a \neq b \neq c\}$, $\{[b,c,a,a] : a,b,c \in X, a \neq b \neq c\}$
5. $g-3$ orbits of patterns with four distinct entries. The reason is this. There are $g(g-1)(g-2)(g-3)$ 4-tuples with four distinct entries and each orbit contains $g(g-1)(g-2)$ 4-tuples as $|L(g-1)| = g(g-1)(g-2)$.

We show an example to explain the method.

Example 1 Let $g = 4$, $k = 15$, and $n = 364$. Then $X = GF(3) \cup \{\infty\}$. For $g = 4$, there are 15 orbits: 1 orbit of type 1, 4 orbits of type 2, 3 orbits of type 3, 6 orbits of type 4 and 1 orbit of type 5. Let $v = (0001102\infty\infty1\infty101\infty)$. Build the following circulant matrix M from v:

$$M = \begin{pmatrix}
0 & \infty & 1 & 0 & 1 & \infty & 1 & \infty & \infty & 2 & 0 & 1 & 1 & 0 & 0 \\
0 & 0 & \infty & 1 & 0 & 1 & \infty & 1 & \infty & \infty & 2 & 0 & 1 & 1 & 0 \\
0 & 0 & 0 & \infty & 1 & 0 & 1 & \infty & 1 & \infty & \infty & 2 & 0 & 1 & 1 \\
1 & 0 & 0 & 0 & \infty & 1 & 0 & 1 & \infty & 1 & \infty & \infty & 2 & 0 & 1 \\
1 & 1 & 0 & 0 & 0 & \infty & 1 & 0 & 1 & \infty & 1 & \infty & \infty & 2 & 0 \\
0 & 1 & 1 & 0 & 0 & 0 & \infty & 1 & 0 & 1 & \infty & 1 & \infty & \infty & 2 \\
2 & 0 & 1 & 1 & 0 & 0 & 0 & \infty & 1 & 0 & 1 & \infty & 1 & \infty & \infty \\
\infty & 2 & 0 & 1 & 1 & 0 & 0 & 0 & \infty & 1 & 0 & 1 & \infty & 1 & \infty \\
\infty & \infty & 2 & 0 & 1 & 1 & 0 & 0 & 0 & \infty & 1 & 0 & 1 & \infty & 1 \\
1 & \infty & \infty & 2 & 0 & 1 & 1 & 0 & 0 & 0 & \infty & 1 & 0 & 1 & \infty \\
\infty & 1 & \infty & \infty & 2 & 0 & 1 & 1 & 0 & 0 & 0 & \infty & 1 & 0 & 1 \\
1 & \infty & 1 & \infty & \infty & 2 & 0 & 1 & 1 & 0 & 0 & 0 & \infty & 1 & 0 \\
0 & 1 & \infty & 1 & \infty & \infty & 2 & 0 & 1 & 1 & 0 & 0 & 0 & \infty & 1 \\
1 & 0 & 1 & \infty & 1 & \infty & \infty & 2 & 0 & 1 & 1 & 0 & 0 & 0 & \infty \\
\infty & 1 & 0 & 1 & \infty & 1 & \infty & \infty & 2 & 0 & 1 & 1 & 0 & 0 & 0
\end{pmatrix}$$

Now the 24 elements of $L(3)$ acting on M produces 24 matrices. We also need to add the following matrix,

$$C = \begin{pmatrix} 0&0&0&0&0&0&0&0&0&0&0&0&0&0&0 \\ 1&1&1&1&1&1&1&1&1&1&1&1&1&1&1 \\ 2&2&2&2&2&2&2&2&2&2&2&2&2&2&2 \\ \infty&\infty&\infty&\infty&\infty&\infty&\infty&\infty&\infty&\infty&\infty&\infty&\infty&\infty&\infty \end{pmatrix}^T$$

to ensure the coverage of orbit 1. By horizontally concatenating the 24 matrices and C, we build a 15×364 testing array with coverage measure 0.834.

2.1 *Choice of Vector v*

Recall that we set M to be the $k \times k$ circulant matrix generated from a starter vector v. Then, any four rows in the matrix M must have at least one element from each of the orbits $2-5$. To determine which vectors will work as starters, we define the sets $d[x, y, z]$ for positive integers x, y, and z as follows:

$$d[x, y, z] = \{(v_i, v_{i+x}, v_{i+x+y}, v_{i+x+y+z}) \mid 0 \leq i \leq k-1\}$$

where the subscripts are taken modulo k. For v to be a starter vector, each set $d[x, y, z]$ must contain a representative from each of the orbits $2-5$. A covering array of strength 4 satisfies the property that for any four distinct rows all possible 4-tuples of g symbols occur at least once as a column. For computational convenience, we divide the collection of $\binom{k}{4}$ choices of four distinct rows from k rows into few equivalence classes. We define an equivalence relation $\sim$ on this collection of choices of four rows out of k rows as

$$(\alpha, \beta, \gamma, \delta) \sim (\alpha', \beta', \gamma', \delta') \text{ if and only if}$$

$\beta - \alpha = \beta' - \alpha' \bmod k$ and $\gamma - \beta = \gamma' - \beta' \bmod k$ and $\delta - \gamma = \delta' - \gamma' \bmod k$. This equivalence relation induces a partition of the set of all choices of four rows into equivalence classes $[x, y, z]$ given by

$$[x, y, z] = \{(i, i+x, i+x+y, i+x+y+z) \bmod k \mid i = 0, 1, ..., k-1\}$$

where $x = \beta - \alpha$, $y = \gamma - \beta$ and $z = \delta - \gamma$ all considered under modulo k. For $k = 9$, the class $[2, 2, 3] = \{(0, 2, 4, 7), (1, 3, 5, 8), (2, 4, 6, 0), (3, 5, 7, 1), (4, 6, 8, 2), (5, 7, 0, 3), (6, 8, 1, 4), (7, 0, 2, 5), (8, 1, 3, 6)\}$. Given k, to generate all equivalence classes without repetition, we give specific choices for x, y, and z: $x = 1, 2, ..., \lfloor \frac{k}{4} \rfloor$, $y = x, x+1, ..., k-1$ and $z = x, x+1, ..., k-1$ such that

1. $x + y + z \leq k - 1 - x$ **:** The reason is this. Let $w = k - x - y - z$. Then w has to be strictly greater than x. If $w = x$, then the classes $[x, y, z]$ and $[x, x, y]$ are the same.
2. **If** $y > \lfloor \frac{k-2x}{2} \rfloor$ **then** $z \geq x + 1$**:** The class $[x, y, x]$ for $y > w$ and the class $[x, w, x]$ for $y < w$ are the same where $x + y + x + w = k$ as $z = x$. Hence when $y \leq \lfloor \frac{k-2x}{2} \rfloor$ we allow $z \geq x$ and for $y > \lfloor \frac{k-2x}{2} \rfloor$ we require $z \geq x + 1$.

3. **If** $k \equiv 0$ **mod** 4 **:** In addition to the above choices of x, y and z we need to consider one more class with $x = y = z = \frac{k}{4}$.

At this stage, we would like to make few remarks about the size of equivalence classes defined by above choices of x, y, and z.

1. $k \not\equiv 0 \bmod 2$:
   If k is an odd integer, then each class contains exactly k distinct choices from the collection of $\binom{k}{4}$ choices and hence there are total $l = \frac{(k-1)(k-2)(k-3)}{24}$ distinct classes of size k.
2. $k \equiv 0 \bmod 2$:
   If k is an even integer, then $\frac{k}{2}$ can be written as sum of two positive integers a and b where $a \leq b$ in $\lfloor \frac{k}{4} \rfloor$ different ways.
   Case 1 : If $k \not\equiv 0 \bmod 4$, then class of the form $[a, b, a]$ contains only $\frac{k}{2}$ distinct choices. So there are total $\lfloor \frac{k}{4} \rfloor$ equivalence classes of size $\frac{k}{2}$ and the remaining classes are of size k.
   Case 2 : If $k \equiv 0 \bmod 4$, then a class of the form $[a, b, a]$ contains only $\frac{k}{2}$ distinct choices and a class of the form $[a, a, a]$ where $a = \frac{k}{4}$ contains only $\frac{k}{4}$ distinct choices. Here we get total $\frac{k}{4} - 1$ equivalence classes of size $\frac{k}{2}$, exactly one class of size $\frac{k}{4}$ and the remaining classes are of size k.

For $k = 9$, the equivalence classes are $[1, 1, 1]$, $[1, 1, 2]$, $[1, 1, 3]$, $[1, 1, 4]$, $[1, 1, 5]$, $[1, 2, 1]$, $[1, 2, 2]$, $[1, 2, 3]$, $[1, 2, 4]$, $[1, 3, 1]$, $[1, 3, 2]$, $[1, 3, 3]$, $[1, 4, 2]$, $[2, 2, 2]$, and each equivalence class is of size 9. Thus $14 \times 9 = \binom{9}{4}$. A vector v is said to be a starter vector if each set $d[x, y, z]$ has a representation from each of the orbits $2-5$. Often, representation from orbit of type 1 is taken care by attaching g constant columns, one for each symbol. Once a starter vector is found, the circulant matrix M is constructed and acted upon by the group $L(g-1)$ and concatenated to form an array of size $k|L(g-1)|$. To this array, g constant columns are added to ensure the coverage of orbit 1 to produce a covering array of size $k|L(g-1)| + g$.

3 Results

The coverage measure of a covering array is always one. For computational convenience, we rewrite the coverage measure of an array A for $t = 4$ in terms of equivalence classes $[x, y, z]$ and $d[x, y, z]$ as follows:

$$\mu_4(A) = \frac{\sum\limits_{x,y,z} |[x, y, z]| \times \text{number of distinct 4-tuples covered by } d[x, y, z]}{\binom{k}{4} g^4}.$$

We use computer search to find vectors v with very high coverage measures. Table 1 shows vectors with high coverage, the number of test cases (n) generated by our

Table 1 A comparison of the number test cases (n) produced by our construction with high coverage measure and best known n for full coverage

(g, k)	Vector v with good coverage	Our Results n (coverage measure)	Best known n [3]
(3, 16)	00001001∞∞011∞1∞	99 (0.828)	237
(3, 17)	0000010∞∞101∞01∞1	105 (0.851)	282
(3, 18)	00010∞0∞1001∞111∞∞	111 (0.864)	293
(3, 19)	000010010∞01∞0∞111∞	117 (0.883)	305
(3, 20)	0000110101∞0∞10∞∞11∞	123 (0.892)	314
(3, 21)	00001010∞1∞∞10∞∞001∞1	129 (0.906)	315
(3, 22)	0000011∞0∞0110∞1∞∞∞01∞	135 (0.913)	315
(3, 23)	0000001∞∞0101∞10∞10∞∞∞1	141 (0.923)	315
(3, 24)	00000001∞∞0101∞10∞101∞∞1	147 (0.924)	315
(3, 25)	0000000011∞0∞011∞01∞0∞11∞	153 (0.930)	363
(4, 18)	00010021∞∞∞21020∞2	436 (0.851)	760
(4, 19)	0000121011∞01∞0∞221	460 (0.866)	760
(4, 20)	0000112101202∞0221∞2	484 (0.878)	760
(4, 21)	0000011021010∞2∞0221∞	508 (0.887)	1012
(4, 22)	0000001102∞02021∞∞01∞1	532 (0.894)	1012
(4, 23)	00000001210210∞∞20112∞1	556 (0.898)	1012
(4, 24)	00000000121∞011∞02∞0∞112	580 (0.899)	1012
(4, 25)	000000000121220∞011∞2012∞	604 (0.901)	1012
(5, 21)	110131300∞30010∞∞3203	1265 (0.834)	1865
(5, 22)	3∞32011200∞∞00∞0∞10010	1325 (0.842)	1865
(5, 23)	0002∞03100∞203021332320	1385 (0.854)	1865
(5, 24)	003∞21022212300032302310	1445 (0.860)	1865
(5, 25)	∞200∞0∞∞31020∞300303∞∞33	1505 (0.869)	2485
(5, 26)	202002211000∞0121031∞∞2300	1565 (0.873)	2485
(5, 27)	∞∞03002030∞000∞11∞0031301∞3	1625 (0.880)	2485
(5, 28)	013333130320∞1∞1003200310300	1685 (0.883)	2485
(5, 29)	00012212∞010∞31100310200031010	1745 (0.891)	2485
(5, 30)	33001∞0∞000330∞∞010012∞1313001	1805 (0.894)	2485
(6, 25)	000403014003033404320∞1∞∞	3006 (0.811)	6325
(6, 26)	∞0∞40021404010013010011444	3126 (0.819)	6456
(6, 27)	433∞∞01∞∞20∞03020∞∞0∞00401∞	3246 (0.826)	6606
(6, 28)	4023031100232200∞21∞∞2020020	3366 (0.829)	6714

For $g = 5$, the elements of $GF(4)$ are represented as 0,1, 2, and 3; here 2 stands for x and 3 stands for $x + 1$

technique, best known n with full coverage. A comparison of our construction with best known covering array sizes show that our construction produces significantly smaller testing arrays with very high coverage measures.

4 Conclusions

In this paper, we have proposed construction of strength four covering arrays with budget constraints. In order to test a software component with 25 parameters each having three values, our construction can generate a test suite with 153 test cases that ensure with probability 0.93 that software failure cannot be caused due to interactions of two, three, or four parameters whereas best known covering array in [3] requires 363 test cases for full coverage. The results show that the proposed method could reduce the number of test cases significantly.

Acknowledgments The first author gratefully acknowledges support from the Council of Scientific and Industrial Research (CSIR), India, during the work under CSIR senior research fellow scheme.

References

1. Chateauneuf, M.A., Colbourn, C.J., Kreher, D.L.: Covering arrays of strength three. Des. Codes Crypt. **16**, 235–242 (1999)
2. Cohen, D.M., Dalal, S.R., Fredman, M.L., Patton, G.C.: The AETG system: an approach to testing based on combinatorial design. IEEE Trans. Softw. Eng. **23**(7), 437–443 (1997)
3. Colbourn, C.: Covering Array Tables for t=2,3,4,5,6. http://www.public.asu.edu/ccolbou/src/tabby/catable.html
4. Covering Arrays generated by IPOG-F, http://math.nist.gov/coveringarrays/ipof/ipof-results.html
5. Hartman, A., Raskin, L.: Problems and algorithms for covering arrays. Discret. Math. **284**, 149–156 (2004)
6. Hartman, A.: Software and hardware testing using combinatorial covering suites. Graph Theory, Comb. Algorithms: Interdisc. Appl. (Kluwer Academic Publishers) **34**, 237–266 (2006)
7. Kaner, C., Falk, J., Nguyen, H.Q.: Testing Computer Software, 2nd edn. Wiley, New York (1999)
8. Maity, S.: 3-Way software testing with budget constraints. IEICE Trans. Inf. Syst. **E-95-D**(9), 2227–2231 (2012)
9. Maity, S., Nayak, A.: Improved test generation algorithms for pair-wise testing. In: Proceedings of 16th IEEE International Symposium on Software Reliability Engineering, pp. 235–244. Chicago (2005)
10. Meagher, K., Stevens, B.: Group construction of covering arrays. J. Comb. Des. **13**(1), 70–77 (2005)
11. Richard, K.D., Wallace, D.R., Gallo, A.M.: Software fault interactions and implications for software testing. IEEE Trans. Softw. Eng. **30**(6), 418–421 (2004)
12. Robinson, D.J.S.: A course in the theory of groups, 2nd edn. Springer, Heidelberg (1995)
13. Yilmaz, C., Cohen, M., Porter, A.: Covering arrays for efficient fault characterisation in complex configuration spaces. IEEE Trans. Softw. Eng. **32**(1), 20–34 (2006)

Amplitude Equation for a Nonlinear Three Dimensional Convective Flow in a Mushy Layer

Dambaru Bhatta and Daniel N. Riahi

Abstract We consider a nonlinear three-dimensional convective flow in a mushy layer. A mushy layer is a partially solidified region formed during solidification of binary alloys. During solidification, fluid flow within the mushy layer can cause vertical chimneys or channels void of solid. These chimneys can generate imperfections in the final form of the solidified alloy. The equations governing the mushy layer system are the continuity equation, heat equation, solute equation, and conservation of momentum which is governed by Darcy's law. A quadratic nonlinear evolution equation satisfied by the amplitude is derived for hexagonal cells. This equation is obtained from the first-order system using the adjoint of the linear system. An explicit solution of the amplitude equation is also presented.

Keywords Evolution equation · Three-dimensional · Convective flow · Solidification · Mushy layer

1 Introduction

In solidification of binary alloys, vertical chimneys or channels void of solid that are typically oriented in the direction of gravity are observed by experimentalists. When a binary alloy is solidified from cooled boundary, due to the temperature difference at the solidification front, the interface becomes unstable. Hence a heterogeneous region of both solid and liquid, famously referred as mushy layer, is formed. This layer is sandwiched between a solid layer at the bottom and a liquid layer at the top. Convection plays a very important role during the solidification process of binary alloys. The convective flow within the mushy layer influences the formation of thread-like structures known as freckles in the final form of the solidified alloy. Convective flows in a horizontal mushy layer are known to produce chimneys during solidification of binary alloys. It is well known that convection in the chimneys causes a thin hair-like

D. Bhatta (✉) · D.N. Riahi
University of Texas-RGV, Edinburg, TX, USA
e-mail: dambaru.bhatta@utrgv.edu

R.N. Mohapatra et al. (eds.), *Mathematics and Computing*,
Springer Proceedings in Mathematics & Statistics 139,
DOI 10.1007/978-81-322-2452-5_27

structure called freckles. Study of hydrodynamic stability was carried by many scientists including Helmholtz, Kelvin, Rayleigh, and Reynolds in the nineteenth century because of its practical importance. Lev Landau [1] proposed an equation to analyze hydrodynamic stability. Various case studies on hydrodynamic and hydromagnetic stabilities have been presented by Chandrasekhar [2].

A fairly large number of, theoretical as well as experimental, studies have been devoted to predicting the chimney formation during the solidification process. Drazin and Reid [3] have presented various studies done by many researchers. They presented methods and results of thermal convection, rotating and curved flows, and parallel shear flows. Development of asymptotic theory of Orr–Sommerfeld equation, applications of linear stability theory, and nonlinear theory of hydrodynamic stability have been presented. Many previous studies have examined in detail about the mechanism of freckle formation during the solidification of multicomponent alloys. A thermodynamically consistent model was proposed by Hill et al [4] to analyze a mushy layer. Fowler [5] developed mathematical analysis of freckle formation to predict the criterion for freckling. This prediction is equivalent to the classical Rayleigh number condition for convective instability. Huppert et al. [6] studied experimentally the six different cases that arise when homogeneous solution is cooled from below and also evaluated the criterion for which solid–liquid interface becomes unstable. Worster [7] developed the model for the dendritic growth that often formed during the solidification of binary alloy by considering the region of mixed phase as continuum.

Several experimental studies concerning the solidification of binary alloy with or without magnetic filed have been reported in the literature. Vives and Perry [8] carried out an experimental investigation of solidification of tin and aluminum alloys under the influence of externally imposed magnetic field. They reported that the stationary magnetic field decreases the superheat and increases the rate of solidification. Chen et al. [9] and Chen [15] carried out experimental studies on directional solidification of aqueous chloride solution. Worster [10, 11] applied linear stability analysis for the two-layer model and concluded that the mushy layer mode is responsible for the development of chimneys. Tait et al. [12] observed the hexagonal pattern of convection just when the system becomes unstable during their experimental work. Amberg and Homsy [13] studied the simplified mushy layer model with constant permeability. They carried out a weakly nonlinear analysis of simplified mushy layer model that was proposed by Worster [7]. A near-eutectic approximation was applied and the limit of large far-field temperature was considered. Such asymptotic limits allowed them to examine the dynamics of mushy layer. A weakly nonlinear analysis of simplified mushy layer model that was proposed in [7] was carried out by Anderson and Worster [14]. A near-eutectic approximation was applied and the limit of large far-field temperature was considered. Such asymptotic limits allowed them to examine the dynamics of mushy layer. They also considered the limit of large Stefan number, which enabled them to reach a domain for the existence of the oscillatory mode of convection.

Study on oscillatory modes of nonlinear compositional convection in mushy layers was carried out by Riahi [16, 19]. In another development by Okhuysen and Riahi [17, 18], a weakly nonlinear analysis of buoyant convection in two-layer model was considered. They predicted subcritical down-hexagonal pattern for the case of reactive mushy layer. Muddamallappa et al. [20] investigated linear marginal stabilities for magnetoconvection cases. Bhatta et al. [21, 22, 24] studied weakly nonlinear convective flow in mushy layer with permeable mush–liquid interface for constant and variable permeability cases. Lee et al. [23] carried out numerical modeling of one-dimensional binary alloy solidification with a mushy layer evolution. Fluxes through steady chimneys in a mushy layer during binary solidification was studied by Rees et al. [25]. Wells et al. [26] analyzed the stability and optimal solute fluxes for nonlinear mushy layer convection. Various numerical methods needed to compute the solutions for the mushy layer system are presented by Cheney and Kincaid [27]. However, analysis of nonlinear convection for three-dimensional mushy layer case had not been undertaken until recently. The objective of this paper is to analyze three-dimensional convective flow in a mushy layer. As a first step, we derive an evolution equation satisfied by the amplitude for hexagonal cells. This amplitude equation is quadratic nonlinear.

2 Governing System for the Mushy Layer

We consider a system governing the mushy layer of thickness d which is cooled from below and the solidification front advances with a constant speed V_0 as shown in the Fig. 1. Derivation and justification of equations governing a mushy layer have been presented various authors [5, 10, 11, 13, 14]. The geometry of the physical system is shown in Fig. 1.

This system is given by

$$\frac{\mu}{\Pi}\overrightarrow{U} = -\nabla p - (\rho - \rho_0)\, g\, \overrightarrow{k}$$
$$\nabla \cdot \overrightarrow{U} = 0$$

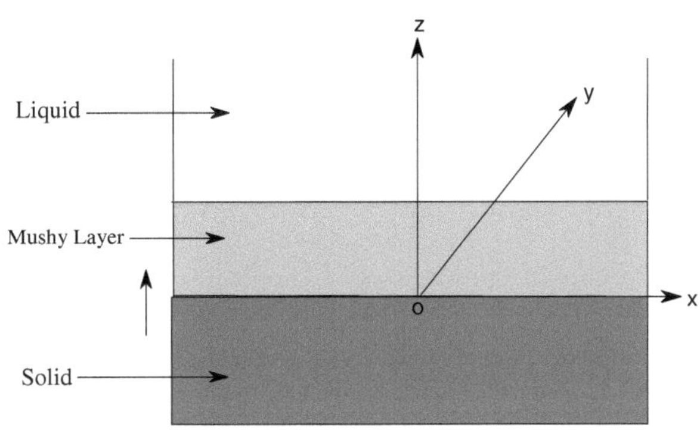

Fig. 1 Geometry of the physical system

$$\frac{\partial T}{\partial t} + \overrightarrow{U} \cdot \nabla T = \kappa \nabla^2 T + \frac{l_h}{\gamma} \frac{\partial \Phi}{\partial t}$$
$$\chi \frac{\partial C}{\partial t} + \overrightarrow{U} \cdot \nabla C = (C - C_s) \frac{\partial \Phi}{\partial t}. \tag{1}$$

Here the equations represent conservation of momentum, conservation of mass, heat equation, and solute equation, respectively. Here t, T, κ, γ, l_h, represent time, temperature, thermal diffusivity of the liquid, specific heat of the liquid, and latent heat per unit mass, respectively. Here $\overrightarrow{U} = U\mathbf{i} + V\mathbf{j} + W\mathbf{k}$ is the liquid flux where U, V are used to denote horizontal components, W denotes the vertical component of $\overrightarrow{U}$, and **i**, **j**, **k** are the unit vectors along x, y, z directions. Also Φ stands for the local solid volume fraction, i.e., $\Phi = 1 - \chi$ where χ is the local liquid volume fraction. C is the composition of the liquid and C_s is the composition of the solid phase. Here μ is used for dynamic viscosity of the liquid, p represents the dynamic pressure, ρ is the density of the liquid, ρ_0 is some reference value of the liquid density, and g denotes the acceleration due to gravity. Also $\rho = \rho_0[1 + \beta(C - C_0)]$ where β is the expansion coefficient of solid and C_0 is some reference value C. Permeability $\Pi = \Pi(\chi)$ is a function of the local liquid volume fraction, χ.

The boundary conditions are

$$T = T_e, \quad W = 0 \qquad \text{at} \quad z = 0$$
$$T = T_0, \quad \Phi = W = 0 \quad \text{at} \quad z = d.$$

Here T_0 denotes the temperature at the mush–liquid interface (at $z = d$), and T_e and C_e represent eutectic temperature and eutectic concentration (at the solid–mush interface, $z = 0$), respectively.

2.1 Nondimensionalization

We nondimensionalize the system in a frame moving with the solidification front at constant speed V_0 and use the following scalings: velocity scale is V_0, i.e., $\overrightarrow{\mathscr{U}} = \frac{\overrightarrow{U}}{V_0}$, length scale is $\frac{\kappa}{V_0}$, time scale is $\frac{\kappa}{V_0^2}$, pressure scale is $\frac{\kappa\mu}{\Pi_0}$, $\Theta = \frac{T-T_0}{\Delta T}$, $\mathscr{K} = \frac{\Pi_0}{\Pi}$ where $\Delta T = T_0 - T_e$, $\Delta C = C_0 - C_e$ and Π_0 is a reference value of Π. The nondimensional constants appearing in the derivation are Rayleigh number, $\mathscr{R} = \frac{\beta g \Pi_0 \Delta C}{V_0 \mu}$, Stefan number, $\mathscr{S} = \frac{l_h}{\gamma \Delta T}$, and concentration ratio, $\mathscr{C} = \frac{C_s - C_0}{\Delta C}$.

Nondimensional system can be expressed as

$$\mathscr{K}\overrightarrow{\mathscr{U}} + \nabla\mathscr{P} + \mathscr{R}\Theta\overrightarrow{k} = \overrightarrow{0}$$
$$\nabla \cdot \overrightarrow{\mathscr{U}} = 0$$

$$\left(\frac{\partial}{\partial t}-\frac{\partial}{\partial z}\right)[\Theta-\mathscr{S}\Phi]+\overrightarrow{\mathscr{U}}\cdot\nabla\Theta=\nabla^2\Theta$$
$$\left(\frac{\partial}{\partial t}-\frac{\partial}{\partial z}\right)[(1-\Phi)\,\Theta+\mathscr{C}\Phi]+\overrightarrow{\mathscr{U}}\cdot\nabla\Theta=0 \tag{2}$$

with boundary conditions:

$$\Theta=-1,\ \mathscr{W}=0 \quad \text{at} \quad z=0$$
$$\Theta=\Phi=\mathscr{W}=0 \quad \text{at} \quad z=\delta$$

where $\mathscr{W}$ denotes the vertical component of $\overrightarrow{\mathscr{U}}$. Also $\delta=\frac{V_0 d}{\kappa}$ is a growth peclet number representing the dimensionless depth of the mushy layer. For this study, we take permeability as constant, i.e., $\mathscr{K}=1$.

3 Solution Procedure

Assuming solutions of the form

$$\Theta(x,y,z,t)=\theta_b(z)+\varepsilon\theta\,(x,y,z,t)$$
$$\Phi(x,y,z,t)=\phi_b(z)+\varepsilon\phi\,(x,y,z,t)$$
$$\overrightarrow{\mathscr{U}}(x,y,z,t)=\overrightarrow{0}+\varepsilon\,\overrightarrow{u}\,(x,y,z,t)$$
$$\mathscr{P}(x,y,z,t)=p_b(z)+\varepsilon p\,(x,y,z,t) \tag{3}$$

where $\theta_b,\ \phi_b,\ p_b$ are solutions to the steady basic state system (system with no flow) and $\theta,\ \phi,\ \overrightarrow{u},\ p$ are perturbation solutions. Here ε is the perturbation parameter.

3.1 Basic State Solutions

Using (3) in (2) and setting $\varepsilon=0$, we obtain steady basic state system as

$$\frac{d^2\theta_b}{dz^2}+\frac{d\theta_b}{dz}-\mathscr{S}\frac{d\phi_b}{dz}=0 \tag{4}$$
$$(1-\phi_b)\,\frac{d\theta_b}{dz}+(\mathscr{C}-\theta_b)\,\frac{d\phi_b}{dz}=0 \tag{5}$$
$$\frac{dp_b}{dz}+\mathscr{R}\theta_b=0 \tag{6}$$

with boundary conditions:

$$\theta_b = -1 \quad \text{at} \quad z = 0$$
$$\theta_b = \phi_b = 0 \quad \text{at} \quad z = \delta.$$

Solutions θ_b and ϕ_b are, respectively, given by

$$z = \frac{r_1 - \mathscr{C}}{r_1 - r_2} \ln\left[\frac{1 + r_1}{r_1 - \theta_b}\right] + \frac{\mathscr{C} - r_2}{r_1 - r_2} \ln\left[\frac{1 + r_2}{r_2 - \theta_b}\right] \tag{7}$$

and

$$\phi_b = \frac{\theta_b}{\theta_b - \mathscr{C}} \tag{8}$$

where $r_1,\ r_2$ are given by

$$r_1 = \frac{\mathscr{C} + \mathscr{S} + \theta_\infty + \sqrt{(\mathscr{C} + \mathscr{S} + \theta_\infty)^2 - 4\mathscr{C}\theta_\infty}}{2}$$
$$r_2 = \frac{\mathscr{C} + \mathscr{S} + \theta_\infty - \sqrt{(\mathscr{C} + \mathscr{S} + \theta_\infty)^2 - 4\mathscr{C}\theta_\infty}}{2}.$$

and θ_∞ is the nondimensional temperature far away from mush–liquid interface. Thickness of the layer can be determined as

$$\delta = \frac{r_1 - \mathscr{C}}{r_1 - \beta} \ln\left[\frac{1 + r_1}{r_1}\right] + \frac{\mathscr{C} - r_2}{r_1 - r_2} \ln\left[\frac{1 + r_2}{r_2}\right].$$

4 Perturbed System

Using (3) in the system (2), the perturbed system can be obtained as

$$\vec{u} + \nabla p + \mathscr{R}_c \theta \hat{k} = -\varepsilon \mathscr{R}_1 \theta \hat{k} \tag{9}$$

$$\left(\nabla^2 + \frac{\partial}{\partial z}\right)\theta - \mathscr{S}\frac{\partial \phi}{\partial z} - \theta_b' w = \varepsilon\left[\frac{\partial}{\partial \tau}(\theta - S\phi) + \vec{u} \cdot \nabla\theta\right] \tag{10}$$

$$\frac{\partial}{\partial z}\{(\theta_b - \mathscr{C})\phi - (1-\phi_b)\theta\} + \theta_b' w = \varepsilon\left\{-\vec{u}\cdot\nabla\theta - \frac{\partial}{\partial z}(\theta\phi) + \frac{\partial}{\partial \tau}\{(\theta_b - \mathscr{C})\phi - (1-\phi_b)\theta\} + \varepsilon\frac{\partial}{\partial \tau}(\theta\phi)\right\} \tag{11}$$

$$\nabla\cdot\vec{u} = 0 \tag{12}$$

with $\theta = w = 0$ at $z = 0$ and $\theta = \phi = w = 0$ at $z = \delta$, $\tau = \varepsilon t$ and ε is the perturbation parameter, given by $\frac{\mathscr{R}-\mathscr{R}_c}{\mathscr{R}_1}$. Here $\mathscr{R}_c$ is the critical Rayleigh number and $\mathscr{R}_1$ is the nonlinear contribution to $\mathscr{R}$ beyond the value for the most critical neutrally stable linear solution. Here $\vec{u} = (u, v, w)$ and θ_b' denote the derivative of θ_b with respect to z. Now, we eliminate the pressure from the Eq. (9) by taking the double curl of that equation. Also using the continuity equation, the third component of $\nabla\times\nabla\times\vec{u}$ becomes $-\nabla^2 w$. Similarly, for the third component of $\nabla\times\nabla\times\left(\mathscr{R}_c\theta\hat{k}\right)$, we have

$$-\mathscr{R}_c\left[\frac{\partial^2\theta}{\partial x^2} + \frac{\partial^2\theta}{\partial y^2}\right] = \mathscr{R}_c(\Delta_2\theta).$$

Here Δ_2 represents two-dimensional Laplacian operator. These allow us to transform Eq. (9) as

$$\nabla^2 w + \mathscr{R}_c(\Delta_2\theta) = -\varepsilon\mathscr{R}_1(\Delta_2\theta) \tag{13}$$

4.1 Linear and Adjoint Systems

Considering

$$\begin{aligned}\theta &= \theta_0 + \varepsilon\theta_1 + \varepsilon^2\theta_2 + \ldots\\ \phi &= \phi_0 + \varepsilon\phi_1 + \varepsilon^2\phi_2 + \ldots\\ w &= w_0 + \varepsilon w_1 + \varepsilon^2 w_2 + \ldots\end{aligned}$$

and denoting the solution of linear perturbed system by $q_0 = \left[w_0,\ \theta_0,\ \phi_0\right]^{T_r}$, where T_r is used to denote the transpose, the linear system can be obtained from Eqs. (10), (11) and (13) by comparing the coefficients of ε^0 as follows

$$\nabla^2 w_0 + \mathscr{R}_c(\Delta_2\theta_0) = 0 \tag{14}$$

$$\left(\nabla^2 + \frac{\partial}{\partial z}\right)\theta_0 - \mathscr{S}\frac{\partial\phi_0}{\partial z} - \theta_b' w_0 = 0 \tag{15}$$

$$\frac{\partial}{\partial z}\left[(\theta_b - \mathscr{C})\phi_0 - (1-\phi_b)\theta_0\right] + \theta_b' w_0 = 0 \tag{16}$$

The boundary conditions at the solidifying front are $\theta_0 = w_0 = 0$ and at the mush–liquid interface are $\theta_0 = \phi_0 = w_0 = 0$.

For the quantities belonging to the adjoint system, we use the sub-index a, i.e., $q_a = \left[w_a \;\; \theta_a \;\; \phi_a \right]^{T_r}$. To obtain the adjoint system, we multiply the Eqs. (14), (15) and (16) by w_a, θ_a and ϕ_a, respectively, add them and integrate and then take the limit as follows:

$$\lim_{L\to\infty} \frac{1}{4L^2} \int_0^{\delta} \int_{-L}^{L} \int_{-L}^{L} \Bigg[w_a \left(\nabla^2 w_0 \right) + \mathscr{R} w_a \left(\Delta_2 \theta_0 \right) + \theta_a \left(\nabla^2 + \frac{\partial}{\partial z} \right) \theta_0 - \mathscr{S} \theta_a \frac{\partial \phi_0}{\partial z} - \theta_a \theta_b' w_0 + \phi_a \left\{ \frac{\partial}{\partial z} \left[\left(\theta_b - \mathscr{C} \right) \phi_0 - \left(1 - \phi_b \right) \theta_0 \right] + \theta_b' w_0 \right\} \Bigg] dV = 0 \tag{17}$$

Here L denotes the length in xy-plane. The boundary conditions satisfied by adjoint solutions at the solidifying front are $\theta_a = w_a = \phi_a = 0$ and at the mush–liquid interface are $\theta_a = w_a = 0$. The adjoint system can be obtained as

$$\nabla^2 w_a - \theta_b' \theta_a + \theta_b' \phi_a = 0$$

$$\mathscr{R} \left(\Delta_2 w_a \right) + \left(\nabla^2 - \frac{\partial}{\partial z} \right) \theta_a + \left(1 - \phi_b \right) \frac{\partial \phi_a}{\partial z} = 0$$

$$\mathscr{S} \frac{\partial \theta_a}{\partial z} + \left(\mathscr{C} - \theta_b \right) \frac{\partial \phi_a}{\partial z} = 0$$

5 Evolution Equation for the Amplitude

First-order system is obtained from Eqs. (13), (10) and (11) by comparing the coefficients of ε^1 as

$$\nabla^2 w_1 + \mathscr{R}_c \left(\Delta_2 \theta_1 \right) = -\mathscr{R}_1 \left(\Delta_2 \theta_0 \right) \tag{18}$$

$$\left(\nabla^2 + \frac{\partial}{\partial z} \right) \theta_1 - \mathscr{S} \frac{\partial \phi_1}{\partial z} - \theta_b' w_1 = \left[\frac{\partial}{\partial \tau} \left(\theta_0 - S \phi_0 \right) + \vec{u_0} . \nabla \theta_0 \right] \tag{19}$$

$$\frac{\partial}{\partial z} \left\{ \left(\theta_b - \mathscr{C} \right) \phi_1 - \left(1 - \phi_b \right) \theta_1 \right\} + \theta_b' w_1 = -\vec{u_0} \cdot \nabla \theta_0 - \frac{\partial}{\partial z} \left(\theta_0 \phi_0 \right) + \frac{\partial}{\partial \tau} \left\{ \left(\theta_b - \mathscr{C} \right) \phi_0 - \left(1 - \phi_b \right) \theta_0 \right\} \tag{20}$$

Using poloidal and toroidal decompositions of a divergence-free vector (Chandrasekhar [2]), we can express

$$\vec{u_0}.\nabla\theta_0 = \frac{\partial^2 u_{P_0}}{\partial x \partial z}\frac{\partial \theta_0}{\partial x} + \frac{\partial^2 u_{P_0}}{\partial y \partial z}\frac{\partial \theta_0}{\partial y} - \left(\Delta_2 u_{P_0}\right)\frac{\partial \theta_0}{\partial z}$$

where u_{P_0} is the poloidal component of $\vec{u_0}$. To derive the evolution equation satisfied by the amplitude, we multiply (18), (19) and (20) by w_a, θ_a, and ϕ_a, respectively, and integrate the result with respect to $x,\ y,\ z$ and take the limit. Left-hand side of this operation becomes

$$\begin{aligned} LHS = \lim_{L\to\infty}\frac{1}{4L^2}\int_0^\delta\int_{-L}^{L}\int_{-L}^{L}\Big[& w_a\left\{\nabla^2 w_1 + \mathscr{R}_c\left(\Delta_2\theta_1\right)\right\} \\ & +\theta_a\left\{\left(\nabla^2 + \frac{\partial}{\partial z}\right)\theta_1 - \mathscr{S}\frac{\partial \phi_1}{\partial z} - \theta_b' w_1\right\} \\ & +\phi_a\left\{\frac{\partial}{\partial z}\left\{\left(\theta_b - \mathscr{C}\right)\phi_1 - \left(1-\phi_b\right)\theta_1\right\} + \theta_b' w_1\right\}\Big] dV \end{aligned} \tag{21}$$

where $dV = dx\, dy\, dz$. Right-hand side is given by

$$\begin{aligned} RHS = \lim_{L\to\infty}\frac{1}{4L^2}\int_0^\delta\int_{-L}^{L}\int_{-L}^{L} & \left[-w_a\mathscr{R}_1\left(\Delta_2\theta_0\right)\right. \\ & +\theta_a\left\{\frac{\partial}{\partial \tau}\left(\theta_0 - S\phi_0\right) + \frac{\partial^2 u_{P_0}}{\partial x \partial z}\frac{\partial \theta_0}{\partial x} + \frac{\partial^2 u_{P_0}}{\partial y \partial z}\frac{\partial \theta_0}{\partial y} - \left(\Delta_2 u_{P_0}\right)\frac{\partial \theta_0}{\partial z}\right\} \\ & +\phi_a\left\{\frac{\partial}{\partial \tau}\left\{\left(\theta_b - \mathscr{C}\right)\phi_0 - \left(1-\phi_b\right)\theta_0\right\} - \frac{\partial}{\partial z}\left(\theta_0\phi_0\right)\right. \\ & \left.\left. -\frac{\partial^2 u_{P_0}}{\partial x \partial z}\frac{\partial \theta_0}{\partial x} - \frac{\partial^2 u_{P_0}}{\partial y \partial z}\frac{\partial \theta_0}{\partial y} + \left(\Delta_2 u_{P_0}\right)\frac{\partial \theta_0}{\partial z}\right\}\right] dV \end{aligned} \tag{22}$$

Integration by parts and use of boundary conditions simplify LHS as

$$\begin{aligned} LHS = \lim_{L\to\infty}\frac{1}{4L^2}\int_0^\delta\int_{-L}^{L}\int_{-L}^{L} & \left[w_1\left\{\nabla^2 w_a - \theta_b'\left(\theta_a - \phi_a\right)\right\} + \theta_1\left\{\mathscr{R}_c\left(\Delta_2 w_a\right)\right.\right. \\ & \left.\left. +\left(\nabla^2 - \frac{\partial}{\partial z}\right)\theta_a - \left(1-\phi_b\right)\frac{\partial \phi_a}{\partial z}\right\} + \phi_1\left\{\mathscr{S}\frac{\partial \theta_a}{\partial z} + \left(\mathscr{C} - \theta_b\right)\frac{\partial \phi_a}{\partial z}\right\}\right] dV \end{aligned}$$

which is zero because by the adjoint property. Now we simplify the RHS by writing as

$$RHS = I_1 + I_2 + I_3 + I_4 \tag{23}$$

where

$$
\begin{aligned}
I_1 &= \lim_{L\to\infty} \frac{1}{4L^2} \int_0^{\delta} \int_{-L}^{L} \int_{-L}^{L} \{-w_a \mathscr{R}_1 (\Delta_2 \theta_0)\}\, dV \\
I_2 &= \lim_{L\to\infty} \frac{1}{4L^2} \int_0^{\delta} \int_{-L}^{L} \int_{-L}^{L} \left\{\theta_a \frac{\partial}{\partial \tau} (\theta_0 - \mathscr{S} \phi_0) + \phi_a \frac{\partial}{\partial \tau} \{(\theta_b - \mathscr{C}) \phi_0 \right. \\
&\quad - (1 - \phi_b) \theta_0\}\}\, dV \\
I_3 &= \lim_{L\to\infty} \frac{1}{4L^2} \int_0^{\delta} \int_{-L}^{L} \int_{-L}^{L} (\theta_a - \phi_a) \left\{ \frac{\partial^2 u_{P_0}}{\partial x \partial z} \frac{\partial \theta_0}{\partial x} + \frac{\partial^2 u_{P_0}}{\partial y \partial z} \frac{\partial \theta_0}{\partial y} \right\} dV \\
I_4 &= \lim_{L\to\infty} \frac{1}{4L^2} \int_0^{\delta} \int_{-L}^{L} \int_{-L}^{L} \\
&\quad \left\{ - (\theta_a - \phi_a) \left(\Delta_2 u_{P_0}\right) \frac{\partial \theta_0}{\partial z} - \phi_a \frac{\partial}{\partial z} (\theta_0 \phi_0) \right\} dV
\end{aligned} \tag{24}
$$

We assume that the linear and adjoint solutions take the following form

$$
f(x, y, z, \tau) = A(\tau)\, \widetilde{f}(z) \eta_1(x, y) \tag{25}
$$

Here A represents the amplitude and $\eta_1(x, y) = Re\left[\sum_{j=1}^{3} e^{i\,\overline{a}_j . \overline{r}}\right]$, $\overline{a}_1 = \left(\frac{\alpha\sqrt{3}}{2}, \frac{\alpha}{2}\right), \overline{a}_2 = \left(-\frac{\alpha\sqrt{3}}{2}, \frac{\alpha}{2}\right), \overline{a}_3 = (0, -\alpha)$, $\overline{r} = (x, y)$ and α is the wavenumber. These yield

$$
\eta_1(x, y) = \cos\frac{\alpha}{2}\left(x\sqrt{3} + y\right) + \cos\frac{\alpha}{2}\left(-x\sqrt{3} + y\right) + \cos(-\alpha y) \tag{26}
$$

Now we can simplify the integrals appearing in (24) as follows:

$$
\begin{aligned}
I_1 &= \left[\alpha^2 \mathscr{R}_1 A^2 \left\{ \int_0^{\delta} \widetilde{w}_a \widetilde{\theta}_0 dz \right\}\right] \left(I_{\eta^{(2)}}\right) \\
I_2 &= \left[A \frac{dA}{d\tau} \int_0^{\delta} \{\widetilde{\theta}_a (\widetilde{\theta}_0 - \mathscr{S} \widetilde{\phi}_0) \right. \\
&\quad \left. + \widetilde{\phi}_a \left[(\theta_b - \mathscr{C}) \widetilde{\phi}_0 - (1 - \phi_b) \widetilde{\theta}_0\right]\} dz \right] \left(I_{\eta^{(2)}}\right) \\
I_3 &= A^3 \left\{ \int_0^{\delta} (\widetilde{\theta}_a - \widetilde{\phi}_a) \widetilde{\theta}_0 \left(D \widetilde{u}_{P_0}\right) dz \right\} \left(I_{\eta^{(xy)}}\right) \\
I_4 &= A^3 \left[\int_0^{\delta} \left\{ \alpha^2 (\widetilde{\theta}_a - \widetilde{\phi}_a) \widetilde{u}_{P_0} \left(D \widetilde{\theta}_0\right) - \widetilde{\phi}_a D \left(\widetilde{\theta}_0 \widetilde{\phi}_0\right) \right\} dz \right] \left(I_{\eta^{(3)}}\right)
\end{aligned} \tag{27}
$$

where $D = \frac{d}{dz}$ and $I_{\eta^{(2)}}$, $I_{\eta^{(xy)}}$, $I_{\eta^{(3)}}$ are given by

$$\begin{aligned} I_{\eta^{(2)}} &= \lim_{L\to\infty} \frac{1}{4L^2} \int_{-L}^{L} \int_{-L}^{L} \eta_1^2 dx\, dy \\ I_{\eta^{(xy)}} &= \lim_{L\to\infty} \frac{1}{4L^2} \int_{-L}^{L} \int_{-L}^{L} \left\{ \left(\frac{\partial \eta_1}{\partial x}\right)^2 + \left(\frac{\partial \eta_1}{\partial y}\right)^2 \right\} \eta_1\, dx\, dy \\ I_{\eta^{(3)}} &= \lim_{L\to\infty} \frac{1}{4L^2} \int_{-L}^{L} \int_{-L}^{L} \eta_1^3 dx\, dy \end{aligned} \tag{28}$$

Carrying out the integrations and taking the limits of (28) and equating LHS and RHS, we obtain an equation satisfied by the amplitude as

$$c_1 \frac{dA}{d\tau} = c_2 A + c_3 A^2 \tag{29}$$

where c_1, c_2, c_3 can be expressed as

$$\begin{aligned} c_1 &= \int_0^{\delta} \left\{ \widetilde{\theta}_a \left(\mathscr{S} \widetilde{\phi}_0 - \widetilde{\theta}_0 \right) + \widetilde{\phi}_a \left[\left(\mathscr{C} - \theta_b \right) \widetilde{\phi}_0 + \left(1 - \phi_b \right) \widetilde{\theta}_0 \right] \right\} dz \\ c_2 &= \alpha^2 \mathscr{R}_1 \int_0^{\delta} \widetilde{w}_a \widetilde{\theta}_0 dz \\ c_3 &= \int_0^{\delta} \left[\alpha^2 \left(\widetilde{\theta}_a - \widetilde{\phi}_a \right) \left\{ \widetilde{u}_{P_0} \left(D\widetilde{\theta}_0 \right) + \frac{1}{2} \widetilde{\theta}_0 \left(D\widetilde{u}_{P_0} \right) \right\} - \widetilde{\phi}_a D \left(\widetilde{\theta}_0 \widetilde{\phi}_0 \right) \right] dz \end{aligned} \tag{30}$$

Solution of the Eq. (29) is obtained as

$$A(\tau) = \frac{A_0}{\left(1 + \frac{c}{b} A_0\right) e^{-b\tau} - \frac{c}{b} A_0} \tag{31}$$

where $b = \frac{c_2}{c_1}$, $c = \frac{c_3}{c_1}$, and c_1, c_2, c_3 are given in (30). Here $A_0 = A(0)$ the initial value of A. The steady-state solution for the amplitude can be obtained as $A(\infty) = -\frac{b}{c}$.

References

1. Landau, L.D.: On the problem of turbulence. C. R. Acad. Sci. U. R. S. S. **44**, 311–314 (1944)
2. Chandrasekhar, S.: Hydrodynamic and Hydromagnetic Stability. Dover Publication, New York (1961)
3. Drazin, P., Reid, W.H.: Hydrodynamic Stability. Cambridge University Press, Cambridge (1981)
4. Hills, R.N., Loper, D.E., Roberts, P.H.: A thermodynamically consistent model of a mushy zone. Q. J. Mech. Appl. Math. **36**, 505–539 (1983)
5. Fowler, A.C.: The formation of freckles in binary alloys. IMA J. Appl. Math. **35**, 159–174 (1985)

6. Huppert, H.E., et al.: Dynamic solidification of a binary melt. Nature **314**, 703–707 (1985)
7. Worster, M.G.: Solidification of an alloy from a cooled boundary. J. Fluid Mech. **167**, 481–501 (1986)
8. Vives, C., Perry, C.: Effects of magnetically damped convection during the controlled solidification of metals and alloys. Int. J. Heat Mass Trasf. **30**(3), 479–496 (1987)
9. Chen, C.F., et al.: Experimental study directional solidification of aqueous chloride solution. J. Fluid Mech. **227**, 567–586 (1991)
10. Worster, M.G.: Natural convection in a mushy layer. J. Fluid Mech. **224**, 335–359 (1991)
11. Worster, M.G.: Instabilities of the liquid and mushy regions during solidification of alloys. J. Fluid Mech. **237**, 649–669 (1992)
12. Tait, S., Jahrling, K., Jaupart, C.: The planform of compositional convection and chimney formation in a mushy layer. Nature **359**, 406–408 (1992)
13. Amberg, G., Homsy, G.M.: Nonlinear analysis of buoyant convection in binary solidification with application to channel formation. J. Fluid Mech. **252**, 79–98 (1993)
14. Anderson, D.M., Worster, M.G.: Weakly nonlinear analysis of convection in mushy layers during the solidification of binary alloys. J. Fluid Mech. **302**, 307–331 (1995)
15. Chen, C.F.: Experimental study of convection in a mushy layer during directional solidification. J. Fluid Mech. **293**, 81–98 (1995)
16. Riahi, D.N.: On nonlinear convection in mushy layers Part 1. Oscillatory modes of convection. J. Fluid Mech. **467**, 331–359 (2002)
17. Okhuysen, B.S., Riahi, D.N.: On weakly nonlinear convection in mushy layers during solidification of alloys. J. Fluid Mech. **596**, 143–167 (2008a)
18. Okhuysen, B.S., Riahi, D.N.: Flow instabilities of liquid and mushy regions during alloy solidification and under high gravity environment induced by rotation. Int. J. Eng. Sci. **46**, 189–201 (2008b)
19. Riahi, D.N.: On oscillatory modes of nonlinear compositional convection in mushy layers. Nonlinear Anal.: Real World Appl. **10**, 209–226 (2009)
20. Muddamallappa, B.D., Riahi, D.N.: Numerical investigation on mariginal stability and convection with and without magnetic field in a mushy layer. Trans. Porous Media **79**, 301–317 (2009)
21. Bhatta, D., Muddamallappa, M.S., Riahi, D.N.: On perturbation and marginal stability analysis of magneto-convection in active mushy layer. Trans. Porous Media **82**, 385–399 (2010a)
22. Bhatta, D., Muddamallappa, M.S., Riahi, D.N.: On weakly nonlinear evolution of convective flow in a passive mushy layer. Nonlinear Anal.: Real World Appl. **11**, 4010–4020 (2010b)
23. Lee, D., Alexandrov, D., Huang, H.-N.: Numerical modeling of one-dimensional binary solidification with a mushy layer evolution. Numer. Math. Theor. Methods Appl. **5**(2), 157–185 (2012)
24. Bhatta, D., Riahi, D.N., Muddamallappa, M.S.: On nonlinear evolution of convective flow in an active mushy layer. J. Eng. Math. **74**, 73–89 (2012)
25. Rees, J., David, W., Worster, M.G.: Fluxes through steady chimneys in a mushy layer during binary alloy solidification. J. Fluid Mech. **714**, 127–151 (2013)
26. Wells, A.J., Wettlaufer, J.S., Orszag, S.A.: Nonlinear mushy-layer convection with chimneys: stability and optimal solute fluxes. J. Fluid Mech. **716**, 203–227 (2013)
27. Cheney, W., Kincaid, D.: Numerical Mathematics and Computing, 6th edn. Thomson Brooks/Cole, Florence (2008)

Effect of Variable Bottom Topography on Water Wave Incident on a Finite Dock

Harpreet Dhillon and Sudeshna Banerjea

Abstract The problem of wave scattering by a finite rigid dock floating in water with variable bottom topography is investigated here. Assuming the variation of the bottom topography to be in the form of small undulations, a simplified perturbation analysis is employed to solve the problem approximately. The first-order corrections to reflection and transmission coefficients are obtained in terms of integrals involving the shape function describing the bottom topography. Two types of shape functions describing a patch of sinusoidal ripples and a Gauss-type curve are considered. For a sinusoidal patch of ripples at the bottom, first-order correction to the reflection coefficient shows a resonating behavior when the wavelength of sinusoidal bottom is half the wavelength of the incident field. It is also observed that when the dock totally shadows the sinusoidal undulations, resonance does not occur.

Keywords Wave scattering · Perturbation analysis · Variable bottom topography

1 Introduction

Wave interaction with a thin floating plate can be used to model a wide range of physical system viz breakwaters, docks, sea floes, very large floating structures etc. For this reason, this is one of the well-studied problems and a number of mathematical methods have been developed to handle these problems. The problem of water wave

H. Dhillon (✉) · S. Banerjea
Department of Mathematics, Jadavpur University, Kolkata 700032, India
e-mail: harpreetdhillon1186@gmail.com

S. Banerjea
e-mail: sbanerjee@math.jdvu.ac.in

R.N. Mohapatra et al. (eds.), *Mathematics and Computing*,
Springer Proceedings in Mathematics & Statistics 139,
DOI 10.1007/978-81-322-2452-5_28

scattering by a finite rigid dock floating on free surface was considered by a number of researchers. Most work on finite dock problems is concerned with infinitely deep water or water of uniform finite depth. An integral equation approach is used in the literature to solve these problems approximately. For example, [7, 8] used an integral equation approach to obtain an asymptotic solution for short waves. [9] also used another integral equation approach to obtain approximate solution for large wave numbers. References of more works can be found in the paper of [5] who considered oblique wave scattering by a finite dock floating on water of uniform finite depth. They formulated the problem as a mixed boundary value problem and reduced into a system of dual integral equations. Linton [10] also solved this problem using the modified residue calculus technique. However, it is unlikely that the water depth will be constant under the entire structure. For this reason, there is a need to study the effect of undulated bottom topography on the ocean waves incident on a rigid dock.

Recently, [4] investigated the problem of oblique wave scattering by a semi-infinite rigid dock with bottom undulations. In this paper, oblique wave scattering by a finite rigid dock in ocean with bottom undulations is investigated. We use a simplified perturbation method directly to the governing partial differential equation, the boundary and infinity conditions satisfied by the potential function describing the fluid motion. Use of perturbation analysis procedure produces two boundary value problems (BVPs) for the potential functions upto first order (cf. [4]). The boundary value problem for the zero-order potential function (BVP-I) is concerned with the problem of water wave scattering by a finite dock in water of uniform finite depth. As mentioned earlier, this problem was studied by Linton [10] who used the residue calculus method of complex variable theory to determine the reflection and transmission coefficients explicitly. In the present paper, we have reproduced the method of [10] to obtain expression for the velocity potential, reflection coefficient $|R_0|$. The BVP-II is a radiation problem in water of uniform finite depth. Without solving BVP-II, the first-order correction to the reflection and transmission coefficients are obtained here by a simple application of Green's integral theorem. Analytical expressions for first-order corrections to these coefficients are obtained in terms of integrals involving the shape function describing the bottom topography and the solution of BVP-I. For two different shape functions of the bottom, the first-order corrections to reflection and transmission coefficients are obtained and depicted graphically against the wave number. It is observed that when the ocean bottom has sinusoidal undulations and the undulated bottom is beneath the dock and its extent exceeds the length of the dock, first-order correction to reflection coefficient $|R_1|$ becomes very large and we say that $|R_1|$ exhibits resonance when the wavelength of the sinusoidal bottom is half the wavelength of the incident field. It is also observed that an increase in the number of ripples in the patch of sinusoidal bottom undulation induces a large amount of wave energy radiation at infinity. Also, an increase in angle of oblique incidence enhances the resonance in the first-order correction to reflection coefficient. The phenomena of occurrence of resonance in the reflection coefficient is observed when the undulations in the bottom is in the form of sinusoidal patch, irrespective of whether a barrier is present or not (cf. [1–3, 6, 11, 13]). However, in our problem, it is interesting to observe that if the extent of undulation in the ocean bottom beneath the dock is within the length of the dock, then such resonance does not occur. This shows that if the dock totally shadows the undulations, then the resonant behavior is removed.

It may be mentioned here that the results in this paper are based on the assumption of small undulations in the bottom topography. However in the literature, [12] used variational approach to study the problem of scattering of water waves by an elastic plate (ice cover) of variable thickness on the surface of ocean with undulating bottom topography. They considered a model of more complexity with arbitrary undulations in the ocean bed. The model in the chapter can be further improved by considering the arbitrary undulations in the bottom instead of small undulations and we propose to study this model using different technique in future.

2 Statement and Formulation

We consider time-harmonic potential flow in an ocean of finite depth having small undulations at the bottom. A rectangular cartesian coordinate system is chosen in which y-axis is taken vertically downward and $y = 0$ corresponds to the undisturbed free surface of water. The bottom of the ocean with small undulation is described by $y = h + \varepsilon c(x)$ where ε is a small nondimensional positive number which gives a measure of smallness of the bottom undulations and $c(x)$ is a bounded continuous function and is such that $c(x) \to 0$ as $|x| \to \infty$ so that far away from the undulations the bottom is of uniform finite depth h below the mean free surface. Let a floating dock of width $2a$ occupy the position $y = 0,\ |x| \leq a$, and a wave train be obliquely incident on the dock from the direction of negative x-axis at an angle θ. Let $\mathrm{Re}\{\phi(x, y)e^{i\nu z - i\omega t}\}$ be the velocity potential describing the irrotational motion in the fluid region where ω is the angular frequency and ν is defined below. Then the mathematical problem under consideration is to solve the following BVP for $\phi(x, y)$ satisfying:

$$(\nabla^2 - \nu^2)\phi = 0 \quad \text{in } 0 < y < h, \ -\infty < x < \infty, \tag{1}$$

the free surface condition

$$K\phi + \phi_y = 0 \quad \text{on } y = 0, \ |x| > a, \tag{2}$$

where $K = \omega^2/g$, g being the gravity,

the condition at the dock

$$\phi_y = 0 \quad \text{on } y = 0, \ |x| \leq a, \tag{3}$$

the bottom condition

$$\phi_n = 0 \quad \text{on } y = h + \varepsilon\, c(x), \tag{4}$$

n denoting the normal derivative,

the edge condition

$$\frac{\partial \phi}{\partial r} \sim A \ln r \quad as\ r = ((x \pm a)^2 + y^2)^{1/2} \to 0, \tag{5}$$

for some constant A at the edges $(\pm a, 0)$ of the dock,

and the infinity conditions

$$\phi(x, y) = \begin{cases} (e^{i\mu(x+a)} + Re^{-i\mu(x+a)})\psi_0(y) & \text{as } x \to -\infty, \\ Te^{i\mu(x-a)}\psi_0(y) & \text{as } x \to \infty, \end{cases} \tag{6}$$

where

$$\psi_0(y) = N_0^{-1} \cosh k_0(y - h) \tag{7}$$

with

$$N_0^2 = \frac{1}{2}\left(1 + \frac{\sinh 2k_0 h}{2k_0 h}\right),$$

k_0 being the unique real positive root of the transcendental equation $k \tanh kh = K$. Here, $\nu = k_0 \sin\theta$ and $\mu = k_0 \cos\theta$, R and T denote, respectively, the reflection and the transmission coefficients to be determined.

3 Method of Solution

The bottom condition (4) can be approximated up to the first order of the small parameter ε as

$$-\frac{\partial \phi}{\partial y} + \varepsilon\left[c'(x)\frac{\partial \phi}{\partial x} - c(x)\frac{\partial^2 \phi}{\partial y^2}\right] = 0 \text{ on } y = h. \tag{8}$$

The form of the approximate bottom condition (8) suggests that ϕ,R, and T have the following perturbational expansions in terms of the small parameter ε:

$$\begin{aligned} \phi(x, y; \varepsilon) &= \phi_0(x, y) + \varepsilon\phi_1(x, y) + O(\varepsilon^2), \\ R(\varepsilon) &= R_0 + \varepsilon R_1 + O(\varepsilon^2), \\ T(\varepsilon) &= T_0 + \varepsilon T_1 + O(\varepsilon^2). \end{aligned} \tag{9}$$

Substituting the expansions (9) in (1–3), (5), (6), and (8), we find after equating the coefficients of identical powers of ε^0 and ε^1 from both sides of the results, that the functions $\phi_0(x, y)$ and $\phi_1(x, y)$ satisfy the following boundary value problems, viz, BVP-I and BVP-II, respectively.

BVP-I: The function $\phi_0(x, y)$ satisfies

$$\begin{aligned}
(\nabla^2 - \nu^2)\phi_0(x, y) &= 0 \quad \text{in } 0 < y < h, \quad -\infty < x < \infty, \\
K\phi_0 + \phi_{0y} &= 0 \quad \text{on } y = 0, \ |x| > a, \\
\phi_{0y} &= 0 \quad \text{on } y = 0, \ |x| \le a, \\
\frac{\partial \phi_0}{\partial r} &\sim A \ln r \quad as \ r \to 0, \\
\phi_{0y} &= 0 \quad \text{on } y = h, \\
\phi_0(x, y) &\sim \begin{cases} (\mathrm{e}^{i\mu(x+a)} + R_0 \mathrm{e}^{-i\mu(x+a)})\psi_0(y) & \text{as } x \to -\infty, \\ T_0 \mathrm{e}^{i\mu(x-a)}\psi_0(y) & \text{as } x \to \infty. \end{cases}
\end{aligned} \tag{10}$$

BVP-II: The function $\phi_1(x, y)$ satisfies

$$\begin{aligned}
(\nabla^2 - \nu^2)\phi_1(x, y) &= 0 \quad \text{in } 0 < y < h, \quad -\infty < x < \infty, \\
K\phi_1 + \phi_{1y} &= 0 \quad \text{on } y = 0, \ |x| > a, \\
\phi_{1y} &= 0 \quad \text{on } y = 0, \ |x| \le a, \\
\phi_{1y} &= \frac{\mathrm{d}}{\mathrm{d}x}\left(c(x)\frac{\partial \phi_0(x, h)}{\partial x}\right) - \nu^2 c(x)\phi_0(x, h) \quad \text{on } y = h, \\
\frac{\partial \phi_1}{\partial r} &\sim A \ln r \quad \text{as } r \to 0, \\
\phi_1(x, y) &\sim \begin{cases} R_1 \mathrm{e}^{-i\mu(x+a)}\psi_0(y) & \text{as } x \to -\infty, \\ T_1 \mathrm{e}^{i\mu(x-a)}\psi_0(y) & \text{as } x \to \infty. \end{cases}
\end{aligned} \tag{11}$$

The BVP-I corresponds to the problem of oblique wave scattering by a rigid dock in water of uniform depth h. Linton [10] solved this problem using the residue calculus technique and obtained the reflection and transmission coefficients explicitly. The BVP-II is a radiation problem in uniform finite depth water having a dock which occupies the position $-a \le x \le a$, $y = 0$ and the bottom condition involves ϕ_0, the solution of BVP-I. Without solving for $\phi_1(x, y)$, R_1 and T_1 can be determined in terms of integrals involving the shape function $c(x)$ and $\phi_0(x, y)$.

Following the methodology of [10], and considering geometrical symmetry, it is possible to split $\phi_0(x, y)$ in this case into its symmetric and antisymmetric parts such that

$$\phi_0(x, y) = \phi_0^s(x, y) + \phi_0^a(x, y) \tag{12}$$

where

$$\phi_0^s(x, y) = \phi_0^s(-x, y), \quad \phi_0^a(x, y) = -\phi_0^a(-x, y). \tag{13}$$

The region $x < 0$ is considered and use of (13) will extend the solution into $x > 0$. Then $\phi_0^{s,a}(x, y)$ satisfy (10) with the additional condition $\phi_{0x}^s(0, y) = \phi_0^a(0, y) = 0$ and the requirement that

$$\phi_0^{s,a}(x, y) \to \frac{1}{2}(\mathrm{e}^{i\mu(x+a)} + R_0^{s,a}\mathrm{e}^{-i\mu(x+a)})\psi_0(y) \text{ as } x \to -\infty \tag{14}$$

where $\alpha_0 = -i\mu = -i(k_0^2 - \nu^2)^{\frac{1}{2}}$, $\alpha_n = (k_n^2 + \nu^2)^{\frac{1}{2}}$ $(n \geq 1)$, $\pm ik_n$ $(n = 1, 2, ..)$ are the purely imaginary roots of the transcendental equation $k \tanh kh = K$ and $\psi_n(y) = N_n^{-1} \cos k_n(y - h)$ with $N_n = \frac{1}{2}\left(1 + \frac{\sin 2k_n h}{2k_n h}\right)$.

In $-a < x < 0$, ϕ_0 has the expansion

$$\phi_0^{s,a} = \sum_{n=0}^{\infty} \frac{\varepsilon_n}{2} B_n^{s,a}(\mathrm{e}^{\beta_n x} \pm \mathrm{e}^{-\beta_n x}) \cos\left(\frac{n\pi}{h}\right)y, \tag{15}$$

where $\varepsilon_0 = 1$, $\varepsilon_n = 2$ for $n \geq 1$ and $\beta_n = \left(\left(\frac{n\pi}{h}\right)^2 + \nu^2\right)^{\frac{1}{2}}$. A_n^s, A_n^a $(n = 1, 2, \ldots)$ and B_n^s, B_n^a $(n = 0, 1, 2, \ldots)$ are unknown constants to be determined.

Using the residue calculus method, the expressions for R_0 and T_0 were obtained analytically by [10] given by

$$R_0 = \frac{1}{2}(\mathrm{e}^{2i\delta^s} + \mathrm{e}^{2i\delta^a})\mathrm{e}^{-2i\theta}\mathrm{e}^{2i\delta_\infty},$$

$$T_0 = \frac{1}{2}(\mathrm{e}^{2i\delta^s} - \mathrm{e}^{2i\delta^a})\mathrm{e}^{-2i\theta}\mathrm{e}^{2i\delta_\infty},$$

where

$$\delta_\infty = \sum_{n=1}^{\infty}\left(\tan^{-1}\left(\frac{\mu}{\beta_n}\right) - \tan^{-1}\left(\frac{\mu}{\alpha_n}\right)\right)$$

and

$$\delta^{s,a} = \arg\left[1 - \sum_{n=0}^{\infty} \frac{C_n^{s,a}}{i\mu + \beta_n}\right]$$

$C_m^{s,a}$, $m \geq 0$ are the solutions to the infinite system of real equations

$$C_m^{s,a} \pm D_m \sum_{n=0}^{\infty} \frac{C_n^{s,a}}{\beta_m + \beta_n} = \pm D_m, \; m \geq 0$$

where

$$D_0 = 2\nu e^{-2\nu a} \prod_{n=1}^{\infty} \frac{(1 - \frac{\nu}{\alpha_n})(1 + \frac{\nu}{\beta_n})}{(1 + \frac{\nu}{\alpha_n})(1 - \frac{\nu}{\beta_n})}$$

and for $m \geq 1$,

$$D_m = \frac{2\beta_m(\nu + \beta_m)(\alpha_m - \beta_m)e^{-2\beta_m a}}{(\nu - \beta_m)(\alpha_m + \beta_m)} \prod_{n=1 n \neq m}^{\infty} \frac{(1 - \frac{\beta_m}{\alpha_n})(1 + \frac{\beta_m}{\beta_n})}{(1 + \frac{\beta_m}{\alpha_n})(1 - \frac{\beta_m}{\beta_n})}$$

The constants A_n^s, A_n^a, B_n^s, B_n^a are obtained numerically from the system of linear equations

$$\sum_{n=0}^{\infty} V_n^{s,a}\left[\frac{1}{\alpha_n - \beta_m} \pm \frac{e^{-2\beta_m a}}{\alpha_n + \beta_m}\right] = \frac{1}{\alpha_0 + \beta_m} \pm \frac{e^{-2\beta_m a}}{\alpha_0 - \beta_m}, \quad m \geq 0 \tag{16}$$

$$\sum_{n=0}^{\infty} V_n^{s,a}\left[\frac{1}{\alpha_n + \beta_m} \pm \frac{e^{-2\beta_m a}}{\alpha_n - \beta_m}\right] = \frac{1}{\alpha_0 - \beta_m} \pm \frac{e^{-2\beta_m a}}{\alpha_0 + \beta_m} \mp \frac{2hN_0\beta_m B_m^{s,a} e^{-\beta_m a} \sinh 2\beta_m a}{x_0 \sin x_0 h}, m \geq 0 \tag{17}$$

after truncation, where

$$x_0 = -ik_0 \quad \text{and} \quad x_n = k_n \ (n \geq 1)$$

and

$$V_0^{s,a} = A_0^{s,a} + 1\,,$$

$$V_n^{s,a} = \frac{A_n^{s,a} N_0 x_n \sin x_n h}{N_n x_0 \sin x_0 h}.$$

In the expression (16), the upper signs are associated with superscript s and lower sign with superscript a.

To obtain R_1, we apply Green's integral theorem to the functions ϕ_0 and ϕ_1 in the regions bounded by the lines $y = 0, \ -X \leq x \leq -a;\ y = 0, \ -a \leq x \leq a;\ y = 0,\ a \leq x \leq X;\ x = \pm X,\ 0 \leq y \leq h;\ y = h, \ -X \leq x \leq X\ (X > 0,\ Y > 0)$. If L denotes the contour of this region then

$$\int_L \left(\phi_0 \frac{\partial \phi_1}{\partial n} - \phi_1 \frac{\partial \phi_0}{\partial n}\right) dL = 0$$

where n is the outward normal to the line element dL. The free surface condition satisfied by ϕ_0 and ϕ_1 ensures that there is no contribution to the integral from the lines $y = 0, \ -X \leq x \leq -a$, and $y = 0, \ a \leq x \leq X$. Also there is no contribution to the integral from the line $y = 0, \ -a \leq x \leq a$ because of the condition on the dock. As both ϕ_0 and ϕ_1 describe outgoing waves as $x \to \infty$, there is no contribution to the integral from the line $x = X, \ (0 \leq y \leq h)$ as $X \to \infty$. The only contributions arise from the integral along the line $x = -X, \ (0 \leq y \leq h)$ as $X \to \infty$ and the integral along the bottom. Making $X \to \infty$, we obtain ultimately

$$2i\mu R_1 = \int_{-\infty}^{\infty} c(x)[\phi_{0x}^2(x,h) + \nu^2\phi_0^2(x,h)]\mathrm{d}x. \tag{18}$$

Thus, R_1 given by (3) can be obtained numerically once $c(x)$ is known.

To obtain T_1, we use Green's integral theorem to the functions $\chi_0(x, y)$ $(= \phi_0(-x, y))$ and $\phi_1(x, y)$ in the same region mentioned above and making $X \to \infty$, we find

$$2i\mu T_1 = -\int_{-\infty}^{\infty} c(x)[\phi_{0x}(x,h)\phi_{0x}(-x,h) + \nu^2\phi_0(x,h)\phi_0(-x,h)]\mathrm{d}x. \tag{19}$$

T_1 given by (19) can be obtained numerically once $c(x)$ is known. It may be noted from (19) that if $c(x)$ is an odd function, then T_1 vanishes identically.

Example 1 The form of shape function $c(x)$ is taken as

$$c(x) = \begin{cases} c_0 \sin\lambda x & -\frac{m\pi}{\lambda} \leq x \leq \frac{m\pi}{\lambda}, \\ 0 & \text{otherwise}, \end{cases} \tag{20}$$

where m is a positive integer. This represents m sinusoidal ripples at the bottom with wave length $\frac{2\pi}{\lambda}$.

We consider $a < \frac{m\pi}{\lambda}$, i.e., when the sinusoidal bottom topography is beneath the dock and its extent exceeds the length of the dock then R_1 is given by

$$\begin{aligned} R_1 =& \frac{c_0}{2i\mu}\Bigg[4(N_0^{-1})^2\Bigg(\frac{(\nu^2-\mu^2)}{\lambda^2-4\mu^2}[(T_0^2-R_0^2)\mathcal{E}^+ + \mathcal{E}^-] + \frac{2R_0(\nu^2+\mu^2)}{\lambda}[\cos\lambda a - \\ &(-1)^m]\Bigg) - 4N_0^{-1}\Bigg[\sum_{n=1}^{\infty}\mathcal{F}_n^-(R_0(A_n^s+A_n^a) - T_0(A_n^s - A_n^a)) + \sum_{n=1}^{\infty}(A_n^s + A_n^a)\mathcal{F}_n^+\Bigg] \\ &+ \int_{-\frac{m\pi}{\lambda}}^{-a}\sin\lambda x\Bigg[\Bigg(\sum_{n=1}^{\infty}\alpha_n\mathcal{G}_n^+\Bigg)^2 + \nu^2\Bigg(\sum_{n=1}^{\infty}\mathcal{G}_n^+\Bigg)^2\Bigg]\mathrm{d}x + \int_a^{\frac{m\pi}{\lambda}}\sin\lambda x\Bigg[\Bigg(\sum_{n=1}^{\infty}\alpha_n\mathcal{G}_n^-\Bigg)^2 \\ &+ \nu^2\Bigg(\sum_{n=1}^{\infty}\mathcal{G}_n^-\Bigg)^2\Bigg]\mathrm{d}x + \int_{-a}^{a}\sin\lambda x\Bigg[\Bigg(\sum_{n=1}^{\infty}\beta_n\mathcal{H}_n^-\Bigg)^2 + \nu^2\Bigg(\sum_{n=1}^{\infty}\mathcal{H}_n^+\Bigg)^2\Bigg]\mathrm{d}x\Bigg] \end{aligned} \tag{21}$$

where

$$\mathscr{F}_n^{\pm} = \frac{N_n^{-1}(\nu^2 \pm i\mu\alpha_n)}{\lambda^2 + (\alpha_n \pm i\mu)^2}[(\alpha_n \pm i\mu)\sin\lambda a + \lambda\cos\lambda a - \lambda(-1)^m e^{(\alpha_n \pm i\mu)(a - \frac{m\pi}{\lambda})}]$$

$$\mathscr{G}_n^{\pm} = (A_n^s \pm A_n^a)N_n^{-1}e^{\alpha_n(\pm x + a)}$$

$$\mathscr{H}_n^{\pm} = \frac{\varepsilon_n}{2}(-1)^n(e^{\beta_n x}(B_n^s + B_n^a) \pm e^{-\beta_n x}(B_n^s - B_n^a))$$

$$\mathscr{E}^{\pm} = -2i\mu\sin\lambda a \pm \lambda\cos\lambda a \mp \lambda(-1)^m e^{-2i\mu(a - \frac{m\pi}{\lambda})}$$

while the first-order transmission coefficient is

$$T_1 \equiv 0.$$

In this case, it is observed that $|R_1|$ becomes very large when $\lambda = 2\mu$, i.e., when the wavelength of sinusoidal ripples in the ocean bottom becomes twice the wavelength of incident field.

Here a limiting process is used to evaluate $|R_1|$ as $\frac{2\mu}{\lambda} = \zeta \to 1$. The limiting value of $|R_1|$ as ζ tends to 1 is given by

$$\begin{aligned}
R_1 =& \frac{-ic_0}{\lambda}\Bigg[\frac{2(N_0^{-1})^2}{\lambda}\Bigg(i(\nu^2 - \mu^2)\sin\lambda a(T_0^2 - R_0^2 + 1) + (-1)^m(m\pi - a\lambda)\\
&[(T_0^2 - R_0^2)e^{-i\lambda(a - \frac{m\pi}{\lambda})} + e^{i\lambda(a - \frac{m\pi}{\lambda})}] + 4R_0(\nu^2 + \mu^2)[\cos\lambda a - (-1)^m]\Bigg)\\
&- 4N_0^{-1}R_0\sum_{n=1}^{\infty}\frac{(A_n^s + A_n^a)N_n^{-1}(\nu^2 - i\frac{\lambda}{2}\alpha_n)}{\lambda^2 + (\alpha_n - i\frac{\lambda}{2})^2}[(\alpha_n - i\frac{\lambda}{2})\sin\lambda a + \lambda\cos\lambda a -\\
&\lambda(-1)^m e^{(\alpha_n - i\frac{\lambda}{2})(a - \frac{m\pi}{\lambda})}] - 4N_0^{-1}\sum_{n=1}^{\infty}\frac{(A_n^s + A_n^a)N_n^{-1}(\nu^2 + i\frac{\lambda}{2}\alpha_n)}{\lambda^2 + (\alpha_n + i\frac{\lambda}{2})^2}[(\alpha_n + i\frac{\lambda}{2})\\
&\sin\lambda a + \lambda\cos\lambda a - \lambda(-1)^m e^{(\alpha_n + i\frac{\lambda}{2})(a - \frac{m\pi}{\lambda})}] + 4N_0^{-1}T_0\sum_{n=1}^{\infty}\frac{(A_n^s - A_n^a)N_n^{-1}(\nu^2 - i\frac{\lambda}{2}\alpha_n)}{\lambda^2 + (\alpha_n - i\frac{\lambda}{2})^2}\\
&[(\alpha_n - i\frac{\lambda}{2})\sin\lambda a + \lambda\cos\lambda a - \lambda(-1)^m e^{(\alpha_n - i\frac{\lambda}{2})(a - \frac{m\pi}{\lambda})}] + \int_{-\frac{m\pi}{\lambda}}^{-a}\sin\lambda x\\
&\left[\left(\sum_{n=1}^{\infty}\alpha_n\mathscr{G}_n^+\right)^2 + \nu^2\left(\sum_{n=1}^{\infty}\mathscr{G}_n^+\right)^2\right]dx + \int_a^{\frac{m\pi}{\lambda}}\sin\lambda x\left[\left(\sum_{n=1}^{\infty}\alpha_n\mathscr{G}_n^-\right)^2 + \nu^2\left(\sum_{n=1}^{\infty}\mathscr{G}_n^-\right)^2\right]dx\\
&+ \int_{-a}^{a}\sin\lambda x\left[\left(\sum_{n=1}^{\infty}\beta_n\mathscr{H}_n^-\right)^2 + \nu^2\left(\sum_{n=1}^{\infty}\mathscr{H}_n^+\right)^2\right]dx\Bigg].
\end{aligned} \tag{22}$$

The expression for R_1 when $\frac{m\pi}{\lambda} < a$ is given by

$$R_1 = \frac{c_0}{2i\mu}\left[\int_{-\frac{m\pi}{\lambda}}^{\frac{m\pi}{\lambda}} \sin\lambda x\left[\left(\sum_{n=1}^{\infty}\beta_n\mathscr{H}_n^-\right)^2 + \nu^2\left(\sum_{n=1}^{\infty}\mathscr{H}_n^+\right)^2\right]\mathrm{d}x\right] \tag{23}$$

and

$$T_1 \equiv 0.$$

In this case when the undulation in sea bed is beneath the dock and its extent is less than the length of the dock, then such resonating behavior in $|R_1|$ is not observed. Thus when the dock totally shadows the undulations, then $|R_1|$ does not exhibit any resonance.

Example 2 Another form of shape function is taken as

$$c(x) = b_0 \mathrm{e}^{-\xi^2 x^2} \quad (\xi > 0), \quad -\infty < x < \infty. \tag{24}$$

This represents a Gauss-type curve.

The expression for R_1 in this case is given by

$$\begin{aligned}
R_1 =& \frac{b_0}{2i\mu}\Bigg[4(N_0^{-1})^2\Bigg((\nu^2-\mu^2)[(R_0^2+T_0^2)\mathscr{K}^+ + \mathscr{K}^-] + \frac{R_0(\nu^2+\mu^2)\sqrt{\pi}erfc(a\sqrt{\xi^2})}{\sqrt{\xi^2}}\Bigg) \\
&+4N_0^{-1}\Bigg(\sum_{n=1}^{\infty}(\nu^2-i\mu\alpha_n)\mathscr{L}_n^+[R_0(A_n^s+A_n^a)+T_0(A_n^s-A_n^a)] + \sum_{n=1}^{\infty}(\nu^2+i\mu\alpha_n) \\
&(A_n^s+A_n^a)\mathscr{L}_n^-\Bigg) + \int_{-\infty}^{-a}\mathrm{e}^{-\xi^2x^2}\left[\left(\sum_{n=1}^{\infty}\alpha_n\mathscr{G}_n^+\right)^2 + \nu^2\left(\sum_{n=1}^{\infty}\mathscr{G}_n^+\right)^2\right]\mathrm{d}x + \int_a^{\infty}\mathrm{e}^{-\xi^2x^2} \\
&\left[\left(\sum_{n=1}^{\infty}\alpha_n\mathscr{G}_n^-\right)^2 + \nu^2\left(\sum_{n=1}^{\infty}\mathscr{G}_n^-\right)^2\right]\mathrm{d}x + \int_{-a}^{a}\mathrm{e}^{-\xi^2x^2}\left[\left(\sum_{n=1}^{\infty}\beta_n\mathscr{H}_n^-\right)^2 + \nu^2(\sum_{n=1}^{\infty}\mathscr{H}_n^+)^2\right]\mathrm{d}x\Bigg]
\end{aligned} \tag{25}$$

and

$$\begin{aligned}
T_1 =& \frac{-b_0}{i\mu}\Bigg[4(N_0^{-1})^2\Bigg(T_0R_0(\nu^2+\mu^2)\mathscr{K}^+ + T_0(\nu^2-\mu^2)\frac{\sqrt{\pi}\,erfc(a\sqrt{\xi^2})}{2\sqrt{\xi^2}}\Bigg) \\
&+2N_0^{-1}\Bigg(\sum_{n=1}^{\infty}(\nu^2+i\mu\alpha_n)\mathscr{L}_n^+[R_0(A_n^s-A_n^a)+T_0(A_n^s+A_n^a)]+ \\
&\sum_{n=1}^{\infty}(\nu^2-i\mu\alpha_n)(A_n^s-A_n^a)\mathscr{L}_n^-\Bigg) - \int_{-\infty}^{-a}\mathrm{e}^{-\xi^2x^2}\Bigg[\left(\sum_{n=1}^{\infty}\alpha_n\mathscr{G}_n^+\right) \\
&\left(\sum_{n=1}^{\infty}(A_n^s-A_n^a)N_n^{-1}\alpha_n\mathrm{e}^{\alpha_n(x+a)}\right) - \nu^2\left(\sum_{n=1}^{\infty}\mathscr{G}_n^+\right)\left(\sum_{n=1}^{\infty}(A_n^s-A_n^a)N_n^{-1}\mathrm{e}^{\alpha_n(x+a)}\right)\Bigg]\mathrm{d}x \\
&+\int_{-a}^{0}\mathrm{e}^{-\xi^2x^2}\Bigg[\Bigg(\sum_{n=0}^{\infty}\beta_n\mathscr{H}_n^-)(\sum_{n=0}^{\infty}\frac{\varepsilon_n}{2}(-1)^n\beta_n[\mathrm{e}^{-\beta_nx}(B_n^s+B_n^a)-\mathrm{e}^{\beta_nx}(B_n^s-B_n^a)]\Bigg)
\end{aligned}$$

$$+\nu^2\left(\sum_{n=0}^{\infty}\mathscr{H}_n^+\right)\left(\sum_{n=0}^{\infty}\frac{\varepsilon_n}{2}(-1)^n[\mathrm{e}^{-\beta_n x}(B_n^s+B_n^a)+\mathrm{e}^{\beta_n x}(B_n^s-B_n^a)]\right)\Bigg]\,\mathrm{d}x\Bigg] \tag{26}$$

4 Numerical Results

We have computed $|R_1|$ for different values of wave number Kh, for two types of shape functions $c(x)$ characterizing the uneveness of the bottom as mentioned earlier.

It is verified that in example 1 for $c(x) = c_0 \sin \lambda x$, $x \,\varepsilon\, [\frac{-m\pi}{\lambda}, \frac{m\pi}{\lambda}]$ (Eq. 20), when the length of the plate is made to tend to zero, then $|R_0| = 0$ and $|R_1|$ coincide with the result given in [3] and [6].

For numerical computation, the value of the nondimensional parameter $\frac{c_0}{h}$ is taken as 0.01 and λh as 1 for the figures (1 and 8) which depicts $|R_1|$ against Kh for $c(x)$ given by (20).

In Figs. 1 and 2, $|R_1|$ is plotted against Kh for $\frac{a}{h}=3$, where $\frac{a}{h} < \frac{m\pi}{\lambda h}$ with number of ripples $m=2$, 3, 5 and 7 and $\theta=\frac{\pi}{9}$ and $\frac{\pi}{3}$, respectively. In this case, the sinusoidal bottom topography is beneath the dock and its extent exceeds the length of the dock.

In both the figures it is observed that $|R_1|$ is oscillatory in nature. As the number of ripples increases, the frequency of oscillation in $|R_1|$ increases, and also more wave

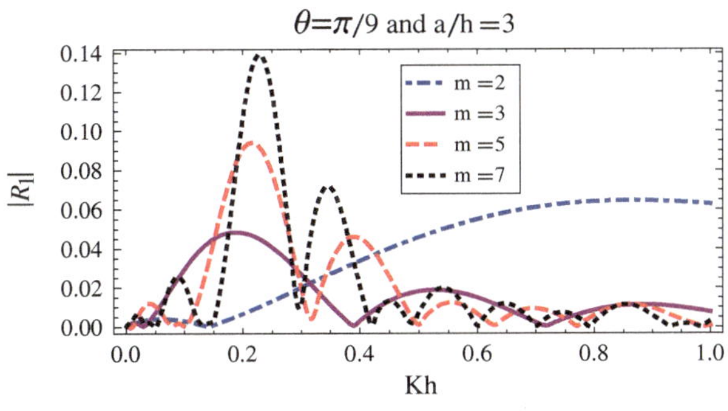

Fig. 1 First-order reflection coefficient $|R_1|$ for $c(x) = c_0 \sin \lambda x$ with $\frac{c_0}{h} = 0.01$ and $\lambda h = 1$

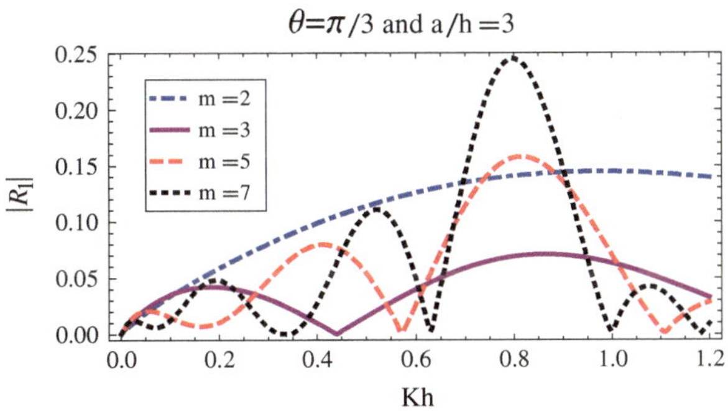

Fig. 2 First-order reflection coefficient $|R_1|$ for $c(x) = c_0 \sin \lambda x$ with $\frac{c_0}{h} = 0.01$ and $\lambda h = 1$

Fig. 3 First-order reflection coefficient $|R_1|$ for $c(x) = c_0 \sin \lambda x$ with $\frac{c_0}{h} = 0.01$ and $\lambda h = 1$

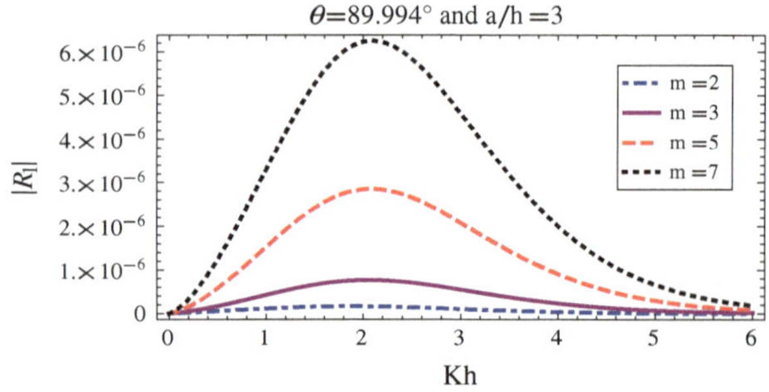

energy is reflected. As the angle of incidence increases from $\frac{\pi}{9}$ to $\frac{\pi}{3}$, it is seen that the frequency of oscillation in $|R_1|$ decreases and the peak value slightly increases. This phenomenon is due to multiple interaction of wave energy with the sinusoidal ocean bed and the dock.

In Fig. 3, θ has been taken as 89.994° (very near to $\frac{\pi}{2}$) with the number of ripples $m = 2$, 3, 5, and 7. We expect that $|R_1|$ should be very small which is indeed the case as is evident from the numerical computation. It is observed $|R_1|$ becomes almost zero, that is, the first-order reflection coefficient vanishes as the angle of incidence is taken very near to $\frac{\pi}{2}$.

In Figs. 4 and 5, $|R_1|$ is depicted against $\zeta = \frac{2\mu}{\lambda}$ for the shape function $c(x)$ given by (20) for $\frac{a}{h} = 1$, $Kh = 0.1$, $\theta = \frac{\pi}{6}, \frac{\pi}{3}$, respectively, and $m = 2, 4$. Here m is the number of sinusoidal ripples in an otherwise flat bottom. These figures show that for $m = 2$ and 4, the graph of $|R_1|$ sharply increases for $\zeta \approx 1$. Here, the value of $|R_1|$ is evaluated from Eq. (21) by a limiting process as $\zeta \to 1$. This behavior in $|R_1|$ is due to resonance which occurs when the wavelength of the ripple is half the wavelength of the incident wave. However, the peak value of $|R_1|$ is much higher for $m = 4$ than for $m = 2$. The peak value also increases as θ increases from $\frac{\pi}{6}$ to $\frac{\pi}{3}$. Therefore, an increase in the number of ripples in the patch of sinusoidal bottom undulation induces a large amount of wave energy radiation at infinity. Also the increase in angle of oblique incidence, enhances the resonance in $|R_1|$.

Fig. 4 First-order reflection coefficient $|R_1|$ against $\zeta = \frac{2\mu}{\lambda}$ for $c(x) = c_0 \sin \lambda x$ with $\frac{c_0}{h} = 0.01$ and $Kh = 0.1$

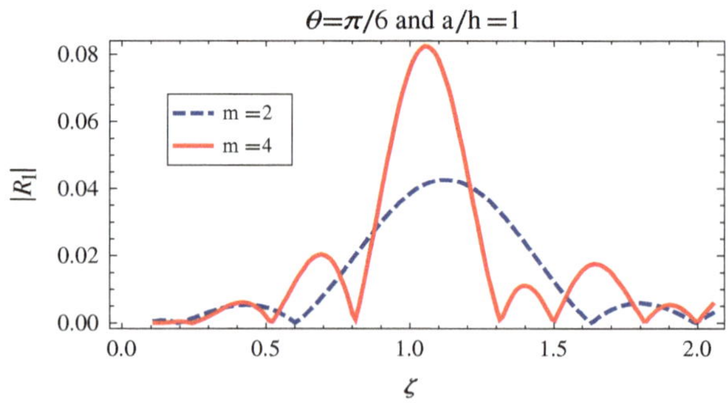

Fig. 5 First-order reflection coefficient $|R_1|$ against $\zeta = \frac{2\mu}{\lambda}$ for $c(x) = c_0 \sin \lambda x$ with $\frac{c_0}{h} = 0.01$ and $Kh = 0.1$

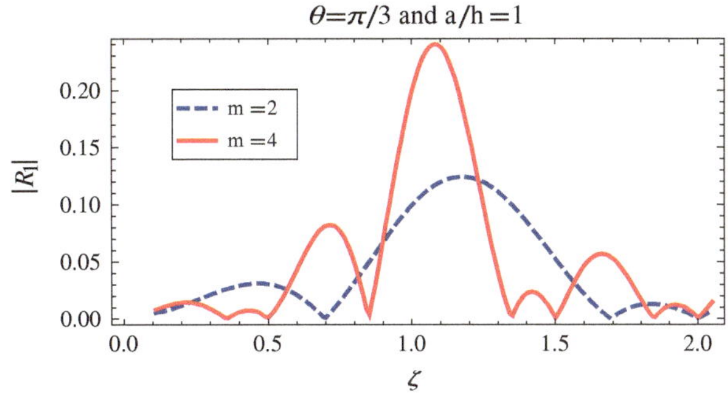

In Figs. 6, 7, and 8, $\frac{a}{h}$ has been chosen to be 18 such that $\frac{m\pi}{\lambda h} < \frac{a}{h}$ with $\theta = \frac{\pi}{9}$, $\frac{\pi}{3}$ and 89.994°. It is observed that $|R_1|$ increases as Kh increases, reaches a peak, and then starts decreasing. The peak value increases as θ increases from $\frac{\pi}{9}$ to $\frac{\pi}{3}$ and also as m increases from 2 to 5. In this case, resonance in $|R_1|$ is not observed. When θ is 89.994°, the value of $|R_1|$ becomes almost negligible. This shows that when the length of the dock overshadows the length of the sinusoidal patch in the bottom, the resonance exhibited by $|R_1|$ can be avoided.

Now, for the shape function given in example 2 by (24) which represents a Gauss-type curve, $|R_1|$ and $|T_1|$ is depicted against Kh for two different values of ξh (Figs. 9, 10, 11, 12) when $\frac{b_0}{h} = 0.1$. It is seen that for each value of ξh, $|R_1|$ and $|T_1|$

Fig. 6 First-order reflection coefficient $|R_1|$ for $c(x) = c_0 \sin \lambda x$ with $\frac{c_0}{h} = 0.01$ and $\lambda h = 1$

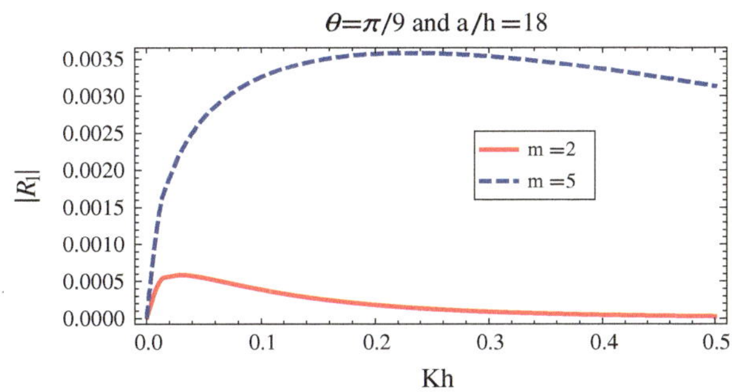

Fig. 7 First-order reflection coefficient $|R_1|$ for $c(x) = c_0 \sin \lambda x$ with $\frac{c_0}{h} = 0.01$ and $\lambda h = 1$

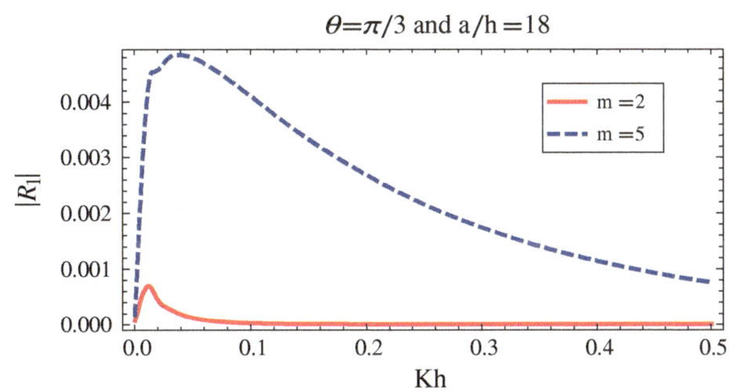

Fig. 8 First-order reflection coefficient $|R_1|$ for $c(x) = c_0 \sin \lambda x$ with $\frac{c_0}{h} = 0.01$ and $\lambda h = 1$

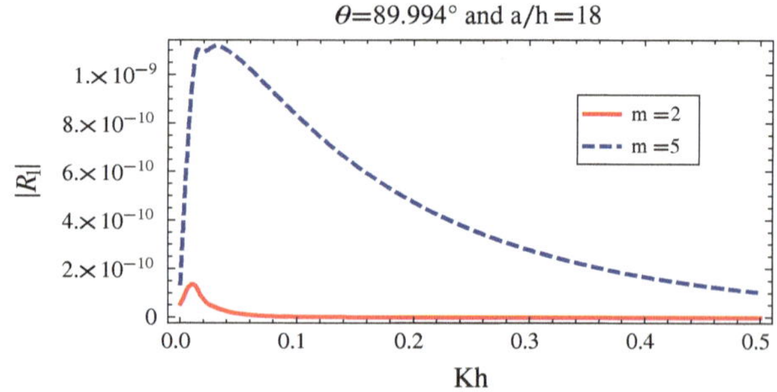

first increases with Kh, attains a maximum, and then decreases as Kh is further increased. For all the figures it is observed that the peak value of $|R_1|$ and $|T_1|$ decreases as ξh increases.

The Figs. 9 and 10 shows the variation of $|R_1|$ and $|T_1|$ against Kh for $\theta = \frac{\pi}{6}$ and $\frac{\pi}{3}$, respectively, and $\xi h = 1$. The Figs. 11 and 12 shows similar variation for $\xi h = 2$. It is observed that the peaks in $|R_1|$ and $|T_1|$ become steeper and sharp with the increase of the incident angle. Also as ξh increases, the reflection and transmission of wave energy diminishes. It is also observed from the figures that transmission coefficient is comparatively smaller than the reflection coefficient, which shows that there is more reflection of wave energy than transmission.

Fig. 9 First-order reflection and transmission coefficients $|R_1|$ and $|T_1|$ for $c(x) = b_0 \mathrm{e}^{-\xi^2 x^2}$ with $\xi h = 1$; $\frac{b_0}{h} = 0.1$

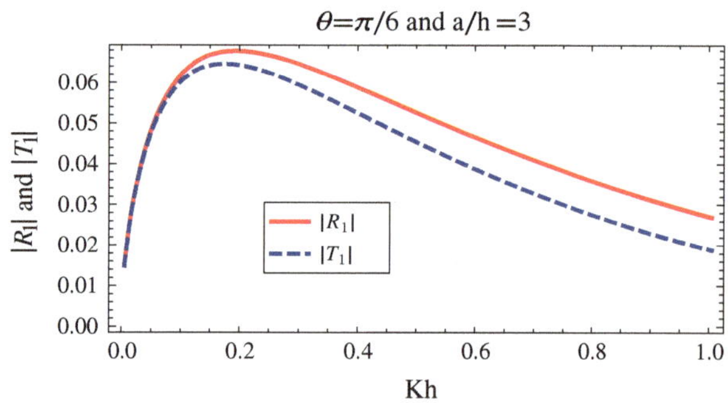

Fig. 10 First-order reflection and transmission coefficients $|R_1|$ and $|T_1|$ for $c(x) = b_0 \mathrm{e}^{-\xi^2 x^2}$ with $\xi h = 1$; $\frac{b_0}{h} = 0.1$

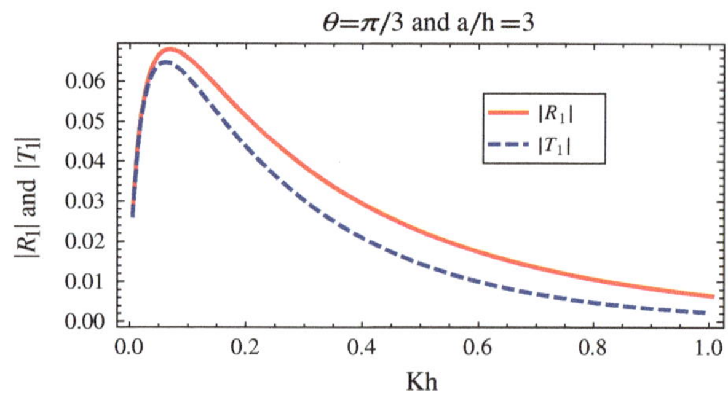

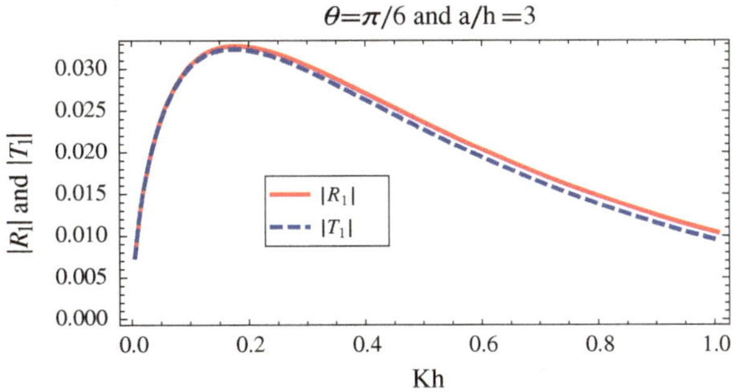

Fig. 11 First-order reflection and transmission coefficients $|R_1|$ and $|T_1|$ for $c(x) = b_0 e^{-\xi^2 x^2}$ with $\xi h = 2$; $\frac{b_0}{h} = 0.1$

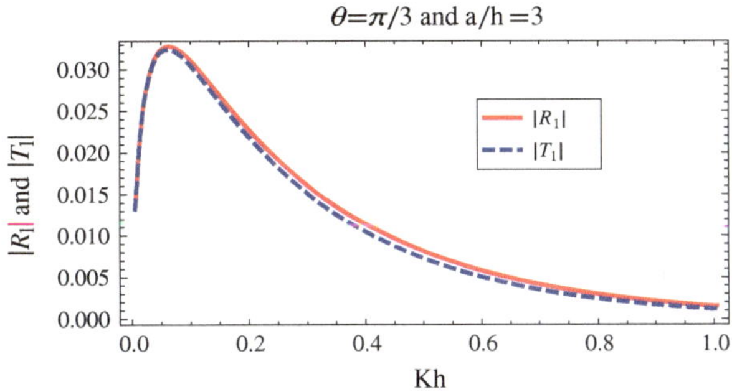

Fig. 12 First-order reflection and transmission coefficients $|R_1|$ and $|T_1|$ for $c(x) = b_0 e^{-\xi^2 x^2}$ with $\xi h = 2$; $\frac{b_0}{h} = 0.1$

5 Conclusion

The problem of oblique water wave scattering by a variable by a finite rigid dock in the presence of variable bottom topography is investigated by employing a simplified perturbation analysis. The first-order corrections to the reflection and transmission coefficients R_1 and T_1 are determined in terms of integrals involving the shape function describing the bottom and the solution of the corresponding scattering problem for uniform finite depth water. For the case of a patch of sinusoidal ripples at the bottom, $|R_1|$ is depicted in a number of figures. When the patch of sinusoidal ocean bed is beneath the dock and its extent exceeds the length of the dock then $|R_1|$ is oscillatory in nature and more over $|R_1|$ exhibits resonance when the wavelength sinusoidal bottom is half the wavelength of the incident wave. This may be attributed to the multiple interactions of the incident wave, sinusoidal bottom topography and the edge of the dock. However, when the patch of sinusoidal bottom is beneath the dock and within the extent of length of dock, $|R_1|$ does not exhibit oscillatory nature and no resonance is found in this case. In this case, the transmission coefficient vanishes identically. For the case of a Gauss-type curve, the first-order corrections to the reflection and transmission coefficients are found to decrease with increase of ξh. Also for a given value of ξh, there is more reflection than transmission of wave energy.

Acknowledgments This work is supported by CSIR, New Delhi, DST research project no. SR/SY/MS:521/08 and DST PURSE scheme and UGC (UPE II) through Jadavpur University. The authors are thankful to Prof. B. N. Mandal for his useful suggestions.

References

1. Basu, U., Mandal, B.N.: Diffraction of water waves by a deformation of the bottom. Indian J. Pure Appl. Math. **22**(9), 781–786 (1991)
2. Davies, A.G., Guazzelli, E., Belzons, M.: The propagation of long waves over an undulating bed. Phys. Fluids A **1**(8), 1331–1340 (1989)
3. Davies, A.G., Heathershaw, A.D.: Surface-wave propagation over sinusoidally varying topography. J. Fluid Mech. **144**, 419–443 (1984)
4. Dhillon, H., Banerjea, S., Mandal, B.N.: Oblique wave scattering by a semi-infinite rigid dock in the presence of bottom undulations. Indian J. Pure Appl. Math. **44**, 167–184 (2013)
5. Dorfmann, A.A., Savvin, A.A.: Diffraction of water waves by a horizontal plate. J. Appl. Math. Phys. (ZAMP) **49**, 805–826 (1998)
6. Heathershaw, A.D., Davies, A.G.: Resonant wave reflection by transverse bedforms and its relations to beaches and offshore bars. Marine Geol. **62**, 321–338 (1985)
7. Holford, R.L.: Short surface waves in the presence of a finite dock. I. Proc. Camb. Phil. Soc. **60**, 957–983 (1964)
8. Holford, R.L.: Short surface waves in the presence of a finite dock. II. Proc. Camb. Phil. Soc. **60**, 985–1011 (1964)
9. Leppington, F.G.: On the scattering of short surface waves by a finite dock. Proc. Camb. Phil. Soc. **64**, 1109–1129 (1968)
10. Linton, C.M.: The finite dock problem. J. Appl. Math. Phys. (ZAMP) **52**, 640–656 (2001)
11. Mandal, B.N., Gayen, R.: Water wave scattering by bottom undulations in the presence of a thin partially immersed barrier. Appl. Ocean Res. **28**, 113–119 (2006)
12. Porter, D., Porter, R.: Approximations to wave scattering by an ice sheet of variable thickness over undulating topography. J. Fluid Mech. **509**, 145–179 (2004)
13. Rakshit, P., Banerjea, S.: Effect of bottom undulation on the waves generated due to rolling of a plate. J. Marine Sci. Appl. **10**, 7–16 (2011)

Electrokinetic Effects on Solute Mixing Near a Conducting Obstacle Within a Microchannel

S. Bera and S. Bhattacharyya

Abstract A numerical study is made on the electroosmotic flow (EOF) near a polarizable obstacle mounted on one of the nonconducting walls of a microchannel. The external electric field induces a Debye layer of nonuniform ζ-potential along the obstacle, which results in a nonlinear electroosmostic flow. The combined effect of surface roughness and nonuniform electric double layer on the polarizable obstacle creates a vortical flow. The form of this vertical flow and its dependence on the bulk ionic concentration is analyzed. Our numerical model is based on the Navier–Stokes equations for fluid flow, Nernst–Planck equations for ionic concentration, and Poison equation for induced electric potential. We have computed the governing nonlinear coupled set of equations by the control volume method over a staggered grid system. Our results show that the form of the vortical flow, which develops in the vicinity of the obstacle, depends on the thickness of the Debye layer along the homogeneous part of the channel. The occurrence of electrical neutrality of fluid outside the Debye layer and recirculating vortex near the obstacle suggests that the fluid flow is influenced by the induced electric field and vice-versa. The vertical flow, which leads to enhanced mixing of solute in the microchannel.

Keywords Induced surface potential · Electroosmosis · Nernst–Planck equations · Finite volume method · Charge density · Micro-vortex

1 Introduction

Electroosmosis offers an alternative means to pressure gradients to drive flow in microchannels. Electrokinetic phenomena provide one of the most popular non-mechanical techniques in microfluidics. This has drawn wide interest due to its

S. Bera (✉) · S. Bhattacharyya
Department of Mathematics, Indian Institute of Technology Kharagpur, Kharagpur 721302, India
e-mail: subrata.br@gmail.com

S. Bhattacharyya
e-mail: somnath@maths.iitkgp.ernet.in

R.N. Mohapatra et al. (eds.), *Mathematics and Computing*,
Springer Proceedings in Mathematics & Statistics 139,
DOI 10.1007/978-81-322-2452-5_29

robustness, no dynamic parts, and easy to operate conditions. Electroosmosis is the preferred mode for manipulating fluids in microdevices, which is gaining increased attention. One of the most promising applications of microfluidics is the lab-on-a-chip device. EOF is the bulk fluid motion driven by the electrokinetic force acting on the net charged ions in the diffuse layer; the outer part of an electrical double layer (EDL). For an EOF in a microchannel with thin EDL (linear EOF), the EOF is often modeled using a simple slip velocity condition known as the Helmholtz–Smoluchowski equation. This boundary condition provides a linear relationship between the slip velocity and local applied electric field. The outside flow is governed by the viscous diffusion and the fluid is considered electrically neutral. The EOF in this case is an irrotational flow and is independent of the depth of the channel.

The nonlinear electrokinetic phenomenon provides a promising alternative mechanism for flow control in microfluidic devices. The vortical flow and mixing through modulation of channel wall potential has been studied by several authors namely, Ghosal [1], Erickson and Li [2], Yariv [3], Fu et al. [4], Bhattacharyya and Nayak [5], Chen and Conlisk [6], Lin and Chen [7]. It is established in those forgoing studies on EOF in a surface-modulated microchannel that the potential patch of opposite sign may induce a vortex, the bulk flow being governed by the induced pressure gradient, viscous diffusion, and electric body force.

The ζ-potential of a nonconducting surface in a classical electroosmosis is taken to be a constant. When an inert conducting surface is embedded in a dielectric medium, an external field will cause the surface to polarize. The electric field drives an ionic current in the electrolyte. These ions cannot penetrate the conducting surface and accumulate in a form of a charged cloud near the surface. This charge cloud grows and expels the electric field lines and the surface behaving like an insulator. A nonuniform charge density develops around the polarizable surface, which results in a nonconstant ζ-potential. The current drives positive ions along the surface where the initial current enters and negative ions where it leaves. Consequently, the variable ζ-potential along the conducting surface has a change of sign. Thus, the electrokinetics around a polarizable conducting surface differ. Squires and Bazant [8, 9] refereed the electroosmosis (ICEO) above a polarizable surface as the induced charge electroosmosis and the electrophoresis of a polarizable particle as induced charge electrophoresis (ICEP). ICEO is a nonlinear phenomenon and many result in the formation of vortices in a microchannel. The circulation and enhancement of species mixing by introducing conducting surface in a microchannel was investigated by Wu and Li [10]. Subsequently, Wu and Li [11] studied the ICEO around conducting hurdles which are embedded with a microchannel. Eckstein et al. [12] made a combined numerical and experimental study on vortex generation phenomena around sharp corners in microfluidic devices through the ICEO mechanism. The recent progress on ICEO and its various applications have been discussed by Bazant and Squires [13].

Microvortices have advantage in species mixing in microdevices. In other situations, the appearance of vortices needs to be suppressed so as to avoid aggregation of suspended particles leading to the eventual jumping of the device. In any case, the study on vortical flow in ICEO is important. Most of the previous studies on ICEO and micromixing are based on the thin Debye layer assumption. If the Debye layer thickness is much less than the geometric length scale of interest, the flows can be captured by prescribing a slip velocity at the outer edge of the Debye layer. The slip velocity is governed by the Helmholtz–Smoluchowski velocity $U_{\text{HS}} = -\varepsilon_e \zeta_i E_0/\mu$, where ε_e is the permittivity of the medium. The slip velocity assumption may not hold if the Debye length is finite or the applied field is strong or both. The electric body force which governs the fluid flow and the convective transport of ions are significant if either of the above conditions hold. In those cases, the momentum equations and ion transport equations are coupled and the distribution of ions are not governed by the equilibrium Boltzmann distribution. It may be noted that the Boltzmann distribution of ions is based on the assumption of thermodynamic equilibrium where convective transport of ions are neglected.

In the present study, we have modeled the ICEO based on the Navier–Stokes equations for fluid flow in which the electric body force effect is included. The transport of ions is governed by the Nernst–Planck equations and the electric field is governed by the Poisson equation. The nonlinear set of governing equations are solved in a coupled manner through a control volume approach over a staggered grid arrangement. We have studied the ICEO in the vicinity of a conducting polarizable hurdle in the form of a block mounted on the lower wall of a microchannel. Our results show that the form of vortical flow induced by the conducting hurdle depends on the ionic concentration of the electrolyte.

2 Mathematical Model

We consider a long rectangular channel of height h and width W with $h \ll W$ be fielded with an incompressible Newtonian electrolyte of uniform permittivity and viscosity. An obstacle of the form of a rectangular block of length (l) and height (d) is considered to be mounted along the lower wall of the channel (Fig. 1). Two types of electric field can be identified with their origin, one is the external electric field and the other is the induced electric field developed due to the movement of ions. The external applied electric field is generated by the electrodes placed at inlet and outlet of the channel. The number $\Lambda = E_0 h/\phi_0$ measures the strength of the external electric field in nondimensional form. The nondimensional equation for the distribution of external potential (ψ) is governed by the following Laplace equation

$$\nabla^2 \psi = 0 \tag{1}$$

The wall and all sides of the block are electrically insulated, i.e., $\nabla\psi \cdot n = 0$, where $\mathbf{n}$ is the unit outward normal. Far upstream ($x \to -\infty$) and downstream ($x \to \infty$) of the block, ψ approaches a linear function of x, i.e., $\psi = -\Lambda x$. Based on the assumption of negligible surface conductivity (Wu and Li [10]), the induced ζ potential along the surface of the obstacle is

$$\zeta_i = -\psi + \psi_c \tag{2}$$

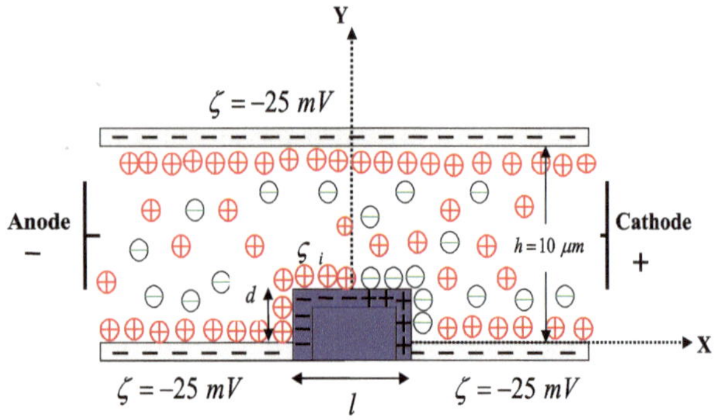

Fig. 1 Schematic diagram of the microchannel with wall-mounted conducting obstacle

where

$$\psi_c = \frac{\int_s \psi \mathrm{d}S}{S} \tag{3}$$

is the constant correction potential imposed for the requirement of electric neutrality of the obstacle surface. Here, S is the total area of the conducting surface. So, the induced zeta potential ζ_i varies with the local value of the externally applied electric field.

The Navier–Stokes equation for electroosmotic flow of a constant property of Newtonian fluid is $\nabla \cdot \mathbf{q}^* = 0$

$$\rho\left(\frac{\partial \mathbf{q}^*}{\partial t} + (\mathbf{q}^* \cdot \nabla)\mathbf{q}^*\right) = -\nabla p^* + \mu \nabla^2 \mathbf{q}^* + \rho_e^* \mathbf{E} \tag{4}$$

where $\mathbf{q}^*(= u^*,\ v^*,\ w^*)$ is the velocity field of the fluid with u^*, v^*, and w^* are the velocity components in the x, y, and z directions, respectively. The electric field $\mathbf{E}$ $(= \mathbf{E}_x, \mathbf{E}_y, \mathbf{E}_z)$ is determined by the superposition of external electric field along with the induced electric field developed due to migration of ions. The charge density ρ_e^* is related to the electric field as

$$\nabla \cdot (\varepsilon_e \mathbf{E}) = -\varepsilon_e \nabla^2 \Phi^* = \rho_e^* = \sum_i z_i e n_i \tag{5}$$

Here Φ^* is the electric potential, z_i and n_i are, respectively, the valance and molar concentration of the i type ion; e is the elementary electric charge and $\varepsilon_e = \varepsilon_0 \varepsilon_r$, where ε_0 is the permittivity of vacuum and ε_r is the dielectric constant of the solution. Total electric potential Φ^* can be written as $\Phi^* = \psi^*(x, y, z) + \phi^*(x, y, z)$, where ϕ^* is the induced electric potential. The variables with superscript $*$ denotes a dimensional quantity.

The transport equation of the ionic species i is governed by the Nernst–Planck equation as

$$\frac{\partial n_i}{\partial t} + \nabla \cdot N_i = 0 \tag{6}$$

where N_i $(= -D_i \nabla n_i + n_i \omega_i z_i F\mathbf{E} + n_i \mathbf{q}^*)$ is the net flux of ionic species. D_i and ω_i $(= D_i/RT)$ are, respectively, the diffusivity and mobility of i type species. Here R is the gas constant F is a Faraday's constant and T is the absolute temperature of the solution. If we consider symmetric electrolyte then $z_i = \pm 1$.

The velocity field is coupled with the mass transfer equation and the Poisson equation for the induced electric potential. We scale the velocity field $\mathbf{q}^*$ by the Helmholtz–Smoluchowski velocity $U_{\mathrm{HS}} = \varepsilon_e E_0 \phi_0/\mu$, electric potential Φ^* by ϕ_0 $(= K_B T/e)$, n_i by the bulk ionic concentration n_0, Cartesian coordinates by $(h,\ h,\ W)$, pressure p by $\mu U_{\mathrm{HS}}/h$ and time t by h/U_{HS}. The parameter $\kappa = [(2e^2 n_0)/(\varepsilon_e K_B T)]^{1/2}$ is reciprocal of the characteristic EDL thickness (λ_d) and $\varepsilon = \lambda_d/h = 1/\kappa h$. We denote the nondimensional concentration of cation by g and anion by f. Here, K_B is Boltzmann constant, μ is the viscosity of the electrolyte, and F is the Faraday constant.

The width (W) of the channel is considered to be on the order of the length of the channel. So, $\varepsilon_1 = h/W << 1$. Thus, all gradients with respect to z can be neglected and the flow can be treated as two-dimensional.

We can either specify the ion concentration or the flux at the surface. We assume a no-slip boundary condition along the walls and all face of the block and constant ζ-potential on the walls. The boundary conditions along the walls ($y=0$ and $y=1$) can be described as

$$u = v = 0;\ \phi = \zeta;\ g = e^{-\zeta};\ f = e^{\zeta} \tag{7}$$

Here, ζ is the surface potential of the channel wall which is constant only along the homogenous parts of the channel wall. On the conducting block, ζ is replaced by the local induced zeta potential ζ_i as defined in Eq. (3). The channel is assumed to be sufficiently long upstream and downstream of the block and the flow is assumed to be fully developed EOF at the far upstream and downstream of the block. We imposed the upstream and downstream conditions at a distance from the origin four times the length of the block. Far upstream of the block $(x = -4)$:

$$u = u^{in};\ v = 0;\ \phi = \phi^{in};\ g = g^{in},\ f = f^{in};\ \frac{\partial p}{\partial x} = 0 \tag{8}$$

Far downstream of the block $(x = +4)$:

$$\frac{\partial u}{\partial x} = 0;\ \frac{\partial v}{\partial x} = 0;\ \frac{\partial \phi}{\partial x} = 0;\ \frac{\partial g}{\partial x} = 0;\ \frac{\partial f}{\partial x} = 0;\ \frac{\partial p}{\partial x} = 0 \tag{9}$$

The values of $u = u^{\mathrm{in}},\ \phi = \phi^{\mathrm{in}},\ g = g^{\mathrm{in}},\ f = f^{\mathrm{in}}$ at the inlet and outlet of the computational domain are due to the $1-D$ fully developed electroosmotic flow over a homogeneous microchannel. The governing equations can be determined by the similar manner as described above. Under the Debye–Huckel approximation with $\kappa h \gg 1$, the electroosmotic flow in plane channel can be obtained as

$$\phi^{\text{in}} = [\zeta_1 \sinh((1-y)h)/\lambda + \zeta_2 \sinh(yh)/\lambda]/\sinh(h/\lambda) - \Lambda x \tag{10}$$

$$u^{\text{in}} = y(\zeta_1 - \zeta_2) - \zeta_1(1 - \phi/\zeta_1) \tag{11}$$

$$g^{\text{in}} = \mathrm{e}^{-\phi^{\text{in}}}; \quad f^{\text{in}} = \mathrm{e}^{+\phi^{\text{in}}} \tag{12}$$

Here, ζ_1, ζ_2 are the zeta potential at lower wall ($y = 0$) and upper wall ($y = 1$), respectively.

2.1 Mass Transport Equation

The electrokinetic transport of the uncharged sample species is considered in the present analysis. In absence of no chemical reaction or absorption of species on the wall, the transport of species are governed by convection and diffusion. The governing equation for species transport in the nondimensional form is

$$\frac{\partial C}{\partial t} + (\mathbf{q} \cdot \nabla)C = \frac{1}{Pe}\nabla^2 C \tag{13}$$

where the advection speed is $\mathbf{q}$. Here $C(x, t)$ is the dimensionless species concentration, scaled by the reference concentration C_0 and D is the diffusion coefficient of the species. Here, Peclet number is based on sample diffusivity D and is defined as $Pe = Uh/D$. No mass flux is assumed along the channel wall. Thus, the boundary conditions of C are as follows:

$C = C_{\text{in}}$ at $x = -L$; $\frac{\partial^2 C}{\partial x^2} = 0$ at $x = L$ and $\frac{\partial C}{\partial y} = 0$ at $y = 0, 1$.

where the value for C_{in} is specified as $C_{\text{in}} = 1$ on the lower inlet ($0 \leq y \leq 0.5$) and $C_{\text{in}} = 0$ on the upper inlet ($0.5 \leq y \leq 1$) of the channel. The nondimensional channel length L is taken to be sufficiently long to establish a steady mixing.

3 Numerical Methods

Our computations start by solving the Laplace Eq. (1) for a given value of Λ to determine the external electric field (ψ) and the induced potential (ζ_i) along the surface. A central difference scheme is used to discretize the Eq. (1). A line-by-line iterative method along with the successive overrelaxation technique (SOR) is used to compute the discretized equations. The grids are considered to be the same as that of the grids used for computing other variables.

We solved the coupled set of governing nonlinear equations for fluid flow and ionic species concentration through a finite volume method on a staggered grid system [14]. The discretized form of the governing equations is obtained by integrating the governing equations over each control volumes. Different control volumes are used to integrate different equations. The equations for fluid flow and ion transport involves first-order derivatives of electric potential. Near the obstacle, both the imposed and induced electric potential undergo a sharp change. In order to capture accurately the sharp change in variable values, we have used the second-order upwind scheme, QUICK (Quadratic Upwind Interpolation Convective Kinematics) [15], to discretize the convective and electromigration terms in both concentration and Navier–Stokes equations. The QUICK scheme uses a quadratic interpolation/exterpolation between the three nodal values of variables to estimate its value at the interface of the control volume. The upwind scheme imparts stability to the numerical solution in the region where a steep gradient in variables occur. An implicit first-order scheme is used for discretizing the time derivative terms. Due to coupling of equations, we solve the system of linear algebraic equations through a block elimination method (Varga [16]). The resulting discretized equations are solved iteratively through the pressure correction-based iterative algorithm SIMPLE [17]. The iteration starts by assuming the induced electric potential ϕ at every cell center.

We considered a nonuniform grid spacing along y-direction and uniform grids along the other axis and δt was taken as 0.001. To check the effects of grid spacing, computations have been performed for three different meshes with Grid 1: 200×240, Grid 2: 400×240, and Grid 3: 400×500 for fully developed EOF and compared with the results due to Mirbozorgi et al. [18]. In Grid 1 and Grid 2, we considered a nonuniform grid size where δy is assumed to vary between 0.005 to 0.01 with δx is either 0.02 (for Grid 1) or $\delta x = 0.01$ (for Grid 2). In Grid 3, we considered $\delta x = 0.01$ and $0.0025 \leq \delta y \leq 0.005$. Figure 2a, b suggests that the results obtained by Grid 2 and Grid 3 agree fairly well with each other and these results are in close agreement with the result due to Mirbozorgi et al. [18]. Thus, we find Grid 2 is optimal.

We have also tested the accuracy of our numerical algorithm by comparing with the results due to Fu et al. [4] when the EOF near a step-jump in ζ-potential is considered

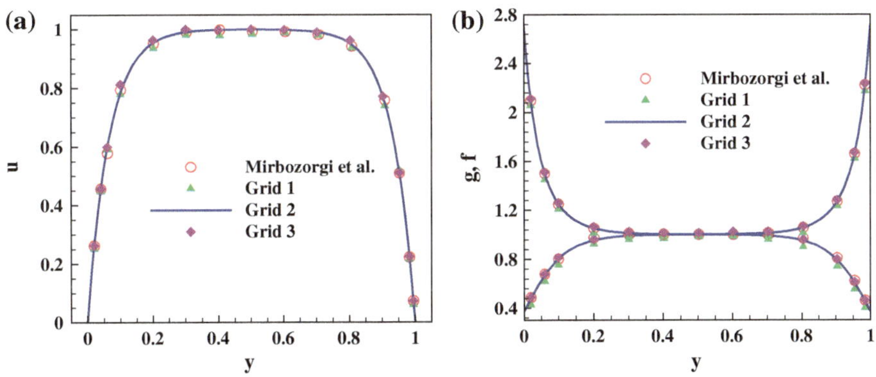

Fig. 2 Comparison of our computed solution with Mirbozorgi et al. [18] and the effects of grid size for fully developed EOF in a plane microchannel when channel height (h) is 10 μm, $\lambda_d = 0.46$ μm ($\kappa h = 21.74$), $\zeta = -25$ mV, and $Re = 0.02$. **a** Velocity; **b** ionic concentrations

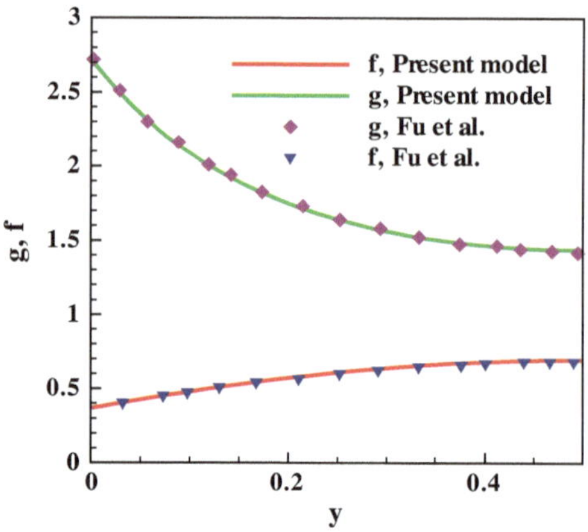

Fig. 3 Comparison of the ionic concentration for cations and anions near a step jump in ζ-potential with Fu et al. [4], when ionic strength is 10^{-2} mol/m^3 ($\kappa h = 1.04$) and external electric field corresponding to 10^5 V/m with height of the channel is 0.1 μm. *Dotted line*, *g*; *dashed line*, *f*

for ionic strength $n_0 = 10^{-4}$ mol/m^3 and external electric field corresponding to 10^5 V/m with height of the channel is 0.1 μm. Figure 3 shows that our result for ionic concentration $(g,\ f)$ is in good agreement with the result due to Fu et al. [4].

4 Results and Discussions

We consider the height of the channel $h = 10\,\mu$m, viscosity $\mu = 0.001$ Kg/m s, density $\rho = 1000$ Kg/m^3, Faraday constant $F = 96500$ Cmol^{-1}, gas constant $R = 8.315$ J/mol K at temperature $T = 300$ K, diffusion coefficient $D_+ = D_- = 2.0 \times 10^{-9}$ m^2/s. The external electric field is assumed 10^4 V/m, thus the $\Lambda = 4.0$ when $h = 10\,\mu$m. In all the computations presented below, the ζ-potential of the homogenous part of the microchannel is -1 and channel height, h is $10\,\mu$m. The Reynolds number based on Helmholtz–Smoluchowski velocity $U_{\mathrm{HS}}(= 1.788 \times 10^{-3}$ m/s) is $Re = 1.78 \times 10^{-3}$ when $\zeta = -1$, Schmidt number, $Sc = 500$, and the Peclet number $Pe = 0.89$. The EDL thickness (λ_d) varies from 0.20 to 0.96 μm when the ionic concentration of the aqueous solution varies from 2.4×10^{-3} to 10^{-4} mol/m^3. This implies that the Debye–Huckel parameter κh varies between 10.42 to 50.12 when $h = 10\,\mu$m.

Figure 4 shows the induced ζ-potential distribution along the top face of the conducting polarizable block. We determined ζ_i on the surface of the conducting block through the Eqs. (2) and (3). When an electric field is applied, the current derives positive ions into a thin diffuse layer on one side of the block and negative ions into the other side, also attracting equal and opposite surface charge on the opposite side surface of the block. After the polarization of the block, it behaves like an insulator and an induced dipolar double layer is formed. The distribution of ζ_i shows a symmetry pattern around the vertical line $y = 0$.

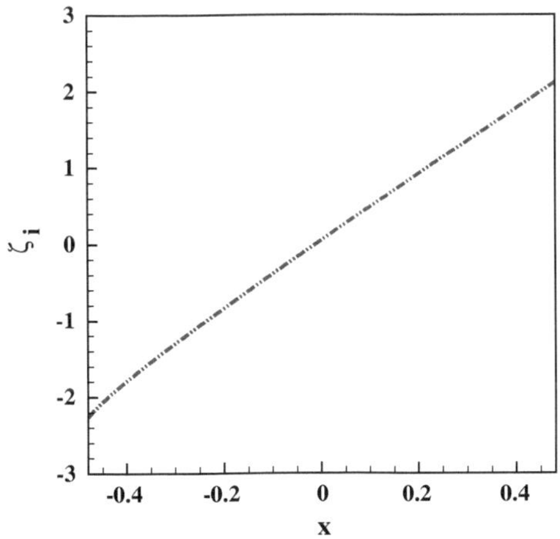

Fig. 4 Distribution of induced zeta-potential (ζ_i) along the *top face* of the block when $\kappa h = 32.84$, $h = 10\,\mu$m and $E_0 = 10^4$ V/m. The ζ-potential along the homogenous part of the channel walls is -1

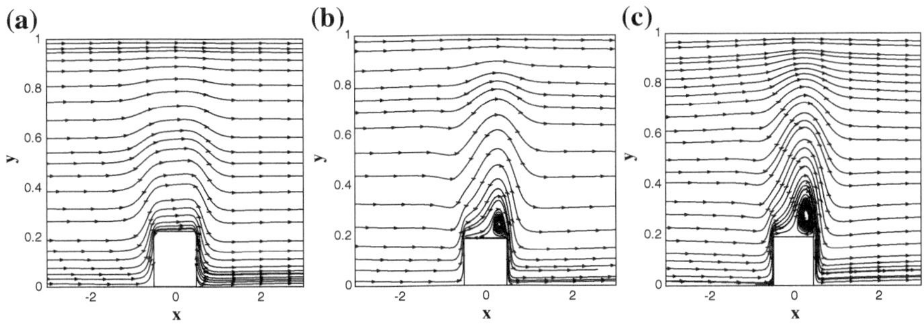

Fig. 5 Streamline patterns in the vicinity of the block when $h = 10\,\mu$m, $\zeta = -1$ and $E_0 = 10^4$ V/m. **a** Nonconducting hurdle with $\lambda_d = 0.30\,\mu$m (κh=23.2); **b** conducting hurdle with $\lambda_d = 0.43\,\mu$m (κh=23.2); **c** conducting hurdle with $\lambda_d = 0.30\,\mu$m (κh=32.84)

The streamline patterns near the block at different EDL thickness is shown in Fig. 5a–c. We have also included the case in which the block is insulated with same ζ-potential as that of the plane wall of the channel. The streamlines of the liquid flow near the polarizable block surface is distorted and micro-vortex is generated because of the nonuniform, induced charge distribution on the conducting hurdle. The flow around the nonconducting channel wall shows no vortex over the block (Fig. 5a). In the presence of an imposed electric field, the surplus ions in the diffused EDL will move. The motion of ions drag the advancement liquids and a electroomsotic flow sets-in. Along the nonconducting surface, the net charge density is uniform; whereas, a nonuniform distribution of net charge occurs along the conducting block. The opposite sign in the net charge density along the conducting block draws flow

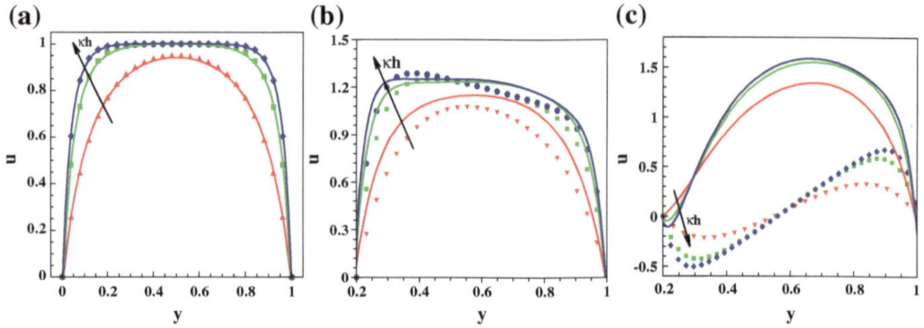

Fig. 6 Streamwise velocity profile at different x-stations when $h = 10\,\mu\text{m}$, $\zeta = -1$, $E_0 = 10^4$ V/m, and κh varies between 10.38 to 32.84. **a** $x = -2.5$ (far upstream of the block); **b** $x = -0.25$ (*top face* of the block where $\zeta_i < 0$); **c** $x = 0.25$ (*top face* of the block where $\zeta_i > 0$). *Symbols* Debye–Huckel solution. *Arrow* indicates the increasing direction of κh

in the opposite direction and a vortex develops close to the top face. The EOF away from the block is, however, independent of κh and the flow is close to the Helmholtz–Smoluchowski velocity U_{HS}.

In Fig. 6a–c, we present the u-velocity profiles far upstream of the block ($x = -2.5$) and on the upper surface of the block ($x = -0.25,\ 0.25$) for different values of the EDL thickness (λ_d) when $E_0 = 10^4$ V/m and $h = 10\,\mu$m. Along the flat surface, the EDL is homogeneous and the flow is primarily along the direction of the electric field. The u-velocity profile is plug-like and independent of the variation of EDL thickness for large κh. Outside the EDL on the plane surface, the flow is determined by the Helmholtz–Smoluchowski velocity U_{HS}. However, at $\kappa h = 10.38$, the EDL is relatively thick and the velocity profile assumes a parabolic shape even at $x = -2.5$. The velocity distribution over the flat surface ($x = -2.5$) is in agreement with the analytic solution based on the Debye–Huckel approximation as given by Eq. (11).

As we move towards the block, the fluid encounters an adverse pressure gradient due to the geometric modification of the channel wall along with the effect of nonhomogeneous EDL. The induced ζ-potential on the upstream half of the obstacle is nonhomogeneous but of the same sign as that of the plane surface. The nonuniformity of the net charge density near the hurdle results in nonuniform EOF velocity, which creates a pressure gradient along the primary flow direction. In addition, an induced pressure gradient develops due to the momentum loss. The Debye–Huckel solution underpredicts the computed solution in this region. The velocity is increased over the block due to the requirement of continuity of the fluid flow.

At $x = 0.25$ (downstream half), the ζ-potential on the upper face of the obstacle is in opposite sign and a flow reversal close to the upper face of the obstacle is evident. The Debye–Huckel solution at $x = 0.25$ shows that the region of flow reversal is much bigger than the computed result. In the Debye–Huckel solution, the influence of induced pressure gradient due to the nonuniform charge density is not considered. Thus, an EOF slip condition in this region will be incorrect. The electric body force in this region influences the flow, which makes the Navier–Stokes equations and the

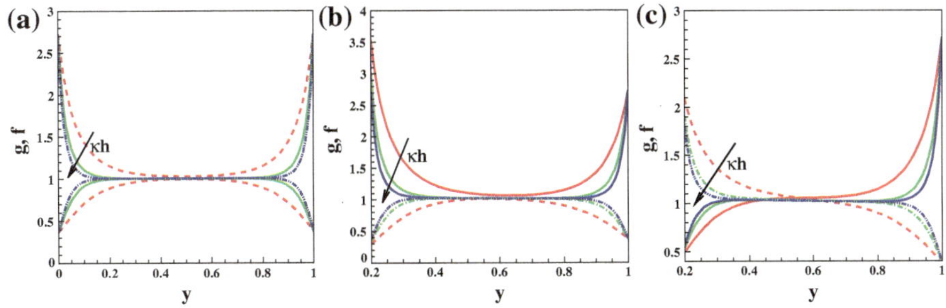

Fig. 7 Profile for ionic species concentration at different x-stations when $h = 10\,\mu\text{m}$, $\zeta = -1$, $E_0 = 10^4$ V/m, and κh varies between 10.38 to 32.84. **a** $x = -2.5$ (far upstream of the hurdle); **b** $x = -0.25$ (*top face* of the hurdle where $\zeta_i < 0$); **c** $x = 0.25$ (*top face* of the hurdle where $\zeta_i > 0$). *Solid lines*, g; *dashed lines*, f. *Arrow* indicates the increasing direction of κh

Nernst–Planck equations for ion transport coupled. It is clear from Fig. 6b, c that u depends on axial location in the region above the obstacle.

The profile for cation and anion along the cross-section of the plane channel ($x = -2.5$) and at two different x-stations over the obstacle upper face, i.e., $x = -0.25,\ 0.25$ is shown in Fig. 7a–c. The effect of fluid motion on the charge distribution along the plane channel can be neglected if the EDL is thin. Outside the Debye layer, the fluid is electrically neutral, i.e., $g - f = 0$. The ion distribution within the Debye layer is governed by the balance of electromigration and diffusion. Thus, the ion density for a thin EDL is governed by the Boltzmann distribution. However, at low κh (thick EDL), the bulk fluid is not electrically neutral and hence, the electric bodyforce outside the EDL is not negligible. For a low value of κh (e.g., $\kappa h = 10.38$), we find that the velocity profile (Fig. 7a) does not assume a plug-like form nor does the ion distribution (Fig. 7a) follow the Guy-Chapman type profile. Along the upstream half of the upper face of the hurdle ($x < 0$), ζ-potential varies but it is of the same sign as that of the homogeneous part of the channel walls. The ion distribution in this region ($x = -0.25$) is similar to the ion distribution on the homogeneous part of the channel ($x = -2.5$). Along the downstream part ($x > 0$) of the hurdle, the ζ-potential is of opposite sign in the lower EDL. We find from Fig. 7c that even for high κh, the net charge density ($g - f$) is nonzero outside the EDL. In this region, the electric body force on fluid flow is nonzero, which makes the governing equations for fluid flow and ion transport coupled. Besides, for lower values of κh (i.e., thick EDL), the bulk fluid is not electrically neutral and thus, the ions do not follow the Boltzmann distribution.

4.1 Mixing of Solute

Species mixing in a plane microchannel arises primarily from diffusion mechanisms and for that, a long mixing length is required in order to achieve a homogeneous mixing of the two sample streams. Figure 8 shows the concentration distribution

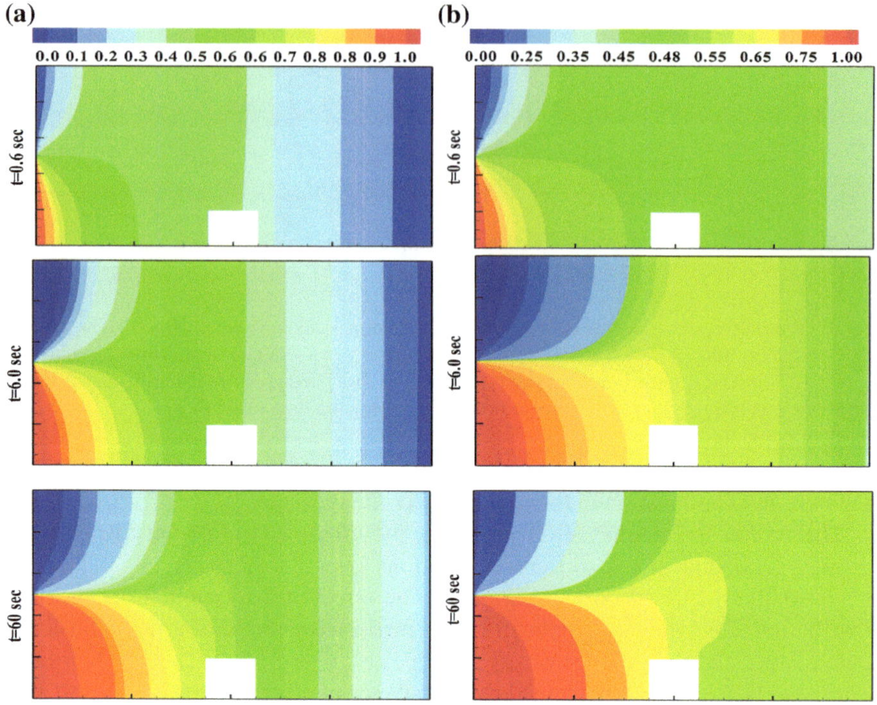

Fig. 8 Distribution of solute concentration within the channel for nonconducting obstacle as well as conducting obstacle when $h = 10\,\mu\text{m}$, $\kappa h = 32.8$, $\zeta = -1$, and $E_0 = 10^4$ V/m **a** nonconducting; **b** conducting. *First row* corresponds to $t = 0.06$ s; *second row*, $t = 0.60$ s; and *third row*, $t = 60$ s

for two cases namely (a) nonconducting block with constant ζ-potential and (b) conducting block. The transport of solute is governed by convection and diffusion. Due to the nonhomogenous surface potential of the conducting block, flow circulation is increased and hence the convection effect is significant compared to diffusion. It is evident from the results that the mixing of solute due to convection mechanism occurs at a faster rate in the downstream region for the conducting block.

5 Conclusions

Based on the nonlinear model, the electroosmotic flow (EOF) due to a conducting obstacle mounted on one wall of a microchannel is studied. A nonuniform, induced surface potential develops over the conducting obstacle and the EOF in the vicinity of the obstacle does not resemble the classical plug-like form. Recirculation vortex develops over the block and the strength of the vortex depends on the Debye layer thickness when the Debye layer is considered finite. In the vicinity of the block, our computed solution for EOF differs from the analytical solution based on the Debye–

Huckel approximation. Our result shows that the bulk fluid in the vicinity of the hurdle is not electrically neutral even for thin EDL case, i.e., $\kappa h > 30$. Thus, the distribution of ions in this region depends on convection and imposed electric field even for the case of thin EDL. For the case of thick EDL, the bulk fluid within the channel is not electrically neutral and the profile for streamwise velocity assumes a parabolic form. We find that the assumption of Boltzmann distribution of ions will be incorrect when EDL is nonhomogeneous or it is thick (low κh). Due to the electrical nonneutrality of the fluid, the electric bodyforce influences the flow outside the EDL. Thus, the assumption of free-slip condition in the region where EDL is nonhomogeneous is incorrect.

References

1. Ghosal, S.: Lubrication theory for electro-osmotic flow in a microfluidic channel of slowly varying cross-section and wall charge. J. Fluid Mech. **549**, 103–128 (2002)
2. Erickson, D., Li, D.: Influence of surface heterogeneity on electrokinetically driven microfluidic mixing. Langmuir **18**, 1883–1892 (2002)
3. Yariv, E.: Electro-osmotic flow near a surface charge discontinuity. J. Fluid Mech. **521**, 181–189 (2004)
4. Fu, L.-M., Lin, J.-Y., Yang, R.-J.: Analysis of electroosmotic flow with step change in zeta potential. J. Colloid Interfaces Sci. **258**, 266–275 (2003)
5. Bhattacharyya, S., Nayak, A.K.: Electroosmotic flow in micro/nanochannels with surface potential heterogeneity: an analysis through the Nernst-Planck model with convection effect. Colloids Surf. A **339**, 167–177 (2009)
6. Chen, L., Conlisk, A.T.: Effect of nonuniform surface potential on electroosmotic flow at large applied electric field strength. Biomed. Microdevices **11**, 251–258 (2009)
7. Lin, T.-Y., Chen, C.-L.: Analysis of electroosmotic flow with periodic electric and pressure fields via the lattice Poisson-Boltzmann method, Appl. Math. Model (2012) (in press)
8. Squires, T.M., Bazant, M.Z.: Induced-charge electro-osmosis. J. Fluid Mech. **509**, 217–252 (2004). doi:10.1017/S0022112004009309
9. Squires, T.M., Bazant, M.Z.: Breaking symmetries in induced-charge electro-osmosis and electrophoresis. J. Fluid Mech. **560**, 65–101 (2006)
10. Wu, Z., Li, D.: Mixing and flow regulating by induced-charge electrokinetic flow in a microchannel with a pair of conducting triangle hurdles. Microfluid Nanofluid **5**, 65–76 (2008). doi:10.1007/s10404-007-0227-7
11. Wu, Z., Li, D.: Micromixing using induced-charge electrokinetic flow. Electrochimica Acta **53**, 5827–5835 (2008)
12. Eckstein, Y., Yossifon, G., Seifert, A., Miloh, T.: Nonlinear electrokinetic phenomena around nearly insulated sharp tips in microflows. J. Colloid Interfaces Sci. **338**, 243–249 (2009)
13. Bazant, M.Z., Squires, T.M.: Induced-charge electrokinetic phenomena. Curr. Opin. Colloid Interface Sci. **15**, 203–213 (2010)
14. Fletcher, C.A.J.: Computational Techniques for Fluid Dynamics, vols. I and II, Springer Series in Computational Physiscs, 2nd edn. Springer, Berlin (1991)
15. Leonard, B.P.: A stable and accurate convective modelling procedure based on quadratic upstream interpolation. Comput. Methods Appl. Mech. Eng. **19**, 59–98 (1979)
16. Varga, R.S.: Matrix Iterative Numerical Analysis. Wiley, New York (1962)
17. Patankar, S.V.: Numerical Heat Transfer and Fluid Flow, Hemisphere Publishing Corporation. Taylor & Francis Group, New York (1980)
18. Mirbozorgi, S.A., Niazmand, H., Renkrizbulut, M.: Electro-osmotic flow in reservoir-connected flat microchannels with non-uniform zeta potential. J. Fluid Eng. T ASME **128**, 1133–1143 (2006)

Distribution of Primitive Polynomials Over $GF(2)$ with Respect to Their Weights

Prasanna Raghaw Mishra, Indivar Gupta and Navneet Gaba

Abstract In this paper, we study the distribution of primitive polynomials over $GF(2)$ with respect to their weights and report some interesting empirical results which can help crypto-designers to select suitable primitive polynomials. We carry out an exhaustive study of primitive polynomials over $GF(2)$ for the degrees up to 30 and figure out the cases where this distribution is symmetrically placed about its mean. We then try to address the issue of effect on variability of primitive polynomials restricted to have certain minimum weight. Further, we propose an empirical lower bound on variability of primitive polynomials when the polynomials are restricted to have at least 40 % taps. We also propose a conjecture on the relationship of degree and the most probable weight of randomly generated primitive polynomials.

Keywords Primitive polynomial · LFSR · Finite fields · Crypto-primitives

1 Introduction

Primitive polynomials over finite fields are the polynomials of great interest for cryptographers. Many of the crypto-primitives like LFSR-based generator, LFG etc. require primitive polynomials for their feedback logic. These polynomials are used in the designs to ensure maximum possible period. However, an arbitrarily chosen primitive polynomial may not be suitable for cryptographic usage. For example, an

---

P.R. Mishra · I. Gupta (✉) · N. Gaba
Scientific Analysis Group, Defence Research and Development Organization,
Metcalfe House, Delhi 110054, India
e-mail: indivar_gupta@yahoo.com; indivargupta@sag.drdo.in

P.R. Mishra
e-mail: prasanna.r.mishra@gmail.com

N. Gaba
e-mail: navneetgaba2000@yahoo.com

R.N. Mohapatra et al. (eds.), *Mathematics and Computing*,
Springer Proceedings in Mathematics & Statistics 139,
DOI 10.1007/978-81-322-2452-5_30

LFSR-based system with a sparse polynomial is susceptible to fast correlation attack [2, 3]. It is therefore, required that the polynomials are not sparse. Though taking the nonsparse polynomial saves crypto-design from various sparseness-based attack, it reduces the number of choices for primitive polynomials of a given degree. In other words, doing so affects the variability of primitive polynomials. For a randomly selected primitive polynomial of degree d, the variability (i.e., number of primitive polynomials available under the given conditions) is $\phi(2^d - 1)/d$ [4, 5]. However, no such explicit formula is available for variability when primitive polynomials are required to have a minimum weight.

We have carried out an empirical study of primitive polynomials of degree up to 30 and tried to find how the variability of primitive polynomials varies with the lower bound on polynomial weight. As far as we know, this type of work has not been done earlier. However, some work on existence of irreducible polynomials over $GF(2)$ of maximum weight can be found in the paper [1].

2 Mathematical Formulation

Let $f(x) = a_d x^d + a_{d-1}x^{d-1} + \cdots + a_1 x + a_0$ be a polynomial in $GF(2)[x]$. We define support of a polynomial Supp(f) as the set of exponents of the nonzero terms, i.e.,

$$\text{Supp}(f) = \{i \,|\, a_i \neq 0\}.$$

We define the weight of a polynomial f, denoted as w_f as the cardinality of the support of f. Mathematically,

$$w_f = \#\text{Supp}(f).$$

For a polynomial of degree d, the polynomial $x^d + x^{d-1} + \cdots + x + 1$ has maximum possible weight $d + 1$. Weight of the polynomial satisfies the obvious inequality $0 \leq w_f \leq d + 1$, where d is the degree of the polynomial f. We define the percentage weight of a polynomial f, denoted as wp_f as the weight of f expressed as a percentage of $d + 1$, where d is the degree of f. In other words,

$$wp_f = \frac{100 w_f}{d + 1} \tag{1}$$

Let S_d denote the set of all primitive polynomial of degree d over $GF(2)$. By variability of primitive polynomials, we mean the number of primitive polynomials of degree d available under some context. Here, we require primitive polynomials to have a lower bound on their weights. We define S_d^B as the set of all primitive polynomials of degree d with a weight bound B.

$$S_d^B = \{f \mid f \text{ is a primitive polynomial over } GF(2), \deg(f) = d, w_f \geq B\}.$$

We define variability of primitive polynomials of degree d with respect to weight lower bound B, denoted by v_d^B as the number of primitive polynomials of degree d with a weight bound B. We have $v_d^B = \#S_d^B$. On the similar lines, we define the percentage variability of primitive polynomials of degree d with respect to percentage weight bound b, denoted by vp_d^b as the number of primitive polynomials of degree d with a percentage weight bound b (where $b = \frac{100B}{d+1}$) expressed as the percentage of total number of primitive polynomials of degree d. As we know that there are $\phi(2^d - 1)/d$ primitive polynomials of degree d, we can write vp_d^b as,

$$vp_d^b = \frac{\#S_d^{\frac{b(d+1)}{100}} \times 100 \times d}{\phi(2^d - 1)} \tag{2}$$

3 The Methodology

To study the distribution of primitive polynomials with respect to weight, we generate all primitive polynomials of a given degree. To do this, we apply the filtering technique, i.e., to filter out primitive polynomials of degree d from the set of all polynomials of degree d. For a given degree d, there are 2^d polynomials but all of them are not required to be generated. We can put some obvious checks at the onset, e.g., for a primitive polynomial $f(x)$ we have $f(0) = f(1) = 1$. This means we are required to generate polynomials with constant term 1 and with odd weights. There are total $\sum_{i=0}^{\lfloor \frac{d}{2} \rfloor - 1} \binom{d-1}{2i+1}$ such polynomials. It can be shown that this number is equal to 2^{d-2}. We generated such 2^{d-2} polynomials for $2 \leq d \leq 30$ and each set was tested for primitivity. After filtering, we got 29 sets of polynomials corresponding to $2 \leq d \leq 30$, each set containing $\frac{\phi(2^d-1)}{d}$ polynomials. For each set, the weight distribution was calculated and the results are compiled in the form of tables. Recall that a primitive polynomial has always odd number of terms. Therefore, frequency of an even weight is always zero. We have, therefore, not listed the frequencies corresponding to even weights. One of the tables (for $d = 10$) is given for example (Table 1).

We have also plotted the number of primitive polynomials with respect to their weights for degrees $2 \leq d \leq 30$. We observe that: *"the distribution is nearly symmetric about mean when degree is of the form* $4k + 1, k > 4$". Plots for degrees 28 and 29 are given in Fig. 1

Table 1 Frequency distribution of weights for degree = 10

S. No.	Weight	Frequency
1	1	0
2	3	2
3	5	20
4	7	28
5	9	10
6	11	0
		Total: 60

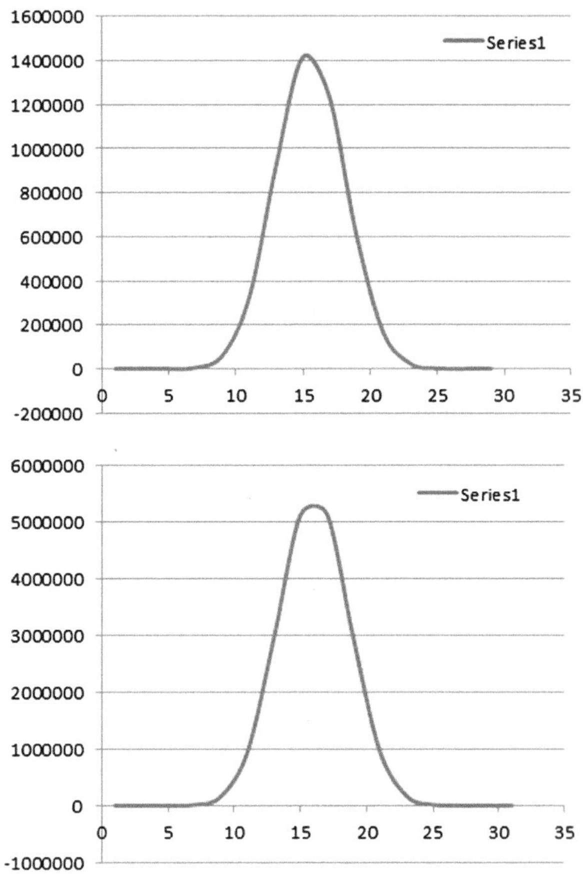

Fig. 1 Plots of number of primitive polynomials versus weight, $d = 28$ and 29

4 Primitive Polynomials with Lower Bound on Weight

As we have discussed in the introduction, for cryptographic applications, the bound on weight of primitive polynomials should be neither be too large nor too small. We have observed that a percentage lower bound 40 % is a reasonably good choice for

Table 2 Table for degree 10

S. No.	b	vp_d^b	S. No.	b	vp_d^b
1	10	100.00	6	60	63.33
2	20	100.00	7	70	16.67
3	30	96.67	8	80	16.67
4	40	96.67	9	90	0.00
5	50	63.33	10	100	0.00

cryptographic applications. To demonstrate this, first we made the table of percent weight bound (b) and the percent variability (vp_d^b) for $b = 10, 20, \ldots, 100$ and $2 \leq d \leq 30$.

Let us denote the frequency of primitive polynomials with weight w by $fr_{d,w}$. Corresponding to the percent lower bound on weight b the variability v_d^B can be given as:

$$v_d^B = \sum_{w=B}^{d+1} fr_{d,w} \tag{3}$$

where, B is the corresponding lower bound on weight connected to b by relation (2).

$$b = \frac{100B}{d+1} \tag{4}$$

Using (3) and (4), vp_d^b can be given as

$$vp_d^b = \frac{d \sum_{w=\lceil (d+1)b/100 \rceil}^{d+1} fr_{d,w}}{\phi(2^d - 1)} \times 100 \tag{5}$$

Using formula (5), the value of vp_d^b can be calculated for a given b and d from the tables created in Sect. 3. We have tabulated the values of vp_d^b against b for degrees 2 to 30. For reference, we give the table for $d = 10$ (Table 2).

For each $2 \leq d \leq 30$, we have plotted graph of percentage of primitive polynomials vp_d^b with respect to percent weight bound b and the overlapped graph[1] is shown in Fig. 2.

We draw a vertical line at 40 % mark on X-axis and also draw a horizontal line at 50 % mark on Y-axis . We see that all the plots remain well above 50 % marks when $b = 40$. Further, to see the variations more precisely, we have tabulated the values of vp_d^b for $2 \leq d \leq 30$ when $b = 40$. See Table 3.

[1]It was originally a color-coded graph which we have converted to monochrome. The actual color-coded graph can be obtained from the authors.

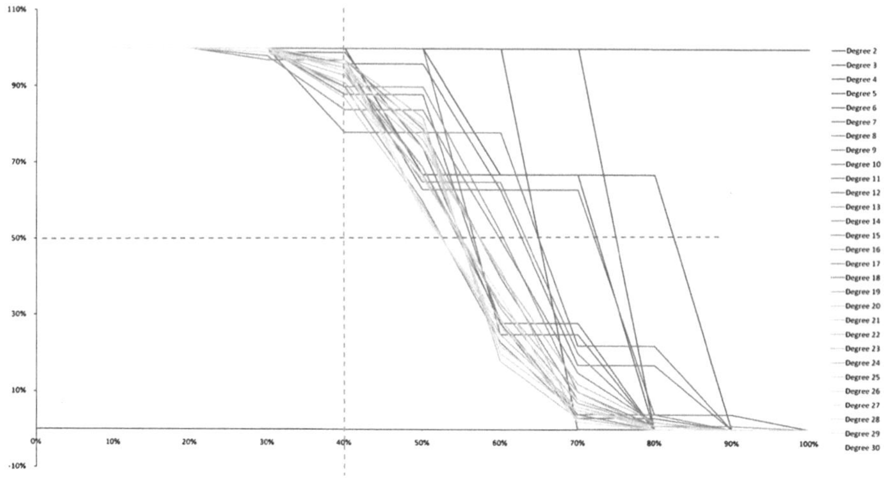

Fig. 2 Plot of percentage variability versus percent weight bound of primitive polynomials

Table 3 Percentage variability with respect to degree for $b = 40$

S. No.	Degree (d)	vp_d^b	S. No.	Degree (d)	vp_d^b
1	2	100	16	17	84.33
2	3	100	17	18	89.81
3	4	100	18	19	93
4	5	100	19	20	94.97
5	6	100	20	21	96.86
6	7	77.78	21	22	87
7	8	100	22	23	90.57
8	9	95.83	23	24	93.29
9	10	96.67	24	25	95.33
10	11	98.86	25	26	96.8
11	12	87.5	26	27	88.53
12	13	89.52	27	28	91.55
13	14	94.44	28	29	93.9
14	15	95.11	29	30	95.64
15	16	97.46	–	–	–

The graphical representation of the above table is given in Fig. 3:

From table and graph it is clear that after $d = 8$, the oscillations of N_{w_p} are confined to the range [84.33, 98.86]. If this trend continues, the percentage of primitive polynomials having at least 40 % weight will never be less than 50 %. So, the lower bound on number of primitive polynomials of degree d having weight at least 40 % is 0.5 times the total number of primitive polynomials of degree d. Precisely, the lower bound is $\frac{\phi(2^d-1)}{2d}$.

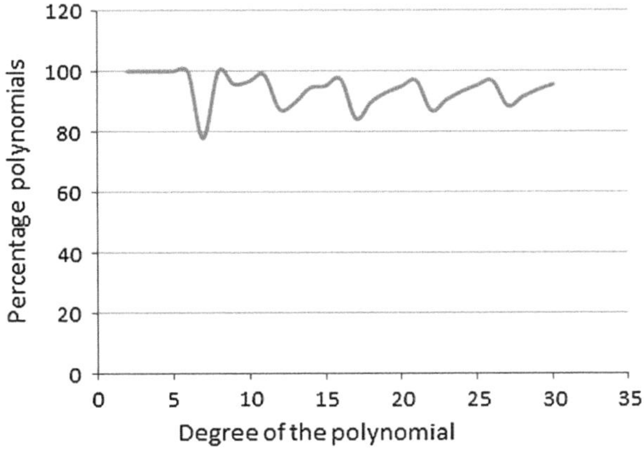

Fig. 3 Plot of percentage variability versus to degree for $b = 40$

5 The Most Probable Weight

We tried to find the most probable weights of a randomly generated primitive polynomial of degree d. The most probable weights of a randomly generated primitive polynomial of a given degree are the weights whose probability is the maximum. Probability of a weight w of a randomly generated polynomial f of degree d is given as

$$= \frac{fr_{d,w} \times d}{\phi(2^d - 1)} \tag{6}$$

Clearly the maximum probability $pr_d^{\max}$ occurs corresponding to the maximum frequency $fr_d^{\max}$, where

$$fr_d^{\max} = \max_w fr_{d,w}$$

Now $pr_d^{\max}$ can be given as,

$$pr_d^{\max} = \frac{d \times fr_d^{\max}}{\phi(2^d - 1)}$$

We define the set of most frequent weight MPW_d of a randomly generated primitive polynomial of degree d as the set

$$\text{MPW}_d = \{wt_f \,|\, fr_{d,wt_f} = fr_d^{\max}\}$$

We used the above relationship to compute the most probable weight for different degree of primitive polynomials varying from 2 to 30. These weights are listed in the following table.

S. No.	Degree (d)	Most probable weight (MPW_d)	S. No.	Degree (d)	Most probable weight (MPW_d)
1	2	3	16	17	11
2	3	3	17	18	11
3	4	3	18	19	11
4	5	5	19	20	11
5	6	5	20	21	13
6	7	5	21	22	13
7	8	5	22	23	13
8	9	7	23	24	13
9	10	7	24	25	15
10	11	7	25	26	15
11	12	7	26	27	15
12	13	9	27	28	15
13	14	9	28	29	17
14	15	9	29	30	17
15	16	9			

We observed that $\#\text{MPW}_d = 1$ for $2 \leq d \leq 30$. This means that for a given $d, 2 \leq d \leq 30$, the most probable weight is unique.

From the above table, we also deduced the relationship between degree and the most probable weight. The relationship is tested and holds good when $2 \leq d \leq 30$ and we believe that this holds good for all $d \in \mathbb{N}, d \geq 2$. So we state the conjecture:

Conjecture 1 *For a given degree d of a random primitive polynomial, the most probable weight* MPW_d *is unique and given by the relation*

$$\text{MPW}_d = 3 + 2\left\lfloor \frac{d-1}{4} \right\rfloor$$

6 Cryptographic Significance of the Work

As discussed earlier in the first section, there is a tradeoff between the weight of a primitive polynomial of a given degree and its variability. For choosing a primitive polynomial, usually filtering technique is used. In this technique, a randomly generated polynomial is checked for primitivism and other desired properties like non-sparseness etc. If the polynomial possesses the requisite properties, it is selected for its use; otherwise, we go on generating random polynomials until we hit the desired one. Success of such a method depends on the probability that a randomly selected polynomial is primitive with certain desired properties for specific applications. In other words, the success depends on the variability of primitive polynomials.

In practice, such a situation arises during formulation of a crypt design especially stream ciphers. Primitive polynomials are used in the crypt designs to ensure maximum possible period. For security reasons, we often require primitive polynomials whose weights are bounded below by a practically derived value. If a filtering algorithm is used for the purpose of selecting such primitive polynomials, it is difficult to predict the success of filtering algorithm for generating the primitive polynomials as no known formula or bound is available in literature which gives the variability of primitive polynomials having a given lower bound on weight.

Results and various observation discussed in the paper will be very useful for crypto-designers. Such formulae will also prove helpful for analysis of designs where even primitive polynomials are not known to attackers. These formulae and bounds can be used to asses the size of key space in such cases.

7 Conclusion and Future Work

We have carried out an empirical study of distribution of primitive polynomials with respect to their weights, and reported two results and proposed one conjecture. We have indicated the special cases when the distribution is symmetric about the mean. We have also tried to address the query: To give a nontrivial lower bound on number of primitive polynomials of a given degree possessing a minimum weight? Our observations also led us to conjecture about the most probable weight of a primitive polynomial of degree d when selected randomly. This analysis can be useful for establishing new results related to distribution of weights of primitive polynomials with respect to their degrees. The results given in the paper can be useful for calculating variability of special types of primitive polynomials required for designing a crypto-algorithm.

We are targeting to conduct a deeper analysis of the results and conjecture proposed in the paper. Apart from this, we will also include study of some primitive polynomial of a specific form with a given weight bound in our forthcoming works.

References

1. Ahmadi, O., Menezes, A.J.: Irreducible polynomials over maximum weight. Utilitas Mathematica **72**, 111–123 (2007)
2. Chepyzhov, V., Johansson, T., Smeets, B.: A simple algorithm for fast correlation attacks on stream ciphers. Fast software encryption' 2000, New York 2000. LNCS, vol. 1978, pp. 181–195. Springer, Berlin (2001)
3. Chepyzhov, V., Smeets, B.: On a fast correlation attack on certain stream ciphers. Eurocrypt' 91. LNCS, vol. 547, pp. 176–185. Springer, Berlin (1992)
4. Lidl, R., Niederreiter, H.: Introduction to finite fields and their applications. Cambridge University Press, Cambridge (1986)
5. Selmer, E.S.: Linear recurrence relations over finite fields (1965 Cambridge University lecture notes). Department of Mathematics, University of Bergen, Norway (1966)

Medial Left Bipotent Seminear-Rings

R. Perumal and P. Chinnaraj

Abstract In this paper, we introduce the concept of medial left bipotent seminear-rings and discuss some of their properties. We have shown that any medial seminear-ring with mate functions is a medial left bipotent seminear-ring. We also obtain a characterization of such a seminear-ring.

Keywords Medial seminear-ring · Mate function · Left (right) ideal · Insertion of factors property

1 Introduction

The concept of seminear-rings was introduced by Willy G. van Hoorn and B. van Rootselaar in 1967 [14]. Seminear-rings are a common generalization of near-rings and semi-rings. However in [14] only a very special type of seminear-rings was considered and the question arose whether it is possible to develop a more general theory of seminear-rings. As a result, seminear-rings came into being as a common generalization of near-rings and semi-rings. Right seminear-rings are algebraic systems $(R, +, .)$ with two binary associative operations, a zero 0 with $x + 0 = 0 + x = x$ and $x0 = 0x = 0$ for any $x \in R$ and one distributive law $(x + y)z = xz + yz$ for all $x, y, z \in R$. This algebraic system lacks subtraction and one distributive law. If we replace the above distributive law by $x(y + z) = xy + xz$, then R is called a left seminear-ring. A natural example of a right seminear-ring is the set $M(\Gamma)$ of all mappings on an additively written semigroup Γ with pointwise addition and composition. More precisely, let $(\Gamma, +)$ be a semigroup with zero ω

R. Perumal (✉)
Department of Mathematics, Kumaraguru College of Technology,
Coimbatore 641049, Tamilnadu, India
e-mail: perumalnew_07@yahoo.co.in

P. Chinnaraj
Department of Mathematics, Park College of Engineering and Technology,
Coimbatore 641659, Tamilnadu, India
e-mail: chinnarajwin@yahoo.co.in

R.N. Mohapatra et al. (eds.), *Mathematics and Computing*,
Springer Proceedings in Mathematics & Statistics 139,
DOI 10.1007/978-81-322-2452-5_31

and $M(\Gamma)$ the set of all maps from Γ into Γ with $\omega x = \omega$ for any $x \in M(\Gamma)$. With the definitions $\alpha(x+y) = \alpha x + \alpha y$ and $\alpha(xy) = (\alpha x)y$ for all $\alpha \in \Gamma$ and $x, y \in M(\Gamma)$, $M(\Gamma)$ is a seminear-ring. Another example of a seminear-ring that generalizes $M(\Gamma)$ is : let $\sum \subseteq End(\Gamma)$, the set of endomorphism on Γ, and define $M_{\Sigma}(\Gamma) = \{f : \Gamma \rightarrow \Gamma / f\alpha = \alpha f, \forall \alpha \in \sum\}$. Then $M_{\Sigma}(\Gamma)$ is a seminear-ring. The theory of seminear-rings has several applications in other domains of mathematics and it is natural that more systems will turn up, that can be subsumed under the theory of seminear-rings. In fact, seminear-rings appear in a natural way in theoretical computer science, mappings of semigroups, linear sequential machines and models of reversible computation. Throughout this paper, by a seminear-ring we mean a right seminear-ring with an absorbing zero. The purpose of this paper is to introduce the concept of medial left bipotent seminear-rings and obtain some of their properties. We write ab to denote the product $a.b$ for any two elements a, b in R.

2 Preliminaries

In this section, we list some basic definitions and results from the theory of seminear-rings that are used in the development of the paper.

Definition 1 (*Definition 1.115, p.41, Pilz* [12] *and van Hoorn W.G.* [14]) A nonempty set R together with two binary operations '+' (called addition) and '.' (called multiplication) is called a right seminear-ring if it satisfies the following conditions:

(i) $(R, +)$ is a semigroup
(ii) $(R, .)$ is a semigroup and
(iii) multiplication distributes over addition on the right, i.e., for all $a, b, c \in R$, $(a+b).c = (a.c) + (b.c)$

Definition 2 (*Shabir M.* [13] *and Weinert H.J.* [15]) A right seminear-ring R is said to have an absorbing zero if,

(i) $a + 0 = 0 + a = a$
(ii) $a.0 = 0.a = 0$, hold for all $a \in R$.

Definition 3 (*Javed AHSAN* [2] *and Shabir M.* [13]) A nonempty subset I of a seminear-ring R is called a left (right) ideal if,

(i) for all $x, y \in I$, $x + y \in I$ and
(ii) for all $x \in I$ and $a \in R$, $ax \in I$ $(xa \in I)$

I is said to be an ideal of R it is both a left ideal and a right ideal of R. We observe that if I is an ideal of a seminear-ring R, then I is a subseminear-ring of R.

Definition 4 A map f from R into R is called a mate function for R if for all x in R, $x = xf(x)x$. ($f(x)$ is called a mate of x).

Definition 5 (*Weinert H.J.* [16]) If S is any nonempty subset of R, then

(i) the left annihilator of S in R is $l(S) = \{x \in R/xs = 0 \text{ for all } s \in S\}$. When $S = \{s\}$, $l(S)$ is denoted by $l(s)$.
(ii) the right annihilator of S in R is $r(S) = \{x \in R/sx = 0 \text{ for all } s \in S\}$. When $S = \{s\}$, $r(S)$ is denoted by $r(s)$.

Lemma 1 *Let R be a seminear-ring. Then $l(S)$ is a left ideal where S is any nonempty subset of R.*

Proof Let $I = l(S) = \{x \in R/xs = 0\}$. For $s \in S, x, y \in I$, $(x+y)s = xs+ys = 0$ implies that $x + y \in I$. Further $x \in I, y \in R$ we have $yxs = y0 = 0$ which implies $yx \in I$. Thus I is a left ideal.

Definition 6 For any two nonempty subsets A, B of R, we denote the subset $\{x \in R/xB \subseteq A\}$ by $(A : B)$.

Lemma 2 *If A is an ideal of a seminear-ring R and B is any subset of R, then $(A : B)$ is always a left ideal.*

Proof Let $I = (A : B) = \{x \in R/xB \subseteq A\}$. For $b \in B$ and $x, y \in I$, $(x + y)b = xb + yb \subseteq A$ implies that $x + y \in I$. For $x \in I, y \in R$ we have $yxb \in yA \subseteq A$ (since A is an ideal) which implies $yx \in I$. Thus I is a left ideal.

Definition 7 (*Javed AHSAN* [2]) An ideal I of R is called

(i) a prime ideal if $AB \subseteq I$ implies $A \subseteq I$ or $B \subseteq I$ holds for all ideals A, B of R.
(ii) a semi prime ideal if $A^2 \subseteq I$ implies $A \subseteq I$ holds for all ideals A of R.
(v) a strictly prime ideal if for left ideals A, B of R, $AB \subseteq I$ implies $A \subseteq I$ or $B \subseteq I$.

Definition 8 (*Definition 9.1, p. 288, Pilz* [12]) A seminear-ring R is said to fulfill the Insertion of Factors Property—IFP for short—if for $a, b, \in R$, $ab = 0 \Rightarrow axb = 0$ for all $x \in R$.

Definition 9 R is said to have DCC (ACC) on left ideals if every descending (ascending) chain of left ideals of R terminates after a finite stage. Similarly, we can define DCC (ACC) on right ideals.

Definition 10 (*Definition 9.30, p. 300, Pilz* [12]) A seminear-ring R is called Boolean if $x^2 = x$ for all $x \in R$.

3 Main Results

In this section, we define the concept of left bipotent and medial left bipotent seminear-ring and furnish examples of these concepts. Also we derive some of the properties of medial left bipotent seminear-rings.

Definition 11 [11] We say that a seminear-ring R is left bipotent if $Ra = Ra^2$ for all $a \in R$.

Example 1 (i) Any Boolean seminear-ring is obviously left bipotent.
(ii) Let $R = \{0, a, b, c, d\}$. We define the semigroup operations "+" and "." in R as follows.

+	0	a	b	c	d
0	0	a	b	c	d
a	a	a	a	a	a
b	b	a	b	b	b
c	c	a	b	c	c
d	d	a	b	c	d

.	0	a	b	c	d
0	0	0	0	0	0
a	0	a	a	a	a
b	0	a	b	b	b
c	0	a	b	c	c
d	0	a	b	c	d

Then $(R, +, .)$ is a left bipotent seminear-ring.
(iii) We consider the seminear-ring $(R, +, .)$ where $R = \{0, a, b, c, d\}$ and the semigroup operations "+" and "." are defined as follows.

+	0	a	b	c	d
0	0	a	b	c	d
a	a	a	b	d	d
b	b	b	b	d	d
c	c	d	d	c	d
d	d	d	d	d	d

.	0	a	b	c	d
0	0	0	0	0	0
a	0	a	a	a	a
b	0	a	b	b	b
c	0	a	b	b	b
d	0	a	d	d	d

Obviously $(R, +, .)$ is a left bipotent seminear-ring. It is worth noting that this seminear-ring is not Boolean.

Definition 12 (*Pellegrini Manara.S.* [7]) A seminear-ring R is said to be medial if $xyzt = xzyt$ for all x, y, z and t in R.

Definition 13 A seminear-ring which is both medial and left bipotent is called a medial left bipotent seminear-ring.

We shall now give an example of a medial left bipotent seminear-ring.

Example 2 Let $R = \{0, a, b, c\}$. We define the semigroup operations "+" and "." in R as follows.

+	0	a	b	c
0	0	a	b	c
a	a	0	c	b
b	b	c	0	a
c	c	b	a	0

.	0	a	b	c
0	0	0	0	0
a	0	b	c	a
b	0	c	a	b
c	0	a	b	c

Obviously $(R, +, .)$ is a medial left bipotent seminear-ring.

Proposition 1 *Let R be a medial seminear-ring with a mate function f. Then R is medial left bipotent seminear-ring.*

Proof Since f is a mate function for R, we have $x = xf(x)x$ for every x in R. Clearly $f(x)x$ is an idempotent. Hence $x = x(f(x)xf(x)x) = xf(x)^2x^2$ (since R is medial). Hence $x \in Rx^2$, so that $Rx \subseteq Rx^2$. Clearly $Rx^2 \subseteq Rx$. Thus $Rx = Rx^2$. Hence R is a medial left bipotent seminear-ring.

Proposition 2 *Let R be a medial left bipotent seminear-ring. Then the following are true.*

(i) *R satisfies IFP.*
(ii) *$ab = 0$ implies $Rba = 0$ for every a, b in R.*

Proof **(i)** Suppose $ab = 0$ for some $a, b \in R$. Since R is left bipotent, for any $x \in R$, $axb = ax'b^2$ for some $x' \in R$. Since R is medial we have $ax'b^2 = ax'bb = abx'b = 0$. So that $axb = 0$. Thus R satisfies IFP.

(ii) Suppose $ab = 0$. Now $(ba)^2 = b(ab)a = 0$. Since R is left bipotent we have, $Rba = R(ba)^2 = 0$. Hence $Rba = 0$.

Proposition 3 *In a medial left bipotent seminear-ring $l(x)$ is an ideal for any $x \in R$.*

Proof By Lemma 1, $l(x)$ is a left ideal. Let us show that $l(x)$ is a right ideal. Let $z \in R$ and let $y \in l(x)$. Then by Proposition 2(i), $yzx = 0$. Hence $yz \in l(x)$ so that $l(x)R \subseteq l(x)$. Hence $l(x)$ is a right ideal. Thus $l(x)$ is an ideal.

Proposition 4 *Let R be a medial left bipotent seminear-ring. If I is any ideal of R, then $(I : S)$ is an ideal for any subset S of R.*

Proof Clearly $(I : S)$ is a left ideal by Lemma 2. Let $x \in (I : S)$ and let $s \in S$. Then $xs \in I$. Since R is left bipotent, $Rs = Rs^2$. Hence for any $y \in R$, $ys = y's^2$ for some $y' \in R$. Hence $xys = xy's^2$. Since R is medial, $xy's^2 = xy'ss = xsy's \in I$ as $xs \in I$. Hence $xys \in I$ so that $(I : S)\, R \subseteq (I : S)$. Hence $(I : S)$ is a right ideal. Thus $(I : S)$ is an ideal.

Definition 14 (*Groenewald N.J.* [4]) An ideal I of a seminear-ring R is said to be strongly prime if and only if for every $x \notin I$, there is a finite subset F of $\langle x \rangle$ such that for all $a \in R$, $Fa \subseteq I$ implies $a \in I$.

Definition 15 A seminear-ring R is called strongly prime if and only if every nonzero ideal I of R contains a finite subset F such that the right annihilator $r\,(F)$ of F is $\{0\}$.

Theorem 1 *Let R be a medial left bipotent seminear-ring without nonzero zero divisors. Then R is prime if and only if for $a, b \in R$, $aRb = 0$ implies $a = 0$ or $b = 0$.*

Proof Assume that R is prime. Suppose that $aRb = 0$. Then $a \in (0 : Rb)$. Hence $\langle a \rangle \subseteq (0 : Rb)$ so that $\langle a \rangle\, Rb = 0$ Hence by Proposition 2 (ii), $Rb\, \langle a \rangle = 0$ so that $\langle Rb \rangle\, \langle a \rangle = 0$. Since R is prime, we have $\langle Rb \rangle = 0$ or $\langle a \rangle = 0$. Hence $Rb = 0$ or $\langle a \rangle = 0$. Since R has no zero divisors, we have $a = 0$ or $b = 0$. The Converse is obvious.

Proposition 5 *Let R be a medial left bipotent seminear-ring without nonzero zero divisors. If R is prime and has D.C.C on right annihilators, then R is strongly prime.*

Proof Let I be any ideal of R and consider the collection of right annihilators $\{r(F)\}$ where F runs over all finite subsets of I. Since right annihilators satisfy D.C.C, by Zorn's lemma there exists a minimal element $M = r(F_0)$. We claim that $M = \{0\}$. For, if $M \neq 0$, then there is $0 \neq m \in M$ such that $F_0 m = 0$. Since R is prime there exists $0 \neq b \in R$ such that $mbm \neq 0$. Hence $bm \neq 0$. Let $S = r(F_0 \cup \{b\})$. Clearly $S \subseteq M$. Now $m \in M$, but $m \notin S$. Hence $S \subset M$, a contradiction. Consequently $r(F_0)(= M) = \{0\}$ and the desired result now follows.

References

1. Ahsan, J.: Seminear-rings characterized by their s-ideals. II. Proc. Jpn. Acad. **71**(A), 111–113 (1995)
2. Ahsan, J.: Seminear-rings characterized by their s-ideals. I. Proc. Jpn. Acad. **71**(A), 101–103 (1995)
3. Balakrishnan, R., Perumal, R.: Left Duo Seminear-rings. Sci. Magna **8**(3), 115–120 (2012)
4. Booth, G.L., Groenewald, N.J.: On strongly prime near-rings. Indian J. Math. **40**(2), 113–121 (1998)
5. Jat, J.L., Choudary, S.C.: On left bipotent near-rings. Proc. Edin. Math. Soc. **22**, 99–107 (1979)
6. Park, Y.S., Kim, W.J.: On structures of left bipotent near-rings. Kyungbook Math. J. **20**(2), 177–181 (1980)
7. Pellegrini Manara, S.: On medial near-rings, near-rings and near fields, Amsterdam, pp. 199–210 (1987)
8. Pellegrini Manara, S.: On regular medial near-rings. Boll. Unione Mat. Ital. VI Ser. D Algebra Geom. **6**, 131–136 (1985)
9. Perumal, R., Balakrishnan, R., Uma, S.: Some special seminear-ring structures. Ultra Sci. Phys. Sci. **23**(2), 427–436 (2011)
10. Perumal, R., Balakrishnan, R., Uma, S.: Some special seminear-ring structures II. Ultra Sci. Phy. Sci. **24**(1), 91–98 (2012)
11. Perumal, R., Balakrishnan, R.: Left bipotent seminear-rings. Int. J. Algebra **6**(26), 1289–1295 (2012)
12. Pilz Günter, Near-rings, North Holland, Amsterdam (1983)
13. Shabir, M., Ahamed, I.: Weakly regular seminearrings. Int. Electr. J. Algebra **2**, 114–126 (2007)
14. van Willy, G., Hoorn, G., van Rootselaar, R.: Fundamental notions in the theory of seminear-rings. Compos. Math. **18**, 65–78 (1967)
15. Weinert, H.J.: seminear-rings. seminearfield and their semigroup theoretical background. Semigroup Forum **24**, 231–254 (1982)
16. Weinert H.J.: Related representation theorems for rings, semi-rings, near-rings and seminear-rings by partial transformations and partial endomorphisms. Proc. Edinburgh Math. Soc. **20**, 307–315 (1976–77)

Subcentral Automorphisms

R.G. Ghumde and S.H. Ghate

Abstract A concept of subcentral automorphisms of group G with respect to a characterstic subgroup M of $Z(G)$ along with relevant mathematical paraphernalia has been introduced. With the help of this, a number of results on central automorphisms have been generalized.

Keywords Central automorphisms · Subcentral automorphisms · Purely nonabelian group

1 Introduction

Let G be a group. We shall denote the commutator, center, group of automorphisms, and group of inner automorphisms of G by G', $Z(G)$, $\mathrm{Aut}(G)$, and $\mathrm{Inn}(G)$, respectively. Let $\exp(G)$ denote the exponent of G.

For any group H and abelian group K, let $\mathrm{Hom}(H, K)$ denote the group of all homomorphisms from H to K. This is an abelian group with binary operation $fg(x) = f(x)g(x)$ for $f, g \in \mathrm{Hom}(H, K)$.

An automorphism α of G is called central if $x^{-1}\alpha(x) \in Z(G)$ for all $x \in G$. The set of all central automorphisms of G, which is here denoted by $\mathrm{Aut}_c(G)$, is a normal subgroup of $\mathrm{Aut}(G)$. Notice that $\mathrm{Aut}_c(G) = C_{\mathrm{Aut}(G)}(\mathrm{Inn}(G))$, the centralizer of the subgroup $\mathrm{Inn}(G)$ in the group Aut(G). The elements of $\mathrm{Aut}_c(G)$ act trivially on G'.

There have been number of results on the central automorphisms of a group. M.J. Curran [2] proved that, "For any non abelian finite group G, $\mathrm{Aut}_z^z(G)$ is isomorphic with Hom $(G/G'Z(G), Z(G))$, where $\mathrm{Aut}_z^z(G)$ is group of all those central

R.G. Ghumde (✉)
Department of Mathematics, Ramdeobaba College of Engineering and Management, Nagpur 440013, India
e-mail: ranjitghumde@gmail.com

S.H. Ghate
Department of Mathematics, R.T.M. Nagpur University, Nagpur 440013, India
e-mail: sureshghate@gmail.com

R.N. Mohapatra et al. (eds.), *Mathematics and Computing*,
Springer Proceedings in Mathematics & Statistics 139,
DOI 10.1007/978-81-322-2452-5_32

automorphisms which preserve the centre $Z(G)$ elementwise." In [3], Franciosi et al. showed that, If "$Z(G)$ is torsion free and $Z(G)/G' \cap Z(G)$ is torsion, then $\mathrm{Aut}_c(G)$ acts trivially on $Z(G)$. It is an abelian and torsion free group". They further proved that,"$\mathrm{Aut}_c(G)$ is trivial when $Z(G)$ is torsion free and G/G' is torsion." In [5], Jamali et al. proved that, "For a finite group G in which $Z(G) \leq G'$, $\mathrm{Aut}_c(G) \cong \mathrm{Hom}(G/G', Z(G))$." They also proved that, "If G is a purely nonabelian finite p-group of class two (p odd), then $\mathrm{Aut}_c(G)$ is elementary abelian if and only if $\Omega_1(Z(G)) = \phi(G)$, and $exp(Z(G)) = p$ or $exp(G/G') = p$," where $\phi(G)$ is Frattini subgroup of G and $\Omega_1(Z(G)) = \langle x \in Z(G) | x^p = 1 \rangle$. Note that, a group G is called purely nonabelian if it has no nontrivial abelian direct factor. Adney [1] proved that, "If a finite group G has no abelian direct factor, then there is a one-one and onto map between $\mathrm{Aut}_c(G)$ and $\mathrm{Hom}(G, Z(G))$."

In this article, we generalize the above results to subcentral automorphisms.

2 Subcentral Automorphisms

Let M and N be two normal subgroups of G.

By $\mathrm{Aut}^N(G)$, we mean the subgroup of $\mathrm{Aut}(G)$ consisting of all automorphisms which induce identity on G/N.

By $\mathrm{Aut}_M(G)$, we mean the subgroup of $\mathrm{Aut}(G)$ consisting of all automorphisms which induce identity on M.

Let $\mathrm{Aut}_M^N(G) = \mathrm{Aut}^N(G) \cap \mathrm{Aut}_M(G)$. From now onward, M will be a characteristic central subgroup, and elements of $\mathrm{Aut}^M(G)$ will be called as subcentral automorphisms of G (with respect to subcentral subgroup M). It can be seen that, $\mathrm{Aut}^M(G)$ is a normal subgroup of $\mathrm{Aut}_c(G)$.

We further, let $C^* = \{\alpha \in \mathrm{Aut}_M(G) | \alpha\beta = \beta\alpha, \forall \beta \in \mathrm{Aut}^M(G)\}$.

Clearly, C^* is a normal subgroup of $\mathrm{Aut}(G)$. Since every inner automorphism commutes with elements of $\mathrm{Aut}_c(G)$, $\mathrm{Inn}(G) \leq C^*$. If we take $M = Z(G)$, then C^* is same as $\mathrm{Inn}(G)$.

Let $K =< \{[g, \alpha] | g \in G, \alpha \in C^*\} >$, where $[g, \alpha] \equiv g^{-1}\alpha(g)$.

If $M = Z(G)$ then $K = G'$. However, in general, G' is a subgroup of K for every central subgroup M.

In the following, K and C^* will always correspond to a central subgroup of M of G as in the above definitions.

Our main results are given by the following theorems.

Theorem 1 *For a finite group G, $\mathrm{Aut}_M^M(G) \cong \mathrm{Hom}\left(\frac{G}{KM}, M\right)$.*

Theorem 2 *Let G be a group with M torsion free and $M/M \cap K$ torsion. Then $\mathrm{Aut}^M(G)$ is a torsion-free abelian group which acts trivially on M.*

Theorem 3 *Let G be a purely nonabelian finite group, then $|\mathrm{Aut}^M(G)| = |\mathrm{Hom}(G, M)|$.*

Theorem 4 *Let G be a purely nonabelian finite p-group (p odd), then* $\mathrm{Aut}^M(G)$ *is an elementary abelian p-group if and only if $exp(M) = p$ or $exp(G/K) = p$.*

Following proposition shows that each element of K is invariant under the natural action of $\mathrm{Aut}^M(G)$.

Proposition 1 $\mathrm{Aut}^M(G)$ *acts trivially on K.*

Proof Consider an automorphism $\alpha \in \mathrm{Aut}^M(G)$. This implies $x^{-1}\alpha(x) \in M$, for all $x \in G$. So $\alpha(x) = xm$ for some $m \in M$. Let $\beta \in C^*$. By definition of C^*, we have $\alpha([x,\beta]) = \alpha(x^{-1}\beta(x)) = (\alpha(x))^{-1}\beta(\alpha(x)) = m^{-1}x^{-1}\beta(xm) = m^{-1}x^{-1}\beta(x)m = x^{-1}\beta(x) = [x,\beta]$. Hence the results follows. □

Proof of Theorem 1 For any $\mu \in \mathrm{Aut}_M^M(G)$, define the map $\psi_\mu \in \mathrm{Hom}\left(\frac{G}{KM}, M\right)$ as $\psi_\mu(gKM) = g^{-1}\mu(g)$.
We first show that ψ_μ is well defined.
Let $gKM = hKM$, i.e., $gh^{-1} \in KM$.
$\therefore \mu(gh^{-1}) = gh^{-1} \Rightarrow g^{-1}\mu(g) = h^{-1}\mu(h) \Rightarrow \psi_\mu(gKM) = \psi_\mu(hKM)$.

For proving ψ_μ is a homomorphism, consider $\psi_\mu(gKMhKM) = \psi_\mu(ghKM) = (gh)^{-1}\mu(gh) = h^{-1}g^{-1}\mu(g)\mu(h) = g^{-1}\mu(g)h^{-1}\mu(h) = \psi_\mu(gKM).\psi_\mu(hKM)$

Now define a map $\psi : \mathrm{Aut}_M^M(G) \longrightarrow \mathrm{Hom}\left(\frac{G}{KM}, M\right)$, as $\psi(\mu) = \psi_\mu$.

We show that ψ is the required isomorphism.
For $f, g \in \mathrm{Aut}_M^M(G)$ and $h \in G$, $\psi(fg)(hKM) = \psi_{fg}(hKM) = h^{-1}fg(h) = h^{-1}f(hh^{-1}g(h)) = h^{-1}f(h)h^{-1}g(h) = \psi_f(hKM)\psi_g(hKM) = \psi_f.\psi_g(hKM)$.
Hence $\psi(fg) = \psi(f)\psi(g)$.

Consider $\psi(\mu_1) = \psi(\mu_2)$, i.e., $\psi_{\mu_1}(gKM) = \psi_{\mu_2}(gKM), g \in G$. This implies $g^{-1}\mu_1(g) = g^{-1}\mu_2(g) \Rightarrow \mu_1 = \mu_2$, as g is an arbitrary element of G. Thus ψ is a monomorphism.

We next show that ψ is onto. For any $\tau \in \mathrm{Hom}\left(\frac{G}{KM}, M\right)$, define a map $\mu : G \to G$ as $\mu(g) = g\tau(gKM), g \in G$.
Now we show that $\mu \in \mathrm{Aut}_M^M(G)$. For $g_1, g_2 \in G, \mu(g_1g_2) = g_1g_2\tau(g_1g_2KM) = g_1\tau(g_1KM)g_2\tau(g_2KM) = \mu(g_1)\mu(g_2)$. $\therefore$ μ is a homomorphism on G.
Further, let $\mu(g) = 1$. This implies $g\tau(gKM) = 1 \Rightarrow \tau(gKM) = g^{-1} \Rightarrow g^{-1} \in M$ $\therefore gKM = KM \Rightarrow \tau(gKM) = 1 \Rightarrow g = 1$. Hence μ is one-one.
As G is finite, μ must be onto. So $\mu \in \mathrm{Aut}(G)$. Further, as $g^{-1}\mu(g) = g^{-1}g\tau(gKM) = \tau(gKM) \in M$, so $\mu \in \mathrm{Aut}^M(G)$. Also if $g \in M$, then $\mu(g) = g(\tau(gKM)) = g\tau(KM) = g$. Thus, $\mu \in \mathrm{Aut}_M^M(G)$ and $\psi(\mu) = \tau$.
Hence the theorem follows. □

Corollary 1 *Let G be finite group with $M \leq K$, then* $\mathrm{Aut}^M(G) \cong \mathrm{Hom}(G/K, M)$.

Proof Since $M \leq K$, $\frac{G}{KM} = G/K$. The result follows directly from Theorem 1 and Proposition 1. □

Proof of Theorem 2 Let $\alpha \in \mathrm{Aut}^M(G)$. If x is an element of M, then by the hypothesis $x^n \in M \cap K$ for some positive integer n. By Proposition 1, we have $x^n = \alpha(x^n) = (\alpha(x))^n$, and hence $x^{-n}(\alpha(x))^n = 1$. Since $x^{-1}\alpha(x) \in M$, this implies $(x^{-1}\alpha(x))^n = 1$. As M is torsion free, this implies that $x^{-1}\alpha(x) = 1$, i.e., $\alpha(x) = x$. Therefore, $\mathrm{Aut}^M(G)$ acts trivially on M.
Let $\alpha, \beta \in \mathrm{Aut}^M(G)$ and $x \in G$. So $\alpha\beta(x) = \alpha(\beta(x)) = \alpha(xx^{-1}\beta(x)) = \alpha(x)x^{-1}\beta(x) = xx^{-1}\alpha(x)x^{-1}\beta(x) = \beta(x)x^{-1}\alpha(x) = \beta(x)\beta(x^{-1}\alpha(x)) = \beta\alpha(x)$. Thus, $\mathrm{Aut}^M(G)$ is an abelian group.
Now, consider $\alpha \in \mathrm{Aut}^M(G)$, and suppose there exists $k \in N$ such that $\alpha^k = 1$. Since $x^{-1}\alpha(x) \in M$ for all $x \in G$, there exists $g \in M$ such that $\alpha(x) = xg$. Further, $\alpha^2(x) = \alpha(\alpha(x)) = \alpha(xg) = \alpha(x)\alpha(g) = xg^2(\because \alpha$ acts trivially on $M)$. Hence, by induction, $\alpha^n(x) = xg^n$. But $\alpha^k = 1 \Rightarrow x = xg^k$, i.e., $g^k = 1$. As M is torsion free, we must have $g = 1$. Thus $\alpha(x) = x$ for every x, i.e., $\alpha = 1$.
Therefore, $\mathrm{Aut}^M(G)$ is torsion free, and the theorem follows. □

Proposition 2 *Let G be a group in which M is torsion free and G/K is torsion, then* $\mathrm{Aut}^M(G) = 1$.

Proof Let $\alpha \in \mathrm{Aut}^M(G)$ and $x \in G$. Then by the assumption, $x^n \in K$ for some $n \in N$. As α fixes K elementwise, we have $(\alpha(x))^n = \alpha(x^n) = x^n$. So $x^{-n}(\alpha(x))^n = 1$. But $\alpha \in \mathrm{Aut}^M(G)$ and hence $x^{-1}\alpha(x) \in M \le Z(G)$. This implies that $(x^{-1}\alpha(x))^n = 1$. Since M torsion free, it follows that $x^{-1}\alpha(x) = 1$, i.e., $\alpha(x) = x, \forall x \in G$. So $\mathrm{Aut}^M(G) = 1$. □

Proof of Theorem 3 For $f \in \mathrm{Aut}^M(G)$, we let $\alpha(f) \equiv \alpha_f$ defined as $\alpha(f)(g) \equiv \alpha_f(g) = g^{-1}f(g), g \in G$. It can be shown that $\alpha_f \in \mathrm{Hom}(G, M)$. We thus have $\alpha : \mathrm{Aut}^M(G) \to \mathrm{Hom}(G, M)$.
One can easily see that α is injective.
It just remains to show that α is onto.
For $\sigma \in \mathrm{Hom}(G, M)$, consider the map $f : G \to G$ given by $f(g) = g\sigma(g)$. f is an endomorphism and also $g^{-1}f(g) = \sigma(g) \in M$, which implies that f is subcentral endomorphism of G, and hence f is normal endomorphism(i.e., f commutes with all inner automorphisms). So, clearly $\mathrm{Im}(f)$ is a normal subgroup of G.
It is easy to see that f^n is also normal endomorphism and hence $\mathrm{Im}\, f^n$ is a normal subgroup of G, for all $n \ge 1$. Since G is a finite group, the two series

$$\mathrm{Ker}\, f \le \mathrm{Ker}\, f^2 \le \ldots\ldots$$

$$\mathrm{Im}\ f \ge \mathrm{Im}\ f^2 \ge \ldots\ldots$$

will terminate.

So there exists $k \in N$ such that

$$\mathrm{Ker}\, f^k = \mathrm{Ker}\, f^{k+1} = \ldots = A$$

$$\mathrm{Im}\, f^k = \mathrm{Im}\, f^{k+1} = \ldots\ldots = B$$

Now, we prove that $G = AB$.

Let $g \in G$, $f^k(g) \in \mathrm{Im}\, f^k = \mathrm{Im}\, f^{2k}$, and so $f^k(g) = f^{2k}(h)$, for some $h \in G$. Therefore $f^k(g) = f^k(f^k(h))$. This implies $f^k(g^{-1})f^k(g) = f^k(g^{-1})f^k(f^k(h))$. Thus $(f^k(h))^{-1}g \in \mathrm{Ker}\, f^k = A$. Thus $g \in AB$ and hence $G = AB$.
Clearly $A \cap B =< 1 >$ and therefore $G = A \times B$. If $f(g) = 1$, then $g^{-1}\sigma(g) = 1$. This implies $\mathrm{Ker}\, f \leq M$. Similarly, if $f^2(g) = 1$, i.e., $f(f(g)) = 1$. Thus $f(g) \in \ker f \leq M$. Therefore, $g\sigma(g) \in M \Rightarrow g \in M$. Hence $\ker f^2 \leq M$. Repetition of this argument gives, $A \equiv \ker f^k \leq M \leq Z(G)$. This implies A is an abelian group. By assumption, G is purely nonabelian and hence, we must have $A \equiv \mathrm{Ker}\, f^k = 1$. This further implies $\mathrm{Ker}\, f = 1$, i.e., f is injective. So $G = B \equiv \mathrm{Im}\ f^k = \mathrm{Im}\, f$. Thus f surjective. Hence, $f \in \mathrm{Aut}^M(G)$. From the definition of α, it follows that $\alpha(f) = \sigma$. α is thus surjective. Therefore, α is the required bijection. Hence the result follows. □

Proposition 3 *Let G be a purely nonabelian finite group, then for each $\alpha \in \mathrm{Hom}(G, M)$ and each $x \in K$, we have $\alpha(x) = 1$. Further $\mathrm{Hom}(G/K, M) \cong \mathrm{Hom}(G, M)$.*

Proof Whenever G is purely nonabelian group, then by Theorem 3, $|\mathrm{Aut}^M(G)| = |\mathrm{Hom}(G, M)|$. For every $\sigma \in \mathrm{Aut}^M(G)$, it follows that $f_\sigma : x \to x^{-1}\sigma(x)$ is a homomorphism from G to M. Further the map $\sigma \to f_\sigma$ is one–one and thus a bijection because $|\mathrm{Aut}^M(G)| = |\mathrm{Hom}(G, M)|$. So every homomorphism from G to M can be considered as an image of some element of $\mathrm{Aut}^M(G)$ under this bijection. Let $\alpha \in \mathrm{Hom}(G, M)$. Since $K = \{[g, \alpha] | g \in G, \alpha \in C^*\}$, a typical generator of K is given by $g^{-1}\beta(g)$ for some $g \in G$, and $\beta \in C^*$. So $\alpha(g^{-1}\beta(g)) = f_\sigma(g^{-1}\beta(g)) = (g^{-1}\beta(g))^{-1}\sigma(g^{-1}\beta(g)) = \beta^{-1}(g)gg^{-1}\beta(g) = 1(\because g^{-1}\beta(g) \in K)$. It follows that $\alpha(x) = 1$, for every $x \in K$.

Now consider the map $\phi : \mathrm{Hom}(G, M) \longrightarrow \mathrm{Hom}(G/K, M)$ such that $\phi(f) = \bar{f}$, where $\bar{f}(gK) = f(g)$ for all $g \in G$. Clearly this map ϕ is an isomorphism. □

Proposition 4 *Let G be a purely nonabelian finite group, then $|\mathrm{Aut}^M(G)| = |\mathrm{Hom}(G/K, M)|$.*

Proof Proof follows directly from Theorem 3 and Proposition 3. □

Proposition 5 *Let p be a prime number. If G is a purely nonabelian finite p-group then $\mathrm{Aut}^M(G)$ is a p-group.*

Proof By the assumption, the subgroup M and hence Hom $(G/K, M)$ are finite p-groups. Hence the result follows directly from Proposition 4. □

Proposition 6 *Let G be a purely nonabelian finite group*
(i) If $gcd(|G/K|, |M|) = 1$, then $\mathrm{Aut}^M(G) = 1$.
(ii) If $\mathrm{Aut}^M(G) = 1$, *then* $M \leq K$.

Proof (i) Follows from Proposition 4.
(ii) Let $|G/K| = a$ and $|M| = b$. Since $\mathrm{Aut}^M(G) = 1$, hence by Proposition 4, $(a, b) = 1$. So there exist integers λ and μ such that $\lambda a + \mu b = 1$. Let $x \in M$. Thus $xK = (xK)^1 = (xK)^{\lambda a+\mu b} = (xK)^{\lambda a}(xK)^{\mu b} = K \Rightarrow x \in K$. □

Remark 1 From Corollary 1, and Proposition 3, we can say that, whenever $M \leq K$, $\mathrm{Aut}^M(G) \cong \mathrm{Hom}(G, M)$. Even when $\mathrm{Im} f \leq K$, for all $f \in \mathrm{Hom}(G, M)$, this result holds. Thus, if G is a purely nonabelian finite group and if for all $f \in \mathrm{Hom}(G, M)$, $\mathrm{Im} f \leq K$, then $\mathrm{Aut}^M(G) \cong \mathrm{Hom}(G/K, M)$.

Remark 2 For every $f \in \mathrm{Hom}(G, M)$, the map $\sigma_f : x \to xf(x)$ is a subcentral endomorphism of G. This endomorphism is an automorphism if and only if $f(x) \neq x^{-1}$ for all $1 \neq x \in G$ (G is finite).

Following lemma has been proved in [4], we shall use it to prove Theorem 4.

Lemma 1 *Let x be an element of a finite p-group G and N a normal subgroup of G containing G' such that $o(x) = o(xN) = p$. If the cyclic subgroup $< x >$ is normal in G such that $ht(xN) = 1$, then $< x >$ is a direct factor of G.*

In the above statement ht denotes height. Height of an element a of a group G is defined as the largest positive integer n such that for some x in G, $x^n = a$.

Proof of Theorem 4 For the odd prime p, let $\mathrm{Aut}^M(G)$ be an elementary abelian p-group. Assume that the exponent of M and G/K are both strictly greater than p. Since G/K is finite abelian, it has a cyclic direct summand $\langle xK\rangle$ say, of order $p^n (n \geq 2)$ and hence $G/K \cong \langle xK\rangle \times L/K$. For $f \in \mathrm{Hom}(G, M)$, consider $f(x) = a$ for any $x \in G$. So $\bar{f}(xK) = a$. Since exp (M) is strictly greater than p, the order of a is p^m, for some m, $2 \leq m \leq n$.
We can use the homomorphism $\bar{f}$ to get corresponding homomorphism (also denoted by same notation) $\bar{f}$ as $\bar{f} : \langle xK\rangle \times L/K \to M$ with $(x^iK, lK) \to a^i$. The map $\bar{f}$ on $< xK > \times L/K$ is well defined, since $o(a)|o(xK)$ (as $m \leq n$).
If $aK = (x^sK, lK)$ then we show that $p|s$. Assume $p \not| s$, then $< xK >=< x^sK >$ and hence $G/K =< aK > L/K$. Now we have $o(a) \geq o(aK) \geq o(x^sK) = o(xK) \geq o(\bar{f}(xK)) = o(a)$. This implies that $o(a) = o(aK)$. Thus $< a > \cap K = 1$. As $o(aK) = o(xK)$, we get $G/K \cong < aK > \times L/K$ and hence $G \cong < a > \times L$. This is a contradiction, as G is a purely nonabelian group. Thus $p|s$.
By Remark 2 and Theorem 3, $\sigma_f \in \mathrm{Aut}^M(G)$ and by assumption $o(\sigma_f) = p$.

Now, we have $\sigma_f(x) = xf(x) = xa$. Since $f(a) = \bar{f}((xK)^s, lK) = a^s$, we have $\sigma_f^2(x) = xa^{s+2} = xa^{\frac{(s+1)^2-1}{s}}$.

Also, $\sigma_f^3(x) = xa^{\left(\frac{(s+1)^3-1}{s}\right)}$.
Generalizing this,
we get $\sigma_f^t(x) = xa^{\left(\frac{(s+1)^t-1}{s}\right)}$, for every $t \in N$.

As the order of σ_f is p, we have $a^{\frac{(s+1)^p-1}{s}} = 1$. Since p is odd and $p|s$, we have $p^2|(\left(\frac{(s+1)^p-1}{s}\right) - p)$.
$\therefore qp^2 + p = \frac{(s+1)^p-1}{s}$ for some $q \in Z$. Thus $(a^p)^{qp+1} = 1$. But $o(a) = p^m \Rightarrow o(a^p) = p^{m-1}$.
Now
(1) if $a^p \neq 1$, then $p^{m-1}|(qp+1)$. But this is impossible as $m \geq 2$.
(2) $a^p = 1$ is also not possible as $o(a) = p^m$ and $m \geq 2$.

So, the assumption that $exp(M)$ and $exp(G/M)$ are stricly greater than p is wrong. Conversly, assume that $exp(G/K) = p$ and $f \in \text{Hom}(G, M)$. Then by proposition 3, $\bar{f} \in \text{Hom}(G/K, M)$. So for $x \in G$, put $\bar{f}(xK) = a$. If $aK \neq 1$, then it follows that $o(aK(G)) = o(a) = p$. Clearly $< a > \leq M(G) \leq Z(G)$ and hence the cyclic subgroup $< a >$ is normal in G. We also have $ht(aK) = 1$. Now by the Lemma 1, the cyclic subgroup $< a >$ is an abelian direct factor of G, and this contradicts the assumption. Therefore $a \in K$. This implies that $\text{Im}(f) \leq K$. Hence by Remark 1 $\text{Aut}^M(G) \cong \text{Hom}(G/K, M)$. But as M is abelian, $\text{Hom}(G/K, M)$ is abelian. Thus $\text{Aut}^M(G)$ is abelian. Since $exp(G/M) = p$, this implies that $\text{Aut}^M(G)$ is an elementary abelian p-group.

Now assume that $exp(M) = p$. Consider $f, g \in \text{Hom}(G, M)$. We first show that $g \circ f(x) = 1$, for all $x \in G$. Assume that $\bar{f}(xK) = b \in M$, for $x \in G$. Since $exp(M) = p$, it implies that $o(b)|p$. If $b = 1$ then $g \circ f(x) = g(\bar{f}(xK(G))) = 1$. Now take, $o(b) = p$. If $b \in K$ then we have $g(f(x)) = g(\bar{f}(xK(G))) = g(b) = 1$. Assume b does not belong to K. As $b^p = 1$, it follows that $o(bK) = p$. Also, as $b \in M \leq Z(G)$, $< b >$ is normal in G. Now if $ht(bK(G)) = 1$, then by the Lemma 1, the cyclic subgroup $< b >$ is an abelian direct factor of G, giving a contradiction. So assume $ht(bk(G)) = p^m$ for some $m \in N$. By the definition of height, there exists an element yK in G/K such that $bK = (yK)^{p^m}$. But $exp(M) = p$. Therefore $g \circ f(x) = g(b) = \bar{g}(bK) = \bar{g}(yK)^{p^m} = 1$. Thus, for all $f, g \in \text{Hom}(G, M)$ and each $x \in G$, $g(f(x)) = 1$. We can similarly show that $f(g(x)) = 1$ and hence $f \circ g = g \circ f$. From Remark 2, $\sigma_f \circ \sigma_g = \sigma_g \circ \sigma_f$. This shows that $\text{Aut}^M(G)$ is abelian.

Now we show that each nontrivial element of $\text{Aut}^M(G)$ has order p. So if $\alpha \in \text{Aut}^M(G)$, then by Remark 2, there exists a homomorphism $f \in \text{Hom}(G, M)$ such that $\alpha = \sigma_f$. Therefore, we have to show that $o(\sigma_f)|p$. Clearly, taking $f = g$ and using $f(f(x)) = 1, x \in G$, we have $x \in G$, we have $\sigma_f^2(x) = \sigma_f(xf(x))$

$= x(f(x))^2$. In general for $n \geq 1$, $\sigma_f^n(x) = x(f(x))^n$. As exp $(M) = p$ and $f(x) \in M$ we have, $\sigma_f^p(x) = x$ which implies $\sigma_f^p = 1_{\text{Aut}^M(G)}$.
Hence $o(\sigma_f)|p$. Thus, $o(\alpha)|p \; \forall \alpha \in \text{Aut}_M(G)$. $\therefore$ $\text{Aut}^M(G)$ is an elementary abelian group.

□

References

1. Adney, J.E., Yen, T.: Automorphisms of a p-group. Illinious J. Math. **9**, 137–143 (1965)
2. Curran, M.-J.: Finite groups with central automorphism group of minimal order. Math. Proc. Royal Irish Acad. **104**(A(2)), 223–229 (2004)
3. Franciosi, S., Giovanni, F.D., Newell, M.L.: On central automorphisms of infinite groups. Commun. Algebra **22**(7), 2559–2578 (1994)
4. Jafri, M.H.: Elementary abelian p-group as central automorphisms group. Commun. Algebra **34**(2), 601–607 (2006)
5. Jamali, A.R., Mousavi, H.: On central automorphism groups of finite p-group. Algebra Colloq. **9**(1), 7–14 (2002)

On Symmetric Laplace Integral of Order n

S. Ray and A. Garai

Abstract A symmetric integral of Perron type is defined using symmetric Laplace derivative, which is more general than other symmetric integrals like SCP integral of Burkill (Proc. Lond. Math. Soc. 3:46–57, 1951, [2]), T^n integral of Mukhopadhyay (Real Anal. Exch. 30:451–494, 2004–2005, [7]). The properties of this symmetric integral are studied.

Keywords Laplace derivative · Symmetric laplace derivative · Laplace smooth · Symmetric integral

1 Introduction

Laplace derivative is a generalization of the Peano derivative. This is a new type of derivative first introduced by Sevtic in [15] in 2000 and the properties of this derivative were recently explored in the articles [8, 10–13], and in the book [6]. The symmetric version of this derivative is studied in [3] and [4] (see also [6] and [9]).

Generally, symmetric integral is a Perron-type integral defined using symmetric derivative. Symmetric integral is a useful tool to handle trigonometric series. In this article we have defined a symmetric integral using symmetric Laplace derivative. This integral is a generalization of the T^n integral introduced in [7]. We study the properties of this integral.

S. Ray (✉)
Visva Bharati, Santiniketan, West Bengal, India
e-mail: subhasis.ray@visva-bharati.ac.in

A. Garai
Memari College, Memari, Burdwan, West Bengal, India
e-mail: garaianupam@gmail.com

R.N. Mohapatra et al. (eds.), *Mathematics and Computing*,
Springer Proceedings in Mathematics & Statistics 139,
DOI 10.1007/978-81-322-2452-5_33

2 Definition and Notation

Let a function $f : \mathbf{R} \to \mathbf{R}$ be integrable in some neighborhood of $x \in \mathbf{R}$. If there are real numbers $\alpha_0, \alpha_2, \ldots. \alpha_{2k}$ such that

$$\lim_{s\to\infty} s^{2k+1} \int_0^\delta e^{-st} \left[\frac{f(x+t)+f(x-t)}{2} - \sum_{i=0}^{k} \frac{t^{2i}}{(2i)!}\alpha_{2i} \right] dt = 0,$$

for some $\delta > 0$, then α_{2k} is called the symmetric Laplace derivative of f at x of order $2k$ and is denoted by $SLD^{2k} f(x)$. Also, if there are real numbers $\alpha_1, \alpha_3, \ldots. \alpha_{2k+1}$ such that

$$\lim_{s\to\infty} s^{2k+2} \int_0^\delta e^{-st} \left[\frac{f(x+t)-f(x-t)}{2} - \sum_{i=0}^{k} \frac{t^{2i+1}}{(2i+1)!}\alpha_{2i+1} \right] dt = 0,$$

for some $\delta > 0$ then α_{2k+1} is called the symmetric Laplace derivative of f at x of order $2k+1$ and is denoted by $SLD^{2k+1} f(x)$. These definitions do not depend on δ, [3]. If $SLD^{2k} f(x)$ exists then $SLD^{2i} f(x)$ exists for $0 < i < k$ also if f is continuous at x then $SLD^0 f(x) = f(x)$ (see Theorem 2.3 of [3]). If $SLD^{2k} f(x_0)$ exists we define

$$\overline{SLD}^{2k+2} f(x_0) = \limsup_{s\to\infty} s^{2k+3} \int_0^\delta e^{-st} \theta_{2k+2}(f, x_0, t) dt,$$

where

$$\theta_{2k+2}(f, x_0, t) = \left[\frac{f(x_0+t)+f(x_0-t)}{2} - \sum_{i=0}^{k} \frac{t^{2i}}{(2i)!} SLD^{2i} f(x_0) \right].$$

The definition of $\underline{SLD}^{2k+2} f(x_0)$ is analogous. Suppose $SLD^{2k-2} f(x)$ exists, then f is said to be Laplace smooth at x of order $2k$ if

$$\lim_{s\to\infty} s^{2k} \int_0^\delta e^{-st} \left[\frac{f(x+t)+f(x-t)}{2} - \sum_{i=0}^{k-1} \frac{t^{2i}}{(2i)!} SLD^{2i} f(x) \right] dt = 0.$$

The definitions for odd cases are similar. It is easy to verify that if $\overline{SLD}^{2k} f(x)$ and $\underline{SLD}^{2k} f(x)$ exist and finite, then f is Laplace smooth of order $2k$ at x (see [6]). Suppose $f : \mathbf{R} \to \mathbf{R}$ be integrable in a right neighborhood of x (respectively left neighborhood of x) if there are real numbers $\alpha_0, \alpha_1, \ldots.. \alpha_r$ such that

$$\lim_{s\to\infty} s^{r+1} \int_0^\delta e^{-st} \left[f(x+t) - \sum_{i=0}^{r} \frac{t^i}{i!}\alpha_i \right] dt = 0$$

(respectively,

$$\lim_{s\to\infty} s^{r+1}\int_0^\delta e^{-st}\left[f(x-t)-\sum_{i=0}^{r}\frac{(-t)^i}{i!}\alpha_i\right]dt=0),$$

then α_r is called the right-hand (left hand) Laplace derivative of f at x of order r and denoted by $LD_r^+ f(x)[LD_r^- f(x)]$. If $LD_r^+ f(x)$ exists it is easy to verify that $LD_i^+ f(x)$ exists and $\alpha_i = LD_i^+ f(x)$ for $0 \le i \le r$. Similar result holds for the left derivative also. If $LD_r^+ f(x)$ and $LD_r^- f(x)$ exist and $LD_i^+ f(x) = LD_i^- f(x)$ for $i = 0, 1, ...r$, then f is said to have Laplace derivative at x of order r and is denoted by $LD_r f(x)$. If $LD_r f(x)$ exists then $SLD^r f(x)$ exists with equal value (Theorem 2.4 [3]).

3 Preliminaries

Lemma 3.1 *Let $F : [a, b] \to \mathbf{R}$. If $F^{(r)}$ exists and is convex in (a, b), then F is $r+2$ convex in (a, b). If moreover $F \in \mathscr{D}$ in $[a, b]$, then F is $r+2$ convex in $[a, b]$.*

This is Lemma 3.5 of [5].

Theorem 3.2 *Let $f : [a, b] \to \mathbf{R}$ be such that*
(i) f is continuous in $[a, b]$,
(ii) $SLD^n f$ exists in (a, b) and $SLD^l f \in \mathscr{D}\cap\mathscr{B}_1^*$ in (a, b) for $l = n-2, n-4, ...$,
(iii) $SLD^n f \in \mathscr{D}\cap\overline{\mathscr{B}_1^*}$ in (a, b),
(iv) $\overline{SLD}^{n+2} f(x) \ge 0$ a.e. in (a, b),
(v) $\overline{SLD}^{n+2} f(x) \ge -\infty$ in (a, b) except on a countable subset $E \subset (a, b)$,
(vi) $f \in \overline{S}_{n+2}(x)$ for $x \in E$.
Then $SLD^n f$ is convex in (a, b) and it is the continuous derivative $f^{(n)}$ in (a, b).

The proof of this theorem follows from Theorem 3.22 of [4] by using standard argument used to prove Theorem 1.1 of [2] or Theorem 16 of [1].

Lemma 3.3 *Let $f : [a, b] \to \mathbf{R}$ be such that*
(i) f is continuous in $[a, b]$,
(ii) $SLD^{n-2} f$ exists finitely in (a, b) and $SLD^l f \in \mathscr{D}$ in (a, b) for $l = n-2, n-4, ...$,
(iii) $\overline{SLD}^n f(x) \ge 0$ a.e in (a, b) ,
(iv) f is Laplace smooth of order n for all $x \in (a, b)$
(v) $\overline{SLD}^n f(x) > -\infty$ except on a countable set in (a, b), Then $f^{(n-2)}$ exists, is continuous and convex in $[a, b]$.

Proof By Corollary 3.3 of [3], $SLD^l f \in \mathscr{B}_1^*(a, b)$ for $l = n-2, n-4, ...,$ and hence by Theorem 3.2, $SLD^{n-2} f$ is convex in (a, b) and it is the continuous derivative

$f^{(n-2)}$ in (a, b). Hence by Lemma 3.1, f is n- convex in $[a, b]$. Hence the $(n-2)$ th order Peano derivative $f_{(n-2)}$ of f exists and is convex in $[a, b]$ (by [1]). Hence $\lim_{x \to a+} f_{(n-2)}(x)$ and $\lim_{x \to b-} f_{(n-2)}(x)$ exists and since $f_{(n-2)}(x) \in \mathscr{D}$ in $[a, b]$ (by [14]), $f_{(n-2)}(x)$ is continuous in $[a, b]$. Therefore by [14], $f_{(n-2)}$ is the ordinary derivative $f^{(n-2)}$ and the result is proved.

Definition 3.4 Let f be an extended real valued function on $[a, b]$. A function Q is said to be an LT^n major function of f if
(i) Q is continuous in $[a, b]$,
(ii) $LD_{n-2}Q$ exists finitely in $[a, b]$,
(iii) $LD_{n-1}Q$ exists finitely in $[a, b]$ except on a set of measure zero in (a, b),
(iv) $LD_r Q(a) = 0$ for $r = 0, 1, \ldots.., n-1$,
(v) $\underline{SLD}^n Q \geq f$ a.e in (a, b),
(vi) $\underline{SLD}^n Q > -\infty$ except on a countable set in (a, b),
(vii) Q is Laplace smooth of order n in (a, b).

Similarly, a function $q : [a, b] \to \mathbf{R}$ is said to be a LT^n minor function of f if $(-q)$ is an LT^n major function of $(-f)$. For LT^n major or LT^n minor function we simply write major or minor function when there is no confusion.

4 Main Results

Lemma 4.1 *If Q and q are, respectively, major and minor functions of f, then for each r, $1 < r \leq n$, $(Q - q)^{(n-r)}$ exists and is k- convex, $0 \leq k \leq r$, in $[a, b]$ and so $LD_{n-r}Q - LD_{n-r}q$ is k- convex in $[a, b]$.*

Proof Let $\phi = Q - q$, then ϕ is continuous and $LD_{n-2}\phi$ exists in $[a, b]$. So by Theorem 17 of [8] $LD_k\phi \in \mathscr{D}$ for $1 \leq k \leq n-2$. Also, ϕ is Laplace smooth of order n in (a, b) and $\underline{SLD}^n\phi \geq \underline{SLD}^n Q - \overline{SLD}^n q \geq 0$ a.e in (a, b), and $\underline{SLD}^n\phi > -\infty$ except on a countable set in (a, b). Hence by Theorem 3.3, $\phi^{(n-2)}$ exists, is continuous and convex in $[a, b]$. So the right-hand derivative of $\phi^{(n-2)}$ exists in $[a, b)$ and left-hand derivative of $\phi^{(n-2)}$ exists in $(a, b]$ and the first order derivative, i.e., $\phi^{(n-1)}$ exists except on a countable set in (a, b) and $\phi^{(n-1)}$ is nondecreasing on the set where it exists. Since $\phi^{(n-1)}(a) = LD_{n-1}Q(a) - LD_{n-1}q(a) = 0$, $\phi^{(n-1)}$ is also nonnegative on the set where it exists. Also, $\phi^{(n-2)}(x) = \int_a^x \phi^{(n-1)}(t)dt$, $x \in [a, b]$. Since $\phi^{(n-1)}$ is nonnegative a.e, $\phi^{(n-2)}$ is nonnegative, nondecreasing, and convex in $[a, b]$. So $LD_{n-2}Q - LD_{n-2}q$ is nonnegative, nondecreasing, and convex in $[a, b]$. Thus the result is true for $r = 2$. Suppose the result is true for any r, $1 < r < n$. Then by induction hypothesis $\phi^{(n-r)}$ exists and is k-convex for $0 \leq k \leq r$ in $[a, b]$, and so

$$\phi^{(n-r-1)}(x) = \int_a^x \phi^{(n-r)}(t)dt, \quad x \in [a, b].$$

Hence $\phi^{(n-r-1)}$ is k-convex, $0 \le k \le r+1$ in $[a, b]$, i.e., $LD_{n-r-1}Q - LD_{n-r-1}q$ is k-convex, $0 \le k \le r+1$. So the result is true for $r+1$. Therefore the proof is completed by induction.

Definition 4.2 Let $\{Q\}$ and $\{q\}$ be the collection of all major and minor functions of f in $[a, b]$. Let
$u = \inf_{Q\in\{Q\}} LD_{n-1}Q(b)$ and $v = \sup_{q\in\{q\}} LD_{n-1}q(b)$.
By Lemma 4.1, $(Q-q)^{(n-2)}$ exists and is k- convex, $0 \le k \le 2$, in $[a, b]$ for any $Q \in \{Q\}$ and any $q \in \{q\}$. So the left-hand derivative of $(Q-q)^{(n-2)}$ exists at b and the right-hand derivative of $(Q-q)^{(n-2)}$ exists at a and $(Q-q)^{(n-1)}$ exists except on a countable set in (a, b) and is nondecreasing on a set where it exists. Hence $LD_{n-1}Q - LD_{n-1}q$ is nondecreasing on a set $S \subset [a, b]$, such that $a, b \in S$ and $\mu(S) = b - a$. Therefore, $LD_{n-1}Q(b) \ge LD_{n-1}q(b)$. Since this is for arbitrary $Q \in \{Q\}$ and $q \in \{q\}$, $u \ge v$. If $u = v \ne \pm\infty$, then f is said to be LT^n integrable on $[a, b]$ and the common value is called LT^n integral of f and is denoted by $(LT^n)\int_a^b f(x)dx$. If there is no confusion we simply write integrable and integral in place of LT^n- integrable and LT^n- integral.

Let f be integrable in $[a, b]$ and let $\varepsilon > 0$ be arbitrary. Then there are $Q \in \{Q\}$ and $q \in \{q\}$ such that for almost all $x \in [a, b]$,

$$0 \le LD_{n-1}Q(x) - LD_{n-1}q(x) \le LD_{n-1}Q(b) - LD_{n-1}q(b) < \varepsilon \tag{1}$$

Since $(Q-q)^{(n-2)} = LD_{n-2}Q - LD_{n-2}q$ is convex,

$$\begin{aligned} LD_{n-2}Q(x) - LD_{n-2}q(x) = (Q-q)^{(n-2)}(x) &= \int_a^x (Q-q)^{(n-1)}(t)dt \\ &= \int_a^x [LD_{n-1}Q(t) - LD_{n-1}q(t)]dt \end{aligned}$$

for $x \in [a, b]$. Hence from (1) and Lemma 4.1,

$$0 \le LD_{n-2}Q(x) - LD_{n-2}q(x) \le LD_{n-2}Q(b) - LD_{n-2}q(b) < \varepsilon(b-a), \quad x \in [a, b] \tag{2}$$

So for each $x \in [a, b]$, since ε is arbitrary,

$$\inf_{Q\in\{Q\}} LD_{n-2}Q(x) = \sup_{q\in\{q\}} LD_{n-2}q(x) = F_2(x)(say).$$

The function F_2 is called second primitive of f.

Suppose the r th primitive F_r, $2 \le r < n$ is defined and the relation

$$0 \le LD_{n-r}Q(x) - LD_{n-r}q(x) \le LD_{n-r}Q(b) - LD_{n-r}q(b) < \varepsilon(b-a)^{r-1}, \quad x \in [a,b], \tag{3}$$

is obtained. Since $LD_{n-r-1}Q(x) - LD_{n-r-1}q(x)$ is convex, by Lemma 4.1, we get

$$LD_{n-r-1}Q(x) - LD_{n-r-1}q(x) = \int_a^x [LD_{n-r}Q(t) - LD_{n-r}q(t)]dt, \qquad x \in [a,b].$$

Hence from (3) and Lemma 4.1 we get

$$0 \le LD_{n-r-1}Q(x) - LD_{n-r-1}q(x) \le LD_{n-r-1}Q(b) - LD_{n-r-1}q(b) < \varepsilon(b-a)^r, \quad x \in [a,b]. \tag{4}$$

and so for fixed $x \in [a,b]$,

$$\inf_{Q \in \{Q\}} LD_{n-r-1}Q(x) = \sup_{q \in \{q\}} LD_{n-r-1}q(x) = F_{r+1}(x)(say).$$

The function F_{r+1} is called the $(r+1)$ th primitive of f. Thus all the primitive F_r, $2 \le r \le n$ of f are defined.

Theorem 4.3 *If f are integrable and if F_k is the k th primitive, $k = 2, \ldots, n$, then there are sequence of major functions $\{Q_\alpha\}$ and minor functions $\{q_\alpha\}$ such that $\{LD_r Q_\alpha\}$ and $\{LD_r q_\alpha\}$ converge uniformly to F_{n-r} in $[a,b]$, $r = 0, 1, \ldots, n-2$.*

Proof If $\varepsilon > 0$ be given, then there is a major function Q and a minor function q such that

$$0 \le LD_r Q(x) - LD_r q(x) < \varepsilon(b-a)^{n-r-1}.$$

Hence for each positive integer α, there is a major function Q_α and minor functions q_α such that

$$0 \le LD_r Q_\alpha(x) - LD_r q_\alpha(x) < \frac{1}{\alpha}(b-a)^{n-r-1}.$$

From the definition of F_{n-r} we have

$$0 \le LD_r Q_\alpha(x) - F_{n-r}(x) \le LD_r Q_\alpha(x) - LD_r q_\alpha(x) < \frac{1}{\alpha}(b-a)^{n-r-1}.$$

This shows that $\{LD_r Q_\alpha\}$ converges uniformly to F_{n-r} on $[a,b]$. The proof for minor functions is similar.

Theorem 4.4 *Let f be integrable on $[a,b]$ and F_r be the r th primitive of f for $2 \le r \le n$. Then for any major function Q and for any minor function q, $LD_{n-r}Q(x) - F_r(x)$ and $F_r(x) - LD_{n-r}q(x)$ are k-convex, $0 \le k \le r$, in $[a,b]$.*

Proof Let $\{q_\alpha\}$ be a sequence of minor functions such that $\{LD_{n-r}q_\alpha\}$ converges uniformly to F_r on $[a, b]$. So for any major function Q, $\{LD_{n-r}Q - LD_{n-r}q\}$ converges uniformly to $\{LD_{n-r}Q - F_r\}$ on $[a, b]$. Since for each α , $\{LD_{n-r}Q - LD_{n-r}q_\alpha\}$ is k-convex, $\{LD_{n-r}Q - F_r\}$ is k-convex in $[a, b]$ for $0 \le k \le r$. The proof for $F_r(x) - LD_{n-r}q(x)$ is similar.

Theorem 4.5 *Let f be integrable in $[a, b]$ and F_n be its n th primitive. $LD_{n-2}F_n$ exists finitely in $[a, b]$ and there is a measurable set $B \subset [a, b]$ with $a, b \in B$ and $\mu(B) = b - a$ such that $LD_{n-1}F_n$ exists finitely in B and $LD_{n-1}F_n(b)$ equal to the integral of f in $[a, b]$ and $LD_{n-1}F_n(a) = 0$.*

Proof Let Q be any major function. Then by Theorem 4.4, the function $\phi = Q - F_n$ is k-convex, $0 \le k \le n$ in $[a, b]$ and so the $(n-2)$ th Peoano derivative $\phi_{(n-2)}$ and hence $LD_{n-2}\phi$ exists everywhere in $[a, b]$, is convex nondecreasing and nonnegative. Since $LD_{n-2}Q$ exists finitely in $[a, b]$, $LD_{n-2}F_n$ exists finitely in $[a, b]$. Also, $\phi_{(n-1)}$ exists in $[a, b]$ except on a countable set in (a, b) and $\phi_{(n-1)}$ is nondecreasing and nonnegative on the set where it exists. Since $LD_{n-1}Q$ exists finitely in $[a, b]$ except on a set of measure zero in (a, b), $LD_{n-1}F_n$ exists finitely in $[a, b]$ except on a set of measure zero in (a, b). Let

$$B = \{x : x \in [a, b]; \quad LD_{n-1}F_n \quad \textit{exists finitely}\}.$$

Then B is the required set. Also, $\phi_{(n-1)}$ being nondecreasing and nonnegative

$$LD_{n-1}Q(x) - LD_{n-1}F_n(x) \ge 0 \tag{5}$$

for those $x \in B$ for which $LD_{n-1}Q(x)$ exists. Similarly,

$$LD_{n-1}F_n(x) - LD_{n-1}q(x) \ge 0 \tag{6}$$

for those $x \in B$ for which $LD_{n-1}q(x)$ exists. Since f is integrable, for any $\varepsilon > 0$ there are Q and q such that $LD_{n-1}Q(b) - LD_{n-1}q(b) < \varepsilon$. Since $LD_{n-1}Q(x) - LD_{n-1}q(x)$ is nondecreasing on the set where it exists, we have

$$0 \le LD_{n-1}Q(x) - LD_{n-1}q(x) < \varepsilon \tag{7}$$

whenever $LD_{n-1}Q(x)$ and $LD_{n-1}q(x)$ exist.
We have from (5)–(7) there is $Q \in \{Q\}$ such that

$$0 \le LD_{n-1}Q(x) - LD_{n-1}F_n(x) < \varepsilon \tag{8}$$

whenever $LD_{n-1}Q(x)$ and $LD_{n-1}F_n(x)$ exist.
From (5) and (8) we get

$$\inf_{Q}\{LD_{n-1}Q(x)\} = LD_{n-1}F_n(x) \tag{9}$$

whenever $LD_{n-1}Q(x)$ and $LD_{n-1}F_n(x)$ exist.

Since $LD_{n-1}Q(b)$ and $LD_{n-1}F_n(b)$ exist, $LD_{n-1}F_n(b)$ is equal to the integral of f on $[a, b]$. Also, (5) and (6) hold for $x = a$ and hence $LD_{n-1}F_n(a) = 0$.

Definition 4.6 The set $B = \{x : x \in [a, b]; \quad LD_{n-1}F_n \quad \text{exists finitely}\}$ defined in the proof of the above Theorem is called the base of the integral of f and $LD_{n-1}F_n$, which exists finitely in B is called the first primitive of f and will be denoted by F_1.

Theorem 4.7 *If Q and q are major and minor functions, then $LD_{n-1}Q$ and $LD_{n-1}q$ exist at each point of B.*

Proof By Theorem 4.4, $Q - F_n$ is n-convex where F_n is the n th primitive. Let $\eta \in B$. Then the one-sided derivatives $(Q - F_n)^{(n-1),+}(\eta)$ and $(Q - F_n)^{(n-1),-}(\eta)$ exist finitely. Again since $LD_{n-1}F_n(\eta)$ exists, $LD^{+}_{n-1}Q(\eta)$ and $LD^{-}_{n-1}Q(\eta)$ exist finitely. Since Q is Laplace smooth of order n at η so $LD_{n-1}Q(\eta)$ exists. Similarly, $LD_{n-1}q(\eta)$ exists on B.

Theorem 4.8 *If f is integrable on $[a, b]$ and F_n is its nth primitive, then*
(i) $LD_{n-r}F_n = F_r$ in $[a, b]$,
(ii) $SLD^n F_n = f$ a.e. in (a, b),

Proof (i) If $r = n$ the case is trivial and if $r = 1$ the proof follows from Theorem 4.5, we suppose $1 < r < n$. By Theorem 4.4 $Q - F_n$ and $F_n - q$ are k convex for $0 \leq k \leq n$ and so $(Q - F_n)^{(n-r)}$ and $(F_n - q)^{(n-r)}$ are k convex for $0 \leq k \leq r$. Hence for $x \in [a, b]$ and for all major functions Q and all minor functions q, we have

$$LD_{n-r}Q(x) - LD_{n-r}(F_n)(x) \geq 0 \quad and \quad LD_{n-r}(F_n)(x) - LD_{n-r}q(x) \geq 0. \tag{10}$$

Let $\varepsilon > 0$ be arbitrary. By (4) there is a major function $\overline{Q}$ and a minor function $\overline{q}$ of f such that

$$0 \leq LD_{n-r}\overline{Q}(x) - LD_{n-r}\overline{q}(x) \leq \varepsilon(b-a)^{r-1} \quad for \quad all \quad x \in [a, b]. \tag{11}$$

From (10) to (11) we get

$$0 \leq LD_{n-r}\overline{Q}(x) - LD_{n-r}F_n(x) \leq \varepsilon(b-a)^{r-1} \quad for \quad all \quad x \in [a, b]. \tag{12}$$

From (10) to (12) we get
$\inf_{Q \in \{Q\}} LD_{n-r}Q(x) = LD_{n-r}F_n(x)$ for all $x \in [a, b]$.
From definition $\inf_{Q \in \{Q\}} LD_{n-r}Q(x) = F_r(x)$, so the proof is completed.

(ii) For any positive integer k define $E_k = \{x : x \in (a, b); \ f(x) > \underline{SLD}^n F_n(x) + \frac{1}{k}\}$. Suppose E_k has positive outer measure p. Choose $0 < \varepsilon < \frac{p}{2k}$. Let Q be a major function such that $LD_{n-1}Q(b) - LD_{n-1}F_n(b) < \varepsilon$, [by (9)]. Let $A \subset [a, b]$ be the set where $LD_{n-1}Q(x)$ exists. Let $R(x) = LD_{n-1}Q(x) - LD_{n-1}F_n(x)$ on $A \cap B$, where B is base of the integral of f. Since $Q - F_n$ is n-convex , $LD_{n-1}Q(x) - LD_{n-1}F_n(x)$ is nondecreasing on $A \cap B$. Extend R on whole of $[a, b]$ such that R remains nondecreasing on $[a, b]$. By Theorem 4.5 , $R(a) = 0$. So

$$\int_a^b R' \leq R(b) < \varepsilon \tag{13}$$

Let $G_k = \{x : x \in A \cap B \cap E_k; 0 \leq R'(x) \leq \frac{1}{2k}\}$ and $H_k = \{x : x \in A \cap B; R'(x) > \frac{1}{2k}\}$. Then H_k is measurable and by (13) $\mu(H_k) < p$. Since $\mu(A \cap B) = b - a$, $A \cap B \cap E_k$ has outer measure p and since R' exists almost everywhere, almost all the points of $A \cap B \cap E_k$ are points of $G_k \cup H_k$ and so G_k has positive outer measure. Since $Q - F_n$ is n- convex, $(Q - F_n)_{(n)}$ and hence $LD_n(Q - F_n)$ exists finitely a.e in (a, b) and $(Q - F_n)_{(n)} = LD_n(Q - F_n) = R'$ a.e in (a, b). So $SLD^n(Q - F_n) = R'$ since R is monotone. Hence almost everywhere in G_k we have

$$f \leq \underline{SLD}^n Q = \underline{SLD}^n F_n + SLD^n(Q - F_n) = \underline{SLD}^n F_n + R' \leq \underline{SLD}^n F_n + \frac{1}{2k}.$$

But this is a contradiction since $G_k \subset E_k$. So $\mu(E_k) = 0$. Since

$$\{x : x \in (a, b); f(x) > \underline{SLD}^n F_n(x)\} = \cup_{k=1}^{\infty} E_k,$$

we have $f \leq \underline{SLD}^n F_n(x)$ a.e in (a, b). Similarly $f \geq \overline{SLD}^n F_n(x)$ a.e in (a, b). Hence $SLD^n F_n$ exists and equals to f a.e in (a, b).

Theorem 4.9 *If f is integrable, then f is measurable and finite almost everywhere.*

Proof Let F_n be the n th primitive of f, then $SLD^n F_n = f$ a.e. Since

$$SLD^n F_n = \lim_{m\to\infty} s_m^{2k+1} \int_0^{\delta} e^{-s_m t} \left[\frac{F_n(x+t) + F_n(x-t)}{2} - \sum_{i=0}^{k-1} \frac{t^{2i}}{(2i)!}\alpha_{2i} \right] dt,$$

when $n(= 2k)$ is even and

$$= \lim_{m\to\infty} s_m^{2k+2} \int_0^{\delta} e^{-s_m t} \left[\frac{F_n(x+t) - F_n(x-t)}{2} - \sum_{i=0}^{k-1} \frac{t^{2i+1}}{(2i+1)!}\alpha_{2i+1} \right] dt,$$

when $n(= 2k + 1)$ is odd.
So $SLD^n F_n$ is measurable. Hence f is measurable. Suppose that $f = \infty$ on a set of positive measure. So by Theorem 4.8 (ii) $SLD^n F_n = \infty$ on a set of positive measure. Let q be any minor function of f. Then $F_n - q$ is n- convex and so $SLD^n(F_n - q)$ exists

finitely almost everywhere in (a, b). Since $\underline{SLD}^n F_n = \underline{SLD}^n q + SLD^n(F_n - q)$, we have $\underline{SLD}^n q = \infty$ on a set of positive measure, which is a contradiction. Thus $f < \infty$ almost everywhere. Similarly, $f > -\infty$ almost everywhere.

5 Generality of LT^n Integral Over T_n integral

For the definitions of d.l.V.P. derivative and ordinary smoothness of a function we refer [7].

Theorem 5.1 *If f is T_n integrable in the sense of [7], then it is LT^n integrable for all $n \geq 2$ and the integrals are equal.*

Proof Let f be T_n integrable in the sense of [7] and M be any major function in the sense of [7]. Then M is also LT^n major function for the following facts: (i) $\underline{D}^n M \leq \underline{SLD}^n M$ [by Theorem 2.5 of [3]]. (ii) If f is ordinary smooth of order n then it is Laplace smooth of order n.[by Theorem 3.2 of [9]]. The case for LT^n minor function is similar.

Theorem 5.2 *If f is integrable and $f \geq 0$ then f is Lebesgue integrable.*

Proof Let Q be any major function. Then $\underline{SLD}^n Q \geq f \geq 0$ a.e. The other properties of Q implies that Q is n- convex. So by Theorem 4.3, F_n is n-convex and F_1 is nondecreasing on B where B is the base of the integral of f. Therefore, $F_1' = LD_1 F_1$ is Lebesgue integrable. By [1] $(F_n)_{(n)}$ exists a.e. and by Theorem 4.26 of [16]; $(F_n)_{(n)} = (F_1)'$ a.e. By Theorem 4.8, f is Lebesgue integrable.

Theorem 5.3 *If f is integrable in $[a, b]$ and $c \in B$, then f is integrable in $[a, c]$ and in $[c, b]$ and*

$$\int_a^b f = \int_a^c f + \int_c^b f.$$

Again if f is integrable in $[a, c]$ and in $[c, b]$ and if $LD_{n-1}^-(G_n(c))$ and LD_{n-1}^+ $(H_n(c))$ exist, where G_n and H_n are the n th primitives of f on $[a, c]$ and $[c, b]$ respectively, then f is integrable in $[a, b]$ and

$$\int_a^c f + \int_c^b f = \int_a^b f.$$

Proof Let f is integrable in $[a, b]$ and let $\{Q\}$ be the collection of all major functions and let F_n be its n th primitive. Then for each $Q \in \{Q\}$ the restriction of Q on $[a, c]$ is a major function of f on $[a, c]$. Then from (8)

$$\inf_{Q \in \{Q\}} LD_{n-1} Q(c) = LD_{n-1} F_n(c) \tag{14}$$

Similarly, $\sup_{q\in\{q\}} LD_{n-1}q(c) = LD_{n-1}F_n(c)$. Hence f is integrable in $[a, c]$ and

$$\int_a^c f = LD_{n-1}F_n(c) \tag{15}$$

Since $c \in B$, by Theorem 4.7, $LD_{n-1}Q(c)$ and $LD_{n-1}q(c)$ exist. For each $Q \in \{Q\}$, let

$$\overline{Q}(x) = Q(x) - \sum_{i=0}^{n-1} \frac{(x-c)^i}{i!} LD_i Q(c).$$

Then $\overline{Q}(x)$ is a major function of f in $[c, b]$.
Also, $LD_{n-1}\overline{Q}(b) + LD_{n-1}Q(c) = LD_{n-1}Q(b)$, and hence

$$\inf_{u\in\{u\}} LD_{n-1}u(b) + \inf_{Q\in\{Q\}} LD_{n-1}Q(c) \leq \inf_{\underline{Q}\in\{Q\}} LD_{n-1}\overline{Q}(b) + \inf_{Q\in\{Q\}} LD_{n-1}Q(c)$$

$$\leq \inf_{Q\in\{Q\}} \left[LD_{n-1}\overline{Q}(b) + LD_{n-1}Q(c)\right] = \inf_{Q\in\{Q\}} LD_{n-1}Q(b) = LD_{n-1}F_n(b),$$

where $\{u\}$ is the class of all major functions of f on $[c, b]$. So by (14)

$\inf_{u\in\{u\}} LD_{n-1}u(b) \leq LD_{n-1}F_n(b) - LD_{n-1}F_n(c)$. Similarly,
$\sup_{v\in\{v\}} LD_{n-1}v(b) \geq LD_{n-1}F_n(b) - LD_{n-1}F_n(c)$, where $\{v\}$ is the class of all minor functions of f on $[c, b]$. Therefore f is integrable in $[c, b]$ and

$$\int_c^b f = LD_{n-1}F_n(b) - LD_{n-1}F_n(c).$$

Hence from (15)

$$\int_a^b f = LD_{n-1}F_n(b) = LD_{n-1}F_n(c) + \int_c^b f = \int_a^c f + \int_c^b f.$$

For the converse part, let Q, q, and u,v be the major and minor functions of f on $[a, c]$ and $[c, b]$, respectively, such that
$LD_{n-1}Q(c) - LD_{n-1}q(c) < \frac{\varepsilon}{2}$ and $LD_{n-1}u(b) - LD_{n-1}v(b) < \frac{\varepsilon}{2}$. Since G_n is n th primitive of f in $[a, c]$, $Q - G_n$ is n convex and so $(Q - G_n)^{(n-2)}$ is convex and continuous in $[a, c]$. So $LD_{n-1}^-(Q - G_n)(c)$ exists. Since $LD_{n-1}^-(G_n)(c)$ exists, $LD_{n-1}^-Q(c)$ exists, and also finite. Since Q is major function, $LD_{n-1}Q(c)$ is finite. Let

$$\overline{Q}(x) = Q(x) \text{ for } x \in [a, c]$$
$$= u(x) - \sum_{i=0}^{n-1} \frac{(x-c)^i}{i!} LD_i Q(c) \text{ for } x \in [c, b].$$

Then $\overline{Q}$ is a major function of $[a, b]$. Similarly, $\overline{q}$ is a minor function of f on $[a, b]$ where

$$\overline{q}(x) = q(x) \text{ for } x \in [a, c]$$
$$= v(x) - \sum_{i=0}^{n-1} \frac{(x-c)^i}{i!} LD_i q(c) \text{ for } x \in [c, b].$$

Since $\left[LD_{n-1}\overline{Q}(b) - LD_{n-1}\overline{q}(b)\right] \leq \left[LD_{n-1}u(b) - LD_{n-1}v(b)\right] + \left[LD_{n-1}Q(c) - LD_{n-1}q(c)\right] \leq \varepsilon$. So f is integrable in $[a, b]$. The rest is clear.

Theorem 5.4

$$(LT^n) \int_a^b f = LD_{n-1}F(b) - LD_{n-1}F(a).$$

Proof Let the function $\phi(x) = F(x) - \sum_{i=0}^{n-1} \frac{(x-a)^i}{i!} LD_i F(a)$ for $a \leq x \leq b$, is both a LT^n major and LT^n minor function of f on $[a, b]$. So f is LT^n integrable on $[a, b]$ and

$$(LT^n) \int_a^b f = LD_{n-1}\phi(b) = LD_{n-1}F(b) - LD_{n-1}F(a).$$

Theorem 5.5 *If f is integrable on $[a, b]$ and k is a finite constant then kf is integrable in $[a, b]$ and*

$$\int_a^b kf = k \int_a^b f.$$

Proof Case-I

Let $k \geq 0$ and let $\{Q\}$ and $\{q\}$ be the collection of all major and minor functions of f in $[a, b]$. Let F_n be the n th primitive of f in $[a, b]$. Since f is integrable

$$\inf_{Q \in \{Q\}} LD_{n-1}Q(b) = LD_{n-1}F_n(b) \quad and \quad \sup_{q \in \{q\}} LD_{n-1}q(b) = LD_{n-1}F_n(b). \tag{16}$$

It is clear that $\{kQ\}$ and $\{kq\}$ be the collection of all major and minor functions of kf in $[a, b]$. Now

$$\inf_{kQ \in \{kQ\}} LD_{n-1}kQ(b) = k \inf_{Q \in \{Q\}} LD_{n-1}Q(b) = kLD_{n-1}F_n(b),$$

and

$$\sup_{kq\in\{kq\}} LD_{n-1}kq(b) = k \inf_{q\in\{q\}} LD_{n-1}q(b) = kLD_{n-1}F_n(b).$$

Hence kf is integrable and

$$\int_a^b kf = \inf_{kQ\in\{kQ\}} LD_{n-1}kQ(b) = kLD_{n-1}F_n(b) = k\int_a^b f.$$

Case-II
Let $k < 0$ and if $\{Q\}$ and $\{q\}$ be the collection of all major and minor functions of f in $[a, b]$, then $\{kQ\}$ and $\{kq\}$ be the collection of all minor and major functions of kf in $[a, b]$. The proof is completed by **case I**.

Theorem 5.6 *Let f and g be integrable on $[a, b]$. If the sum $f + g$ is defined on $[a, b]$ then $f + g$ is integrable in $[a, b]$ and*

$$\int_a^b (f+g) = \int_a^b f + \int_a^b g.$$

Proof If $\{Q_1\}$, $\{q_1\}$ and $\{Q_2\}$, $\{q_2\}$ be the collection of all major and minor functions of f and g in $[a, b]$. Then $\{Q_1 + Q_2\}$ and $\{q_1 + q_2\}$ be the collection of all major and minor functions of $f + g$ in $[a, b]$. Then the proof is same as the above proof.

References

1. Bullen, P.S.: A criterion for n-convexity. Pacific J. Math. **36**, 81–98 (1971)
2. Burkill, J.C.: Integral and trigonometric series. Proc. Lond. Math. Soc. **3**(1), 46–57 (1951)
3. Garai, A., Ray, S.: On the Symmetric Laplace derivative. Acta Mathematica Hungarica **133** (1–2), 166–184 (2011)
4. Garai, A., Ray, S.: Convexity conditions for higher order Symmetric Laplace derivative. Bull. Allahabad Math. Soc. **28**(2), 131–152 (2013)
5. Mukhopadhayay, S.N., Ray, S.: Convexity conditions for approximate symmetric $d.l.V.P.$ derivable functions. Indian J. Math. **49**(1), 71–92 (2007)
6. Mukhopadhayay, S.N.: Higher order derivatives. Chapman and Hall/CRC, Monographs and surveys in pure and applied mathematics **144**, (2012)
7. Mukhopadhyay, S.N.: An n-th order integral and its integration by parts with applications to trigonometric series. Real Anal. Exch. **30**(2), 451–494 (2004–2005)
8. Mukhopadhyay, S.N., Ray, S.: On laplace derivative. Anal. Math. **36**, 131–153 (2010)
9. Ray, S., Garai, A.: On higher order laplace smooth functions, Bull. Allahabad Math.Soc. **29**(2), 153–172 (2014)
10. Ray, S., Garai, A.: The Laplce Derivative, The Math.Student, 81. Nos. **1–4**, 171–175 (2012)
11. Ray, S., Garai, A.: The Laplce Derivative II, The Math.Student, 81. Nos. **1–4**, 177–184 (2012)
12. Ray, S., Garai, T.K.: On Laplace Continuity, Real analysis Exchange, 37(2), 2011/2012, 299–310

13. Ray, S., Ghosh, S.: Some mean value theorems for the Laplace Derivative, IJMA, 4. Nos. **1**, 65–70 (2012)
14. Oliver, H.W.: The exact peano derivative. Trans. Amer. Math. Soc. **76**, 444–456 (1954)
15. Sevtic, R.E.: The Laplace Derivative. Comment. Math. Univ. Carolinae **42**(2), 331–343 (2001)
16. Zygmund, A.: Trigonometric Series. Cambridge University Press, London (1968)

A Sequence Space and Uniform (A, φ)—Statistical Convergence

Ekrem Savaş

Abstract In this, we introduce and study some properties of the new sequence space that is defined using the φ—function and de la Valée-Poussin mean. We also study some connections between $V_\lambda((A, \varphi))$—strong summability of sequences and λ—strong convergence with respect to a modulus.

Keywords Modulus function · φ-function · λ—strong convergence · Matrix transformations · Sequence spaces · Statistical convergence

1 Introduction and Background

Let s denote the set of all real and complex sequences $x = (x_k)$. By l_∞ and c, we denote the Banach spaces of bounded and convergent sequences $x = (x_k)$ normed by $||x|| = \sup_n |x_n|$, respectively. A sequence $x \in l_\infty$ is said to be almost convergent if all of its Banach limits coincide. Let $\hat{c}$ denote the space of all almost convergent sequences. Lorentz [6] has shown that

$$\hat{c} = \left\{ x \in l_\infty : \lim_m t_{m,n}(x) \text{ exists uniformly in } n \right\}$$

where

$$t_{m,n}(x) = \frac{x_n + x_{n+1} + x_{n+2} + \cdots + x_{n+m}}{m+1}.$$

The space $[\hat{c}]$ of strongly almost convergent sequences was introduced by Maddox [7] and also independently by Freedman et al. [3] as follows:

E. Savaş (✉)
Istanbul Commerce University, Department of Mathematics, Sütlüce, Istanbul, Turkey
e-mail: ekremsavas@yahoo.com

R.N. Mohapatra et al. (eds.), *Mathematics and Computing*,
Springer Proceedings in Mathematics & Statistics 139,
DOI 10.1007/978-81-322-2452-5_34

$$[\hat{c}] = \left\{x \in l_\infty : \lim_m t_{m,n}(|x - L|) = 0, \text{ uniformly in } n, \text{ for some } L\right\}.$$

Let $\lambda = (\lambda_i)$ be a nondecreasing sequence of positive numbers tending to ∞ such that

$$\lambda_{i+1} \leq \lambda_i + 1, \lambda_1 = 1.$$

The collection of such sequence λ will be denoted by Δ.

The generalized de la Valée-Poussin mean is defined as

$$T_i(x) = \frac{1}{\lambda_i} \sum_{k \in I_i} x_k$$

where $I_i = [i - \lambda_i + 1, i]$. A sequence $x = (x_n)$ is said to be (V, λ)—summable to a number L, if $T_i(x) \to L$ as $i \to \infty$ (see [9]).

Recently, Malkowsky and Savaş [9] introduced the space $[V, \lambda]$ of λ—strongly convergent sequences as follows:

$$[V, \lambda] = \left\{x = (x_k) : \lim_i \frac{1}{\lambda_i} \sum_{k \in I_i} |x_k - L| = 0, \text{ for some } L\right\}.$$

Note that in the special case where $\lambda_i = i$, the space $[V, \lambda]$ reduces the space w of strongly Cesàro summable sequences which is defined as

$$w = \left\{x = (x_k) : \lim_i \frac{1}{i} \sum_{k=1}^{i} |x_k - L|) = 0, \text{ for some } L\right\}.$$

More results on λ- strong convergence can be seen from [12, 20–24].

Ruckle [16] used the idea of a modulus function f to construct a class of FK spaces

$$L(f) = \left\{x = (x_k) : \sum_{k=1}^{\infty} f\left(|x_k|\right) < \infty\right\}.$$

The space $L(f)$ is closely related to the space l_1, which is an $L(f)$ space with $f(x) = x$ for all real $x \geq 0$.

Maddox [8] introduced and examined some properties of the sequence spaces $w_0(f)$, $w(f)$, and $w_\infty(f)$ defined using a modulus f, which generalized the well-known spaces w_0, w and w_∞ of strongly summable sequences.

Recently, Savas [19] generalized the concept of strong almost convergence using a modulus f and examined some properties of the corresponding new sequence spaces.

Waszak [26] defined the lacunary strong (A, φ)—convergence with respect to a modulus function.

Following Ruckle [16], a modulus function f is a function from $[0, \infty)$ to $[0, \infty)$ such that

(i) $f(x) = 0$ if and only if $x = 0$,
(ii) $f(x + y) \leq f(x) + f(y)$ for all $x, y \geq 0$,
(iii) f increasing,
(iv) f is continuous from the right at zero.

Since $|f(x) - f(y)| \leq f(|x - y|)$, it follows from condition (iv) that f is continuous on $[0, \infty)$.

If $x = (x_k)$ is a sequence and $A = (a_{nk})$ is an infinite matrix, then Ax is the sequence whose nth term is given by $A_n(x) = \sum_{k=0}^{\infty} a_{nk}x_k$. Thus we say that x is A-summable to L if $\lim_{n\to\infty} A_n(x) = L$. Let X and Y be two sequence spaces and $A = (a_{nk})$ an infinite matrix. If for each $x \in X$ the series $A_n(x) = \sum_{k=0}^{\infty} a_{nk}x_k$ converges for each n and the sequence $Ax = A_n(x) \in Y$ we say that A maps X into Y. By (X, Y) we denote the set of all matrices which maps X into Y, and in addition if the limit is preserved then we denote the class of such matrices by $(X, Y)_{reg}$.

A matrix A is called regular , i.e., $A \in (c, c)_{reg}$. if $A \in (c, c)$ and $lim_n A_n(x) = lim_k x_k$ for all $x \in c$.

In 1993, Nuray and Savas [14] defined the following sequence spaces:

Definition 1 Let f be a modulus and A a nonnegative regular summability method. We let

$$w(\hat{A}, f) = \left\{ x : lim_n \sum_{k=1}^{\infty} a_{nk} f(|x_{k+m} - L|) = 0, \text{ for some L, uniformly in } m \right\}$$

and

$$w(\hat{A}, f)_0 = \left\{ x : lim_n \sum_{k=1}^{\infty} a_{nk} f(|x_{k+m}|) = 0, \text{ uniformly in } m \right\}.$$

If we take $A = (a_{nk})$ as

$$a_{nk} := \begin{cases} \frac{1}{n}, & \text{if } \quad n \geq k, \\ 0, & \text{otherwise.} \end{cases}$$

Then the above definitions are reduced to $[\hat{c}(f)]$ and $[\hat{c}(f)]_0$ which were defined and studied by Pehlivan [15].

If we take $A = (a_{nk})$ is a de la Valée poussin mean, i.e.,

$$a_{nk} := \begin{cases} \frac{1}{\lambda_n}, & \text{if } k \in I_n = [n - \lambda_n + 1, n], \\ 0, & \text{otherwise.} \end{cases}$$

Then these definitions are reduced to the following sequence spaces which were defined and studied by Malkowsky and Savas [9].

$$w(\hat{V},\lambda,f)=\left\{x: lim_j\frac{1}{\lambda_j}\sum_{k\in I_j}f(|x_{k+m}-L|)=0, \text{ for some L, uniformly in } m\right\}$$

and

$$w(\hat{V},\lambda,f)_0=\left\{x: lim_j\frac{1}{\lambda_j}\sum_{k\in I_j}f(|x_{k+m}|)=0, \text{ uniformly in } m\right\}$$

When $\lambda_j=j$ the above sequence spaces become $[\hat{c}(f)]_0$ and $[\hat{c}(f)]$.

By a φ-function we understand a continuous nondecreasing function $\varphi(u)$ defined for $u\geq 0$ and such that $\varphi(0)=0$, $\varphi(u)>0$, for $u>0$ and $\varphi(u)\to\infty$ as $u\to\infty$, (see, [26]).

A φ-function φ is called non-weaker than a φ-function ψ if there are constants $c,b,k,l>0$ such that $c\psi(lu)\leq b\varphi(ku)$, (for all large u) and we write $\psi\prec\varphi$.

A φ-function φ and ψ are called equivalent and we write $\varphi\sim\psi$ if there are positive constants b_1,b_2,c,k_1,k_2,l such that $b_1\varphi(k_1u)\leq c\psi(lu)\leq b_2\varphi(k_2u)$, (for all large u), (see, [26]).

A φ-function φ is said to satisfy (Δ_2)-condition, (for all large u) if there exists constant $K>1$ such that $\varphi(2u)\leq K\varphi(u)$.

In this paper, we introduce and study some properties of the following sequence space that is defined using the φ- function and de la Valée-Poussin mean and some known results are also obtained as special cases.

2 Main Results

Let $\Lambda=(\lambda_j)$ be the same as above, φ be given φ-function, and f be given modulus function, respectively. Moreover, let $\mathbf{A}=(a_{nk}(i))$ be the generalized three-parametric real matrix. Then we define

$$V_\lambda^0((A,\varphi),f)=\left\{x=(x_k):\lim_j\frac{1}{\lambda_j}\sum_{n\in I_j}f\left(\left|\sum_{k=1}^{\infty}a_{nk}(i)\varphi(|x_k|)\right|\right)=0, \text{ uniformly in } i\right\}.$$

If $\lambda_j=j$, we have

$$V_\lambda^0((A,\varphi),f)=\left\{x=(x_k):\lim_j\frac{1}{j}\sum_{n=1}^{j}f\left(\left|\sum_{k=1}^{\infty}a_{nk}(i)\varphi(|x_k|)\right|\right)=0, \text{uniformly in } i\right\}.$$

If $x \in V_\lambda^0((A, \varphi), f)$, the sequence x is said to be λ—strong (A, φ)—convergent to zero with respect to a modulus f. When $\varphi(x) = x$ for all x, we obtain

$$V_\lambda^0((A), f) = \left\{ x = (x_k) : \lim_j \frac{1}{\lambda_j} \sum_{n \in I_j} f\left(\left| \sum_{k=1}^{\infty} a_{nk}(i)(|x_k|) \right| \right) = 0, \text{ uniformly in } i \right\}.$$

If $f(x) = x$, we write

$$V_\lambda^0(A, \varphi) = \left\{ x = (x_k) : \lim_j \frac{1}{\lambda_j} \sum_{n \in I_j} \left(\left| \sum_{k=1}^{\infty} a_{nk}(i)\varphi(|x_k|) \right| \right) = 0, \text{ uniformly in } i \right\}.$$

If we take $A = I$ and $\varphi(x) = x$ respectively, then we have

$$V_\lambda^0(I, f) = \left\{ x = (x_k) : \lim_j \frac{1}{\lambda_j} \sum_{k \in I_j} f\left(|x_k| \right) = 0 \right\}.$$

If we take $A = I$, $\varphi(x) = x$ and $f(x) = x$ respectively, then we have

$$V_\lambda^0((I)) = \left\{ x = (x_k) : \lim_j \frac{1}{\lambda_j} \sum_{k \in I_j} |x_k| = 0 \right\},$$

which was defined and studied by Savaş and Savaş [18].

If we define the matrix $A = (a_{nk}(i))$ as follows: for all i

$$a_{nk}(i) := \{ \begin{matrix} \frac{1}{n}, & \text{if} \quad n \geq k, \\ 0, & \text{otherwise.} \end{matrix}$$

then we have,

$$V_\lambda^0(\mathbf{C}, \varphi, f) = \left\{ x = (x_k) : \lim_j \frac{1}{\lambda_j} \sum_{n \in I_j} f\left(\left| \frac{1}{n} \sum_{k=1}^{n} \varphi(|x_k|) \right| \right) = 0 \right\}.$$

If we define

$$a_{nk}(i) := \{ \begin{matrix} \frac{1}{n}, & \text{if } i \le k \le i+n-1, \\ 0, & \text{otherwise.} \end{matrix}$$

then we have,

$$V_\lambda^0(\hat{c}, \varphi, f) = \left\{ x = (x_k) : \lim_j \frac{1}{\lambda_j} \sum_{n \in I_j} f\left(\left| \frac{1}{n} \sum_{k=i}^{i+n} \varphi(|x_k|) \right| \right) = 0, \text{ uniformly in } i \right\}.$$

We now have:

Theorem 1 *Let* $\mathbf{A} = (a_{nk}(i))$ *be the generalized three parametric real matrix and let the* φ*—function* $\varphi(u)$ *satisfy the condition* (Δ_2). *Then the following conditions are true:*

(a) If $x = (x_k) \in w((\mathbf{A}, \varphi), f)$ *and* α *is an arbitrary number, then* $\alpha x \in w((\mathbf{A}, \varphi), f)$.
(b) If $x, y \in w((\mathbf{A}, \varphi), f)$ *where* $x = (x_k)$, $y = (y_k)$ *and* α, β *are given numbers, then* $\alpha x + \beta y \in w((\mathbf{A}, \varphi), f)$.

The proof is a routine verification by using standard techniques and hence is omitted.

Theorem 2 *Let* f *be a modulus function.*

$$V_\lambda^0(A, \varphi) \subseteq V_\lambda^0((A, \varphi), f).$$

Proof Let $x \in V_\lambda^0(A, \varphi)$. For a given $\varepsilon > 0$ we choose $0 < \delta < 1$ such that $f(x) < \varepsilon$ for every $x \in [0, \delta]$. We can write for all i

$$\frac{1}{\lambda_j} \sum_{n \in I_j} f\left(\left| \sum_{k=1}^{\infty} a_{nk}(i) \varphi(|x_k|) \right| \right) = S_1 + S_2,$$

where $S_1 = \frac{1}{\lambda_j} \sum_{n \in I_j} f\left(\left| \sum_{k=1}^{\infty} a_{nk}(i) \varphi(|x_k|) \right| \right)$ and this sum is taken over

$$\sum_{k=1}^{\infty} a_{nk}(i) \varphi(|x_k|) \le \delta$$

and

$$S_2 = \frac{1}{\lambda_j} \sum_{n \in I_j} f\left(\left| \sum_{k=1}^{\infty} a_{nk}(i)\varphi(|x_k|) \right| \right)$$

and this sum is taken over

$$\sum_{k=1}^{\infty} a_{nk}(i)\varphi(|x_k|) > \delta.$$

By definition of the modulus f we have $S_1 = \frac{1}{\lambda_j} \sum_{n \in I_j} f\left(\delta\right) = f(\delta) < \varepsilon$ and moreover

$$S_2 = f(1)\frac{1}{\delta}\frac{1}{\lambda_j} \sum_{n \in I_j} \sum_{k=1}^{\infty} a_{nk}(i)\varphi(|x_k|).$$

Thus we have $x \in V_{\lambda}^{0}((A, \varphi), f)$.
This completes the proof.

3 Uniform (A, φ)—Statistical Convergence

The idea of convergence of a real sequence was extended to statistical convergence by Fast [2] (see also Schoenberg [25]) as follows: If $\mathbb{N}$ denotes the set of natural numbers and $K \subset \mathbb{N}$ then $K(m, n)$ denotes the cardinality of the set $K \cap [m, n]$, the upper and lower natural densities of the subset K are defined as

$$\overline{d}(K) = \lim_{n \to \infty} \sup \frac{K(1, n)}{n} \text{ and } \underline{d}(K) = \lim_{n \to \infty} \inf \frac{K(1, n)}{n}.$$

If $\overline{d}(K) = \underline{d}(K)$ then we say that the natural density of K exists and it is denoted simply by $d(K)$. Clearly $d(K) = \lim_{n \to \infty} \frac{K(1, n)}{n}$.

A sequence (x_k) of real numbers is said to be statistically convergent to L if for arbitrary $\epsilon > 0$, the set $K(\epsilon) = \{k \in \mathbb{N} : |x_k - L| \geq \epsilon\}$ has natural density zero. Statistical convergence turned out to be one of the most active areas of research in summability theory after the work of Fridy [4] and Šalát [17].

In another direction, a new type of convergence called λ-statistical convergence was introduced in [13] as follows.

A sequence (x_k) of real numbers is said to be λ- statistically convergent to L (or, S_λ-convergent to L) if for any $\epsilon > 0$,

$$\lim_{j \to \infty} \frac{1}{\lambda_j} |\{k \in I_j : |x_k - L| \geq \epsilon\}| = 0$$

where $|A|$ denotes the cardinality of $A \subset \mathbb{N}$. In [13] the relation between λ-statistical convergence and statistical convergence was established among other things.

Recently, Savas [20] defined almost λ-statistical convergence using the notion of (V, λ)-summability to generalize the concept of statistical convergence.

Assume that A is a nonnegative regular summability matrix. Then the sequence $x = (x_n)$ is called statistically convergent to L provided that, for every $\varepsilon > 0$, (see, [5])

$$lim_j \sum_{n:|x_n - L| \geq \varepsilon} a_{jn} = 0.$$

Let $\mathbf{A} = (a_{nk}(i))$ be the generalized three parametric real matrix and the sequence $x = (x_k)$, the φ-function $\varphi(u)$ and a positive number $\varepsilon > 0$ be given. We write, for all i

$$K_\lambda^j((A, \varphi), \varepsilon) = \{n \in I_j : \sum_{k=1}^{\infty} a_{nk}(i)\varphi(|x_k|) \geq \varepsilon\}.$$

The sequence x is said to be uniform (A, φ)—statistically convergent to a number zero if for every $\varepsilon > 0$

$$lim_j \frac{1}{\lambda_j} \mu(K_\lambda^j((A, \varphi), \varepsilon)) = 0, \text{ uniformly in } i$$

where $\mu(K_\lambda^j((A, \varphi), \varepsilon))$ denotes the number of elements belonging to $K_\lambda^j((A, \varphi), \varepsilon)$. We denote by $S_\lambda^0((A, \varphi))$, the set of sequences $x = (x_k)$ which are uniform (A, φ)—statistical convergent to zero.

If we take $A = I$ and $\varphi(x) = x$ respectively, then $S_\lambda^0((A, \varphi))$ reduce to S_λ^0 which was defined as follows, (see, Mursaleen [13]).

$$S_\lambda^0 = \left\{ x = (x_k) : lim_j \frac{1}{\lambda_j} |\{k \in I_j : |x_k| \geq \varepsilon\}| = 0 \right\}.$$

Remark 1 (i) If for all i,

$$a_{nk} := \{ \begin{matrix} \frac{1}{n}, & \text{if} \quad n \geq k, \\ 0, & \text{otherwise.} \end{matrix}$$

then $S_\lambda((A, \varphi))$ reduce to $S_\lambda^0((C, \varphi))$, i.e., uniform (C, φ)— statistical convergence.
(ii) If for all i, (see, [1]),

$$a_{nk} := \{ \begin{matrix} \frac{p_k}{P_n}, & \text{if} \quad n \geq k, \\ 0, & \text{otherwise.} \end{matrix}$$

then $S_\lambda((A, \varphi))$ reduce to $S_\lambda^0((N, p), \varphi))$, i.e., uniform $((N, p), \varphi)$— statistical convergence, where $p = p_k$ is a sequence of nonnegative numbers such that $p_0 > 0$ and

$$P_i = \sum_{k=0}^{n} p_k \to \infty (n \to \infty).$$

We are now ready to state the following theorem.

Theorem 3 *If $\psi \prec \varphi$ then $S_\lambda^0((A, \psi)) \subset S_\lambda^0((A, \varphi))$.*

Proof By our assumptions we have $\psi(|x_k|) \leq b\varphi(c|x_k|)$ and we have for all i,

$$\sum_{k=1}^{\infty} a_{nk}(i)\psi(|x_k|) \leq b \sum_{k=1}^{\infty} a_{nk}(i)\varphi(c|x_k|) \leq K \sum_{k=1}^{\infty} a_{nk}(i)\varphi(|x_k|)$$

for $b, c > 0$, where the constant K is connected with properties of φ. Thus, the condition $\sum_{k=1}^{\infty} a_{nk}(i)\psi(|x_k|) \geq \varepsilon$ implies the condition $\sum_{k=1}^{\infty} a_{nk}(i)\varphi(|x_k|) \geq \varepsilon$ and in consequence we get

$$\mu(K_\lambda^j((A, \varphi), \varepsilon)) \subset \mu(K_\lambda^j((A, \psi), \varepsilon))$$

and

$$lim_j \frac{1}{\lambda_j} \mu\Big(K_\lambda^j((A, \varphi), \varepsilon)) \leq lim_j \frac{1}{\lambda_j} \mu(K_\lambda^j((A, \psi), \varepsilon))\Big).$$

This completes the proof.

Theorem 4 *(a) If the matrix A, functions f, and φ are given, then*

$$V_\lambda^0((A, \varphi), f) \subset S_\lambda^0(A, \varphi).$$

(b) If the φ- function $\varphi(u)$ and the matrix A are given, and if the modulus function f is bounded, then

$$S_\lambda^0(A, \varphi) \subset V_\lambda^0(A, \varphi), f).$$

(c) If the φ- function $\varphi(u)$ and the matrix A are given, and if the modulus function f is bounded, then

$$S_\lambda^0(A, \varphi) = V_\lambda^0(A, \varphi), f).$$

Proof (a) Let f be a modulus function and let ε be a positive number. We write the following inequalities:

$$\frac{1}{\lambda_j}\sum_{n\in I_j} f\left(\left|\sum_{k=1}^{\infty} a_{nk}(i)\varphi(|x_k|)\right|\right)$$
$$\geq \frac{1}{\lambda_j}\sum_{n\in I_j^1} f\left(\left|\sum_{k=1}^{\infty} a_{nk}(i)\varphi(|x_k|)\right|\right)$$
$$\geq \frac{1}{\lambda_j} f(\varepsilon)\sum_{n\in I_j^1} 1$$
$$\geq \frac{1}{\lambda_j} f(\varepsilon)\mu(K_\lambda^j(A,\varphi),\varepsilon),$$

where

$$I_j^1 = \left\{n \in I_j : \sum_{k=1}^{\infty} a_{nk}(i)\varphi(|x_k|) \geq \varepsilon\right\}.$$

Finally, if $x \in V_\lambda^0((A,\varphi), f)$ then $x \in S_\lambda^0(A,\varphi)$.

(b) Let us suppose that $x \in S_\lambda^0(A,\varphi)$. If the modulus function f is a bounded function, then there exists an integer M such that $f(x) < M$ for $x \geq 0$. Let us take

$$I_j^2 = \left\{n \in I_j : \sum_{k=1}^{\infty} a_{nk}(i)\varphi(|x_k|) < \varepsilon\right\}.$$

Thus we have

$$\frac{1}{\lambda_j}\sum_{n\in I_j} f\left(\left|\sum_{k=1}^{\infty} a_{nk}(i)\varphi(|x_k|)\right|\right)$$
$$\leq \frac{1}{\lambda_j}\sum_{n\in I_j^1} f\left(\left|\sum_{k=1}^{\infty} a_{nk}(i)\varphi(|x_k|)\right|\right)$$
$$+ \frac{1}{\lambda_j}\sum_{n\in I_j^2} f\left(\left|\sum_{k=1}^{\infty} a_{nk}(i)\varphi(|x_k|)\right|\right)$$
$$\leq \frac{1}{\lambda_j} M\mu(K_\lambda^j((A,\varphi),\varepsilon) + f(\varepsilon).$$

Taking the limit as $\varepsilon \to 0$, we obtain that $x \in V_\lambda^0(A,\varphi, f)$.
The proof of (c) follows from (a) and (b).
This completes the proof.

In the next theorem we prove the following relation.

Theorem 5 *If a sequence $x = (x_k)$ is $S(A, \varphi)$—convergent to L and*

$$\lim inf_j \left(\frac{\lambda_j}{j}\right) > 0$$

then it is $S_\lambda(A, \varphi)$ convergent to L, where

$$S(A, \varphi) = \{x = (x_k) : lim_j \frac{1}{j}\mu(K(A, \varphi, \varepsilon)) = 0\}.$$

Proof For a given $\varepsilon > 0$, we have, for all i

$$\{n \in I_j : \sum_{k=0}^{\infty} a_{nk}(i)\varphi(|x_k - L|) \geq \varepsilon\} \subseteq \{n \leq j : \sum_{k=0}^{\infty} a_{nk}(i)\varphi(|x_k - L|) \geq \varepsilon\}.$$

Hence we have,

$$K_\lambda(A, \varphi, \varepsilon) \subseteq K(A, \varphi, \varepsilon).$$

Finally the proof follows from the following inequality:

$$\frac{1}{j}\mu(K(A, \varphi, \varepsilon)) \geq \frac{1}{j}\mu(K_\lambda(A, \varphi, \varepsilon)) = \frac{\lambda_j}{j}\frac{1}{\lambda_j}\mu(K_\lambda(A, \varphi, \varepsilon)).$$

This completes the proof.

Theorem 6 *If $\lambda \in \Delta$ be such that $lim_j \frac{\lambda_j}{j} = 1$ and the sequence $x = (x_k)$ is $S_\lambda(A, \varphi)$—convergent to L then it is $S(A, \varphi)$ convergent to L,*

Proof Let $\delta > 0$ be given. Since $lim_j \frac{\lambda_j}{j} = 1$, we can choose $m \in N$ such that $|\frac{\lambda_j}{j} - 1| < \frac{\delta}{2}$, for all $j \geq m$. Now observe that, for $\varepsilon > 0$

$$\begin{aligned}
&\frac{1}{j}\left|\left\{n \leq j : \sum_{k=0}^{\infty} a_{nk}(i)\varphi(|x_k - L|) \geq \varepsilon\right\}\right| \\
&= \frac{1}{j}\left|\left\{k \leq j - \lambda_j : \sum_{k=0}^{\infty} a_{nk}(i)\varphi(|x_k - L|) \geq \epsilon\right\}\right| \\
&+ \frac{1}{j}\left|\left\{n \in I_j : \sum_{k=0}^{\infty} a_{nk}(i)\varphi(|x_k - L|) \geq \epsilon\right\}\right| \\
&\leq \frac{j - \lambda_j}{j} + \frac{1}{j}\left|\left\{n \in I_j : \sum_{k=0}^{\infty} a_{nk}(i)\varphi(|x_k - L|) \geq \epsilon\right\}\right|
\end{aligned}$$

$$\leq 1 - (1 - \frac{\delta}{2}) + \frac{1}{j} \left| \left\{ n \in I_j : \sum_{k=0}^{\infty} a_{nk}(i)\varphi(|x_k - L|) \geq \epsilon \right\} \right|$$

$$= \frac{\delta}{2} + \frac{1}{j} \left| \left\{ n \in I_j : \sum_{k=0}^{\infty} a_{nk}(i)\varphi(|x_k - L|) \geq \epsilon \right\} \right|,$$

This completes the proof.

References

1. Edely, O.H.H., Mursaleen, M.: On statistical A-summability. Math. Comput. Model. **49**(3), 672–680 (2009)
2. Fast, H.: Sur la convergence statistique. Colloq. Math. **2**, 241–244 (1951)
3. Freedman, A.R., Sember, J.J., Raphel, M.: Some cesaro-type summability spaces. Proc. London Math. Soc. **37**, 508–520 (1978)
4. Fridy, J.A.: On statistical convergence. Analysis, **5**, 301–313 (1985)
5. Kolk, E.: Matrix summability of statistically convergent sequences. Analysis. **13**, 1877–83 (1993)
6. Lorentz, G.G.: A contribution to the theory of divergent sequences. Acta. Math. **80**, 167–190 (1948)
7. Maddox, I.J.: Spaces of strongly summable sequences. Quart. J. Math. **18**, 345–355 (1967)
8. Maddox, I.J.: Sequence spaces defined by a modulus. Math. Proc. Camb. Philos. Soc. **100**, 161–166 (1986)
9. Malkowsksy, E., Savaş, E.: Some λ- sequence spaces defined by a modulus. Arch. Math. **36**, 219–228 (2000)
10. Moricz, F.: Tauberian conditions under which statistical convergence follows from statistical summability $(C, 1)$. J. Math. Anal. Appl. **275**, 277–287 (2002)
11. Moricz, F., Orhan, C.: Tauberian conditions under which statistical convergence follows from statistical summability by weighted means. Stud. Sci. Math. Hung. **41**(4), 391–403 (2004)
12. Mursaleen, M., ÇakanC., Mohiuddine, S.A., Savaş, E.: Generalized statistical convergence and statistical core of double sequences. Acta Math. Sin. (Engl. Ser.) **26**, 2131–2144 (2010)
13. Mursaleen, M.: λ-statistical convergence. Math. Slovaca **50**, 111–115 (2000)
14. Nuray, F., Savaş, E.: Some new sequence spaces defined by a modulus funcion. Indian J. Pure. Appl. Math. **24**(11), 657–663 (1993)
15. Pehlivan, S.: sequence space defined by a modulus function. Erc. Univer. J. Sci. **3**, 875–880 (1989)
16. Ruckle, W.H.: FK spaces in which the sequence of coordinate vectors in bounded, Cand. J. Math. **25**, 973–978 (1973)
17. Šalát, T.: On statistically convergent sequences of real numbers. Math. Slovaca **30**, 139–150 (1980)
18. Savaş, E., Savaş, R.: Some λ- sequence spaces defined by Orlicz functions. Indian J. Pure. Appl. Math. **34**(12), 1673–1680 (2003)
19. Savaş, E.: On some generalized sequence spaces defined by a modulus. Indian J. Pur. Appl. Math. **30**(5), 459–464 (1999)
20. Savaş, E.: Strong almost convergence and almost λ-statistical convergence.Hokkaido Math. J. **24**(3), 531–536 (2000)
21. Savaş, E.: Some sequence spaces and statistical convergence. Inter. Jour. Math. Math. Sci. **29**, 303–306 (2002)
22. Savaş, E., Kiliçman A.: A note on some strongly sequence spaces. Abstr. Appl. Anal. Article ID 598393, 8 (2011)

23. Savaş, E.: On some sequence spaces and A- statistical convergence, 2nd Strathmore International Mathematics Conference 12–16 August. Nairobi, Kenya (2013)
24. Savaş, E.: On asymptotically λ-statistical equivalent sequences of fuzzy numbers. New Math. Nat. Comput. **3**(3), 301–306 (2007)
25. Schoenberg, I.J.: The integrability of certain functions and related summability methods. Amer. Math. Mon. **66**, 361–375 (1959)
26. Waszak, A.: On the strong convergence in sequence spaces. Fasciculi Math. **33**, 125–137 (2002)